Applied Economic Analysis for Technologists, Engineers, and Managers

Michael S. Bowman

Prentice Hall
Upper Saddle River, New Jersey   Columbus, Ohio

Library of Congress Cataloging-in-Publication Data

Bowman, Michael S.

   Applied economic analysis for technologists,
engineers, and managers / Michael S. Bowman.

     p. cm.

   Includes index.

   ISBN 0-13-375932-6 (case)

   1. Engineering economy. 2. Manufacturing costs.

3. Finance—Case studies. I. Title.

TA177.B69 1999

658.1—dc21   98-35687

                                     CIP

Editor: Stephen Helba
Production Editor: Patricia S. Kelly
Production Coordinator: Clarinda Publication Services
Design Coordinator: Karrie M. Converse
Text Designer: The Clarinda Company
Cover Designer: Tom Mack
Production Manager: Deidra M. Schwartz
Illustrations: Academy Artworks
Marketing Manager: Frank Mortimer, Jr.

This book was set in Times Roman and Optima by The Clarinda Company and was printed and bound by R.R. Donnelley & Sons Company. The cover was printed by Phoenix Color Corp.

©1999 by Prentice-Hall, Inc.
Simon & Schuster/A Viacom Company
Upper Saddle River, New Jersey 07458

Printed in the United States of America

10 9 8 7 6 5 4 3 2 1

ISBN: 0-13-375932-6

Prentice-Hall International (UK) Limited, *London*
Prentice-Hall of Australia Pty. Limited, *Sydney*
Prentice-Hall of Canada, Inc., *Toronto*
Prentice-Hall Hispanoamericana, S. A., *Mexico*
Prentice-Hall of India Private Limited, *New Delhi*
Prentice-Hall of Japan, Inc., *Tokyo*
Simon & Schuster Asia Pte. Ltd., *Singapore*
Editora Prentice-Hall do Brasil, Ltda., *Rio de Janeiro*

Preface

This book is about money. It is about economic analysis, engineering economy, financial calculations, cost reduction, and profit improvement. It is written for students and for those working in design engineering, process engineering, manufacturing engineering, purchasing, and financial analysis in both manufacturing and service organizations; for members of financial improvement teams; and for technical and senior managers. It is also for individuals, proprietors, and small organizations.

No longer are an organization's financial planning and improvements the sole responsibility of a small number of technical, financial, or management staff. Today, organizations implement economic decisions using teams of operating employees, engineers, managers, office and manufacturing personnel, technical and marketing employees, accounting and nonaccounting staff, and others. This book is for those who are or will be members of those teams.

Today, approximately half of the people in the U.S. work force are responsible for funding their own retirement without employer contributions, and this segment is growing. Also, government retirement plans such as Social Security are becoming a smaller proportion of retirees' benefits. Individuals face more complex financial decisions than in the past as they manage their current finances and plan for future retirement. This book is for individuals and families who are managing their money decisions.

At work, employees are more involved in the financial decisions and improvements of the organization than they were in the past. At home, they face financial decisions about investments and retirement. With this increased financial responsibility at work and at home, individuals are becoming more financially knowledgeable. This book is for those who want to learn about economic analysis, apply financial tools, understand financial statements, and monitor and continuously improve their financial results.

TOPICS

The four parts of this book are based on the **financial or money cycle.** The sections are **historical recording and presentation** of the financial results and data; **economic analysis** of projects and investments; **application of economic tools and selection** of projects and investments; and **monitoring and improvement** of the products, projects, and processes.

- **Part I** provides the basis for economic analysis, decisions, and improvement. It includes definitions, financial statements, the Accounting Equation, depreciation, and cash flow. This section provides important terminology and lays the financial foundation for economic analysis and improvement.

- **Part II** presents the traditional approaches of engineering economic analysis, including interest calculations, time value of money, project and investment analysis and selection techniques, economic alternatives, taxes, and asset replacement analysis. This is the central section of the book.

- **Part III** extends economic analysis to the applications of minimum cost analysis, breakeven, personal applications, and a variety of contemporary applications including just-in-time, constraint theory, inventory investment and analysis, and cycle time economics. This section demonstrates important contemporary financial concepts in design and process engineering, purchasing, and production, as well as personal economic topics.

- **Part IV** completes the financial cycle with traditional and contemporary financial monitoring techniques of economics of quality, continuous financial improvement, and team approaches to financial improvements. This section is a vital component for technicians, engineers, managers, and teams in today's organizations.

GOALS AND THEMES

This book presents a set of basic financial decision-making tools that can be applied to many different financial situations. It includes the mathematics as well as the qualitative concepts, and provides a balance between quantitative analysis and qualitative factors. The book presents a mixture of traditional engineering economy concepts, financial and accounting principles, and contemporary continuous cost reduction and profit improvement techniques necessary for today's financial success. Additional themes include team-based economic analysis and improvement, applications to both organizations and individuals, and continuous monitoring and improvement vital to economic success.

Team-Based Decisions

One basic theme is derived from the change in the typical American organization's culture over the past 150 years. Traditional organizational culture and authoritarian decision making was based on the belief that management is "all knowing" about the customer, the product, and the process and always makes the correct decision. Because of competition, customer needs and requirements have become more important. The technical complexity of products, processes, and customer requirements has increased so that no single person can have total knowledge of the complete process or product. Organizations are becoming "leaner" with fewer staff specialists. As a result of these changes, financial decisions now are made at many locations in the organization, and often by teams. Technical, management, operating, and financial personnel are all involved. Technical knowledge, techniques of accounting, time value of money, decision models, monitoring, and continuous improvement techniques are combined to arrive at successful financial results. This book's examples, problems, and cases demonstrate team-based financial concepts applied in different parts of typical organizations. The case problems at the end of each chapter provide an opportunity for students to analyze, discuss, and resolve actual financial situations. These cases provide for the application of the basic concepts and principles in a realistic environment.

Universal Applications

A second theme of the text is the overlap and complexity of the financial tools and concepts that apply to individuals as well as to large corporations, small proprietorships, and service and manufacturing organizations. The financial attitudes, techniques, calculations, and applications are similar regardless of the size of the organization or its products and services. Individuals find that the same concepts can be applied to the family. The book presents applications, questions, problems, and cases based on small and large organizations, individuals, and technical and nontechnical companies.

Continuous Monitoring and Improvement

The third theme is that financial analysis and decisions are continuous activities for the organization and the individual. This continuous improvement mode requires paying attention to financial monitoring and improvements, not just to the initial project or investment analysis. Supervisors, managers, engineers, technicians, operating, purchasing, accounting, marketing, and other personnel are all responsible for the organization's financial success. Traditional periodic "cost cutting" techniques are replaced by the application of continuous monitoring and improvement tools. The book includes many economic tools for monitoring and improving financial results.

USING THE CHAPTERS

- Each chapter begins with **key terms, learning concepts, and an introduction.** These help the reader focus on chapter topics. Chapters end with a **summary, questions, problems, and discussion cases.** This end-of-chapter material is important to learning and applying the techniques and topics.

- Chapters present the **concepts** and **problems** with **solutions, calculations, tables, diagrams,** and **discussion** of the material.

- Numerous **individual, team, and organization-based examples** of financial analysis and decisions using traditional time value of money calculations, economic models, and tools are included.

- More than **380 review questions, 300 problems, and 60 discussion cases** covering all topics are included. The questions, problems, and cases are drawn from traditional and contemporary situations in many different areas to demonstrate the traditional principles as well as new viewpoints. These provide both quantitative and qualitative application of the material. Discussion cases can be used as team analysis problems with either formal participant presentation or informal group discussion. They are derived from real situations and give the participants experience in combining quantitative calculations with realistic situations. These cases represent different sizes and types of organizations.

- Only a calculator and knowledge of college-level algebra are needed for the examples and problems. Financial calculators, programmable calculators, or spreadsheets are helpful, but not required.

■ An **Instructor's Resource Manual,** including **solutions to questions, problems, and discussion cases,** as well as other material, is available.

Flexible Chapter Sequencing

Readers may choose to read or teach the materials in a sequence different from the one presented in the text. For example, financial statement topics may be omitted and the reader could start with the chapters on interest calculations and time value analysis. This approach would be valid in a curriculum in which accounting is a prerequisite. Or the last section, on measurement and continuous financial improvement, may be omitted without revising the applications of the first three sections.

The basics of interest calculations in Part II can be studied without covering Part III on tools and application of financial decision models. Part IV topics, monitoring and improvement techniques, can be introduced after Part II.

When used in in-service training courses of engineering, operating, management, and improvement teams, the parts and chapters may be selected individually as they apply to the organization's needs. For example, the organization's actual financial statements and cost information can be used to supplement Part I financial statement material. Specific chapters from Parts II and III can then be selected. If the organization is just beginning or continuing team approaches to financial improvements, topics from Part IV can be emphasized.

Specific focus on engineering, manufacturing, or service organizations can be obtained by assigning and discussing those examples, questions, problems, and cases that correlate to the size and type of organization.

Individuals interested in using the text for their personal financial understanding and improvement can focus on Parts I and II with specific emphasis on the examples, questions, problems, and cases that relate to the individual and the family.

The topics of the book can be used to prepare for the state engineering exam. Topics in Part II and sections of Part I on depreciation and of Part III on breakeven and minimum costs would be useful.

ACKNOWLEDGMENTS

Many students' suggestions have guided me in the refinement of this material. Engineers, managers, workshop participants, manufacturing organizations, clients, financial personnel, and members of a variety of financial improvement teams have encouraged, supported, and advised me and made suggestions. Their efforts have been very important and valuable to the generation of this book. Peggy Ammerman's suggestions, editing, advice, and help were invaluable. Much appreciation must be given to Steve Helba, Sylvia Huning, Nancy Kesterson, Patty Kelly, Cindy Miller, and Gail Gavin. Without their encouragement, advice, and help, the work could not have been done. Thanks also to Sheryl Rose, who copyedited the text; Jerry Ratchford, who checked the problems and mathematical examples, and Joy McComb, who did the proofreading. The many reviewers of the material, including Roger H. Clarke of Temple University, Charles E. Mobley of University of North Carolina–Charlotte, and Joseph Thompson, were extremely helpful. Their recommendations were very important. The assistance of Christiana Halam with her careful review of calculations has been most helpful. Thanks to all!

THE FINANCIAL JOURNEY

The journey of learning, analysis, implementation, and improvement of financial decisions is a never-ending activity. Whether the reader is new or experienced, in a large organization or a one-person firm, on a financial improvement team or using this material for personal finances, it is hoped that this book will provide a guide and map for the journey. It is a career-long and lifelong trip for the organization and individual. Enjoy the journey and have a profitable trip!

The author is available at bowman@tech.iupui.edu.

> There is a tide in the affairs of men,
> Which, taken at the flood, leads on to fortune;
> Omitted, all the voyage of their life
> Is bound in shallows and in miseries.
> On such a full sea are we now afloat,
> And we must take the current when it serves,
> Or lose our ventures.
>
> —Shakespeare, *Julius Caesar*

Contents

List of Discussion Cases

FINANCIAL CONCEPTS

Chapter 1   Introduction to Financial Decisions

KEY TERMS

Agricultural and craft manufacturing period Economic period in United States before the factory system began in the 1800s. Output was primarily agricultural, with some products produced by small craft shops. During this period there were limited money management techniques, analysis, or accounting.

Bond Borrowing by the corporation, usually in increments of $1,000. A debt (rather than ownership) at the prevailing interest rate that is paid during the life of the debt. Principal is repaid at the end rather than during the bond's life.

Common stock Shares of a corporation that represent ownership in the corporation with variable dividends. The shareholder has voting authority proportional to the amount of shares owned. The shares may be, but are not required to be, repurchased by the corporation.

Contemporary period Economic period beginning in mid-1900s to present that is based on knowledge, information, international and total quality management. Service output exceeds manufactured product output, and extensive economic analysis occurs.

Corporation A business that is formed by a charter granted by a state. The owners may be separate from management, and the company may sell securities and share profits with owners or retain them. This type of business has unique tax requirements.

Cross-functional team Team composed of members from many different functions or departments.

Factory system Economic period that began in the 1800s. The use of the steam engine, mechanical inventions, and labor specialization produced large quantities of similar products. During this period, management science, industrial engineering, and economic analysis of investments and projects began.

Financial journey Continuous improvement of financial results by an organization or individual during its life from collecting, analyzing, making economic decisions, monitoring, and improving financial results and benefits.

Money (Financial) Cycle The collection and generation of historical accounting data; the calculation of interest and other analysis; the application of decision-making tools; and the monitoring and improvement of financial results, all of which result in new historical accounting data to complete the cycle.

Partnership A business that is formed by contract between individuals and has more than one owner. The life of the business is limited to the life of an individual partner. The partners share control, profits, and liability.

Preferred stock Shares of a corporation that are similar to common stock except ownership in the corporation has a fixed dividend rate.

Profits A source of funds for the organization that results from the money remaining after costs are deducted from revenues.

Proprietorship An easy-to-form business that is completely controlled by a single owner. The life of the business is limited to the life of the owner, and the liability of the business and the owner are concurrent.

Total Financial Management The use of work-group and cross-functional teams to analyze and make decisions about financial improvements, new projects, and investments based on accounting data, interest calculations, and other techniques. Customer-driven analysis and decisions are supported by senior management, and results are continuously measured and improved.

Work-group team Team composed of members from a related work group or department.

LEARNING CONCEPTS

The following concepts are emphasized in this chapter:

- This book is about money.
- The organization and individual have many money responsibilities and constituents, which include customers, employees, shareholders, suppliers, lenders, government taxing agencies, and the community.
- Today's organization relies on teams to analyze and make financial decisions using financial tools, technical information, and team techniques. A primary purpose of this book is to help teams, individuals, and organizations understand and apply financial tools and techniques. Tools and techniques for financial analysis and decisions are used by everyone in the organization. The tools are not limited to accounting, finance, senior management, or other personnel.

- Contemporary total financial management is the team approach to continuous financial analysis, decision making, monitoring, and improvement of processes, products, projects, and investments. Teams, led by management, improve financial results using data from accounting system and engineering economy techniques.
- The economic tools and techniques presented here are applicable to both individuals and organizations.
- The steps of economic analysis and decisions can be applied to financial projects, investments, and improvements.
- The money cycle or financial cycle is the continuous process of collecting historical accounting data, record-keeping, applying contemporary financial tools, and continuously improving the financial results. It also includes economic analysis using time value of money techniques.
- Current economic techniques, methods, and tools of financial analysis work best in our economic system, although they may be applied in other economic systems.
- Our current economic system and financial analysis techniques evolved from our economic history, which includes the era of agricultural and craft manufacturing, the factory system era, and the current eras of knowledge and information and international and total quality management.
- The legal forms of the organization influence the sources of capital, the analysis of financial projects, and the economic decisions made by the organization.
- Sources of money include profits, loans, bonds, preferred stocks, and common stocks.
- Money attitudes are brought by individuals to the organization and can change when individuals receive financial training and adopt the beliefs and attitudes of the organization.
- The money journey is continuous.

INTRODUCTION

This chapter introduces some of the concepts and viewpoints expressed throughout the book. The primary topic of this book is money. Financial data collection and presentation, analysis, decision making, monitoring, and performance improvements are the main topics of this book.

The topics, financial techniques, and tools are presented with the viewpoint that the organization's teams implement and apply many of the techniques discussed in the book. The teams may be cross-functional and include members from different work areas, such as engineering design, process manufacturing, office personnel, or other departments. (The financial topics and tools are equally applicable to individuals.) The term *organization* is used to include profit and not-for-profit organizations; manufacturing and service organizations; and software, banking, insurance, health care, educational, and government organizations. The techniques are universal.

The money cycle or financial cycle is another way of describing data collection, analysis, decision making, and measurement and improvement of financial results. The specific steps of these processes are presented in this chapter.

Economic success of the organization and the individual depends on the ability to analyze, record, monitor, and improve financial performance. The sum of the financial successes of all the organizations and all the individuals is the economic success of the country. The United States is economically successful because of its people, its economic system, and its political system. The techniques and tools presented in this book seem to work best in a capitalistic economy, even though they are useful in other systems. Our present techniques, tools, and decision-making methods have evolved during our economic history. These periods of evolution are described briefly in this chapter.

Sources of funds and types of legal forms of organizations are reviewed briefly in this chapter. The legal form of the organization, proprietorship, partnership, or corporation influences how the organization makes its financial analysis and decisions and how the organization obtains the funds. Sources of money include past and current profits, loans, sale of bonds, preferred stock, and common stock.

Finally, the chapter concludes with a discussion of the financial attitudes, beliefs, and values that an individual brings to the organization and the influence of the organization on an individual's beliefs and attitudes. This interchange is an evolutionary process. The financial performance of the individual, department, division, organization, and society is a continuous economic journey that does not end.

MONEY

This book is about money.

The journey and quest for money is continuous for organizations and individuals. This book presents the concepts, techniques, and tools that can be used to make the journey easier, fun, more successful, and profitable.

Money can be as controversial as politics. Often, attitudes about money are as embedded as attitudes about religion. Money, like weather and sports, is a common and familiar topic of conversation. Is there anyone in the United States, or even in the world, who does not recognize the dollar sign ($)?

This book is about the organization's and the individual's money: money of the large and small organization, the profit and not-for-profit organization, the start-up company and the century-old organization. It is about the money of the corner grocery store, the giant software firm, the multiple plant manufacturing company, the health-care facility, the engineering consulting firm, and the neighbor kid's lemonade stand.

Money is not only something for which we work; it is something that works for us. Both individuals and organizations, no matter what size, try to manage their money in order to maximize the results from spending and investing their money. This book is about money working for us.

> Money wasn't always in its current form as paper, coin, credit card, or image on a computer screen. In the past, it has taken the form of beaver pelts, tobacco, gold, bullets, horses, sea shells, and whatever else was in short supply at the time.

Money and Organizations

Organizations invest money into assets such as equipment, land, buildings, research and development projects, new products, services, advertising programs, employee training, and related projects, in the hope that they will receive the maximum benefits from their investments. Earning and improving profits, obtaining revenues, reducing costs, improving efficiencies, and selling more products provide returns and rewards for the many financial constituents of the organization. The economic analysis of these projects and investments is a major topic in this book.

Organizations have many money constituents. Customers, shareholders, employees, lenders, suppliers, the community, and others are constituents of the organization. Each of these constituents has expectations from the organization. The organization attempts to balance and optimize the benefits to constituents in the following ways:

- Serving the **customer's** expectations and needs requires services and products of high quality and competitive price. Managing finances to provide these customer benefits requires skilled financial analysis, decision making, and methodology.

- **Shareholders and owners** expect a return on their investments. They expect financial returns in the form of dividends and/or appreciation of their investment over time. The organization's use of financial tools, techniques, and analysis, and how it makes decisions must lead to the owners receiving their appropriate rewards.

- **Employees** expect, of course, to be paid for the services they deliver. They also desire a financially well-managed organization that provides a stable environment for their personal growth and development.

- **Suppliers** of materials and services and **lenders** of money expect to be paid on time.

- **Taxes and community** responsibilities must be met by the organization.

To meet these obligations, the financial requirements of the organization are continuous and relentless. This book considers these and other topics throughout the chapters.

Money and Individuals

At the start-up of an organization is an individual with an idea, a dream. Steve Jobs and Steve Wozniak began Apple Computer, Inc. in Wozniak's parents' garage. Henry Ford's and Eli Lilly's dreams began in a small way. Bill Gates started Microsoft at his computer. Individuals create the organization. Their dreams, skills, and abilities become the organization's.

Individuals try to maximize the financial benefits and rewards from their money just as an organization does. Individuals make investments into their retirement accounts, buy a home, have money deducted from their paychecks for services, start businesses, save for education, invest in mutual funds, and make other financial decisions.

The word *economics* comes from the Greek *oikos,* meaning house or household. Economics begins at home with the individual, then transfers to the organization and finally to society. When this personal knowledge and skill of individual and family financial management is amplified and combined with the knowledge of other individuals on an organization's improvement team, significant financial success can result, especially with the help of the organization's resources. There is a strong relationship between an individual's financial attitudes and the behavior of that individual at home and at work. Learning and applying financial analysis and decision-making tools in the organization, on the improvement team, and at home has many reciprocal benefits for organizations and individuals. The many tools and techniques presented in this book will help individuals and families, who in turn help the organization reach its financial goals.

Money and Teams

In the past, an organization's financial decisions began in engineering, manufacturing, purchasing, marketing, or other functional areas. The money requests moved through the organization, and the final decisions were made by management, accounting, and finance departments. Today, more organizations are placing responsibility for operating financial decisions in the hands of teams. The teams may be *cross-functional,* composed of personnel from different work areas, or a *work-group team,* composed of individuals from closely related work areas. Some teams even include customers, suppliers, or other individuals from outside of the organization. Teams have the responsibility to analyze financial data and spend money. For some teams, the magnitude of the expenditure may be large. Senior management continues to make the strategic, long-term investment decisions, but teams are frequently involved in the tactical and operating financial decisions. Part of the reason for this shift in financial responsibility is the complexity and sophistication of today's financial and technical information. Team-based money decisions are often faster, better, more efficient, and more productive. Financially well-trained employees from all parts of the organization, who have a command of financial tools and techniques, are valuable assets.

Using teams requires that the organization change not only its way of decision making, but also its behavior and thinking. Organizational culture, its beliefs, values, and attitudes, has changed to accommodate these new ways of making money decisions. Centralized, slow, and autocratic decisions are replaced with participation, involvement, and "ownership" of the financial decisions and their results. "It isn't my job. Those guys in the executive office do that," is being replaced with, "We understand the problems. We can correct them. We can do the analysis. We can make the decisions, and share responsibility for the financial successes of the organization."

The team approach to financial analysis and decision making is not a replacement of the traditional accounting and financial functions of the organization. However, with the support of the financial personnel, the contemporary organization can combine finance, marketing, manufacturing, technical, and management personnel so that financial analysis and decisions can be most effective. By providing data and support, accounting and other financial personnel can be of great help to the operating, technical, and other personnel that are making the improvements in the financial performance of the organization. The historical financial data provide the basis for analysis, decisions, and financial-based improvements in operations, products, processes, and customer service.

We want our money to work as hard for us as we work for it. Or, as the fireworks salesperson tells us, "We want the most bang for our buck."

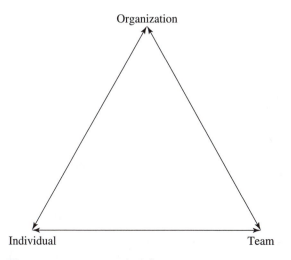

Figure 1–1   Team Triangle

Well-trained financial improvement teams and individuals that participate in the organization's financial analysis and decision making are a very important part of the organization's culture and resources. The relationship of the team, the individual, and the organization forms an interrelated triangle (Figure 1–1). This relationship becomes a powerful force for financial improvement. This book is about giving individuals and teams the financial tools and techniques necessary to improve both the organization's and the individual's financial success.

THE MONEY TOOLS

This book is like a "box of tools" to help engineering, management, technical, financial, and nonfinancial team members assess and fix financial problems. To meet the requirements of all the financial constituents and obtain money rewards, the organization must be financially successful. Understanding and applying these many different money tools, techniques, methods, analyses, and concepts are the central themes and purposes of this book. This continuous improvement of products and processes, measurements, analysis, and decisions must be made by the team members. Below is a summary of the topics, tools, and techniques presented throughout the book that are useful for analyzing, monitoring, and improving the organization's financial results.

Understanding financial statements

Introduction to the accounting system

Depreciation

Manufacturing costs

Interest calculations

Time value of money

Personal applications of projects and investments

Breakeven analysis

Minimum costs analysis

Techniques to analyze alternatives

Techniques to select projects

Analyzing replacement of equipment and buildings

Tax effects on the organization

Measurement of money results

Economics of quality

Continuous financial improvement

Total Financial Management techniques

These tools help us understand, explore, and learn to use financial concepts. Within each chapter, there are many separate techniques and special tools. If we can learn to apply these tools, it will help us to improve our financial decisions as team members and individuals.

The techniques and tools are grouped into four subcategories or sections in the book:

Section 1	Historical accounting and record-keeping techniques
Section 2	Interest calculations and time value of money analysis
Section 3	Decision and analysis techniques
Section 4	Financial improvement tools

These four sections present the concepts and techniques for effective data analysis, measurement, decision making, methods, and understanding of financial improvement efforts.

Section 1 considers financial statements, accounting basics, cost accounting, depreciation, and cash-flow concepts that provide the background from which financial decisions are made. Understanding of these topics permits economic analysis and decisions to be integrated in the overall financial structure of an organization. Nonfinancial personnel, including operating employees, can easily understand these concepts and apply them to their work areas. Being able to read and understand financial statements and to comprehend the basic accounting system helps employees understand the organization's financial health. These statements, the accounting system, and other historical reports and data are closely connected to the process of evaluating equipment, manufacturing processes, product design, financial improvements, taxes, depreciation, and other economic analysis topics.

Approximately two-thirds of the financial tools and techniques are devoted to traditional analysis and decision making. Sections 2 and 3 cover the interest calculations and the decision-making techniques.

Section 4 presents many of the techniques and tools used for measuring and improving the processes and products to enhance financial results. Improvements in productivity, quality, delivery times, customer response times, and related areas can be analyzed using the combined techniques presented in the other sections. These improvements are both technically and financially based. They require both technical knowledge and financial techniques and understanding. The combination of techniques can be applied by the design or process engineer, technician, supervisor, manager, team member, or others in the organization. Understanding the concepts of continuous improvement and total financial management is vital to today's organization if it is to maintain its financial success in international competition.

The combination of interest analysis and other tools presented in the book, coupled with an understanding of basic financial statements, accounting, continuous financial improvements, and total financial management, can be a powerful financial combination for today's organization. This combination can be implemented by technical, management, office, and operating personnel.

Who Uses the Tools?

Since this is a book about finances, one might expect that all the readers are accountants or financial mangers. The contrary is true. These techniques and tools are intended for nonfinancial personnel. It is the manufacturing, technical, scientific, engineering design, process engineering, marketing, purchasing, sales, office, operating, supervisory, management, and other nonfinancial employees that spend, invest, and improve the firm's finances. The sales and marketing people bring in money from selling the product and services of the organization; manufacturing, design, purchasing, and other personnel spend and invest the money; and the accounting department reports, perhaps to the nearest penny, sources and uses for the money. It does the "scorekeeping."

This book is intended for:

design engineers	technicians
technologists	middle and upper managers
purchasing personnel	quality engineers and technicians
operating employees	hospital administrators and
software designers	medical staff
marketing and sales representatives	supervisors
retail store managers	all financial improvement teams
office personnel	research and development staff
process and manufacturing engineers	

Financial and Nonfinancial People

Applying the financial tools by individuals or teams is done primarily by the nonfinancial members of the organization because they have the knowledge of the processes, products, customers, equipment, projects, and investments. The accounting and financial people keep financial history and records of the income, spending, and investment made by the others in the organization. They provide data and assist the teams and individuals with their money decisions and analysis.

THE MONEY CYCLE

Another way of looking at the process of financial analysis and decision is to consider the money cycle. The **accounting system** accumulates the historical financial information. Financial accounting data are combined with **analysis** of the identified problem or improvement. **Tools and techniques** including interest calculations are used to **evaluate the alternatives** economically as well as to consider other benefits and disadvantages. After the decision is made, the results are measured and monitored. Finally, the problem is reviewed in order to make additional **improvements** on a continuous basis. The outcomes of the decision produce **financial results** that are recorded by the accounting system and the cycle is repeated. Figure 1–2 shows the cycle in diagram form.

As Figure 1–2 illustrates, the book's sections and chapters are presented in the same order as the money cycle. We begin with historical accounting data, move on to analysis, then to tools and decisions, and finally to continuous improvements. In actual practice, the decision process may start at any point. For example, a product design or manufacturing process may already be established and is being improved, which will result in the team beginning its study with the improvement part of the cycle.

TODAY'S ORGANIZATION AND MONEY

The process of economic analysis and decision making has evolved over time. Below is a summary of the old and new techniques.

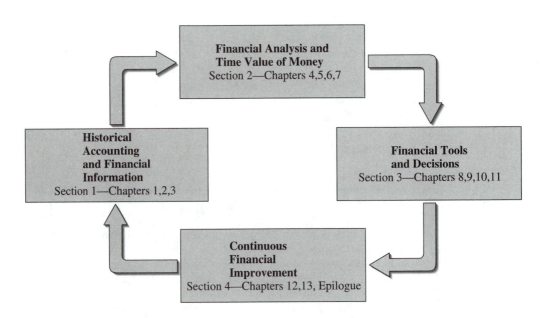

Figure 1–2   The Money Cycle

Traditional Techniques	Contemporary Techniques
Top-down decisions	Manager is team leader
Manager is decision maker	Customer-driven quality
Design- and process-driven quality	Flatter organization
Layered organization	Fast, decentralized decisions
Slow, central decisions	Longer-term financial goals
Short-term profit goals	Continuous financial improvement
Quality: "It is good enough"	Quality, profit, and customer service first
Profit and productivity first	Globally directed
Domestic focus	Decentralized financial decisions
Centralized financial decisions	Shorter, more timely deliveries
Long lead time deliveries	Partnering with suppliers
We vs. they supplier relationships	Reducing costs of quality
High costs of poor quality	Quality based on prevention
Quality based on inspection	Work group financial improvements
Centralized financial decisions	Team-based financial analysis and decisions
Centralized financial analysis and decisions	

Contemporary Total Financial Management uses a team of financially knowledgeable individuals that represent many different functions, who are supported by senior management making financial recommendations and decisions, with the customers' needs and expectations considered on a continuous basis. These individuals and team members often have direct participation in the financial results through profit sharing, bonuses, or other reward systems. This can be a very productive approach to continuing and improving the financial performance of the organization.

The organizations in which many of us work today are different from those of the past. Accounting-driven financial decisions have been replaced with analysis and decisions that are customer directed. In many organizations, the emphasis is on financial decisions made by improvement teams. Senior management still makes the longer-term, strategic financial plans and decisions, but the tactical, shorter-term financial improvements are frequently made by teams. These may be work-group teams with members from related work areas, or cross-functional teams with members representing different departments. The financial improvements are made continuously with emphasis on customer requirements and expectations while focusing on reducing costs, improving quality, and achieving faster response times. Other money constituents, employees, managers, shareholders, suppliers, and others are also part of the team's financial improvement targets.

> When someone says, "It's not the money, it's the principle" they usually really mean it's the money.
>
> —Anon.

Money Rewards

Money rewards for both organizations and individuals are vital, not only paychecks, but other forms of financial rewards and benefits too. Organizations have discovered reward techniques in addition to the paycheck. A significant motivator is the participation in financial decisions and the receiving of financial rewards in the form of stock ownership, bonuses, team or individual incentives, and profit or gain sharing based on the financial successes of the organization or department. As the financial effort taps into the financial intelligence, skills, and experience of the personnel, and their financial skills are expanded with training, financial successes are rewarded. As employees, team members, and individuals we generally want to contribute to the financial success of the organization. If we can also receive financial rewards in addition to our paycheck, we are furthered motivated to contribute to the organization's success. Although senior managers have frequently received a large portion of their compensation in the form of incentives, bonuses, or stock ownership plans, these are now available to many others in the organization.

As a strategy to ensure that both managers and workers are on the side of the company and "think and act like owners," many firms have developed extensive management and employee stock ownership plans. Generally these shares are purchased, sometimes at a discount, or sometimes given to the employees as a stock option incentive plan or outright gift. Ford Motor Co. chairman Alex Trotman says, "I'd surely want this kind of program for everyone from the chairman to the guy who sweeps the floors. It's great for team spirit, and that is one of the most difficult challenges for companies today."

Steps of Money Analysis and Decision Making

We make many personal decisions as we go through the day. Many of the decisions are about money. This book is about the money part of our decisions, mostly job and organization decisions, but the topics apply to personal financial decisions as well.

Much of our waking hours, we are doing financial analysis and making financial decisions. Should I buy that new dress? Continue to rent my apartment or buy a house? Put a new roof on the house or wait another year? Buy the sale items at the supermarket or the brand we usually purchase? Put our savings in the credit union or the mutual fund? Hire someone to cut the lawn or buy a new mower and do it myself? At the manufacturing company, from which supplier should we purchase material? Should we make a component or buy it from a supplier? Which of the alternative designs of the product should be selected? Should we buy the robot? Borrow funds from the bank or sell common stock? Ship the finished product by rail or truck? At the hospital, should we purchase the new machine? Hire more technicians for faster blood analysis in the lab? Build an employee and patient parking garage or expand the surface lots? At the insurance company, should we keep the old computer or buy a new one? The examples are almost limitless.

By combining physical data, financial information, accounting records, observation, customer and supplier information, product and process design calculations, and other information sources, we can analyze and select financial solutions. Then measuring and monitoring processes, customers, products, and financial results can be continuously improved.

Typically, the steps of making economic analysis, financial improvements, and money decisions are:

1. *Identify* the problem or improvement to be made.
2. Select the desired *result, outcome, target, or goal* to be achieved.
3. *Collect data* about financial information, customer feedback, product and process information, quality, delivery, productivity, etc.
4. Determine *alternate solutions* using team-based techniques.
5. *Analyze* using related accounting data, interest calculations, analysis tools, marketing, product, and process information.
6. *Select and approve* an alternative.
7. *Predict outcomes;* measure, audit, and monitor outcomes.
8. *Improve* again and again.
9. Move to other projects and investments.

These steps are typically performed by the financial improvement team and other individuals in the organization. Sometimes, suppliers and customers are added to internal teams to assist with the design of new products, processes, or services or the improvement of existing products.

> The evil money monsters, Greed and Fear, hang around whenever we are making financial decisions. Fortunately, the good money fairies, Hopes and Dreams, are also present.

Money Analysis and Decision Topics

This book is about the bookkeeping/accounting techniques of money, the analysis techniques and tools of money, making money-based decisions by the individual and organization, and the management and improvement of financial results.

- **Record-keeping and bookkeeping** can be divided into two main parts, managerial and financial accounting. Financial accounting is accounting and reporting for external purposes including for shareholders, customers, taxes, and others outside the organization. Internal or managerial accounting is accounting for costs, inventory, assets, and related data for reporting the organization's internal financial system. Both are historical records that present financial results in a standard and consistent form. A brief introductory survey of the accounting and record-keeping system is presented so that nonfinancial members of the organization can understand the basic concepts and vocabulary of historical accounting and record-keeping systems.

- **Analysis techniques and tools** of money include interest calculations, breakeven analysis, minimum cost analysis, replacement, taxes, and related techniques. This mathematical portion is the heart of the book. These tools are presented and supplemented with examples and calculations. The tools and techniques are applicable to individuals, small and large organizations, and to service or manufacturing firms. Although the techniques are quantitative, basic algebra is the only mathematical requirement; and, in practice, usually must be mixed with judgment, experience, and other nonquantitative factors to arrive at a final decision.

- **Decision making** about money covers the analysis of projects, investments, equipment purchase or replacement, and other allocations of money by the organization and individuals. By combining the data from the record-keeping function with economic analysis techniques, decisions can be made that will provide significant financial results for the firm or individuals. In today's organization, the emphasis is on operating and tactical financial decisions that are customer driven and usually implemented by a cross-functional team.

- **Continuously improving** the financial results is the final topic of the section. By involving personnel in the organization and giving them the knowledge and skills to monitor and improve the financial performance of the organization, significant financial results can be realized. Engineering, manufacturing, operating, marketing, purchasing, office support staff, and other personnel can be combined with financial personnel to improve processes, products, facilities, and other components of the organization to measure, monitor, and improve financial results while meeting customers' requirements and expectations. Financial measurements, quality economics, continuous improving techniques, and organization-wide total financial management are key topics in this section.

These four areas are often considered together by teams. There is overlap in the data, analysis, decision, and improvement activities. For our purposes, the material is presented separately and in linear order. Once the individual concepts are understood, they can be applied together in whatever order fits the specific problem.

As Figure 1–3 shows, the analysis and decision process, in practice, contains the overlap of all parts of the financial cycle. In addition, noneconomic factors may be part of the decision process. These include design criteria for the product or service, customer requirements,

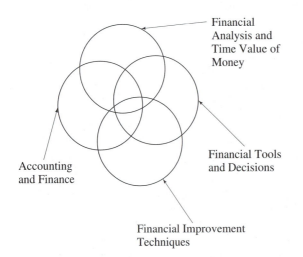

Figure 1–3   Overlap of Financial Functions

equipment capabilities, material characteristics, quality requirements, and many other technical, physical, and nonfinancial criteria. An attempt is made to convert these qualitative characteristics to financial data for the analysis, but these nonfinancial and nonquantitative components of the problem must be considered in addition to the financial characteristics.

What are the reasons for making financial analysis and decisions? Decisions made in organizations or by individuals can be grouped into categories. These include, for example, economic analysis and decisions to:

Meet the customer's requirements and expectations.

Make the organization more profitable.

Improve products.

Improve processes.

Improve services.

Improve quality.

Increase productivity.

Reduce costs.

Improve competitiveness.

Meet environmental standards.

Provide new services or products.

Improve safety of work areas.

Increase returns to shareholders and employees.

Meet legal or ethical requirements.

Simplify work, processes, and methods.

Assist in the personal financial decisions of employees and team members including investments, mortgages, loans, retirement funds, auto lease or purchase, and related decisions.

This list suggests the many areas in which economic and financial analysis is applied.

Money and Capital Investment Decisions

Another way to categorize economic decisions is between current money in the present accounting period and capital investment that includes both present and future periods. Current money is spent for services and materials that are consumed quickly. Items that have a life of a year or more, such as equipment and buildings, are capital investments. Most financial analysis and decisions contain both types of expenditures.

Examples of current money decisions include the purchase of office supplies and materials, housekeeping services from outside the organization, labor services, electric power, and other items that are used within the annual financial period. Current money decisions also include deciding from which supplier to purchase materials by analyzing the physical properties of the material, its quality, the terms of delivery, warranties, and other characteristics. When the organization purchases manufacturing equipment, computers, buildings, or long-lived assets, it is making a capital investment decision. Rather than consuming the equipment's value within the year, the services are provided over a number of years, and it is assumed that only part of the equipment's value is used in any single year. Analysis of this financial decision, for example, includes the equipment's initial cost, operating costs including maintenance, output, efficiency, productivity, safety, reliability, lifespan, value at the end of its useful life, and many other economic and physical factors.

The economic analysis is similar for both types of decisions. Costs, revenues, benefits, and disadvantages of each are considered and evaluated. But, since the capital investment lasts for a number of years, an additional analysis using interest calculations is made.

AMERICA'S ECONOMIC SUCCESS

Before going further, we can look at the economic and financial environment in which we make our money decisions. We live in a country with a political and economic system that permits and encourages us to be financially successful. Although the techniques and methods presented here are applicable to any economic system, they work best in an economically free and capitalistic system. In the U.S., we are fortunate to have an economic system that rewards work, ingenuity, creativity, good analysis, risk taking, sound financial decisions, and meeting the customer's requirements and expectations.

The United States has had unsurpassed economic success. Any standard of living measurement would rank the U.S. economy very near the top of the economic systems in the world. Why has the U.S. been so economically successful? There is no single answer, but there are a number of factors. When groups of students, working engineers, senior managers, and others in organizations are asked this question, they give similar answers. Their suggestions are summarized, not necessarily in order of importance:

Economic freedom—Open economy, freedom to choose services and products to manufacture, sell, or purchase, a cornerstone of capitalism.

Competitive system—The customer can choose the best when there are multiple producers.

Political stability—Stable and flexible government.

Natural resources—Mining, agriculture, and other resources and the technology to utilize them.

Management science—Innovative evolution of management techniques.

Education system—Public education available at all levels.

Legal system—Constitution and common law–based system.

Private property—Allocate land, money, and resources to maximize benefits to customers, a key component of capitalism.

Technology and science—Technical creativity and inventions.

Comprehensive monetary system—Extensive banking, financial markets, and government money system.

Strong military—Protection of property and economic system.

"Yankee ingenuity"—Creativity, innovation, and talents of citizens.

The financial success that U.S. organizations and individuals enjoy is a result of not only applying financial techniques successfully, but also being able to function in an economic system that promotes and encourages economic success.

How Did We Get Here?

Much of the basis for today's current financial practices have their roots in our economic, technical, and business history. To understand today's money techniques and management, it is helpful to look briefly at our financial past. Past events have had significant long-term impact on the present approach. They are the foundation of today's organization, its culture, and its decision-making process. These important periods of economic and financial development are the agricultural and craft periods; the factory system era; and the contemporary era, which includes the knowledge and information period, the international era, and the development of Total Quality Management. We will look at the characteristics of each period and its effect on money and finances of the individual and the organization.

Agricultural and Crafts Manufacturing

This period of our history existed into the 1800s. Its economic and financial characteristics included the following:

- Economy based on family farms (more than 90% of the population), with some crafts manufacturing such as blacksmithing.
- Organizations were family or one-person firms.
- Most of the output from the farm was consumed for its own use. Exceptions were cash crops of tobacco and cotton.
- Little mechanization or work specialization; the use of hand or animal-powered tools.
- Little, if any, external financing of farms or businesses such as loans or sale of stock.
- Barter and exchange rather than cash or credit.
- Financial system of farm or business was brief record-keeping, if at all. The bookkeeping system was "coins in the box or bowl," and there was no financial or economic analysis of projects or investments.
- Worker/owner was in close contact with customers and knew their needs.

Factory System

Frederick Taylor–Type Training

The young drugstore clerk's only training by the owner the first day on the job was how to make change from the customer's payment, and the instruction by the owner, "You do the working, I'll do the thinking."

In the 1800s, with the invention of the steam engine and other mechanical manufacturing equipment, money was invested to purchase equipment and build factories. Employees moved from rural colonial towns to work in the factories and immigrants were hired. Work was hard, but the pay was good for the times and skills required. Characteristics of this period were:

- Larger and more formal organizations patterned after the military.
- Outside investment in factories by owners, partners, and shareholders.
- Banks loaned money to organizations.
- With outside investors and purchase of equipment and facilities came the beginnings of economic analysis of fixed assets and investment returns.
- High output due to steam engine, mechanization, and labor specialization.
- Factory system evolved over most of two centuries to the present, rather than an abrupt "industrial revolution."
- Transportation, banking, and government taxing systems expanded.
- Unions and antitrust legislation appeared toward the end of the 1800s. Oil, rail, and tobacco were the targets.
- Management science emerged in the early 1900s based on military and plantation experiences, and specialization. A few management scientists such as Frederick Taylor, Frank and Lillian Gilbreth, Harrington Emerson, and Henry Gantt were recognized.
- Organizations were formal. Engineering design, manufacturing, accounting, marketing emerged as specialties and departments within organizations.
- Corporations became a popular form of organization because they could sell securities. Stock exchanges raised funds in the form of common and preferred stock and industrial bonds for new firms.
- Owners were separate from managers and workers. Owners and workers were removed from customers.
- Most economic analysis and financial management was based on historical accounting records and analysis. Asset analysis using time value concepts began.

Customer Requirements?

They can have any color they want, as long as it's black.

—Henry Ford

Rules of This Establishment
(From an 1878 Carriage Shop)

Employees working here shall dust the furniture, clean their desks and sweep the floor daily.

All windows shall be cleaned once a week.

Each employee shall bring his own bucket of water and scuttle of coal for the day's work.

Lamps shall be trimmed and chimneys shall be cleaned daily.

Working hours shall be 7:00 A.M. to 8:00 P.M. every evening but the Sabbath. On the Sabbath, everyone is expected to be in the Lord's house.

Employees are expected to trim their own pen nibs to suit themselves.

It is expected that each employee shall participate in the activities of the church and contribute liberally to the Lord's work.

All employees must show themselves worthy of their hire.

All employees are expected to be in bed by 10:00 P.M. *Except:* Each male employee may be given one evening a week for courting purposes and two evenings a week in the Lord's house.

After an employee has been with the firm for 5 years, he shall receive an added payment of 5 cents per day, providing the firm has prospered in a manner to make it possible.

It is the bounden duty of each employee to put away at least 10% of his wages for his declining years so that he will not become a burden upon the charity of his betters.

Any employee who is shaven in public parlors, frequents pool rooms, or uses tobacco shall be brought before management to give reasons why he should be continued in employment.

Money and the Organization's Legal Form

The legal form of an organization affects the way the organization manages its finances, pays its taxes, distributes its profits to owners, and makes financial decisions. Below is a summary of the similarities and differences between legal forms of organizations that have evolved in the U.S. These have impacted the financing, economic analysis, and decisions of organizations. Until the rise of the factory system, the **proprietor** form of organization was the primary organization form. The proprietorship has the following characteristics:

single owner

quick decisions

complete control

easy to form

life limited to the natural life of the individual

capital limited to individual's capacity to borrow and make profits

no limit to potential size

largest number of organizations in U.S.

liability of individual and organization overlap

individual is taxed for the proprietorship

The **partnership** was important in the provision of capital for the factory. Characteristics of the partnership include:

two or more owners

decisions affect all partners

defused control

formed by contract between partners

life limited to natural life of partner

capital available based on partners' capacity to borrow and on profits

may be large

shared profits between partners

shared liability between partners

partners pay taxes as individuals

The **corporation** is uniquely American. The corporation's characteristics are:

formed by charter from state and can conduct its business as an "individual"

owned by shareholders

shareholders having votes proportional to ownership elect directors who appoint managers

unlimited potential life

may sell securities including bonds, common and preferred stock

liability of shareholders limited to their investment

profits may be retained or shared with owners as dividends

special tax requirements for corporation and owners pay tax on their dividends from the corporation

There are other special forms of these three primary types of organizations such as cooperatives, special organizations for tax purposes, and limited partnerships. The organization may make different financial decisions depending on its legal form, so it is important to understand the basic differences.

Contemporary Era

In 1960 approximately 50% of the workers in the world's industrialized countries were involved in making things. It is estimated by 2000, approximately 15% of the workers will be involved in making and moving products. In 1995 it was estimated that approximately 65% of the U.S. work force was working in the services industries. Information and knowledge is now our largest product.

—Peter F. Drucker, *Post-Capitalist Society*

Since the middle of the 1900s, events have evolved rapidly to change the financial and economic environment of the United States. Three important concepts have shaped this economic period for organizations and individuals during the past 50 years: rapid information growth coupled with computer processing of information; the rise of international competition; and the cultural changes brought by the application of Total Quality Management. Our recent, rapid growth and hunger for information can be seen in the ways we communicate. R. Tetseli tells us that between 1983 and 1993 the number of cellular phones in the U.S. increased from zero to more than 16,000,000. More than 25,000,000 people carry individual pagers, and in 1993 more than 12,000,000,000 messages were sent to voice mail boxes (R. Tetseli, "Surviving Information Overload," *Fortune*, July 1994).

The characteristics of the contemporary period include:

- Knowledge, information, and services, rather than manufactured products, are the majority of products produced and sold.
- Over half of the economic output is based on services.
- The computer is to contemporary information what the steam engine was to the development of the factory system.
- International competition is a significant economic force on U.S. organizations. Europe and Asia have developed into strong competitors.
- American firms have expanded globally.
- Total Quality Management evolved into an economic weapon and caused important cultural and philosophical changes in U.S. organizations.
- Financial analysis of the organization and economic analysis of its internal projects and investments became extensive. Analysis of financial decisions is no longer based primarily on historical accounting data.
- Automation and computer control of processes are widespread.
- Design, manufacturing, and financial analysis have "reconnected" with customers, and organizations have become more customer driven.

The characteristics of Total Quality Management include the following:

- **Senior management**—Emphasis on senior management leadership and participation rather than attempting to be all-knowing experts directing and delegating to subordinates.
- **Culture**—Changes are made in organization's core values, belief, and attitudes rather than cosmetic changes.
- **Customer driven**—Organization performance, philosophy, and techniques are based on customer needs and expectations.
- **Prevention**—Organization attempts to prevent rather than detect problems, defects, and mistakes.
- **Continuous improvement**—Small frequent improvements in internal and external customer services and products are made.

- **Employee empowerment**—Consensus, team, cross-functional decisions and improvements rather than top-down "all-knowing" decisions are made.
- **Training**—Coupled with empowerment, people continuously participate in training.
- **Fact and statistic–based decisions**—Emphasis on teams utilizing data, statistics, and customer satisfaction data for decisions.

Sources of Money

Without external financing, factories would be less likely to be created. Until banking and the security exchanges became available to small organizations, funds had to come from individuals, families, and past profits. Corporations were able to sell securities, both stocks and bonds. Briefly, a review of the sources of funds available to organizations are:

- **Profits**—Past profits provide the individual, partnership, or corporation with a source of funds. When a service or product is sold for more than it cost to produce and deliver, a profit is made. Even the blacksmith was able to obtain a profit when horseshoes were sold for more than they cost to make. The profits could be reinvested into new tools, inventory, or other assets to produce other services and products. For the modern corporation, profits can be reinvested and/or paid to the shareholders as dividends.
- **Loans**—The terms and conditions of loans are unique for each borrower and lender. The main difference between profits and loans is that the loan is repaid with interest to the lender.
- **Bonds**—Bonds can be issued by the federal government, state agencies, municipalities, and corporations. They represent a loan by the lender to the borrower. Usually they are in $1,000 multiples and their interest rate is based on the prevailing market interest and the borrower's financial ability. Interest is paid during the life of the bond and the principal is repaid at the end of the bond's life. Typically a bond is for 10 or more years. Bonds are marketable on exchanges and their price fluctuates in the market.
- **Preferred stock**—A corporation can sell preferred stock, which has a fixed dividend rate, similar to bonds. It represents ownership, not debt. The price of the security fluctuates in the market, and since they represent ownership, securities do not need to be repaid as bonds are.
- **Common stock**—A corporation may sell common stock, which represents ownership and does not have a fixed dividend rate. If profits are high, the organization may pay large dividends; if they are low, small or no dividends are paid. Common stock may be repurchased by the organization, and it may be marketable in the security exchanges for a publicly held corporation.

Few historical or contemporary organizations could grow and expand only on their profits. The factory system could not have grown with such rapidity and longevity were it not for external funds.

MONEY ATTITUDES

This book is about money, and the approach used here is generally quantitative, factual, and mathematical. What is missing is the discussion of personal attitudes about money, and the behavior patterns that people demonstrate concerning money. Families, individuals, and organizations have many different attitudes and beliefs about money. They take these attitudes and beliefs with them to the organization when they go to work, and the employees take the organization's attitudes and beliefs about money home with them. The analysis of money must be balanced with the personal attitudes we have about money. Behavior about money often comes from the family as well as from our experience, and these money attitudes are formed early in life.

Individuals bring both good and bad attitudes about money to the organization. It helps both the individual and the organization if the attitudes are appropriate. It is a difficult task to change embedded attitudes. Many individuals in the corporate environment and those managing their own money have difficulty with financial decisions, not because of a lack of knowledge, but often because of their personal attitudes. We don't use one set of financial attitudes at home and another set at the office. Underlying the analysis and calculations are attitudes, beliefs, and values about money.

This book assumes that the reader has healthy attitudes toward money and wants to learn the decision-making techniques that are discussed. These principles of the analysis of money are always much easier to apply (and more successful) when money attitudes are healthy.

After receiving financial rewards, organizations can direct money to the needs and requirements of the financial constituents: lowering prices and improving service and products for customers, paying shareholders' dividends, paying wages and other employee benefits, paying suppliers, paying lenders and bond holders, paying taxes, and making gifts to the community. The organization is also able to reinvest for growth and expansion. Although this book is not about how to allocate and balance these financial rewards, this is certainly an important topic. Our goal is to understand financial record-keeping, analysis, decisions, measurement, and continuous improvement for the organization and individual.

A Lunch About Money

An old friend called and offered to buy lunch. It turned out that this was a celebration lunch: a celebration of reaching a personal net worth of one million dollars! This same person hadn't been able to go to college full time because of limited money. When asked about the secret of financial success, answers like, "Work hard. Have talent and luck. Join a good company and work your way up. Start your own business," or even "Marry the right person," might be expected. Instead the answer was, "*Want* to have a net worth of one million dollars, and have a *good money attitude*."

The Financial Journey

Organizations and individuals continually attempt to improve their financial condition. Firms want to be more profitable, employees want bigger paychecks, shareholders want higher dividends and stock prices, customers want lower prices and higher quality, sales departments want to sell more services and products, production departments want higher quality and lower costs; and individuals want lower mortgage payments, larger retirement accounts, and more savings. The financial journey for the organization and the individual is an unending quest!

To help us improve the organization's and our own personal finances, this book presents the methods, tools, and techniques. Not all of the tools in our tool box are used all the time, but they are available when we need them.

Before we start the lifelong process of making financial decisions to improve the organization's and our personal finances, we need to look at the tools in our tool box and learn how to use them. That is the purpose of this book: to help the organization, its financial improvement teams, and the individual improve their financial situations.

The sequence of the book is from accounting and financial history topics to interest calculations to decision techniques and then to continuous financial improvement techniques. This sequence works for many. But, depending on the reader's needs and interests, the topics may be learned in other sequences. The sections, chapters, and tools are designed to stand alone or together. Ideally, the reader will begin at the first chapter and continue to the last, but this is not necessary.

Let's begin the money journey.

Financial Journey

A student graduates from college at age 22 and opens an individual retirement account, adding $2,000 each year until age 65, and is able to earn 8 percent compound annual return on the money during that time. Upon retirement, this account is worth approximately $650,000. Only $86,000 has been deposited into the account.

SUMMARY

This chapter has presented the following topics:

- Money.
- Money analysis and decisions in the organization, teams, and individuals.
- The team triangle.
- Topics in the money tool box and four sections of the book.
- Typical users of the money tools.
- Contemporary and traditional management of finances.
- Steps of money analysis and decisions.
- The financial cycle.
- Capital investment decisions.
- Characteristics of America's economic success and major economic periods in the U.S.—the agricultural and crafts period, the factory system, and the contemporary period including information services, international competition, and Total Quality Management.
- Legal forms of the organization.

- Sources of funds for organizations.
- Individual and organizational attitudes about money.
- Looking forward and the financial journey.

QUESTIONS

1. What are the topics of this book?
2. Who are the money constituents of the organization?
3. What is meant by the statement, "Economics begins at home"?
4. What is the difference between a financial team composed of work-group members and one of cross-functional members?
5. Why are teams used for financial improvements of existing projects and analysis of new investments?
6. What are changes that have occurred to the employee financial reward system in recent years?
7. What is the team triangle?
8. What is the money tool box? What does it contain?
9. Who in the organization uses the financial tools?
10. Why is it that even though an employee's primary job is not financial, he or she also needs to understand the organization's financial system, financial analysis, and how to participate in money decisions?
11. What are the steps to money analysis and decisions? Apply them to a personal financial decision you have.

12. What are the four parts to the financial cycle?
13. What is meant by "financial functions overlap"?
14. What are some typical financial decisions that the organization makes? The individual?
15. What are some reasons for America's economic success?
16. What are some characteristics of the following economic periods in the U.S.:
    a. agricultural and crafts manufacturing period
    b. factory system
    c. contemporary period
17. What are the characteristics of Total Quality Management?
18. What are the legal and financial differences between a proprietorship, a partnership, and a corporation?
19. To which of the three types of organizations are individuals and families related?
20. How are the sources of funds different for each of the three legal forms of organizations?
21. Why are money attitudes important to the organization's and individual's financial success?
22. What is meant by the "financial journey"?

DISCUSSION CASES

River Software Company

Al and Nancy Roberts are both software engineers. They have begun River Software, a software consulting company, at home while they still maintain their other jobs. They are currently doing custom software work for two companies in the evenings and weekends. Business is picking up and they are each spending approximately 10 to 15 hours each week on the work. They bill the clients each month based on an hourly charge. Recently they have been approached by two other firms about doing some work. If these inquiries become actual contracts, they think they will need to hire some part-time helpers. Al has some ideas for a custom software package that he thinks they would be able to sell to PC users. He is thinking about marketing it over the World Wide Web or by direct mail.

Both Al and Nancy are excited about the possibilities of their company becoming a full-time job for both of them so they can work at home and quit their other jobs. Currently, they are reporting the income from River Software along with the income from their full-time jobs on their tax return. They wonder if they should become a formal partnership when they hire other part-time programmers, or stay as a proprietorship. Nancy suggests that they become a corporation. Al says that is not necessary and could cause a lot of lawyer fees that they cannot afford to pay right now. He thinks the taxes would be higher also. Both are thinking about some kind of bonus or profit sharing for all of the employees if the organization grows. Each has a profit-sharing and stock plan where they work now and they say that it is a strong motivator for employees.

Now, the clients are coming to them by word of mouth, but Nancy wonders if they will need to hire salespeople someday. Al wonders if they will need to have a complex accounting system. Both Al and Nancy like the idea of having their own business, but they are not sure if they want all the administrative and management problems that go along with it.

1. What are the advantages and disadvantages of having your own business?
2. What are the advantages of staying a proprietorship? Formalizing a partnership agreement? Becoming a corporation?
3. Is it expensive and difficult to form a corporation?
4. How might a bonus or profit-sharing plan be established?
5. If the organization grows, how can they solve the marketing and accounting problems?
6. What other things should Nancy and Al consider?

Wood River Products Corporation

As a medium-size manufacturer of modular homes, Wood River is growing rapidly. Sales are growing at more than 12 percent annually, and earnings are growing rapidly also. This expansion is causing "growing pains," according to Jim Forbes, the president of Wood River. They are having difficulty finding qualified employees at all levels, and are having to do more training to help the new employees gain the necessary skills. They are working two shifts in most departments and three shifts in a few departments. Equipment is operating at almost full capacity, and new equipment is being added regularly. They are currently adding on to both the warehouse and manufacturing areas. When these are completed, the design engineering, process engineering, purchasing, and accounting offices will be expanded into the old employee parking lot.

Eighteen months ago, Jim established quality improvement teams in areas where they were having critical quality problems. The teams have the responsibility for making process and operating changes to ensure that the product quality is not compromised as the organization grows rapidly. The teams are typically composed of key operating employees, including experienced and new employees, process engineers, quality technicians, supervisors, design engineers when needed, maintenance supervisors, purchasing, field crew representatives, and senior management as needed. The time and cost of training have been more than anticipated, but the feedback from the field crews that set up the modular homes and the final purchasers of the units has improved significantly. Customer complaints are way down in the past six to eight months. The time and expense is worth it, Jim and others believe.

Many requests are for new and replacement of old equipment and facilities. Process engineers usually write a formal request, which goes to the plant manager; if approved, it then goes to the accounting and finance departments for analysis and recommendations, back to the plant manager, and then to the board of directors for final approval. The process typically takes between 6 and 14 weeks. Approvals are inconsistent and are usually based on the analysis by the financial department. A number of quality teams and manufacturing staff believe that a major bottleneck to quality and productivity is the slow and inconsistent approval of the requests. The manufacturing and engineering staff feel that the accounting and finance people don't understand their critical equipment requirements and base their analysis on the accounting numbers of the past.

In a recent meeting about the problem, Bill Johnson, the treasurer, stated, "All these expenditures for new equipment and facilities have to be looked at from the impact they have on our overall finances. We have responsibilities to the shareholders. Sometimes the requests can't be justified because they don't contribute to increased output. And we have to be careful to balance our long-term capital equipment expenditures with our profits and loan availability from the banks. The reason it takes so long for approval is that my staff is so busy. There are only so many hours in the day, you know!"

Margaret Harris, the plant manager, said, "We're doing a good job in the shop. Quality is greatly improved. The quality teams are very effective. But we are falling behind in our deliveries, and the sales department is complaining about our lead times. We have to get our new equipment and expansion on line faster. Purchasing and suppliers are cooperating with their schedules, but our approval process is slow. Also, some critical equipment requested is often turned down for reasons that I'm not sure are correct. Quality, deliveries, equipment, space, and costs are all critical areas. We're in trouble."

For the next 30 minutes, the discussion was intense. Tempers were getting short. Bill Johnson defended the present system and wanted to hire more finance and accounting personnel. Margaret suggested that she should be given an annual and quarterly capital equipment dollar amount and be allowed to spend it as needed. The analysis could be done using existing teams if they could get some additional training in financial analysis and cost accounting reports. "They are the ones that understand the critical nature of the needed equipment; let them make the decisions. Finance and accounting can sit in on the team meetings and also help with the initial training. Some of the supervisors and engineers have done cost analysis at other jobs they had in the past," she said.

Bill was very angry at the suggestion. "It will never work. I'm against it!"

1. If you were Jim Forbes, what would you do?

2. Is Bill right?

3. How could Margaret be sure that her suggestion will work?

4. To improve the speed of the equipment approval process and to make sure that the needed equipment is purchased, what would you recommend?

5. What other things should be considered?

Chapter 2   Introduction to Financial Statements

KEY TERMS

Accounting Equation Assets = liabilities + owners equity + revenue − costs.

Asset Something of value that is owned.

Balance Sheet (B/S) This statement shows assets equal to liabilities plus owners equity at a point in time.

Cash Flow Statement This statement shows the receipts and disbursements to the cash account.

Costs Expenses for materials, labor services, overhead, and other expenses. Costs are deductions from the Income Statement.

Direct Labor Adding value to the product by changing the material, assembly, or other operation.

Direct Material Direct material is found in the finished products.

Dividends Paid to shareholders after expenses and taxes have been paid.

Financial Statements Reports derived from historical accounting records. Primary financial statements are the Income Statement (I/S), the Balance Sheet (B/S), and the Cash Flow Statement. They show the financial health of the organization.

Income Statement (I/S) This statement, also called a profit and loss statement (P&L), shows revenue less costs, which equals profit or net income over a period of time.

Indirect Labor Assists direct labor to perform their tasks. Indirect labor does not add value directly to the product. Includes material handling operators, inspectors, and maintenance employees.

Indirect Material Necessary for making the product, but not found in the finished product. Examples are cutting fluid for machine tools, bins and trays for moving and storing material, or a restaurant using wax paper or aluminum foil to prepare food.

Journal A chronological record of financial events.

Ledger A classification of financial events by individual accounts or categories.

Liability Debt or obligation to pay for services or products.

Net Profits or Net Income Revenue less costs or expenses for a period of time. This may be a loss if costs are greater than revenues for the period.

Owners Equity or Net Worth The net value of an organization measured by subtracting liabilities from assets at a point in time.

Retained Earnings Funds kept in the organization after dividends are paid.

Revenue Income from the sale of products or services. Other revenue may include interest and dividend income, sale of assets, or other sources.

LEARNING CONCEPTS

- Reading an Income Statement and understanding the account names.
- Reading a Balance Sheet and learning the account names.
- Understanding types of revenues, costs, assets, liabilities, and owners equity.
- Using the Accounting Equation.
- Generating simple financial statements using the Accounting Equation.
- Generating examples of income statements and balance sheets.
- Studying related financial statements.

WHAT ARE FINANCIAL STATEMENTS?

Financial statements represent the financial history of the organization, showing its past successes and failures. These statements assist in financial planning, budgeting, and investment analysis.

Managers, engineers, and technicians have many responsibilities associated with their jobs, but frequently the least understood aspect of an organization is its financial responsibilities. Often, the attitude is, "The financial area is the responsibility of the accounting department. They do their job and I'll do mine." This attitude is changing in many organizations as they begin to use financial improvement teams to analyze projects and investments, and to make improvements in processes, products, and services.

In a small organization, employees are much closer to customers and to seeing the income and expenses of the company. Financial planning and analysis are better understood. At a large firm employees may find it difficult to see how individual actions can affect overall corporate finances and profits. However, at successful firms, both small and large, employees have a sense of the effect of their actions on company finances. Some organizations go to great lengths to provide the proper training to educate everyone in the organization about company finances.

Part of the attitude about employees' financial understanding is created by the belief that "Everyone has their own finances to deal with, so they probably understand the finances of the company." This is not necessarily so. Some individuals have better personal financial management habits than others. Personal financial behavior seems to carry over to the workplace. We don't usually change financial attitudes when we come to work. Positive attitudes of managing personal funds carry over into positive management of the firm's funds.

As employees gain experience and abilities, they are promoted in the organization. With this upward movement usually comes more financial responsibility. Therefore, it is helpful to acquire financial skills, including an understanding of financial statements, along with related work skills. Learning about budgets, accounting techniques, investing the firm's money, estimating costs, and related concepts is as important as other job responsibilities.

This chapter describes the basic format and vocabulary of financial statements, how they are generated, and how to analyze them. No attempt is made here to compress complete

accounting and financial management understanding into a single chapter. Rather, this information describes the basic techniques of a record-keeping system, the foundation for understanding financial management.

To understand financial statements, we will first look at definitions and the format of the
statements. Then we will study the terminology of the statements. Finally, we will see how the
financial statements are generated by the accounting system as it records the financial events of
the organization. Although the examples and discussion are based on the organization, the techniques and vocabulary apply to individuals' financial statements too.

INCOME STATEMENT

The **Income Statement** (I/S) covers a period of time such as a year, quarter, or month. The
units on the Income Statement accounts are dollars per year. In physical terms, the I/S can be
thought of as a flow variable with units of $/year. This is similar to the flow measurement of
physical materials such as gallons/minute or BTU/hour.

The typical format for the Income Statement is:

$$\begin{array}{l} \text{Revenue} \\ -\text{ Costs} \\ \hline \text{Profit} \end{array}$$

The Income Statement shows revenue minus costs, which equals profit over a period of
time. Or, in equation form, the definition of the I/S is:

$$\text{Revenue} - \text{Costs} = \text{Profit}$$
$$\text{for a period of time}$$

For example:

Income Statement for XYZ Inc. for Year 19xx	
Revenue	$300,000
Operating Expenses	250,000
Gross Profit	50,000
Administrative Costs	10,000
Other Income	5,000
Net Income Before Tax	45,000
Tax	20,000
Net Income or Net Profit After Tax	25,000
Dividends	10,000
Retained Earnings	$15,000

The Income Statement is sometimes called the profit and loss statement. It shows the flow of income from sales into the organization and the costs of producing those sales. The residual is net income or profit. After taxes and dividends are paid, the remainder is retained earnings. These are the funds available for reinvestment in the organization.

Noncash accounts appearing on the Income Statement are sometimes misunderstood. Revenues from sales are recorded on the I/S even though the customer gives an IOU rather than paying cash. It is assumed that the customer eventually will pay the money owed, so the sale is recorded on the I/S as an actual sale even though cash is not given at the time. Other noncash expenses on the I/S include depreciation. As equipment and other fixed assets decline in value, the noncash decline in value called depreciation expense is recorded on the I/S. Also, taxes that are calculated from the organization's profits are shown on the I/S even though they are not yet paid. These noncash revenues and expenses are combined with cash revenues and expenses on the Income Statement. This approach to accounting record-keeping is called *accrual accounting,* rather than cash accounting. The reason for using this form of accounting is to show the revenues and costs when they occur rather than showing only cash transactions. To maintain balance in the Accounting Equation, noncash entries such as depreciation must be shown. More details concerning cash and noncash record-keeping are presented in later chapters.

Generally, the format of the Income Statement includes three sections:

1. The initial section shows operating income less operating expenses. This is the income from the sale of the services or products of the firm, less the cost to produce and deliver the services or products.
2. The second section shows other income and expenses of the company. This section shows income and costs of a nonrecurring nature, such as income from securities and investments, and nonoperating costs.
3. The third section includes the taxes and dividends of the organization. Taxes for the corporation are calculated based on the firm's profit, so taxes cannot be calculated until profit is determined. Whatever funds remain after the taxes are paid belongs to the shareholders. The directors of the organization may distribute these funds as dividends to the shareholders or part of the funds may be retained in the organization.

Revenue

Revenue, or income from sales, can be from the sale of products or services, or from other sources. The primary source is the sale of products or services to customers. Both cash and non-cash revenues are shown on the Income Statement.

Other types of revenue or income include the sale of old, used equipment; the sale of defective product as scrap; and the sale of services that the firm does not usually try to sell. An example of this last-mentioned type would be selling computer services or software that the company has developed in the course of its business, even though its primary goal is to sell a different product. Finally, perhaps the firm has invested in securities of other firms or in government investments that yield dividends or interest. These other types of revenue would be included on the I/S.

Costs and Expenses

The term *expenses* is often used as a synonym for costs. Both refer to the outflow of funds in exchange for services or products received by the organization. Costs may be divided into three main categories: material, labor, and overhead expense. **Costs** are an outflow of funds. In exchange, the firm receives products or services from vendors. Purchases of raw material, expenditures for labor, management salaries, advertising expenses, rent payments to the landlord, and donations to charities are all expenses. Both cash and noncash expenses, such as depreciation, are shown on the Income Statement.

Material Costs. Material costs can be further divided into indirect and direct costs. **Direct material** is found in the product. For a television set, the cathode ray tube is direct material; for an automobile, the engine is direct material; and for a fast-food hamburger, the bun is direct material.

Indirect material is material that is necessary for production, but that is not found in the finished product. For an automobile, the cutting fluid to machine the engine parts is indirect material and the engine component is direct material. For fast food, the aluminum foil used in preparation is indirect material and the food sold to customers is direct material. For house painters, the solvent used to clean paint brushes is an indirect material and the paint used on houses is direct material.

Labor Costs. Direct labor actually adds value to the product. Examples of direct labor would be an operator machining a product's components, a chef cooking a hamburger in a restaurant kitchen, or a worker assembling components of a product.

Indirect labor is a person who is necessary for production of the product, but does not actually add value to the product. He or she assists the direct labor employee. Examples of indirect labor are the forklift driver who delivers material to the manufacturing department, the person who sweeps the floor or clears the tables in a restaurant, the quality control inspector who determines if the manufactured parts meet specifications, and the maintenance employee who repairs the production equipment so production can be resumed.

Overhead. Costs that are neither material nor labor are called *overhead* or *burden.* These include all of the other costs that are necessary to run the business. Overhead is sometimes further divided into manufacturing overhead and administrative overhead.

Manufacturing overhead includes rent for the manufacturing facilities, power costs, taxes on the building and inventory, manufacturing supervisors' wages and benefits, and expenses for maintenance and repairs. Administrative overhead includes advertising expenses, corporate taxes, management salaries and benefits, legal costs, accounting expenses, and such things as the company picnic expenses.

Profit or Net Income

Revenue can be thought of as funds coming into the firm and costs as funds leaving the organization. The remainder is **profit.** This applies to both organizations and individuals. For individuals, *costs* such as those for housing, food, entertainment, taxes, etc. are subtracted from *income* to give the difference, *profit.* A person who has an income of $300 per week and expenses of $275 would have a "profit" of $25 per week. This could be put in the bank or in a retirement account, invested in securities, used to purchase a home, deposited in a credit union, or whatever the individual desires. The $25 are the *retained earnings*. At the firm, profits can be reinvested into inventory, paid to shareholders as dividends, used to purchase equipment, or applied to other investments desired by the firm.

For the firm, profits are necessary to expand and provide more products or services to customers. Profits from a good year can be used to offset a year of low sales, high costs, and losses (negative profits). Employees may share in the profits if the firm has a profit-sharing plan. Shareholders may receive dividends from the corporation's profits as a return on their investment in the company. Profits are the driving force of the free economy and the reward for successful companies. Other organizations and individuals that see a company making profits will be attracted to that type of business and become competitors. More producers of similar products give customers a wider selection.

Internally, the organization may have different products, and the relative success of each product is typically measured by its contribution to the overall profits of the firm. Some products that are no longer purchased by customers, that have low or negative profits, may be dropped from the firm's product line. This will allow refocusing on more production of successful products. As a measure of success, profits are very important both to firms and to individuals.

Taxes

Taxes are outflows of funds from a firm to local, state, and federal governments. A percentage of the profits from products or services, wealth, sales, or other income is exchanged for government support of schools, highways, military protection, and all the other services that governments provide. There are taxes on income, "wealth" taxes based on ownership such as property tax, and consumption taxes such as sales tax. Taxes for a corporation are calculated differently from those for an individual or partnership. Corporate taxes are calculated on the profits of the organization after all costs have been deducted from the income. For individuals, taxes are based on total income less some permitted deductions, giving a net income. Tax rates are variable, being an increasing percentage as profits or income increase. In a later chapter, we will look in detail at the calculation and impact of taxes on economic analysis and decisions for both individuals and organizations. For now, we understand that taxes are shown on the Income Statement for the corporation or individual and calculated after gross profits are determined.

Dividends

Dividends are a payment to the shareholders, the owners, for their investment of funds in the company. The Board of Directors of a corporation is elected by and represents the shareholders. The board decides on the amount of dividends to be paid to shareholders. This decision is typically based on the organization's profits for the most recent accounting period. Dividends are shown on the Income Statement as a reduction net profit, which is calculated after taxes are deducted from the pretax gross profit. As the firm grows, it may need to obtain more funds from shareholders. Seeing a successful history of the firm paying dividends will indicate to future shareholders that buying the company's stock could be a good investment of their funds.

Retained Earnings

After dividends have been paid and subtracted from the net income or net profit, the remainder is called **retained earnings.** These funds are the residual of the sales income after all expenses, taxes, and dividends. These funds are a source of money for the firm and can be invested into inventory, plant and equipment, research and development, or other projects that will benefit the organization. Retained earnings belong to the shareholders, but have not been paid as dividends. Rather, these funds have been retained for expansion of the firm.

   To summarize, the Income Statement shows the revenues or income minus the costs or expenses, which gives gross profit. From this amount the taxes are calculated and deducted, leaving net profit or net income. Dividends are deducted from the net profit and the remainder is retained earnings. We add to this definition *the period of time* over which the flow of dollars occurs. For example, it would not be complete to say, "Jones Corp. had sales of $100,000 and net profits of $5,000." We would have to add that the sales and profits occurred over a period of time such as a year.

BALANCE SHEET

The next important financial statement to consider is the Balance Sheet (B/S). The **Balance Sheet** shows, at a point in time, assets, liabilities, and net worth (also called owners equity). Assets are items that the firm owns and that have value. Liabilities are debts of the organization. Net worth or owners equity is the difference between assets and liabilities. It is a liability to the shareholders/owners rather than to suppliers of material or services, but unlike other liabilities, it is owed but not usually paid to them. The retained earnings could be distributed to the shareholders should the company be dissolved.

   As we will see, the assets always equal the sum of the liabilities plus the owners equity—thus the name Balance Sheet. The reason is that the accounting system used to generate both the Income Statement and the Balance Sheet, called a *double-entry system,* always keeps the equation in balance.

   The typical format for the Balance Sheet shows the assets in the left column, usually listed from the most liquid or cash to the least liquid such as buildings. The liabilities are shown in the right column with owners equity shown below the liabilities. Sometimes the assets are shown above and the liabilities and owners equity below on the page. The Balance Sheet is shown below in equation form:

$$\text{Assets} = \text{Liabilities} + \text{Net Worth}$$
$$\text{at an instant in time}$$

   The Balance Sheet is an instant financial photograph of the firm's condition at a specific time, such as the close of business at the end of the month or the year. In physical terms, it is a state variable such as temperature or pressure. Recall that the Income Statement is a measurement of flow.

An example of the Balance Sheet for XYZ Inc. is shown below.

Balance Sheet for XYZ Inc.
as of December 31, 19xx

Assets		Liabilities and Owners Equity	
Cash	$50,000	Accounts Payable	$30,000
Securities	5,000	Note Payable	20,000
Accounts Receivable	5,000	Taxes Payable	10,000
Inventory	90,000	Common Stock	300,000
Equipment	200,000	Retained Earnings	90,000
Buildings	100,000		
		Total Liability and	
Total Assets	**$450,000**	**Owners Equity**	**$450,000**

Assets

Looking first at the assets side of the Balance Sheet, **assets** are those items to which the organization has legal title and that have some value. Assets may be further subdivided into current assets and fixed assets. A current asset typically can be converted into cash within one year. Fixed assets have a life greater than one year. Assets are generally listed in the order of liquidity with cash and most cashlike items first, down to the least liquid such as buildings, equipment, and land. There is an asset category that is even less liquid than buildings: intangibles. This category includes such intangible items as patents and goodwill.

Cash. Cash is not just money on hand, but represents checking accounts, savings accounts, and other demand deposits that are convertible to cash easily and quickly. An organization many have many cash-type accounts, especially if it has facilities in many states or other countries. It is likely to have accounts in each geographic area or different currency location.

Securities. Securities are investments in stocks, industrial and government bonds, and subsidiaries. Securities may be marketable and can typically be converted into cash in a short time.

Accounts Receivable and IOUs from Customers. Accounts receivable is included as a current asset on the asset side of the Balance Sheet. These are debts of other firms to the organization. They are created when a customer takes possession of the services or product, but gives an IOU to the supplier. The customer will pay later based on the terms of the contract. The sale is recorded even though it is a credit sale in most accounting systems. This is an accrual accounting system that assumes that the debt will be paid by the customer in the future. Until then, the IOU is held as an asset on the Balance Sheet.

Inventory. Inventory represents the purchased raw materials that the company will use to make final products; work-in-process, partially completed products; and finished goods ready for sale. Supplies, repair parts, and other indirect material are included in this inventory account also. Another way of looking at the three inventory categories is that raw material has *no* direct labor applied to it yet; in-process inventory has direct labor *being added* to it currently; and finished goods have *all the direct labor added* that is necessary for it to be sold to the customer—no more labor will be added.

Equipment. The equipment account on the Balance Sheet includes such items as manufacturing equipment, computers, copy machines, desks, lawn mowers, material handling equipment, and delivery trucks. It is usually separate from inventory since equipment is used to produce the product or service and is not sold to the customer. Only the owned equipment is shown on the Balance Sheet. Leased equipment is not shown as an asset since it is not owned by the organization. Equipment usually has a life of a number of years. Annually, or in some other time period, the equipment is reduced in value to indicate wear or technological obsolescence. This is called *depreciation,* and will be covered in detail in another chapter.

Buildings. Buildings represent the value of the structures and improvements on the land, including offices, manufacturing facilities, warehouses, farm sheds, or storage facilities. Like equipment, buildings may be reduced in value over their lives.

Land. Land is also a fixed asset. It may be used for parking lots, for warehouses and offices, or it may be idle, waiting for future expansion by the organization. Land is purchased and normally kept on the Balance Sheet at its original value since it is thought not to decrease in value as do equipment and buildings. Land is generally not depreciated unless it is being used for the production of oil, gas, minerals, or timber. Then its reduction in value is called *depletion.* This, along with depreciation, is covered in detail in a later chapter.

Other Assets—Intangibles. There may be other assets, intangibles, which the organization owns, that can be shown on the asset side of the balance sheet, such as research and development, patents, goodwill, prepaid expenses, and other such assets. Generally, these are small in value compared to the other current and fixed assets. Many organizations do not show any of these assets.

Liabilities

On the liabilities side of the Balance Sheet, the two major categories of accounts are liabilities and owners equity. **Liabilities** are debts the firm owes to suppliers, employees, government units for taxes, and others. These debts could include money owed for past wages, money borrowed, material delivered and in the inventory account but not yet paid for, and any other debts. Owners equity, sometimes called net worth, represents the money owed to the shareholders. This includes money invested by the shareholders and the past or accumulated earnings of the firm that have been retained and reinvested.

On the liabilities section of the Balance Sheet, liabilities are further divided into current liabilities and term or long-term liabilities. Current liabilities are those that are normally paid within a year. Term liabilities will be paid later than during the next twelve months.

Current Liabilities. Debts or IOUs owed by the organization that will be paid in one year are called current liabilities. Typical types of current liabilities are money owed to suppliers, or accounts payable; money owed in taxes, or taxes payable; this year's portion of borrowed funds, or notes payable; and other costs and expenses that are due within the year.

Term Liabilities. Money owed later than in the current year is called term liabilities. These may be long-term loans or notes payable that are due sometime in the future. Bonds that are due in the future are also term liabilities. Although the interest on loans and bonds may be due currently, the principal may not be due until many years later.

Accounts Payable. Accounts payable represents money owed to suppliers for material, services, equipment, or other assets received, but not yet paid for. The organization has given the suppliers an IOU for the items and they are on the Balance Sheet as an asset. As an example of an individual's accounts payable, you may have purchased a new stereo with a credit card for $300. The item is listed on the asset side as Stereo, and there is also an accounts payable entry of Credit Card Payable $300 on the liability side of your Balance Sheet.

Notes Payable. Notes payable is usually money owed to financial institutions such as banks or insurance companies, or money owed to individuals. The part that is owed during the current year is a current liability and future payments are listed as a term liability. It is different from accounts payable because it represents money borrowed rather than payment for purchased goods or services.

Taxes Payable. Taxes payable, of course, are owed to local, state, or federal governments. Taxes are paid annually, quarterly, or monthly.

Owners Equity

Owners equity or **net worth** is money owed to the shareholders. Normally, the debt is not repaid to them, but listed on the right side of the Balance Sheet with liabilities. Owners equity includes the value of the capital paid in by the shareholders when the common or preferred stock was initially sold to them. It also includes the accumulated retained earnings of the organization.

Common Stock. The common stock account represents the money that has been paid to the firm by shareholders. It is a debt owed to the shareholders by the firm. The difference between this debt and other payables is that the money owed to the stockholders will not be paid back in normal day-to-day operations. The shareholders *may* be repaid if the organization chooses to buy back some of the stock. The shareholders would also receive money back if the organization goes out of business.

Common stock has a *variable dividend* that is paid annually or quarterly to the owners. The directors, who are elected by the shareholders, decide on dividend payments based on current profits and the organization's dividend policy. Typically a start-up company, such as a small technological organization, will pay small or no dividends, choosing instead to reinvest the profits in the firm. The owners hope to be rewarded with higher share prices rather than dividends. A well-established, stable organization, such as a utility, will pay a high proportion of its earnings to the owners on a regular basis.

Preferred Stock. Preferred stock, unlike common stock, has a *fixed dividend* rate. The owners of preferred securities know when the stock is purchased what their quarterly or annual dividends will be. Like common shares, preferred shares are not normally repurchased by the corporation. They may be repurchased, but the organization is not normally obligated to do so.

Retained Earnings. Retained earnings represent the sum of all of the firm's historical profits and losses that have not been distributed to the shareholders and have been retained and reinvested into the organization.

Going out of Business and Owners Equity. To further understand the concept of owners equity or net worth, let's look at the situation of a firm going out of business and see what would happen to the organization's funds.

Consider the example of XYZ Inc. discussed above. Let's assume that the firm decides to go out of business. Theoretically, the assets are worth and could be sold for $450,000. This money would be placed in the cash account and used to pay all of the outstanding liabilities of $60,000, leaving $390,000 in the checking account. This equals the value of the owners equity, $300,000 + $90,000 owed to the shareholders. When this amount is distributed to the shareholders, the Balance Sheet is "blank" and the firm has no remaining assets or liabilities.

Although it would be unusual for the firm to go out of business if it is successful, this example demonstrates the relationship between assets and the owners equity account. The owners equity account belongs to and is owed to the shareholder owners of the organization even though the funds are retained in the company. Bankruptcy does occur, however. If the organization fails to make profits over a period of time and lets liabilities exceed assets, resulting in zero or negative owners equity, it may fail. Sometimes, it can seek assistance from the courts under the bankruptcy laws and recover financially to become profitable once again.

MORE FINANCIAL STATEMENTS

Now let's look at another set of financial statements. They are a little more complex than the previous ones, and are typical of the type that could be found in an organization's annual report to its shareholders.

Barmac Inc.
Income Statement
12 Months Ending Dec. 31, 19xx

Net Sales	$1,650,000
Less: Cost of Goods Sold	1,155,000
Gross Profit	495,000
Depreciation Expense	52,000
Selling, Admin. Expense	242,000
Operating Profit	201,000
Investment Income	16,000
Interest Expense	(36,000)
Profit Before Tax	181,000
Federal Income Tax	58,000
Net Income (Profit After Tax)	123,000
Dividends Paid	61,000
Retained Earnings	$62,000

Barmac Inc. Financial Statements

Look at the Income Statement for Barmac Inc. There are some new terms that require explanation:

Barmac Inc.
Balance Sheet
Dec. 31, 19xx

Assets			Liabilities	
Current Assets			**Current Liabilities**	
Cash		$35,000	Accounts Payable	$145,000
Marketable Securities		80,000	Notes Payable	60,000
Inventories		610,000	Federal Tax Payable	30,000
Prepaid Expenses		9,000	**Total Current Liabilities**	**$235,000**
Total Current Assets		**$734,000**	**Term Liabilities**	
Fixed Assets			Debentures 8% 2010	260,000
Buildings	300,000		**Total Liabilities**	**$495,000**
Equipment	490,000		**Owners Equity**	
Office Equipment	30,000		Preferred Stock	$160,000
Total Plant and Eq.	820,000		Common Stock	300,000
Less: Accumulated			Accumulated Retained Earnings	492,000
Depreciation	(180,000)		**Total Owners Equity**	**$952,000**
Net Fixed Assets		$640,000		
Land		$70,000		
Other Assets				
Intangibles		3,000		
Total Assets		**$1,447,000**	**Total Liabilities + Owners Equity**	**$1,447,000**

- Net sales are the sales of the organization's products and services after returns and adjustments. Sometimes the gross sales are not the final sales because of customers' returns.

- Cost of goods sold (CGS) is the next account on the Income Statement. CGS is the total cost of manufacturing the product. It includes direct and indirect materials, direct and indirect labor, and the overhead expenses that occurred during manufacturing such as direct supervision, equipment expense, building costs, utilities, and any other costs directly associated with production.

- Depreciation is an estimate of the cost of the fixed assets' decrease in value. When equipment is purchased, it becomes an asset. This asset decreases in value over time because of wear and technological obsolescence. This reduction in value reduces the dollar value of the asset on the Balance Sheet and the resulting deduction in value on the Balance Sheet becomes an expense on the Income Statement for that period of time. The Income Statement shows the *current* year's depreciation, while the Balance Sheet shows all the *accumulated* depreciation of the fixed assets over all years. This transaction will become clearer when we discuss the generation of the financial statements.

- Selling and administrative expense is also known as administrative overhead, marketing, and administrative cost. The account includes the overhead items that are not related to manufacturing, such as office, management, administrative, and marketing costs. Overhead can be divided into two parts: overhead for the production of the product and overhead applicable to administrative or nonmanufacturing areas.

- Operating profit, like many accounting terms, is not universal. Here, it implies the difference between the net sales and the costs associated with the sale and production of products or services. This is an attempt to identify profits that are related to the manufacture and sale of the product and does not include other income from investments or other costs such as interest expenses on loans.

- Investment income is typically from subsidiaries, dividends on stock owned, or other sources. It is separate from the money received from the sale of products or services.

- Interest expense is the money paid during the year for interest on loans and bonds. In this case, parentheses rather than a negative sign are used to show a cost.

- Profit before tax is the difference between all income and costs during the period. For corporations, taxes are calculated using this figure. Income tax is deducted from the profit before tax and the difference is called profit after tax. Often the terms net profit or net income are used.

- After taxes are deducted, dividends distributed to shareholders are determined. This, by the way, results in the profits of the corporation being taxed twice. First the organization pays income taxes, and then when the dividends are distributed to the shareholders, they must report the dividends as income and pay their individual income tax on the dividends.

- Funds left over after dividends are paid are kept by the firm. They are retained earnings.

Looking at the Balance Sheet of Barmac Inc., we can see that the format is expanded to include more accounts than in the previous examples. Some of these new terms need explanation.

- The cash, marketable securities, and inventories accounts are similar to those on the statements of XYZ Inc.

- An account that we have not seen before is prepaid expenses. These are costs paid *before* they are due. An example is an organization that pays rent not yet due. If the October rent is paid in August, then the amount paid would be a current asset until it normally comes due. When it is due, the asset is reduced and an expense is created on the Income Statement.

- Land, buildings, plant, and equipment are fixed assets, similar to those seen previously. This statement shows the original price paid for the fixed assets and subtracts the accumulated depreciation to determine the net fixed asset value. On the Income Statement, only the current year's depreciation is shown as an expense.

- An example of an intangible asset would be the value of patents or trademarks.

On the liabilities side of the Balance Sheet, the two major categories are liabilities and owners equity. Within the liabilities section, the accounts are divided into current liabilities and term or fixed liabilities.

- Accounts payable, notes payable, and taxes payable have been discussed. Only the notes payable portion for the current year is shown as a current liability. Other notes payable would be shown as term liabilities.

- The account called debentures represents bonds that are owned by individuals and other organizations. A bond is a loan. Usually a bond is marketable and can be transferred to others, which is different from a loan from a financial institution, which is usually not marketable. The bond listed here is an 8% bond, which means that the investor receives an 8% payment each year during the life of the bond. Most bonds are sold for a value of $1,000. So 8% of $1,000 would be $80 per year interest paid to each bondholder each year. At the end of the life of the bond, in this case the year 2010, the principal of $1,000 would be repaid by the firm. In the case of limited profits for a year, the bondholders typically are paid before the shareholders. It is said that the bonds are "senior" to the stock.

- Stock types have been discussed previously. Preferred stock is "senior" to the common stock and is paid before common stock if the organization has limited funds.

- Accumulated retained earnings is the sum of all the historical profits that the organization has earned (minus losses) since it was started. This amount, like the other owners equity accounts, is owed to the shareholders even though it is not normally paid to them. It is the difference between assets minus liabilities and is the shareholders' equity in the firm.

The two financial statement examples for XYZ and Barmac demonstrate the basic format and terminology of income statements and balance sheets. Each organization's financial statements are unique. The statements often reflect the differences of the business or industry in which they operate. A manufacturing organization, for example, uses different terminology than a telephone, pharmaceutical, hotel, or not-for-profit organization. However, with a little practice, you will be able to understand the financial statements for most organizations.

PERSONAL FINANCIAL STATEMENTS

The above discussion presented a sample of financial statements for corporations. The same concepts can be applied to an individual's finances. For example, when an individual goes to the bank or other financial institution to obtain a loan, one of the important items of information for the bank is the individual's income and expenses for the past year or a period of years. Usually, copies of tax returns are submitted as a summary of these revenues and costs. In addition, the loan officer might request a list of the person's assets and liabilities. The difference between a person's assets and liabilities is his or her net worth.

For example, consider two individuals applying for a loan at a bank. The first has assets of $1,000,000, but also debts of $990,000. The difference of $1,000,000 − $990,000 is $10,000, which is the net worth of the individual. A second person has only $50,000 of assets and has $5,000 in debts or liabilities. The net worth of the second person is $45,000. Looking only at net worth, the millionaire would not qualify for as large a loan as the second individual. Of course the lender would look at income and other items as well. Large assets are not the only measure of financial strength.

Below is a typical Income Statement and Balance Sheet for an individual:

Income Statement of J. Smith for 19xx

From Employment—Wages, Profit Sharing, Bonus	$45,000	
Interest and Dividend Income	1,500	
	$46,500	
Mortgage Interest		$3,800
Auto Expense		1,100
Medical Expense		550
Children's College Expenses		7,500
Food and Clothing		9,500
Insurance Expense—Home, Car, Medical, Etc.		3,000
Vacation and Entertainment		2,500
Miscellaneous		2,450
Total Expenses		30,400
Net Income Before Taxes		16,100
Local, State, and Federal Tax + FICA		14,100
Earnings Retained After Income, Expenses, and Taxes (these funds are available for saving and investment)		$2,000

Balance Sheet for J. Smith for Dec 31, 19xx

Assets		Liabilities	
Cash,	$4,000	Mortgage	$21,000
Savings and Investments	23,000	Credit Cards	400
IRA/Retirement	34,000	Total Liabilities	$21,400
Cash Value of Insurance	1,000	Net Worth of J. Smith (found by subtracting total liabilities from total assets)	
Automobile	5,000		
Home	65,000		$121,600
Personal Property	11,000		
Total Assets	$143,000	Total Liabilities and Net Worth	$143,000

From the above example we can see that J. Smith's statements are similar to the corporation's financial statements we discussed. The J. Smith Income Statement shows the family has $2,000 left at the end of the year from earnings of $46,500. This could be invested into savings, an IRA, or any other investments. This is approximately 4.3% residual earnings or "profit" on income of $46,500 for the year.

Financial statements for organizations and individuals are similar in format and presentation. Many of the concepts we will discuss for corporations' financial statements and bookkeeping systems can be applied to those of individuals. An exception is the selling of securities by the corporation. With the growth of home computers and user-friendly software for managing a family's money, many people are applying the financial concepts discussed here. Certainly, someone with a family business would find it necessary to perform basic accounting functions and generate financial statements on a regular basis to measure the business' financial progress.

FINANCIAL STATEMENT GENERATION

We have seen the general format and terminology of income statements and balance sheets. Now, let's look at how these statements are generated.

The accounting system is responsible for collecting and recording all of the financial events that occur in the organization, organizing this financial data into usable information, and presenting internal financial reports and financial statements. Our next step is to gain understanding of this historical data collection and organization using the Accounting Equation. The Accounting Equation combines the two categories of the Income Statement, revenue and costs, and the three categories of the Balance Sheet, assets, liabilities, and owners equity, to collect, categorize, and record the financial events in the organization.

Although a complete demonstration of all the entries in the accounting system would require more detail than is contained here, we can still look at the basic entries and gain an understanding of simple accounting techniques. The more understanding that engineers, managers, technicians, designers, purchasing personnel, and marketing people have of the accounting system, the better they can implement the technical and financial goals of the organization.

Financial Statements for ABC Inc.

Let's assume that a firm is just beginning, and that it has made no financial transactions yet. Its initial financial events are shown below in a *journal format*. A **journal** is a chronological listing of financial events. It is derived from the French word *jour,* meaning day. For the following discussion, we will be using some manual techniques to generate the Income Statement and Balance Sheet. In actual practice, most accounting systems are now computerized. Even for home bookkeeping purposes there is good software available for personal computers. These manual demonstrations will show the basic techniques of generating simple financial statements.

The Accounting Equation

Now let's discuss the **Accounting Equation,** a combination of the Income Statement and Balance Sheet accounts.

$$\text{Assets} = \text{Liabilities} + \text{Owners Equity} + \text{Revenue} - \text{Costs}$$

This equation is the framework for all accounting transactions, and since it is an equation, it must always be kept in balance. Note that it is a combination of the Income Statement (revenue − costs) and the Balance Sheet (assets = liabilities + owners equity). By making entries into this equation, accounts under the five main headings can be changed as financial events occur. At the end of the accounting period, these accounts can be summarized, and this resulting summary becomes the Income Statement, Balance Sheet, and other financial statements.

The Double-Entry System

The technique applied to the Accounting Equation is the *double-entry system.* That is, for every financial event, a minimum of two entries are made into the Accounting Equation. When only one entry is made, as into a personal checkbook register, it is a "single-entry system." The individual records the date, check number, check recipient, and amount in the check register, and then deducts the amount from the previous balance to get a new balance. No other entries are made in the individual's bookkeeping system.

For double entry, the first entry and then a parallel entry are made into the equation. For example, if a check is for "Rent" then the double entry is deducted from the checkbook and the cash account under assets in the equation, and also entered into the rent expense account under costs in the equation. As we will see, this double entry will maintain the Accounting Equation in balance after all entries are made.

We will be using the double-entry system requiring a minimum of two and sometimes additional entries for each financial event. Each event is taken from the chronological listing of the financial events, the journal, and recorded into the Accounting Equation. Each account category in the Accounting Equation is called a **ledger** account.

Journal for ABC Inc.

a. Sell common stock, $5,000

b. Buy materials, $1,000

c. Pay rent, $200

d. Buy equipment on credit, $2,000

e. Sell all inventory, $1,500

f. Pay wages, $200

g. Buy material, $800

h. End of accounting period, generate financial statements

We will make additions and subtractions to the Accounting Equation for the given financial events from the journal of ABC Inc., and we will always be sure to keep the equation in balance.

The Accounting Equation for ABC Inc. is shown below along with the entries from the journal into the appropriate accounts under each of the five main equation headings. The lowercase letter notation identifies the journal entries as they are transferred to the categories of the Accounting Equation.

A	=	L	+	OE	+	R	–	C
Cash + 5,000 a				Stock + 5,000 a				
– 1,000 b								
– 200 c								Rent – 200 c
+ 1,500 e						Sales + 1,500 e		
– 200 f								Wages – 200 f
– 800 g								
Inv. + 1,000 b								
– 1,000 e								Cost of
+ 800 g								Goods
								Sold – 1,000 e
Equip. + 2,000 d		Accts. Pay. + 2,000 d						

An approach that is helpful in understanding the placement of the double entries from the journal into the Accounting Equation is to think in terms of exchanging money for some service or product. For example, when raw material is purchased, cash is given to the supplier and the supplier gives the customer material. The two accounts that are affected are cash, which is an asset, and material or inventory, which is also an asset. This exchange causes cash to decrease and inventory to increase. Rather than trying to memorize all of the possible combinations of accounts, think of the process of exchange or contractual relationship and almost automatically the accounts affected will be identified.

Looking at the above Accounting Equation entries for ABC Inc. in detail:

a. When common stock is sold, cash is exchanged for ownership. The cash comes into the firm from the shareholders and the opposite transaction is that the shareholders are owed the amount they paid. In this case, cash increases by $5,000 and the owners equity account called stock increases by $5,000. Note that this transaction adds $5,000 to both sides of the Accounting Equation and the equation is in balance before and after this transaction.

b. When materials are purchased, cash is exchanged for materials. In this case, cash is reduced by $1,000 and another asset account called inventory is created and increased by $1,000. The equation is in balance after the transaction. This transaction increases as well as decreases the asset side of the Accounting Equation by $1,000. The equation remains in balance.

c. When rent is paid, cash is reduced, and the rent expense under the cost column is also increased. Cash is reduced by $200 and rent is increased by $200. This transaction *subtracts* $200 from both sides of the Accounting Equation. Note that increasing a cost is increasing a *negative amount.* The equation is in balance after the transaction.

d. Equipment is received from the vendor, and ABC Inc. gives the vendor an IOU, accounts payable, for the amount owed. The equipment account under assets increases by $2,000 and the liabilities account called accounts payable increases by $2,000. The Accounting Equation is still in balance because $2,000 has been added to *both sides* of the equation.

e. A product sales transaction is a little different from the previous transactions because it needs four entries. The customer gives money in exchange for the product. In this case the customer gives $1,500 to ABC. Cash increases by $1,500 and the revenue account, sales, increases by $1,500. In addition, the inventory has been reduced to 0, so the inventory account must be reduced by $1,000, the cost of the inventory. The other half of this entry is to create an account called cost of goods sold, a negative, under the cost column in the Accounting Equation. A total of four entries were needed to record this transaction. Note that the equation is in balance: $1,500 has been added to both sides of the equal sign and $1,000 has been subtracted from both sides of the equal sign. Note also that a profit has been generated since the cost of the inventory was $1,000 and the revenue was $1,500, giving a profit of $1,500 − $1,000 = $500.

f. When wages are paid, cash is reduced and the expense called wages is entered. Cash is reduced by $200 and $200 in wages is entered under the cost (negative) column. The equation is still in balance because the same amount was *subtracted* from both sides of the equation.

g. This entry is similar to entry b. Cash is reduced and the inventory is increased by $800. The assets side of the equation is both reduced and increased by the same amount, and it is in balance after the entries.

h. The final step in generating the financial statements is to obtain the total amounts in each account. The totals in each account are found by adding and subtracting algebraically the amounts in each account as a result of transferring the information from the journal to the Accounting Equation ledger accounts.

The summary of account totals is shown below.

A	=	L	+	OE	+	R	−	C
Cash $4,300		A/P $2,000		Stock $5,000		Sales $1,500		Rent $200
Inv. 800				Ret. Earn. 100*				Wages 200
Equip. 2,000								CGS 1,000

*Note: This final adjusting entry occurs *after* the Income Statement is generated by algebraically summing the revenues and costs ($1,500 − $1,400) and a profit of $100 is calculated for the month. Since this is the shareholders' property, but not distributed, it will be included under the owners equity account (owed to the shareholders) on the Balance Sheet.

Other than the retained earnings entry on the balance sheet, no new entries are made. The values for the Income Statement and Balance Sheet are from the totals in the Accounting Equation accounts.

Income Statement for ABC Inc.

Next, let's generate the Income Statement by formatting the totals from the Accounting Equation into the typical Income Statement format.

Now a final or closing entry must be made. As soon as the retained earnings are calculated on the Income Statement, this amount, $100, must be put under the owners equity account in the Accounting Equation and on the Balance Sheet. The reason for this entry is that the retained earnings belong to the shareholders.

Income Statement for ABC Inc.
January 31, 19xx

Sales		$1,500
Cost of Goods Sold		1,000
Gross Profit		500
Expenses		
Rent	200	
Wages	200	
Profit Before Tax		100
Tax		0
Profit After Tax—Net Income		100
Dividends		0
Retained Earnings		$100

Balance Sheet for ABC Inc.

We are now ready to generate the balance sheet. We look at the totals of the accounts at the end of the month and format them into the standard Balance Sheet format.

Balance Sheet
ABC Inc.
January 31, 19xx

Assets		Liabilities and Owners Equity	
Cash	$4,300	Accounts Payable	$2,000
Inventory	800	Common Stock	5,000
Equipment	2,000	Retained Earnings	100
		Liabilities and	
Total Assets	$7,100	Owners Equity	$7,100

On the Balance Sheet, total assets equal the sum of total liabilities and owners equity, which leads to the name Balance Sheet. The double-entry system requires a minimum of two entries in the ledger/Accounting Equation and the equation must be kept in balance for each entry into it.

The above example is simplified, of course. In actuality, accounting entries number in the hundreds and thousands per accounting period, and the arithmetic is performed by computer software. In its simplified form, however, the example demonstrates the principles of the double-entry system using the Accounting Equation.

Balance Sheet and Income Statement for Jones Co.

As another example of financial statement generation, let's assume that Jones Co., a small firm, has been in business for some time. The B/S for the end of July is shown below. The ending values of the July B/S will be the beginning values for August.

Jones Co., July 31, 19xx
Balance Sheet

Assets		Liabilities	
Cash	$39,000	Accounts Payable	$78,000
Accounts Rec.	75,000	Notes Payable—Bank	20,000
Securities—T-Bills	35,000	Taxes Payable	25,000
Inventory	120,000	**Current Liabilities**	**$123,000**
Current Assets	**$269,000**	**Term Liabilities**	
Fixed Assets		Mortgage on Bldg.	340,000
Equipment	45,000	**Term Liabilities**	**$340,000**
Building	670,000	**Total Liabilities**	**$463,000**
Fixed Assets	**$715,000**	**Owners Equity**	
		Common Stock 1,000 shares	$5,000
		Retained Earnings	516,000
Total Assets	**$984,000**	**Total Liabilities and O.E.**	**$984,000**

Jones Co. Journal Entries for August 19xx

a. Sold on credit, $35,000 in material to South Side Building. (Note: This is the selling price for the material. The value of the inventory is unknown at this time. At the end of the month, a physical count of the inventory will be made and the CGS entry can be made then.)

b. Pay $15,000 on accounts payable.

c. Advertising in trade journal, paid $3,000.

d. Wages paid, $2,500.

e. Received $22,000 from customer for account receivable.

f. Received bill from supplier for repair of equipment, $1,000

g. Paid sales commission of $1,500.

h. Sold $28,000 of material for cash to customer (see note for entry a).

i. Billed Simpson Co. $8,000 for installation services.

j. Paid Mortgage Co. $17,000 of which $15,000 was toward principal and $2,000 was interest.

k. Paid part of tax owed, $2,000.

l. Declared and paid dividends of $1,000.

m. Depreciation of equipment is $1,000 and the building is $2,000.

n. Physical count of inventory at the end of the month is $80,000. (Beginning inventory − removal during month = ending inventory)

o. End of month; generate Income Statement and Balance Sheet.

Since there are ending amounts in the accounts from last month, we will use them as starting values for the current month's entries into the Accounting Equation. The Accounting Equation with entries is shown below:

A	=	L	+	OE	+	R	−	C
Cash		**A/P**		**Com. Stock**		**Sales**		**Ad. Exp.**
+ 39,000		+ 78,000		+ 5,000		+ 35,000 a		− 3,000 c
− 15,000 b		− 15,000 b				+ 28,000 h		
− 3,000 c		+ 1,000 f		**R.E.**		_____		**Wage Exp.**
− 2,500 d		_____		+ 516,000		63,000		− 2,500 d
+ 22,000 e		64,000		+ 17,000				
− 1,500 g				_____		**Instal. Income**		**Repair Exp.**
+ 28,000 h		**N/P**		533,000		+ 8,000 i		− 1,000 f
− 17,000 j		+ 20,000						
− 2,000 k		**Tax Payable**						**Sales Comm. Exp.**
− 1,000 l		+ 25,000						− 1,500 g
_____		− 2,000 k						
47,000		_____						**Interest Exp.**
		− 23,000						− 2,000 j
A/R								
+ 75,000		**Mortgage Payable**						**Div. Exp.**
+ 35,000 a		+ 340,000						− 1,000 l
− 22,000 e		− 15,000 j						
+ 8,000 i		_____						**Depreciation**
_____		325,000						− 3,000 m
96,000								
Securities								**Cost of Goods Sold**
+ 35,000								− 40,000 n
Inventory								
+ 120,000								
− 40,000 n								
_____								
80,000 n								
Equipment								
+ 45,000								
− 1,000 m								
_____								
44,000								
Building								
+ 670,000								
− 2,000 m								
_____								
668,000								

a. Only part of the entries into the accounting equation can be made now. The inventory and CGS entries can be made when the physical inventory is made at the end of the month.

b. Cash and A/P both decrease.

c. Cash decreases and an advertising expense is created. The same amount is subtracted from both sides of the equation.

d. Cash decreases and a wage expense is created. The same amount is deducted from both sides of the equation.

e. Cash increases, A/R is reduced. The same amount is added to and subtracted from the same side of the equation.

f. Received, but *did not pay* invoice. Increase A/P and create a repair expense. The same amount is added to and subtracted from the right side of the equation.

g. Cash decreases and sales commission expense is created. Subtract the same amount from both sides of the equation.

h. Similar to entry a.

i. Sent invoice that was *not yet paid*. A/R increases and revenue increases. No inventory involved, only service.

Jones Co.
Income Statement, August 19xx

Sales and Installation	$71,000
Cost of Goods Sold	40,000
Operating Profit	31,000
Expenses	
Advertising	3,000
Wages	2,500
Repairs	1,000
Commission	1,500
Depreciation	3,000
	20,000
Less: Interest Exp.	2,000
Profit Before Tax	18,000
Tax*	0
Profit After Tax	18,000
Dividends Paid	1,000
Retained Earnings	$17,000

*Note: No tax is calculated for the month. This will be done for the quarter at the end of September. The tax paid this month was due from the previous quarter.

j. Principal reduces the mortgage payable and decreases cash. Decrease the equation by $15,000 on both sides. Interest payment reduces cash and creates an expense. Reduce both sides of the equation by the same amount, a total of three entries.

k. Cash decreases and tax payable decreases. The same amount is deducted from both sides of the equation.

l. Cash decreases and dividend expense is created. Reduce the equation by the same amount on both sides.

m. Equipment and building values decrease and depreciation expense is created. Reduce both sides of the equation by the same amount. Depreciation is a noncash expense.

n. Starting inventory of $120,000 less unknown deductions must equal $80,000. The deductions for the month must be equal to $40,000. Reduce inventory by $40,000 and create a cost of goods sold of $40,000. Both sides of the equation decrease by $40,000.

o. By taking the net ending values (sum the positive and negative values) of each of the accounts in the five categories, the Income Statement can be generated.

The ending retained earnings of $17,000 are added into the retained earnings under the Accounting Equation. This "disconnects" the Income Statement from the Balance Sheet, and is a closing entry for the Income Statement. Now, when the account totals are placed into the standard Balance Sheet format, the statement will balance. The ending Balance Sheet is shown below:

Jones Co.
Balance Sheet, August 31, 19xx

Assets		Liabilities and Owners Equity	
Cash	$47,000	Accounts Payable	$64,000
Accounts Receivable	96,000	Notes Payable	20,000
Marketable Securities	35,000	Taxes Payable	23,000
Inventory	80,000	**Current Liabilities**	**107,000**
Current Assets	**258,000**	Mortgage on Bldg.	325,000
Equipment	44,000	**Total Liabilities**	**432,000**
Building	668,000	Common Stock	5,000
Fixed Assets	**712,000**	Retained Earnings	533,000
		Total Owners Equity	538,000
Total Assets	**$970,000**	**Total Liabilities and O.E.**	**$970,000**

Related Financial Statements

Although the income statements and the balance sheets are the primary financial statements generated by the accounting system, other reports might be relevant to companies such as Jones Co. and ABC Inc.

- The Cash Flow statement is a summary of the increases and decreases in the cash account for the period. This statement shows the available cash for investment in equipment, buildings, inventory, R&D, payment of dividends, repayment of debt, and other investments. Unlike the Income Statement, the Cash Flow statement shows only the cash income and expenses, not noncash flows such as depreciation, amortization, depletion, and credit sales. Cash flow is very important to engineers, managers, and other nonfinancial personnel as they analyze projects and investments. Cash flow is discussed in later chapters.

- The cost of goods sold on the Income Statement can be expanded to show the details of the CGS account. These details would be the listing of the material costs, labor costs, and manufacturing overhead. This statement would be useful to manufacturing management and would assist in measuring the product costs in the manufacturing process.

- Inventory on the Balance Sheet is frequently shown as raw material, in-process, and finished goods. The manufacturing, purchasing, and inventory control personnel would need to know the details of inventory costs on a regular basis to manage the inventory. In a computerized system this could be a real-time statement that could show all the additions and subtractions to the inventory accounts as they occur.

- The accounts receivable accounts can be expanded to show the details of those who owe the firm money, as well as those who are late in paying their accounts. This would be helpful to the sales department.

- Sales broken down by customer and product type would be important to the marketing department.

- A complete listing of the equipment account and the individual value of the assets would be important to production management. This report could show the initial purchase price of the equipment and the accumulated depreciation for each piece of equipment.

- In general, every account shown on the financial statements contains more details than are shown on the statements. Other reports could be generated. Each account in the ledger can be reported separately. In the days of manual accounting systems, this would require simply copying and reformatting pages from the ledger that represent the account of interest. With today's computerized accounting systems, the individual accounts may be accessed from the accounting system database.

Accounting Audits and Standards

Shareholders want to be assured that the organization is managed honestly. Shareholders annually appoint outside accounting firms to audit the organization's accounting system. Not only must the accounting system be free of errors, it also must meet accounting standards and conventions with respect to internal practices and generation of the financial statements. Normally, the internal accounting system also does periodic internal audits of itself. By meeting the accounting standards and the internal and external audits, the organization assures shareholders and other readers of the financial statements that the presentation is accurate and fairly represents the financial results and health of the organization.

> Figures don't lie, but sometimes liars do the figuring.
> —Anon.

Now that historical financial record-keeping and financial statement generation have been introduced, we will explore depreciation and cash flow calculation. The financial history of the organization has been recorded. The next question is how to use these statements as performance measures to assist with project analysis and financially evaluate the organization.

SUMMARY

- This chapter introduced and defined financial statements. The format and contents of the Income Statement and Balance Sheet were presented, along with examples of typical financial statements. There is little difference between the techniques and principles used to generate statements for individuals and those used for corporations.

- The Accounting Equation was introduced and examples of generating income statements and balance sheets from a simple journal were demonstrated. Using a double-entry system, the Accounting Equation's balance is maintained. The financial statements are generated from the Accounting Equation.

- Accounting terminology, including new definitions, was introduced. By understanding the terminology and techniques that the accounting system uses, management and technical staff can communicate effectively and better understand the accounting role in the firm.

- By "keeping the financial score," an accounting system provides financial statements that detail the financial history of the firm and also provide a financial database for use in decision making by engineers, technicians, managers, and others in the organization.

> Some financial statements are like sausage-making. The more you know about it, the less likely you are to swallow it.

QUESTIONS

1. Define Income Statement and Balance Sheet.
2. Give examples of types of revenues.
3. What are three basic types of costs? What are examples of each type?
4. What is the difference between direct and indirect labor? Between direct and indirect material?
5. Identify the following as indirect or direct material, direct or indirect labor, or manufacturing or administrative overhead costs.
   a. Steel used to make a car
   b. Gas used in a heat-treating furnace
   c. Federal income taxes
   d. Salespersons' commissions
   e. Wages of the maintenance foreman
   f. Training costs of accountants
   g. Firm's donation to the Boy Scouts
   h. Drafting paper
   i. Depreciation of computer
   j. Travel expense of quality control engineer visiting a supplier
6. What are three classifications of inventory?
7. Describe what would happen to the financial statements if the firm went out of business.
8. Define or identify the following:
   a. Double-entry system
   b. Accounts payable
   c. Retained earnings
   d. Assets
   e. Net worth
   f. Single-entry system

9. What four accounts are affected when a sale of product is made for cash? What would be different if the sale was a credit sale?
10. Why is the Income Statement generated before the Balance Sheet?
11. Explain why the owners equity includes common stock as a liability even though the shareholders will not normally be repaid their investment.
12. What is the Accounting Equation?
13. Why does the Balance Sheet always balance?
14. State the entries in the Accounting Equation if the following events occur:
    a. Old equipment is sold for scrap.
    b. A customer returns product for a refund.
    c. A debt is paid.
    d. Part of the inventory is damaged and must be thrown away.
    e. Taxes are paid.
    f. The firm buys back some of its own stock in the open market.
    g. A customer returns a defective product and receives a cash refund.
    h. Interest is paid on a debt.
    i. Defective products are sold to a scrap dealer.
    j. Employee withholding tax is paid to the IRS.
    k. Bonds are redeemed by the company.
    l. Funds are loaned to a subsidiary.
    m. Inventory is transferred from raw material to in-process.
15. What account appears on both the Income Statement and the Balance Sheet?
16. How does the conventional manual system of keeping account records differ from computerized methods?
17. What is the difference between the terms "current" and "fixed"?
18. How can a manager, designer, or manufacturing engineer use the financial statements in their individual jobs in the firm?
19. What are examples of subsidiary statements that can be derived from the accounting records that can assist the technical person or manager in the firm?
20. What is the meaning of the following terms?
    a. Depreciation
    b. Retained earnings
    c. Indirect material
    d. Profits
    e. Intangibles
    f. Common and preferred stock
    g. Accounts receivable
21. What would be the benefits to an individual of using one of the currently available software accounting products for keeping personal and family financial records?

PROBLEMS

For the following problems, entries can be made into the accounting equation using pencil and paper, but it is also possible to use a spreadsheet such as Lotus or Excel. A spreadsheet template of the five columns of the Accounting Equation can be used for all the problems.

1. Obtain an annual report of a public company from their shareholder relations department, a brokerage office, or from the Web. Read the financial statement section and notes to financial statements.
2. Generate your own personal Income Statement by listing all of the revenues and costs for either a month or a year. Use the format presented in this section.
3. Generate your own personal Balance Sheet by listing all the assets and liabilities; the difference will be the net worth. Use the format presented in this section.
4. Widget Inc. is just starting operations. Below are the first week's financial transactions. Set up the ending B/S and period I/S for Widget Inc.
    a. Sell $10,000 worth of common stock.
    b. Buy 10 widgets for $100 each.
    c. Purchase $1,500 worth of equipment.
    d. Sell 5 widgets at $150 each to Smith on credit.
    e. Buy 10 widgets from Jones Supply for $100 for credit each.
    f. Borrow $500 from National Bank and Trust.
    g. Pay wages of $100.
    h. Pay federal tax of $20.
    i. Pay rent of $50.
    j. Pay dividend of $30.
    k. End of month.
5. A small one-person business is just starting. It will provide housekeeping and lawn-care services for other firms. Following are the startup financial events for the company.

a. Place $5,000 into a checking account for the company by the proprietor.

b. Buy cleaning equipment and lawn equipment for cash, $1,500.

c. Buy supplies on credit from Janitor Supply Co., $200.

d. First job pays $450 for lawn services.

e. Bill R and J Company for housekeeping services, $500.

f. Pay wages of $300.

g. Pay truck expenses at Shell Service Station, $120.

h. R and J pays their bill to company.

i. Purchase supplies for $200.

j. Pay Janitor Supply Co. in full.

k. Generate financial statements.

6. For Ajax Inc., set up the initial Balance Sheet, make the entries for the month using the Accounting Equation, and present the Income Statement for the next month and the ending Balance Sheet. At the end of this month, the following balances are in the accounts: cash $5,000; inventory $10,000; equipment $3,000; land $10,000; owe Jones $10,000; common stock $10,000; retained earnings $8,000. (Note that the value of the inventory sold is unknown at the time of sale. However, the value of the remaining inventory is known at the end of the month. The value sold is the difference between the inventory received and the inventory remaining.)

a. Buy inventory from Jones on credit, $10,000.

b. Sell goods for $4,000.

c. Sell common stock, $1,000.

d. Pay wages, $2,000.

e. Pay tax, $1,000.

f. Sell goods to Smith on credit for $15,000.

g. Pay rent, $100.

h. Pay Jones $2,000.

i. Depreciate equipment $200.

j. Inventory at the end of the month is worth $7,000.

k. End of month.

7. For Giswhiz Inc., set up the March Income Statement and the Balance Sheet at the end of March. Giswhiz is an organization of manufacturers' representatives that sell products for other firms and do not manufacture any of their own products. The company's income is derived from sales commissions. Giswhiz also has some equipment that it rents to customers.

Giswhiz Manufacturers' Representatives Inc.
February 28, 19xx

Cash	$1,100	A/P Industrial Advertising	$120
A/R Hammil Inc.	220	A/P ABC Metals	1,000
A/R Michaels Inc.	680	Common Stock	5,000
Equipment	1,500	Retained Earnings	1,380
Land	4,000		
	$7,500		$7,500

March transactions:

a. Received $300 from Michaels Inc.

b. Paid Industrial Advertising in full.

c. Received bill from Wilson Co. for equipment, $320.

d. Received $100 from Thomas Inc. for rentals of equipment.

e. Paid bill from Metal News advertising, $70.

f. Received $180 from Hastings Co. for commission earned on the sale of goods for Hastings.

g. Billed Hammil $200 for commission charges.

h. Paid Wilson Inc. $150 on account.

    i. Received bill from Industrial Advertising for $100.
    j. Billed Michaels for commission charges $50.
    k. Paid D. Rose, a salesperson, her share of commissions, $200.
    l. End of month.

8. The following is the ending Balance Sheet for Part-Time Programmers Co. The company provides temporary computer and data processing help to organizations. It is a small, new organization.

Part-Time Programmers			
Oct. 31, 19xx			
Cash	$4,500	Accounts Payable	$1,800
Gov. Securities	5,000	Wages Payable	2,000
Supplies	1,000	Tax Payable	500
Equipment	8,000	Net Worth	14,200
Total Assets	**$18,500**	**Total Liabilities + N.W.**	**$18,500**

    a. Receive $800 from customer for services.
    b. Buy supplies for cash, $500.
    c. Bill customers for services, $4,000.
    d. Pay consulting wages owed of $2,000.
    e. Pay taxes owed of $500.
    f. Purchase equipment on credit, $2,000.
    g. Received payments from previously billed customers, $2,500.
    h. Pay wages of $1,500.
    i. End of month, generate I/S and B/S.

9. As of the end of November the Technology Co. has the following assets and liabilities: Quincy owes the firm $400; cash totals $2,690; outstanding common stock is worth $5,000; Stone owes $800 to the company; Technology owes Harco $1,200; retained earnings are $900; 16 widgets are in inventory at a value of their cost of $150 each; and 9 type X cables are in inventory at $90 each. Using a manual format or computerized spreadsheet, generate the beginning B/S, ending B/S, and I/S for December.

December transactions:
    a. Paid Harco in full.
    b. Billed Quincy $30 for installation charges.
    c. Although not yet due, paid the January rent of $180 (show as an asset until due, then expense the item).
    d. Sold 6 widgets for $250 each.
    e. Purchased 10 widgets at $150 each and 6 cables at $90 each from Harco on credit.
    f. Paid $40 to XYZ Printing for advertising.
    g. Sold 5 widgets to Stone for $200 each on credit.
    h. Received $120 from Quincy on account.
    i. Paid commissions to salesman of $270.
    j. Sold on credit to Sears, a widget for $250, X cable for $150, and installation charges of $40.
    k. Received $500 from Stone.
    l. Sold 4 cables at $150 each.
    m. Received bill from *Tribune* for advertising, $80.
    n. Paid $140 to trucking co. for delivery service.
    o. Received bill from East Co. for $60 for installation service.
    p. Paid Harco $700.
    q. Paid commissions, $280.
    r. End of month.

DISCUSSION CASES

Abby Pet Products Inc.

Jim and Ann Stuart founded a pet products company eight years ago. The firm sells to pet shops such items as pet toys, collars, travel kennels, food dishes, and related items. They manufacture some of the items and also buy from foreign suppliers. They have 12 manufacturers' representatives who cover the U.S. and part of Canada. The firm has grown rapidly. There are 43 employees, many of whom have been with the firm from the beginning.

Ann believes that in training new employees and holding periodic meetings with current employees, the company should give more financial information so that employees will have a better understanding of the firm's financial situation. She also thinks that employee ownership in the firm should be considered. She and Jim hold all of the common stock of the firm.

Jim isn't sure about publishing the firm's financial information or employee stock ownership. The company has a full-time bookkeeper who does all of the financial record-keeping. An outside accounting firm assists with tax preparation. Jim feels that it's too easy for financial information to be misunderstood or gossiped about and that such information could even get in the hands of competitors, customers, and suppliers. This could hurt the company's competitive position, he feels.

1. What are the advantages and disadvantages of sharing financial information with employees?
2. How could employee stock ownership be an advantage to the firm?
3. Is offering employees stock ownership and sharing financial information a positive motivation for employees? Or is the risk of gossip too great for this idea to work?

Mountain Electronics Co.

The company is located in Colorado and employs 250. Currently, there are 18 in the accounting department. They handle all of the day-to-day bookkeeping functions including payroll, invoicing, taxes, internal cost accounting reports and other necessary accounting functions. An outside CPA firm advises, audits, and assists with tax matters.

The design and process engineering departments are responsible for all design and manufacturing decisions. Lately, there have been some difficult and heated discussions between the engineering staff and the accounting managers about financial decisions. The process and design engineers say that too often their recommendations for a supplier are overruled, and a less-expensive supplier is selected, because of the financial manager's role in advising purchasing. The process engineers also complain that their decisions about tooling, equipment purchases, make or buy decisions, and other activities face frequent interference by the accounting department.

The vice president of finance, Harry Snoad, says that the financial success of the firm is *his* responsibility and that he should be involved in the important financial decisions. He further believes that the design and process engineering people are good engineers, but have no financial training and don't understand financial matters. "They just make decisions on technical criteria, including the quality of suppliers' material and equipment. They should make the technical decisions, and we should make the financial decisions," Harold states.

Hank Walton, the plant manager, is trying to "keep peace in the family" but he isn't sure of the best approach.

1. What recommendations do you have for Hank Walton?

2. What are the "boundaries" of the typical accounting department and of the design and process engineering departments?

3. Are the technical employees normally involved in making some of the financial decisions of the firm? What about the comment that "the technical people don't have any financial training"?

The Millionaires Club (A)

A group of students is interested in learning more about investments, personal money management, and saving for retirement. They have formed the "Millionaires Club" to meet weekly and discuss investments. They feel that the best way to learn is to learn by doing and to "invest on paper" by assuming the group has $15,000 to invest, analyzing various investment alternatives, discussing alternatives, and then placing orders on paper. They can then evaluate their results and learn as they invest without any real dollar outlay. They have divided potential areas of investment into the following categories:

1. Money market funds and savings accounts

2. Mutual funds

3. Common and preferred stocks

4. Industrial and government bonds

They have asked one of their instructors to assist them with their club. The instructor will serve as the "broker" for their orders, assist with questions, refer them to sources of information, and help them bring speakers to their meetings. The group consists of eight students, and they are not sure if they should try to include more students. Some of the members are working and have some experience with their companies' investment programs.

The students are aware that Individual Retirement Accounts, 401K plans, and other retirement investments have great potential, but they don't know the differences between these plans.

1. How could joining a group such as this be better than learning about investments individually?

2. Is it realistic to think that the members can learn some investment techniques or should they rely on an expert such as a broker or financial advisor for advice?

3. Why is investing on paper a better technique than reading about investments?

4. What sources of information would be helpful to the group?

5. The group plans to keep financial records and prepare monthly, quarterly, and annual reports. What types of information and techniques would be recommended for the financial reports?

6. Apply the above situation to a group of which you are a member and get the instructor's assistance in developing the group and making investments "on paper" over a semester or longer.

7. Is it too soon for the students to begin thinking about investing in retirement accounts? Should they wait until later?

8. Can the club eventually convert to using members' actual money?

Sticks and Bricks Construction Inc.

Ben Furguson is the owner of a small, but successful commercial construction firm. The projects are in the $500,000 to $2 million range. He has been in the construction business for 23 years and for the last 9 years has had his own firm. He understands the construction business and is respected by his customers and architects. The ups and downs of business are one of the most difficult problems to manage for Ben. Most of the labor is seasonal and he maintains a small office staff on a full-time basis. Sometimes he has lots of income and a large bank balance and other times when things are slow, he has to dip into his personal savings to pay the rent and the staff.

He has always done most of his own bookkeeping with the help of his wife. An accounting firm assists with the company's taxes at the end of the year. Ben is considering expanding and knows that he will need more funds to run the business. He wants to explore obtaining bank loans to help him expand. Also, he has had an offer from a friend to help finance larger projects that Ben may not be able to handle financially.

He wants to place his financial management on a more professional basis and knows that he will need to submit financial information to the bank and to the investor on a regular basis as he expands and projects progress. He knows that he needs to do some financial planning and analysis, but isn't sure what to do.

1. What financial records and other information should Ben prepare to get ready for his meeting with the bank?

2. What are the advantages and disadvantages of expanding financially? Should the outside investor be considered as a source of funds or only the bank?

3. How are problems different in the construction business compared to a manufacturing or service firm? What differences do these cause in accounting and financial management?

Computer Tech Co.

A small, but successful software company is owned by two programmers that started the firm four years ago. They are able to obtain contracts for custom-designed software. Their advantage is being small and able to give fast deliveries, which, however, require full-time and part-time staff to work long hours when a contract is obtained, sometimes all night and weekends. In addition, their prices are lower than the competition. Many customers are repeat customers.

The two owners are considering developing some software that they feel could be sold to the general public to assist users in working with the Internet. They know that some investment and time would be required to bring the product to market. They have had preliminary discussions with a bank and the bank's reaction is positive. They also feel that they can contact existing software distributors to assist in selling the product.

They estimate that approximately five programmers will be required for six to eight months to do the work. The investment in the proprietary software would be in the range of $250,000 to $400,000. Assuming the loan is approved, there is a question as to how to handle the development costs of the software. One alternative is to expense the cost as it occurs. To do this, as checks are written to the programmers and for project expenses, cash would decrease in the Accounting Equation and development costs would increase in the income/costs accounts. This would have the effect of showing high costs for the coming year's Income Statement. Profits, as a result, would be low for the year, but as the software starts to sell, profits would be increasing over the next two to four years.

Another alternative is to consider the development costs in the same way as a purchased computer. That would be to decrease cash in the Accounting Equation and increase an asset account called Internet software. Then over the next few years, Internet software can be "depreciated" with part of the expenses going to the Income Statement accounts each year, similar to depreciating a computer. Profits shown on the Income Statement would then be a result of the income from sale of the software less the "depreciation" and other distribution costs.

Knowing that the bank will require regular financial updates as well as repayments on the loan, the owners are not sure which approach would be best as far as the bank is concerned. Showing good financial results from this activity is important for future bank relationships.

1. Which approach seems to be the best from a financial viewpoint?

2. How would taxes be affected by each of these two approaches?

3. If the firm were publicly held, would there be any advantage to either method?

Chapter 3   The Accounting Equation—Depreciation, Inventory, and Cash Flow

KEY TERMS

Accelerated Cost Recovery System (ACRS) A depreciation method based on a schedule created by Congress in the early 1980s to replace previous accelerated methods such as SOYD and DB.

Accounting Equation Assets = Liabilities + Owners Equity + Revenue − Costs

Accrual System Recognizing sales and expenses when they occur even though they are credit transactions.

Amortization The reduction in value over time of intangible assets such as patents, copyrights, leasehold improvements, franchise rights, or other nonphysical assets.

Appraised Value The estimated market value of an asset.

Book Value The value of the asset determined by subtracting the accumulated depreciation expenses from the asset's initial value.

Capitalizing Expenditures are first recognized as assets and then reduced in value over time. The value reduction creates an expense on the Income Statement.

Cash Flow Increases or decreases in the cash account in the Accounting Equation. They may be categorized as operating, investment, or financial cash flows. Cash flow can be determined indirectly from the Income Statement.

Cash System Recording the financial event only when cash is received or disbursed.

Declining Balance (DB) A mathematical depreciation method that creates high depreciation expense in the early years and lower amounts in the later years by multiplying the straight line depreciation amount by a factor between one and two times.

Depletion The reduction in value of natural resources occurring with land such as oil, minerals, gas, or timber.

Depreciation Fixed assets such as equipment or buildings are reduced in value over time. The reduction in value is entered as an expense in the cost category of the Income Statement.

Expensing Expenses are created in the cost category of the Income Statement as soon as the expenditures are made.

Group Depreciation A combination of similar assets to calculate total depreciation expenses.

Intangible Assets Economic and legal rights of long-life, nonphysical assets including copyrights, patents, franchises, and leasehold improvements.

Market Value The actual price/value when an asset is sold.

Matching Attempting to have expenses and associated revenues for products and services sold appear on the financial statements during the same period.

Modified Accelerated Cost Recovery System (MACRS) Replaced the ACRS in the 1986 Tax Code revisions. It is based on a schedule of lives and categories of assets to determine the annual depreciation expense values.

Salvage Value The estimated value of the asset when it is no longer used by the organization. Approximates the future market value of the asset.

Straight Line A depreciation method that reduces the asset value at a constant rate over its life.

Sum-of-Years-Digits (SOYD) A mathematical depreciation method that creates high depreciation expense in the early years of the asset's life and low depreciation expense in the later years.

Unit-of-Output or Use Method Depreciation expenses calculation based on the fractional use of the asset's estimated total output or capacity during its life.

LEARNING CONCEPTS

- Understanding the Accounting Equation entries applied to capitalizing and expensing and their impact on the Balance Sheet and Income Statement.

- Calculating and graphing depreciation expense and book values of assets using the accepted methods, including straight line, sum-of-years-digits, declining balance, unit-of-output, accelerated cost recovery system, and modified accelerated cost recovery system.

- Understanding the Accounting Equation entries when assets are disposed of earlier or held longer than originally estimated, and the entries when the asset is sold for less and for more than its book value.

- Understanding the recent trends in selecting depreciation methods, and the impact of the methods on profits, taxes, and cash flow.

- Learning the concepts of noncash entries into the Accounting Equation of depreciation, amortization, and depletion, as applied to land, patents, copyrights, goodwill, franchise and leasehold improvements; and their impact on asset values, expenses, profits, and taxes.

- Applying three types of inventory accounting to manufacturing, wholesale, and retail organizations using the Accounting Equation.

- Understanding the concept of cash flow.

- Relating the concepts and types of operating, investing, and financing cash flows to the Accounting Equation and differentiating between operating cash flow and profits shown on the Income Statement.

- Calculating operating cash flow from the financial statements.
- Understanding the relationship of depreciation and cash flow to the design, manufacturing, and technical decision-making concerning products, processes, and investment decisions.

> Profit is today a fighting word. Profits are the lifeblood of the economic system, the magic elixir upon which progress and all good things ultimately depend.
>
> —Paul Samuelson, economist, 1976

INTRODUCTION

Whenever an analysis and an investment decision are made, there will be financial results in the financial statements. Financial statements are often the basis for measuring the financial success of the organization or individual. Therefore, it is important to understand the effects of investment decisions on the Accounting Equation and on financial statements. This chapter looks at important components of economic analysis, investment decisions, and their effects on the financial statements.

Chapter topics include timing the recording of financial events so that revenues and costs match, credit and cash transactions, capitalizing versus expensing financial events, noncash expenses including depreciation, and measuring cash flow. Understanding these topics will provide a base for understanding the effects of equipment, projects, buildings, and other investment decisions on the financial statements and financial results.

We have looked at the basic Accounting Equation concepts in the previous chapter. This same basic reasoning applies to the more complex financial events introduced in this chapter as well. Whenever a new financial event is encountered, we can apply the same Accounting Equation–based reasoning.

Engineers, managers, and technical personnel in the organization often have financial responsibility for the organization's equipment, buildings, inventory, research and development, and land-based assets that may include mining and timber resources. This chapter relates these financial responsibilities and their impact on the Accounting Equation by looking at the following topics:

Timing, matching, and capitalizing the recording of financial events.

Depreciation of buildings, equipment, tooling, and related assets.

Amortization of patents and R&D costs.

Depletion of mining and timber resources.

Expensing of inventory.

Analysis of cash flow.

The techniques for handling the above topics are closely related. Depreciation, amortization, and depletion are treated the same way in the Accounting Equation and they have similar effects on the financial statements. Cash flow and depreciation have an important impact on the way investment analysis and decisions are made. Both cash flow and depreciation are based on the timing and matching of key financial events so that the resulting financial statements are synchronized and present true and fair financial measurements.

Traditionally, process engineers, design engineers, and manufacturing managers performed the economic analysis of processes, products, and equipment and then made a request for funding. The information would be sent to the finance department, which along with senior management, would make the final decisions. In contemporary organizations, a cross-functional team composed of management, financial, and technical personnel is often the most effective approach to making investment decisions. This team approach requires that the participants bring their special knowledge and abilities to the team and also understand the other areas. Team members need to understand the effect of their decisions on the overall financial system as well as on the individual product, services, facilities, and projects. This chapter will help provide understanding of the accounting approach to capitalizing expenditures, depreciation, amortization, inventory, depletion, cash flows, and related entries into the Accounting Equation.

> The rich get richer and the poor get children.
> —George Bernard Shaw

CAPITALIZING OR EXPENSING OF CASH OUTFLOWS

Two matching and timing techniques are the capitalizing and expensing of cash outflows. The term **expensing** refers to an expenditure that is reduced in the cash account and an expense that is immediately created under the cost category of the **Accounting Equation.** For example, if office supplies are purchased, the entries into the Accounting Equation are to reduce cash, an asset, and to create an expense under the cost category. Both sides of the Accounting Equation are decreased. The immediate expensing of the cash outflow is based on the judgment that the benefit from the use of the supplies to generate revenue will occur in the current period.

The term **capitalizing** refers to an expenditure that causes cash to be reduced and another asset created, rather than an immediate expense. An example of capitalizing is the purchase of equipment that will provide service during future years. Cash, an asset, is reduced and equipment, an asset, is increased. This adds and subtracts the same amount to the asset side of the Accounting Equation. It is assumed that the benefits from the expenditure for equipment will be received during future accounting periods. Each year of the equipment's life part of the equipment's value will be reduced and made an expense under the cost portion of the Accounting Equation. To summarize, the equipment is purchased with cash, made an asset, and part of the value is expended over the equipment's life. This is the process of *depreciating the asset.*

The equipment continues to provide service and generate revenues in the future, and the costs of the equipment are **matched** with the future period's revenues. By dividing the value of the equipment over future periods, the costs are better matched with the revenues that the equipment generates.

Guidelines for Expensing or Capitalizing

> When there is an outflow of cash from the organization, it is sometimes an immediate expense or cost. This is called *expensing*. The cash may be exchanged for another asset. This is called *capitalizing*. Later, as that asset is consumed, it becomes an expense.

After the economic analysis is complete and the decision is made, the next step is to determine the entries that will be made into the Accounting Equation. The determination of expensing or capitalizing must be made. This decision has a very important impact on the Income Statement and Balance Sheet. If the choice is made to expense, the period's costs may be overstated. If the decision is made to capitalize the expenditure, the costs are understated. It is important to attempt to match costs with revenues at the appropriate time.

Decisions to expense or capitalize cash expenditures are based in part on the following criteria.

1. *Life of the expenditure.* The shorter the life of the expenditure, the more likely it should be expensed. Tools, supplies, and materials that are consumed rapidly are usually expensed.

2. *Value of the expenditure.* Low-priced items including tools and supplies are usually expensed. Expensive equipment is capitalized and depreciated. Some organizations arbitrarily expense or capitalize if the expenditure is above or below an arbitrary amount, such as $500 or $1,000.

3. *Matching considerations.* Attempts are made to expense and capitalize items so that the costs of production occur in the same time period as the revenues.

4. *Accounting conventions.* The method that an organization has decided to use to capitalize or expense items in the past sets a precedent for future decisions. The professional accounting societies may establish accepted procedures that are recommendations for accounting systems to follow.

5. *Special situations.* The acquisition or sale of organizations, purchase of patents, internal creation of patents, and write-off of obsolete equipment, material, and R&D may be unique, one-time events that require special expensing or capitalizing procedures.

> ### Accounting Equation Entries
>
> Entries into the Accounting Equation must *always* maintain the equation's balance. Adding the same amount to both sides of the equation or adding and subtracting on the same side of the equation will maintain balance.

Cash Outflows for Printer Supplies at Consulting Engineers Inc. Consider an example of accounting entries at a medium-size consulting firm, Consulting Engineers Inc. The engineering design department decides to buy $500 worth of supplies, including paper for the printer, printer pens, and other computer supplies. What Accounting Equation entries should be made?

The cash account is reduced by $500 when the check is written to the supplier, and a printer supplies expense under the cost (a negative) column in the Accounting Equation is made. The amount of $500 is subtracted from both sides of the Accounting Equation, and the equation is in balance after this financial event. The generated Balance Sheet would show $500 less cash and the Income Statement shows $500 as expenses. Profits for the period would be reduced by $500.

The decision to purchase the supplies was made by the engineering design department. The recording of the financial event was made by the accounting department. And it is likely that the purchasing department was responsible for contacting the supplier and entering into the contract to purchase the supplies as specified by the design department (size, quality, quantity, and other specifications). The technical result of this decision will permit drawings and specifications of products to be made using the supplies that were purchased. The financial result is that cash decreased and costs were incurred. The simple decision to purchase printer supplies has both technical and financial implications for the organization.

Looking at the Income Statement, it would appear that costs are higher and profits are lower than if the purchase had not been made. From the viewpoint of matching, the computer supplies are likely to be consumed in the current period. From the value viewpoint, the supplies have small total value. These considerations lead to the conclusion that expensing rather than capitalizing is the correct decision. The Accounting Equation of these entries is shown in Figure 3–1.

Many similar examples occur in the organization. It is such a common occurrence that we usually don't think about the process or the results of the decision. Now, what if the design department decided to purchase a new computer printer rather than printer supplies? How does this decision affect the financial statements of the organization? Below is an example that shows the difference between expensing supplies and capitalizing equipment, the new printer.

Cash Outflows for a New Printer at Consulting Engineers Inc. Consulting Engineers purchases a new computer printer for $5,000. What are the resulting entries into the Accounting Equation?

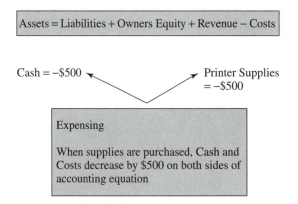

Figure 3–1 Expensing

The printer has a life greater than a year. It has a relatively high price. In order to match its costs with the services it produces, its cost should be spread over more than one year. The printer value is capitalized (an asset is created) when it is purchased, and then over the next three years the value of the asset is reduced with the the entry of depreciation expense on the Income Statement.

Assuming the purchase is for cash, not credit, cash in the Accounting Equation decreases by $5,000 as money is given to the supplier, and an asset, computer equipment, is increased by $5,000. If the purchase were for credit, accounts payable would increase and the equipment account would also increase. The diagram of this transaction is shown in Figure 3–2.

In the Accounting Equation, cash is reduced by $5,000 and computer equipment increased by $5,000. The same amount has been added and subtracted on the left side of the equation. There is no effect on the costs on the Income Statement accounts. Only the Balance Sheet is affected.

The printer value becomes an expense, but it occurs gradually over time. This process of reducing the value of the printer on the asset side of the equation and entering the same amount under the cost portion of the equation is called *depreciation*. This entry is repeated for the assets over time until the asset is disposed of or its value reaches the estimated scrap or salvage value.

What if the value of the printer is incorrectly entered as a cost immediately when it is purchased? In this case, the cash would still be reduced by $5,000, and costs of $5,000 would appear on the Income Statement for that period. It would assume that all of the printer value is used up the first year. This would have the effect of *understating asset value* on the Balance Sheet since the printer still has value past the first year, and *overstating expenses* on the Income Statement for the first year. Then no printer expenses would be incurred for the second and third years even though the printer is delivering service. This would result in misleading conclusions, due to mismatching, about the organization's assets and expenses for the period.

Timing of Financial Events and Cash Flow

An organization earns revenue by selling a product or service. There are many events that occur in the selling process: meetings with customers, proposals, advertising, quotations, purchase orders, discussions, design and manufacture of the product or service, delivery to the customer, and finally payment by the customer to the supplier. When should the seller record the financial results of the sale?

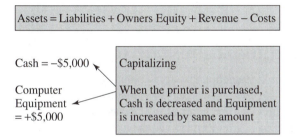

Figure 3–2   Capitalizing

One method would be to deliver the product, wait for the payment, and record the sale in the Accounting Equation when the money is received. The difficulty with this cash method is that sometimes there is a time delay between delivery of the product and payment by the customer. Timing of the record keeping is important. If the inventory is delivered in January, the reductions in cash for material, labor, and overhead costs are recorded in the Accounting Equation as costs when the product is shipped in January, but no revenue is shown, costs and profits are overstated. If the customer pays in April, the revenues and profits will be overstated then. Costs of the goods were already recorded in January. This *cash system* of recording financial transactions leads to revenues and costs that are not synchronized or matched. A system that will yield better timing is the **accrual system** of recording the costs and revenues.

The costs and revenues can be more closely matched if they are recognized in the Accounting Equation at approximately the same time. In the above example, if the sale as well as the costs are recorded in the Accounting Equation when the goods are shipped, timing is improved. When the shipment is made, inventory and cost of goods sold entries are made in the Accounting Equation, reducing both left and right sides of the equation by equal amounts. The entry is made to increase sales as a revenue and to increase accounts receivable, an IOU from the customer, by the sales amount. This makes the revenue and costs appear on the Income Statement accounts at the same time.

This accrual method assumes that the customer will pay for the goods at the agreed-upon terms and time. When this payment is received, cash is increased and accounts receivable, an asset, is reduced. The Accounting Equation is in balance throughout the process. This method of accruing the revenues as a temporary asset, accounts receivable, permits the matching and synchronization of the revenues and costs from the transaction. The profits or net income from the transaction will occur in a more correct time relationship.

An Individual's Cash Accounting System

An example of a **cash accounting system** is an individual's checking account. When a check is written, the date, check number, payee, and amount are recorded in the check register. The amount is deducted from the previous cash balance. When a deposit is made, the balance is recalculated. This is a single-entry cash system. No Accounting Equation is used. No second entry is necessary to balance the addition or reduction of cash to the cash account. Costs are deducted from the checkbook balance when they occur, and revenues are recorded as deposits when they occur. There is no matching or synchronization of costs and revenues. A small business could use this cash accounting system, but for the generation of financial statements, a double-entry and accrual system is more effective than a cash system.

DEPRECIATION, FIXED ASSETS, AND THE ACCOUNTING EQUATION

Process engineers, design engineers, production managers, managers, and project engineers recommend the purchase of buildings, equipment, and other assets. They perform much of the technical and financial analysis on which the purchases are based. They can easily understand the financial impact of their decisions by understanding depreciation and capitalization.

Some confusion surrounds the concept of depreciation. For example, it is sometimes thought to be a creation of the Internal Revenue Service. Although it is true that Congress and the IRS specify the acceptable mathematical methods for calculating depreciation expense for tax purposes, the concept of depreciation was not created by the IRS. **Depreciation** entries in the Accounting Equation are made to reduce the value of a fixed asset such as a building or equipment over time, and to make the expenses match with the revenues that the asset produces.

Since the depreciation expense is a pretax expense that affects taxes paid, Congress and the IRS specify the procedure for calculating the expense. However, this calculation may or may not parallel what is actually happening to the market or physical value of the asset. The IRS specifies the procedure to use for tax purposes, but it doesn't "cause or create" depreciation.

Another misconception about depreciation is that when an asset declines in value, depreciation expense is used to establish a fund to replace the asset when it is depleted. Although depreciation expense is a pretax entry and thus reduces the amount of taxes paid for the period, the tax savings do not go into a "replacement fund." The organization can choose to establish a fund that could be used for asset purchase, but this fund is not a result of depreciation expense.

An analogy may help to explain why depreciation entries are made. Assume that $5,000 is placed in the company's safe at the end of the day. The cash account would show this amount as an asset. During the night a thief breaks in and steals the $5,000. What is the appropriate entry in the Accounting Equation when the theft is discovered the next day? On the Balance Sheet an entry that reduces cash by $5,000 and the paired entry of "loss due to theft" or "theft expense" would be made in the expense category. A $5,000 expense was created. This results in higher costs, which reduces profits before taxes and the amount of taxes for the period. Should the owners and employees of the company be happy that the cash was stolen so that taxes would be lower? Of course not. A similar process occurs when equipment decreases in value because of wear and tear or technological obsolescence. The value of the asset decreases and the paired entry in the Accounting Equation is to create a "loss due to equipment declining in value expense" or simply "depreciation expense." The additional depreciation expense lowers costs, taxes, and assets, but the owners and employees would prefer that the equipment maintained its value and productivity forever, and that it did not decline in value.

Depreciation, as an expense on the Income Statement, affects the calculation of the profit before tax. For this reason, the Internal Revenue Service requires specific depreciation methods to be used. Over the years, Congress has changed the Tax Code frequently, especially the acceptable depreciation methods for tax calculations. For this reason, the different depreciation methods presented here are likely to change in the future. The basic depreciation concepts will not change.

It should be noted that corporations, partnerships, and individual proprietorships are permitted by the Tax Code to depreciate equipment and other fixed assets. Individuals who are not proprietorships, with business income and expenses, are not permitted to deduct depreciation expense from their personal tax returns. For example, individuals generally are not permitted to depreciate personal assets such as an automobile or a house unless they are used for business purposes and provide services to produce revenues from customers.

Calculations of straight line, sum-of-years-digits, declining balance, modified accelerated cost recovery system, and output methods are considered here. Although it is likely that there will be changes in the Tax Code in the future, understanding these basic methods will be a sound basis for understanding any future depreciation methods.

Determining Depreciation

A machine, tool, or building can decrease in value because of use, wear, time, or obsolescence. A computer may work as well after many years of operation as it did when purchased. However, newer and faster computers may make the old computer less desirable and its value to the user and potential purchasers decreases, which is reflected in a lower market value. Its mechanical and electrical functions are as good as when it was purchased, but faster speed, new hardware, and improved software at a lower cost make newer models more valuable. Depreciation calculations and entries into the Accounting Equation are an attempt to reflect the older computer's decline in value.

After the decision to purchase an asset such as a building or equipment, the cash account is reduced and the asset account, equipment or building, is increased. Physically, the asset generally decreases in value. Financially, the asset must be reduced in value on the Balance Sheet to reflect what is happening physically and in the marketplace. Assume that a new piece of equipment is purchased for $100,000 and is published in the current Balance Sheet as having a value of $100,000. A year later, the Balance Sheet is published again. Should the equipment be shown as having a value of $100,000? If not, how should the value be determined? Three methods that could be used to determine the value of the year-old equipment are market value, appraised value, and mathematical calculation of depreciation.

Market Value. One method to determine the value of the equipment is to sell it. The **market** determines the value when the asset is sold. As the asset is sold, the entries into the Accounting Equation would be to increase cash and revenue and to reduce equipment value to zero and create a cost of equipment sold. Of course, the organization cannot sell its equipment and buildings each time it wants to know an asset's value.

Appraised Value. Another method to determine the value of the equipment would be to have an expert **appraise** it and give an estimate of the value. This method has the advantage of not requiring the sale of the asset, but if the firm has many different items, many different appraisals would be required and would be expensive and time consuming to obtain. If this method were used, the amount the equipment has declined in value during the year would be subtracted from the equipment's value in the Accounting Equation, and in the cost column, depreciation expense of the same amount would be entered. This would be repeated each year until the equipment is disposed of or sold for its salvage value. This method is practical only for very special assets, since it is time consuming and costly to appraise every asset periodically.

Mathematical Value. Usually, neither the sale nor appraisal method of determining value is used. Rather, a schedule or formula is applied to the asset's value to determine the amount of depreciation expense each year. This method is easier to apply to many assets. Although this method appears to be precise it may or may not reflect the true market value of the asset. The advantage of this method is that it can be done quickly, using tables, a calculator, or a computer.

The asset's value and depreciation expense may be overstated or understated during the asset's life. When all of the firm's assets are considered, there may be a significant difference between the *actual value* of the assets and the *estimated value* using depreciation formulas or schedules. This variation is important to remember when reading financial statements. Also, this can result in different tax amounts that are calculated and paid by the organization. In equipment-intensive organizations that have many fixed assets, the depreciation method used may have a large impact on the financial statements and tax payments.

Methods to measure value of fixed asset:
1. Sell it.
2. Appraise it.
3. Use a formula or schedule to calculate an estimate of it.

Depreciation Calculation Methods

- Prior to 1954, the Tax Code required using the straight line (SL) method for depreciating fixed assets.

- In 1954, the sum-of-years-digits (SOYD) and declining balance (DB) methods were approved for use. These are two of the methods that give high depreciation in the early stages of the asset's life. They are called *accelerated depreciation* methods because they show a high depreciation amount in the early years of the asset's life and lower amounts in later years. The straight line method was still permitted.

- In 1981, the accelerated cost recovery system (ACRS) was adopted by Congress. It was based on the double declining balance method.

- In 1986, the ACRS method was revised to the current modified accelerated cost recovery system (MACRS).

- Currently, the straight line method is required for buildings, and MACRS or straight line may be used for equipment.

(Source: 1995 U.S. Tax Code, Internal Revenue Service)

These mathematical methods do not give equal results. Straight line gives constant depreciation expenses over the life of the asset, and others (SOYD, DB, and MACRS) give high depreciation expenses in the early years of the asset's life and lower depreciation expense values in the asset's later life.

Since employees, managers, engineers, technicians, shareholders, suppliers, and customers rely on the published financial statements to make financial decisions, they should understand different depreciation methods. Footnotes in the financial statements usually describe the type of depreciation used. The expenses for the accounting period, which include depreciation expense, are used to determine the product's unit cost, and the costs of products and processes is partially determined by the depreciation expense. These costs provide a basis for measuring the financial success of the organization. Cost measurements affect the decisions to make new investments in projects and fixed assets. Permitted depreciation methods are a function of the interpretation of the current Tax Code. A qualified tax practitioner can be consulted to determine the applicability of different depreciation methods to a particular situation.

Let's look at the calculation of depreciation and its impact on costs, asset values, profits, taxes, and financial investment decisions.

Estimating Asset Life and Salvage Value

When an asset is purchased and its depreciation amounts calculated, some mathematical depreciation formulas or schedules require the use of **salvage value** and life estimates. Straight line is one of these methods that requires both of these estimates. The other acceptable method of depreciation for assets placed in service since 1986 is the MACRS method, which specifies the life to be used for particular types of assets but assumes no salvage value. Considerations in estimating the life and salvage values are summarized below.

1. **Physical Considerations**—Each asset has quality, reliability, durability, and expected use or service, which together determine its expected life. History, experience, vendor recommendations, and related data may assist in making this estimate.

2. **Economic Considerations**—Economic life is often different from physical life. It may not be economical to keep the asset, even though it is still physically functional, because of maintenance and repair costs compared to other alternatives.

3. **Replacement Policy**—Some assets are replaced because of company policy, which may or may not reflect physical or economic effects. Delivery trucks may be replaced every three years, for example, or salespersons' cars every two years, because of the image the company wants to present to customers.

4. **Technological Considerations**—Life and salvage are based on technical characteristics. Computers may be upgraded or replaced in the engineering department every two years even though the old ones are still functional.

Straight Line Method of Depreciation Calculation

The simplest and most common of the mathematical depreciation methods is **straight line.** The life of the equipment and its residual value are estimated when it is purchased. The formula for calculating straight line depreciation is:

$$\text{Annual Dep}_{SL} = \frac{\text{First Cost} - \text{Salvage Value}}{\text{Life}} = \frac{P - S}{n}$$

where P = the initial first cost, which is the purchase price

S = the net salvage or residual value, which is the estimated revenue from the sale of the used asset

n = the estimated life of the asset in years

Using Straight Line at Fire Equipment Manufacturing Co. The company purchased a new small robot for $100,000. It has an estimated life of 10 years and an estimated net salvage or residual value at that time of $10,000. The annual depreciation is:

$$\text{Annual Dep}_{SL} = \frac{\$100,000 - \$10,000}{10 \text{ years}} = \$9,000 \text{ per year}$$

When the robot is purchased, the entries into the Accounting Equation reduce the cash account by $100,000 and increase the robot equipment account by $100,000. (The same amount is subtracted *and* added to the same side of the equation.)

Then, each year, the robot equipment account is reduced by $9,000 and the cost (a negative) category of the Accounting Equation is reduced by a $9,000 depreciation expense. This entry is repeated every year over the estimated life of the asset. The reported value of the robot on the Balance Sheet is called the **book value.** This value, of course, is a function of the depreciation method and the assumptions concerning life and salvage value.

The ending book value of the robot after the first year is:

$$\text{Initial Value} - \text{1st year Depreciation Expense} = \text{Book Value at end of 1st Year}$$
$$\$100,000 - \$9,000 = \$91,000$$

The book value of the robot at the end of the second year is:

$$\text{Book Value at end of 1st year} - \text{2nd year Depreciation Expense} = \text{Book Value at end of 2nd Year}$$
$$\$91,000 - \$9,000 = \$82,000$$

These calculations and annual entries into the Accounting Equation continue until the end of the tenth year.

An advantage of a mathematical method of depreciation expense estimates is that it can be calculated using a calculator or spreadsheet rather than by making annual appraisals of the assets. But a disadvantage of a mathematical depreciation method is that it might not reflect the true market value of the asset. Table 3–1 is a spreadsheet-generated table and graph for the robot's straight line depreciation.

The end of the year book value declines to the end of the tenth year when the book value is $10,000, the estimated net salvage value. (For graphical purposes, the start of year 1 is also the end of year 0 when the robot is purchased.) When the accounting system follows this depreciation schedule for the 10-year period, the annual depreciation expenses that appear on the Income Statement for this robot are $9,000 each year. On the Balance Sheet the asset will have a book value that declines by $9,000 each year. Did the asset really go down in value by $9,000 during the year? Probably not exactly. The only way to answer this question accurately would be to sell the asset.

The user of financial statements must understand that the depreciation entries are based on the assumptions of (1) the depreciation method, (2) the life, and (3) the estimated residual net salvage value. The cumulative results of the depreciation calculations that appear on the financial statements may be representative or may be misleading. Depreciation calculations are estimates of value rather than historical fact. The financial statements contain both actual historical financial values and estimated values such as depreciation.

Different Salvage Value

Using the previous example of depreciation of the robot at Fire Equipment Manufacturing, what would occur if the asset is sold for $20,000 at the end of the tenth year rather than at the original salvage value estimate of $10,000? We can return to the Accounting Equation to answer this question.

In this case, the book value of the robot on the Balance Sheet at the end of the tenth year is $10,000 based on the original calculations. If it is sold for $20,000, then the cash account would increase by $20,000 and the revenue account, called sale of equipment (or similar title), would increase by $20,000 ($20,000 is added to both sides of the equation). The other entries

Table 3–1   Straight Line Depreciation for Robot

Year	Book Value at Start of Year	Annual Depreciation $(P - S)/n$	Book Value at End of Year
0			$100,000
1	$100,000	$9,000	$91,000
2	$91,000	$9,000	$82,000
3	$82,000	$9,000	$73,000
4	$73,000	$9,000	$64,000
5	$64,000	$9,000	$55,000
6	$55,000	$9,000	$46,000
7	$46,000	$9,000	$37,000
8	$37,000	$9,000	$28,000
9	$28,000	$9,000	$19,000
10	$19,000	$9,000	$10,000

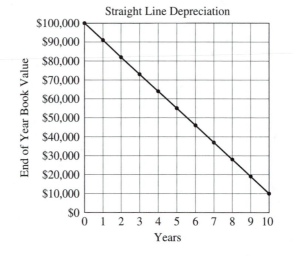

would reduce the equipment account by $10,000 and create a cost of equipment sold (or similar title) account of $10,000 under the cost category. (The entry for this second part of the transaction is the subtraction of $10,000 from both sides of the equation.) A diagram of these accounting equation entries is shown in Figure 3–3.

Figure 3–3 shows the following:

1. The revenue and cash accounts are each increased by $20,000.

2. Robot equipment (asset) is reduced by $10,000, bringing the book value of the robot equipment to zero, and cost of equipment sold (negative) is entered as $10,000 in the cost category.

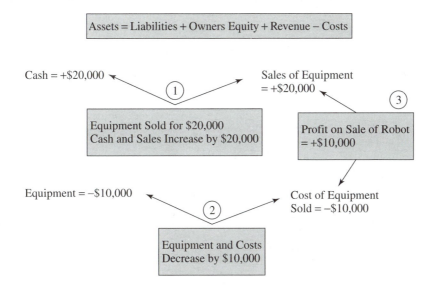

Figure 3–3   Sale of Robot for a Different Salvage Value

3. The result of selling the equipment for more than the estimated salvage value results in a profit. (This profit has tax implications, but is usually added to the operating profits of the organization. This should be determined by a tax practitioner.)

4. The Accounting Equation is in balance after these entries.

If the asset had been sold for less than the estimated salvage value, a loss on the transaction would have occurred. During the life of the asset, its book value and its depreciation expenses are based on the mathematical formula, estimates of life, and net salvage value, not market value. When the asset is sold, the above entries are made to bring the historical mathematical values in agreement with actual market conditions.

Using Different Life Calculations

Another situation can occur. What if the robot in the above example is disposed of at the end of the seventh year for $42,000, rather than kept for 10 years as originally estimated?

The original estimates of life and salvage value produce a book value of $37,000 at the end of the seventh year ($100,000 initial cost − 7 years × $9,000 per year = $37,000). This book value is different from the actual selling price of $42,000, and the entries made when it is sold bring these two values into agreement, as shown in Figure 3–4.

Figure 3–4 shows the steps in the entries to the Accounting Equation for this reduced life of seven years. At the end of the seventh year, the accounting system will make the following entries into the Accounting Equation:

1. Cash and revenue accounts are increased by $42,000. (Adding the same amount to both sides of the equation.)

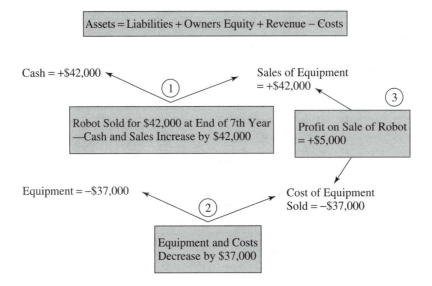

Figure 3–4   Sale of Robot with a Different Life

2. Robot equipment and cost of equipment sold accounts are decreased by the book value of the robot of $37,000, bringing the book value to zero. (Subtracting the same amount on both sides of the equation.)

3. Selling the robot for more than its book value at the end of the seventh year results in a profit of $5,000 on the transaction.

4. The Accounting Equation is in balance after these entries.

What If Life Exceeds the Original Estimate? The opposite case from selling before the estimated life is completed is that the asset lasts longer than originally estimated. If the robot's life exceeds the estimated 10 years, the book value of the asset would be decreased down to the estimated salvage value of $10,000 at the end of the tenth year. This amount would normally be kept as the book value on the Balance Sheet and not reduced any further until the asset is retired or sold.

Installation and Removal Costs

After it is purchased, equipment must be installed, which may be a significant cost. Utilities might have to be changed, and space might have to be prepared to accept the new equipment. At the end of the asset's life, removal costs may be high. Often the cost of installation is included along with the initial cost of the building or equipment. Removal costs are often included as a deduction against the revenue from the sale of the removed equipment.

Repairs and Maintenance

Normal repairs and routine maintenance costs are usually expensed when they occur. A major overhaul of equipment and buildings that extend the life of the asset would normally be depreciated over time. Capitalization, for example, of a new roof on a building would be depreciated separately from the original structure and have a depreciation life of its own. Replacing a large motor would likely extend the life of a machine, and the motor could be depreciated separately with its own life. Generally, the important criterion is whether the repair or maintenance extends the life of the asset. If so, the cost should be capitalized and depreciated. When the repair or maintenance is small in value, routine and repetitive, and does not significantly extend the asset's life, then the cost should be expensed immediately. Certainly, this is an area requiring judgment. Engineering and maintenance personnel can often advise the accountants about the expected repair and maintenance of the equipment or project.

> The buyer drove his new car home from the showroom, and it went down in value by $1,000!

Sum-of-Years-Digits Method of Depreciation Calculation

A popular depreciation method for assets acquired in the past is the **sum-of-years-digits** method (**SOYD**). It was introduced in the 1950s and used through the mid-1980s. Although it has since been replaced by MACRS for assets acquired after 1986, it might still exist in the accounting system for older equipment. SOYD was a popular method because it gave high depreciation values in the early years and lower values later in the equipment's life. SOYD is similar to the current MACRS method, which replaces it.

The straight line method applies the same depreciation expense per year, and the asset declines at a constant rate. Many assets decline more rapidly in their early years and more slowly in later years. An automobile is an example of this rapid initial decline in value and slower loss of value later. The sum-of-years-digits method approximates this decline in value. It generates high depreciation expense in the early years and thus lowers the tax rate compared to the straight line method. Later in the asset's life, the depreciation expense per year is lower, and the tax is higher. SOYD was one of the early "accelerated" depreciation methods. When an accelerated depreciation method is used, it has an impact on the annual cash flow for the organization.

Again, using the robot at Fire Equipment Manufacturing, we can use an initial cost of $P = \$100,000$, salvage = $\$10,000$, and an estimated life of 10 years. The SOYD calculation uses the sum on the years of life of the asset by adding digits, $1 + 2 + 3 + \ldots + 10 = 55$. This value becomes the denominator in a fraction. The numerator of the fraction for the first year's depreciation is 10. This fraction, 10/55, is multiplied times the difference between the first cost and the salvage value, $P - S$. For the second year's calculation, the fraction multiplier becomes 9/55; for the third year, 8/55, and so on.

Generally, to determine the sum of the years,

$$\text{Sum of Years} = 1 + 2 + 3 + \ldots n = \frac{n(n+1)}{2} = \frac{10(10+1)}{2} = 55$$

The annual depreciation for the first year is calculated by:

$$\text{SOYD} = \left(\frac{n+1-j}{SY}\right)(P-S) = \left(\frac{10+1-1}{55}\right)(\$100,000 - \$10,000) = \$16,364 \text{ rounded}$$

where j is the year for which depreciation is being calculated
SY is the denominator, 55
P is the first cost of \$100,000
S is salvage or residual value of \$10,000

This amount, \$16,364, is deducted from the equipment's value at the end of the first year. The other part of this entry is a depreciation expense of \$16,364 in the cost category in the Accounting Equation. (The same amount is deducted from both sides of the equation.)

The second year's depreciation is calculated using $j = 2$:

$$\text{SOYD} = \left(\frac{n+1-j}{SY}\right)(P-S) = \left(\frac{10+1-2}{55}\right)(\$100,000 - \$10,000) = \$14,727 \text{ rounded}$$

Depreciation for succeeding years is calculated in a similar manner.

The end of the year book value is determined by subtracting the annual depreciation expense from the value of the asset at the beginning of the year.

Book Value for the end of the first year = Initial Value − Depreciation amount =
\$100,000 − \$16,364 = \$83,636
Book Value at end of 1st year

Note that the first-year depreciation of \$16,364 is greater than the \$9,000 calculated using straight line depreciation. Also, the second-year SOYD depreciation expense of \$14,727 is less than the first year's SOYD depreciation. The end of the first year's book value for the SOYD is \$83,636 compared to the straight line value of \$91,000. The complete schedule of depreciation expenses and a graph of book values is shown in Table 3–2. Values on the graph are rounded to the nearest whole dollar. It can be seen that book value declines rapidly in the early years compared to straight line, and declines more slowly in the later years.

Compared to SL, SOYD gives higher expenses which result in lower taxes in the early years. In the later years, larger taxes result because of the lower depreciation expenses. SOYD has the effect of postponing the highest taxes into the later years. Many depreciation decisions attempt to postpone taxes rather than to reflect the true physical or economic life of the asset. By postponing the higher taxes, cash outflow from the cash account is minimized in the early years. More cash is then available for investment into other projects. Engineers, technicians, and managers who are aware of the impact of depreciation can structure their investment plans to take advantage of depreciation methods and the cash available for investment.

The entries into the accounting equation for the SOYD method are similar to those for the SL method, except that the annual depreciation amounts for SOYD would decrease over time.

Declining Balance Method

A depreciation method that was used for more than 30 years until the Tax Code was revised in 1981 is the **declining balance (DB)** method. Since there may be older assets that are still being depreciated using this method, it is discussed here. It is related to the straight line

Table 3–2   Sum-of-Years-Digits Depreciation for Robot

Year	Book Value at Start of Year	Fraction Multiplier	P – S	Annual Depreciation	Book Value at End of Year
0					$100,000
1	$100,000	10/55	$90,000	$16,364	$83,636
2	$83,636	9/55	$90,000	$14,727	$68,909
3	$68,909	8/55	$90,000	$13,091	$55,818
4	$55,818	7/55	$90,000	$11,455	$44,363
5	$44,363	6/55	$90,000	$9,818	$34,545
6	$34,545	5/55	$90,000	$8,182	$26,363
7	$26,363	4/55	$90,000	$6,545	$19,818
8	$19,818	3/55	$90,000	$4,909	$14,909
9	$14,909	2/55	$90,000	$3,273	$11,636
10	$11,636	1/55	$90,000	$1,636	$10,000
			Total	$90,000	

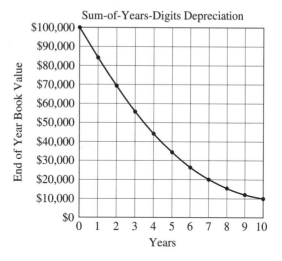

method in that the annual calculated straight line depreciation expense is multiplied by a factor of 1.25, 1.5, 1.75, or a maximum of 2.0 (SL)(DB factor). When a factor of 2.0 is used, the method is called the *double declining balance* method (DDB). This depreciation amount is then used for the first year's depreciation. The new book value for the coming year is determined by subtracting the annual depreciation expense from the original cost. For the second year's depreciation expense, this new book value is multiplied by the factor to obtain the second year's depreciation expense. This process is continued over the life of the asset.

For example, if the robot at Fire Equipment Manufacturing Co. is purchased for $100,000, with an estimated life of 10 years and a salvage value of $10,000, and a declining balance factor of 1.5 is used, the first year's depreciation expense is calculated as follows:

$$\text{Depreciation Expense} = \left(\frac{\text{DB Factor}}{\text{Estimated Life}}\right)(\text{First Cost} - \text{Accumulated Depreciation})$$

$$\text{Depreciation Expense} = \left(\frac{1.5}{10}\right)(\$100,000 - \$0) = \$15,000$$

Note that the salvage value is neglected using this method.

The book value of the equipment at the end of the first year using this method is:

$$\text{First Cost} - \text{1st Year Depreciation} = \text{Book Value at end of 1st Year}$$
$$\$100,000 - \$15,000 = \$85,000$$

The second year's depreciation is:

$$\text{Depreciation Expense} = \left(\frac{\text{DB Factor}}{\text{Estimated Life}}\right)(\text{First Cost} - \text{Accumulated Depreciation})$$

$$\text{Depreciation Expense} = \left(\frac{1.5}{10}\right)(\$100,000 - \$15,000) = \$12,750$$

The book value at the end of the second year is:

$$\text{Book Value at end of 1st Year} - \text{2nd Year Depreciation} = \text{Book Value at end of 2nd Year}$$
$$\$85,000 - \$12,750 = \$72,250$$

The accumulated depreciation at the end of the second year is:

$$\$15,000 + \$12,750 = \$27,750$$

This amount is used to determine the third year's depreciation. This calculation continues over the life of the robot.
The third year's depreciation expense is:

$$\text{Depreciation Expense} = \left(\frac{1.5}{10}\right)(\$100,000 - \$27,750) = \$10,838$$

This DB method is used to give accelerated depreciation expenses similar to the SOYD and MACRS methods. Note that the resulting annual depreciation decreases over time similar to the SOYD method. Cost and profit effects in the financial statement for DB are similar to those of the SOYD method.

Unit-of-Output Method

Equipment's depreciation expense can also be estimated based on use or output rather than time or life. Automobiles, aircraft engines, dies for presses, and office copiers are examples of this application. An automobile that has an estimated total mileage of 150,000 miles that is driven 15,000 miles in one year could be assumed to consume 10% (15,000 miles/150,000 miles) of its value that year using this method.

An office copy machine that will produce an estimated 500,000 copies over its life decreases in value by 20% (100,000 copies/500,000 copies) in a year when 100,000 copies are made. During another year, perhaps only 25,000 copies are made and the asset would decline in value by 5% of its value for that year.

If the **units-of-output** depreciation method is chosen, the disadvantage is that a detailed record of the asset's output must be maintained in order to calculate the percentage of use. This method has logic on its side when compared to time depreciation methods, if a reasonable estimate of total output can be made. As always, the current Tax Code should be checked for updates and changes in the details of depreciation procedures.

Modified Accelerated Cost Recovery System

The **MACRS,** established in 1986, is the current rapid or accelerated depreciation system specified in the Tax Code. It replaces the **accelerated cost recovery system** of 1981. This method, straight line, and unit-of-output are the common techniques used today for assets acquired after the change in the Tax Code in 1986. The MACRS method, like the SOYD and DB techniques, gives higher depreciation expenses in the early years and lower amounts in the asset's later life. Because of its current importance and application (although the Tax Code is subject to change by Congress at any time), the MACRS method will be considered here in detail.

MACRS depreciation classifies assets into categories and specifies the length of lives to be used in the calculations. These lives might not always represent the actual physical or economic life of the asset. For example, a piece of equipment can have a life of seven years under the MACRS method, but the equipment is used three shifts a day, seven days a week. This intense use might mean the equipment only lasts five years before it must be replaced. Important, too, in this system is the fact that no residual or net salvage values are considered. They are assumed to be zero for calculation purposes. Table 3–3 gives the general classification of asset types and lives.

Table 3–3   MACRS Major Classes and Lives

Class Life	Asset Type
3 years	Special tooling and devices such as dies and fixtures, semi-trailer trucks.
5 years	Autos, small trucks, computers, office copiers and office equipment, construction equipment.
7 years	Industrial equipment, fixtures and office furniture.
10 years	Longer life industrial equipment, petroleum pressure vessels, pumps, water transportation equipment, fruit-bearing trees.
15 years	Roads, steam and electric generation equipment, communication equipment, sewers, fences, landscaping, sidewalks and paving.
20 years	Farm and ranch structures.
27.5 years*	Rental apartments and residential real property.
39 years*	Commercial and industrial real property purchased after May 1993 (31.5 years life if purchased prior to May 1993)

*Straight line method applies to real property.
Source: 1995 U.S. Tax Code, Internal Revenue Service.

The MACRS depreciation method specifies annual percentages that are multiplied by the initial cost to determine the annual depreciation amount in each asset class. Using the straight line method with no estimated salvage value is an option also. Straight line *must* be used with real property, and can be elected for other assets. For commercial, industrial, and rental property and buildings, straight line with no salvage value is required by the 1986 tax law that established the MACRS technique. Also, the MACRS table shown here assumes that the asset is placed in service at the middle of the first year and that service ends at the mid-year point. Other tables are based on quarter-year assumptions. A five-year property, for example, is depreciated into the sixth year using the mid-year convention. The table of percentages is shown in Table 3–4.

The MACRS method assumes no salvage value; 100% of the asset cost is depreciated using this method. Final entries would be made in the Accounting Equation if the asset is sold for some salvage value at the end of its life using this method. Also, using Table 3–4, the method assumes that it is acquired at the midpoint of the first year.

Using the example of the robot at Fire Equipment Manufacturing Co. and using a first cost of $100,000, net salvage of $10,000, and an estimated life of 10 years, the MACRS method of depreciation can be applied. For this example, assume that the 10-year life would apply.

Table 3–4   Modified Accelerated Cost Recovery System (Half-Year Convention)

Year of Ownership	3-Year Asset	5-Year Asset	7-Year Asset	10-Year Asset	15-Year Asset	20-Year Asset
1	33.33%	20.00%	14.29%	10.00%	5.00%	3.750%
2	44.45%	32.00%	24.49%	18.00%	9.50%	7.219%
3	14.81%	19.20%	17.49%	14.40%	8.55%	6.677%
4	7.41%	11.52%	12.49%	11.52%	7.70%	6.177%
5		11.52%	8.93%	9.22%	6.93%	5.713%
6		5.76%	8.92%	7.37%	6.23%	5.285%
7			8.93%	6.55%	5.90%	4.888%
8			4.46%	6.55%	5.90%	4.522%
9				6.56%	5.91%	4.462%
10				6.55%	5.90%	4.461%
11				3.28%	5.91%	4.462%
12					5.90%	4.461%
13					5.91%	4.462%
14					5.90%	4.461%
15					5.91%	4.462%
16					2.95%	4.461%
17						4.462%
18						4.461%
19						4.462%
20						4.461%
21						2.231%

Source: 1996 U.S. Tax Code, Internal Revenue Service.

Depreciation for the first year is calculated as:

$$\text{Dep}_1 = (\text{Initial Cost})(\text{MACRS Factor})_1 = (\$100,000)(0.10)_1 = \$10,000 \text{ per year}$$

Book value at the end of the first year is:

$$\text{Book Value at end of 1st Year} = \text{Initial Cost} - \text{1st Year Dep}$$
$$= \$100,000 - \$10,000 = \$90,000$$

Depreciation for the second year is:

$$\text{Dep}_2 = (\text{Initial cost})(\text{MACRS})_2$$
$$= (\$100,000)(0.18) = \$18,000$$

$$\text{Book Value at end of 2nd Year} = \text{Book Value at end of 1st Year} - \text{Dep for 2nd Year}$$
$$= \$90,000 - \$18,000$$

This calculation continues for the rest of the life of the equipment. Note that the salvage value is not part of the calculations.

Table 3–5   MACRS Depreciation Method for Robot

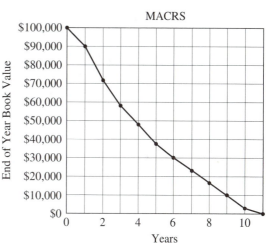

Year	End of Year Book Value	MACRS Factor	Annual Depreciation
0	$100,000		
1	$90,000	0.1000	$10,000
2	$72,000	0.1800	$18,000
3	$57,600	0.1440	$14,400
4	$46,080	0.1152	$11,520
5	$36,860	0.0922	$9,220
6	$29,490	0.0737	$7,370
7	$22,940	0.0655	$6,550
8	$16,390	0.0655	$6,550
9	$9,830	0.0656	$6,560
10	$3,280	0.0655	$6,550
11	$0	0.0328	$3,280
		Total	$100,000

Using the MACRS method for the robot, the annual depreciation expenses and book values are generated as shown in the table and graph in Table 3–5. Note that 11 years are used because of the half-year convention of this method. The end-of-the-year book value declines to zero since the method assumes no net salvage value. If the robot is sold any time during the 11 years, the entries into the Accounting Equation would be made to adjust the selling price and the book value of the equipment, similar to the straight line example. This method gives high depreciation expense in the early years with low depreciation expense in the asset's later years, similar to the older SOYD method.

Compare the straight line depreciation with the MACRS depreciation for the robot at Fire Equipment Manufacturing Co. The straight line method shows a constant depreciation expense over the life of the robot, while the MACRS method shows a high depreciation expense in the early years and a low depreciation expense in the later years. In other words, the straight line method results in constant taxes over the life of the robot, while the MACRS method results in an increase in taxes over the life of the robot.

Depreciation and Economic Life

It would be coincidental for any of the mathematical depreciation methods that have been discussed to agree exactly with the actual decrease in market value of the asset during its life. Values found using the mathematical calculation of depreciation or the assumed life may not necessarily agree with the actual physical or economic life, or the market value of the building or equipment.

The *economic life* of an asset is determined by the period of time the asset will provide the lowest cost of service compared to other alternatives. This concept recommends replacing the equipment if the cost of the service that the asset delivers begins to increase due to higher operating costs and/or when an alternate asset is less costly. We will see later, in the discussion of replacement analysis, that economic life is a critical determination in the decision to keep or to replace an asset. Replacement analysis is based on the time value of money, economic life analysis, and the impact of a depreciation method on tax expenses. It is sometimes confusing when dealing with conflicts between depreciation life and expenses for tax purposes, and economic analysis based on the physical and economic life and the market value of the asset. Both depreciation techniques and economic analysis should be understood by engineering and manufacturing personnel in order to make a complete analysis of the product, process, and asset's cost.

Recent Trends in Depreciation Methods

Accelerated depreciation methods such as MACRS have the advantage of generating high depreciation costs in the early years of an asset's life. At a high rate of inflation, which causes replacement asset values to increase, organizations are more likely to use accelerated depreciation methods. Conversely, in low-inflation periods, more firms would be likely to use constant or straight line methods. Accelerated methods permit the recovery of the cash earlier, which can be used for other inflation-driven investments.

In the 1970s, inflation increased to above 10% in the United States. In the 1980s and 1990s, however, inflation has been in a moderate range of 2% to 4%. General inflation applies to the cost of the replacement of fixed assets as well as other costs. An asset that has a 10-year life, if replaced with an identical asset, would likely cost 25% to 35% or more than it did originally if inflation is 3% per year. Over time, the organization would have to spend more funds annually to replace assets than their total depreciation amount. For this reason, the higher the inflation, the more advantageous an accelerated depreciation method becomes in making funds available sooner for reinvestment. Also, the application of interest and the return that can be earned on cash encourage the use of accelerated depreciation methods. An accelerated method may provide for 60% to 70% of the asset's value to be depreciated in five years of its 10-year life. This larger cash flow, caused by reduced taxes in the early years, permits earlier reinvestment of the cash into other projects, products, and profit-producing equipment and buildings.

In the current low inflation economy, however, large firms are not using the accelerated methods as much as they did 25 years ago when inflation was high. The trend has been away from accelerated methods (sum-of-years-digits, MACRS, and other accelerated methods) and toward straight line methods. A survey by the American Institute of Certified Public Accountants indicates that of the 600 largest firms in the U.S. in 1969, 250 used some type of accelerated depreciation methods. As of 1994, of the 600 largest firms, only *85* were using accelerated methods. Of the firms in the Dow Jones Industrial Average, large firms representing 30 industries, only 6 of the 30 indicated in their recent annual reports that they were using accelerated methods. Recalling that inflation percentages were above 10% in the 1970s and fell back to below 5% per year in the 1980s, firms switched to the straight line method because inflation was a reduced financial factor and the organizations could show higher and more consistent net profits in the early years of an asset's life than if accelerated methods were used. Consistent reporting of annual depreciation values was a major consideration (Tom Burnett, "Twin Towers," *Barron's*, 15 July 1996).

The straight line approach gives a steady decline in fixed asset values and a constant annual depreciation expense. From the standpoint of earnings consistency, the straight line method of depreciation is preferred. In an ideal, "zero inflation" situation, new assets are purchased for the same price as the old assets. As inflation increases, the replacement asset becomes more expensive and the cost to replace is greater than the total depreciation expense of the older equipment or building. Accelerated methods of depreciation partially offset this difference in the early years with proportionally large depreciation expense and lower tax payments. So long as inflation is moderate in the 2–3% range, it is likely that firms will continue to use straight line depreciation so they will not present exaggerated reported earnings on the Income Statement. Although it is more financially conservative to use the accelerated methods of depreciation when equipment replacement is considered, earnings consistency is also important to organizations when publishing the Income Statement.

The magnitude of depreciation and amortization expenses can often have a major effect on the Income Statement. General Motors Corporation, for example, had larger depreciation and amortization expenses than its net income from operations in the years of 1994 and 1995. Below is an abbreviated summary of the Income Statement data for General Motors Corp. as shown in their annual reports of 1994 and 1995.

From General Motors Annual Reports

(dollars in millions)	1995	1994
Net Sales and Revenues	$168,828	154,951
Costs and Expenses	147,031	136,347
Depreciation of Real Estate, Plant and Equipment*	8,554	7,124
Amortization of special tools*	3,212	2,901
Amortization of intangible assets*	255	226
Total Costs and Expenses	$159,052	$146,598
Income Before Income Taxes	9,776	8,353
Taxes	2,895	3,452
Net Income After Taxes	$6,881	$4,901
*The totals of depreciation and amortization for the two years are:		
Depreciation and amortization	$12,021	$10,251

Source: General Motors Annual Reports, 1994, 1995.

Looking at GM's annual reports, we can see that depreciation and amortization have a significant impact on net income and taxes for the two periods. In 1994, depreciation and amortization were approximately 7% of the total costs and expenses; 7.5% in 1995. In both years, depreciation plus amortization is greater than net income after taxes. By including depreciation and amortization in costs and expenses, the required tax payments are reduced significantly. The taxes that would have been paid without the consideration of depreciation and amortization for these two years would have been more than double the tax paid after including depreciation and amortization. The method and amount of depreciation and amortization for General Motors and other organizations are very important components of the financial process. Engineers, managers, and technicians involved in equipment and building investment decisions should have a sound understanding of depreciation concepts, its calculation, application, and overall impact on the organization's finances (Tom Burnett, "Twin Towers," *Barron's,* 15 July 1996; GM Annual Reports 1994, 1995).

Group Depreciation

Sometimes the organization owns a large number of identical or very similar assets. When this occurs, **group depreciation** can be calculated. For example, a utility that installs electrical or telephone service in a new subdivision may use many utility poles, fixtures, and lines. Rather than depreciating each line, transformer, and utility pole, all the poles, fixtures, and lines can be depreciated as a group. Although it is unlikely that all the poles will have the same exact physical life, the group depreciation method would be applied to them collectively. The actual physical lives would be distributed over time even though some would be replaced earlier, and some later, than the group life estimate. The *average estimated life* is used for group depreciation. Steady-state replacement of individual items in the group may occur over time for the assets.

In manufacturing, material handling, production, and storage equipment can be purchased in multiple units and depreciated as a group as well. Office furniture and computer workstations are other examples where group depreciation could be applied.

Partial Year Depreciation

The half-year convention of the MACRS method assumes that the asset is acquired at the midpoint of the first year and is kept until the midpoint of the year after its assumed life. For example, a 10-year-life asset is assumed to be purchased at the middle of the first year and kept until the middle of the 11th year. MACRS depreciation schedules are available for quarter-years also. When using straight line methods, fractions of years can be used to calculate depreciation expense. If the asset is purchased on October 1, three months (October, November, and December) or three-twelfths of the annual depreciation is applied for that year.

OTHER NONCASH OUTFLOWS

Cash that flows out of the cash account, which reduces the asset side of the Accounting Equation, may be offset by an expense under costs, *expensing,* or may be balanced by an entry in the asset side of the Accounting Equation, *capitalizing.* Depreciation is an example of capitalizing and then noncash expensing. Other noncash items that can be capitalized or expensed are:

1. Research and development
2. Amortization of intangible assets
3. Land
4. Prepayments of future expenses
5. Inventory

These noncash expenses are important to engineers, scientific personnel, manufacturing engineers, and other staff in the organization. They affect product and services costs, cost allocation, equipment justification, and related investment analysis and decisions.

Research and Development Expenditures

Many organizations conduct research and development on a continuous basis, and engineering and scientific personnel may be involved in the cost and benefit estimation of the R&D effort. Since it is difficult to estimate the value and the life of the benefits that are produced by these development efforts, R&D expenditures are generally expensed when they occur, rather than capitalized and expensed over time. However, when patents and other R&D are purchased from outside the organization, they are usually capitalized and then expensed over time, similar to fixed assets.

When the purchase of the R&D is made, cash decreases and R&D, an asset, increases (capitalization). The life of the R&D is estimated and the asset is reduced in value over its estimated life (expensed). From the Accounting Equation viewpoint, this is similar to depreciating equipment or buildings.

Amortization of Intangible Assets

The organization may acquire intangible assets. **Intangible assets** are nonphysical and long-lived; they may be legal or economic rights. When these assets are decreased in value over time, the process is called **amortization.** Examples of amortized assets are:

1. **Copyrights**—legal rights to written or other creative works.
2. **Trademarks**—legal rights to names and logos.
3. **Patents**—legal rights to inventions, designs, and processes.
4. **Goodwill**—economic rights to the reputation and profitability of an acquired business.
5. **Franchise**—contractual rights to manufacture and/or sell a product or service with some exclusive privileges such as geographical territory or type of customer.
6. **Leasehold improvements**—Improvements made by the tenant of leased property.

These intangible assets are purchased by the organization or they can be created within the organization. The purchase of a patent, software, or a franchise to sell a particular product in a geographic area is a purchase of an intangible asset. Such an asset is capitalized and then expensed over time. The entries in the Accounting Equation would be to reduce the cash account and to increase the intangible asset. Intangible assets are then assigned an estimated life, decreased in value, and expensed over this period of time. Amortization is similar to depreciation of a physical asset, except amortization applies to intangible assets.

Goodwill may be established when a business or asset is purchased for more than its fair market value. Goodwill is the difference between the price paid and the fair market value. For example, if a business is purchased for $100,000 but its fair market value is $75,000, the additional $25,000 is paid as incentive for the seller to sell the business quickly. Cash is reduced by $100,000, the $75,000 is entered as an asset, and $25,000 is an asset, goodwill. An estimate of the goodwill life is made and then goodwill is amortized over the estimated period of time. Normally straight line methods are used to amortize intangibles.

Land and Depletion

When land is purchased, to build an office or manufacturing facility, for example, the cash account is reduced and the asset account, land, is increased. However, rather than depreciating the land asset, as is the case with equipment and buildings, the land is maintained on the Balance Sheet at its original cost until it is sold, since land is assumed to have an unlimited life. Although the land may actually increase or decrease in value over time, this change in value usually is not recognized on the financial statements until it is sold. When it is sold, the cash and revenue accounts increase and the land is reduced to zero with the other half of the entry being cost of land sold (or similar title), entered under the cost category.

There can be a profit or loss on the sale with tax consequences when the land is sold. If the land's market value has increased, the assets on the balance sheet are *understated* by the difference between the land's cost and its current market value. For some organizations, this understatement may be large. For example, railroad companies that purchased land right-of-ways in the 1800s have sold that land for prices that are many times the original price. Manufacturing organizations that established plants on the edge of towns in farming communities in the early 1900s now find themselves surrounded by urban developments and land values that are very high compared to the amounts shown on the Balance Sheet.

Equipment and buildings may be *depreciated.* Land that is mined may be *depleted.* Land that is not mined for its natural resources is maintained at its value when purchased. Intangible assets are *amortized* over time.

An exception to land being carried at cost on the Balance Sheet until it is sold would be the case of land purchased to remove natural resources from it such as timber, minerals, gas, or oil. In this case, the value of the property decreases as the material is removed from the land. Reducing productive land in value is called **depletion.** Engineers, managers, and technicians in the mining, petroleum, and timber industries need to understand depletion calculations. This decrease in value is handled in the same way as buildings and equipment are depreciated. The land value on the asset side of the Balance Sheet is reduced as the minerals are removed. The other half of the entry is to create a cost called depletion expense under costs in the Accounting Equation. Depletion of land is similar to depreciation of equipment. The timber, minerals, oil, or natural gas value is estimated and then reduced over time as an expense on the Income Statement.

There are many specific Tax Code provisions for the application of depletion expense. Depletion, like depreciation, affects the profits and the taxes of the organization. A knowledgeable tax practitioner can be consulted for current tax information details since depletion methods are a frequent political topic and the Tax Code has often been revised by Congress.

Prepayment of Future Expenses

Another example of capitalizing and expensing is when expenses are paid early. If, for example, the firm pays rent in advance of when it is due, or advertising expenditures in advance before they are required, the expenditure is capitalized and maintained as an asset until it would normally be paid. The Accounting Equation entries would be to decrease the cash account and increase an asset account called, for example, prepaid rent. This prepaid rent is kept as an asset until the time comes when the rent is actually due. At that point, the prepaid rent account is reduced to zero and the other half of the entry is in the cost category as rent expense. A temporary asset is created, then expensed.

INVENTORY MANAGEMENT AND ACCOUNTING

Inventory costs, product costs and prices, and the resulting financial statements are dependent on the inventory accounting system. Many manufacturing managers, process engineers, operating personnel, and financial improvement teams are responsible for improving product costs and developing budgets; they need to have an understanding of inventory accounting. This section gives an introduction.

The purchase of materials that will be converted into a product that is to be sold can be accounted for in different ways. One method of making entries into the Accounting Equation is to *decrease* the cash account by the amount of the purchase while *increasing* cost of goods sold (increasing a negative account) under the cost category. This method would apply to supplies and materials that are consumed in a short period of time and have a small dollar value. This immediate expensing method was demonstrated with the example of computer supplies entries at Consulting Engineers Inc., at the beginning of the chapter.

A second, more typical method is to *reduce* the cash account while *increasing* an asset account called inventory. In this method, cash is decreased and inventory is increased when the material is purchased (subtracting and adding the same amounts to the asset side of the equation). Then, when the inventory is sold, inventory is decreased and cost of goods sold is increased (increasing a negative value). Cash and sales are increased by the selling price of the inventory. This second type of inventory accounting is shown in Figure 3–5.

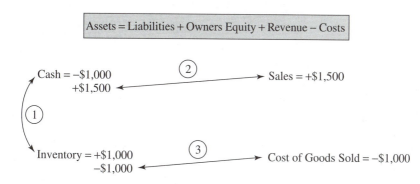

Figure 3–5   Inventory Purchase and Sale for a Retail or Wholesale Distribution Company

Inventory in a Retail or Wholesale Distribution Company

1. The material is purchased. Cash decreases and inventory increases by $1,000. The Accounting Equation is in balance because $1,000 is subtracted and added to the same side of the equation.

2. Inventory is sold to a customer for $1,500. Cash increases and sales increase by the $1,500.

3. The inventory account is decreased by $1,000 (the cost of the sold inventory) and the offsetting entry, cost of goods sold, is $1,000. The resulting profit on this sale is: $1,500 − $1,000 = $500.

This method of accounting for inventory and materials is applicable to merchandising, wholesale, retail, and distribution organizations. In these organizations, no significant change to the purchased materials is made, other than repackaging, before it is sold to the customer.

Inventory Accounting for Manufacturing

Normally, changes in the raw material are made as the inventory physically moves through the manufacturing area. Labor and overhead expenses are incurred to convert the raw material into a finished product. When the raw material is purchased, cash is decreased and inventory, an asset, is increased in value as the material moves through the system. As labor and overhead costs are expended, cash decreases and inventory increases. The value of the inventory, unlike in the other inventory accounting methods, *increases* as the product moves through the "manufacturing pipeline." Labor and overhead is added to the value of the raw material as it becomes a finished product.

When the inventory is finished and sold to the customer, the accounting entries are made to decrease the inventory asset account and to offset this decrease by entering an equal amount as cost of goods sold. See Figure 3–6.

Inventory in a Manufacturing Company

1. The cash is spent for material, labor, and overhead. For example, as the check is sent to the material supplier, cash is decreased and the value of the inventory on the asset side of the Accounting Equation is increased. The same accounting steps are applied to labor and overhead expenditures until the inventory becomes a finished product.

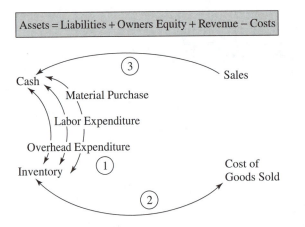

Figure 3–6   **Inventory Purchase and Sale for a Manufacturing Company**

2. Inventory is then decreased and the cost of goods sold entry is made (an increase in costs) when the product is sold.

3. Finally, the customer's payment for the product is recorded by increasing cash and revenue or sales by the selling price of the product.

Figure 3–7 shows the manufacturing pipeline for inventory. Raw material is added to the inventory account, other components and supplies are added to the original inventory, and material, labor, and overhead costs are expended and added to the inventory, which further increases its value. When the finished product is shipped to the customer, the total value of the product includes material, labor, overhead costs, and other materials and supplies.

Three Types of Inventory Accounting Entries

1. Material is expensed immediately as it is purchased. This would apply to supplies that are consumed immediately or have small dollar value.
2. Material is capitalized in an inventory asset account until sold to the customer. Labor and overhead costs are expensed. This is applicable to merchandising, distribution, wholesale, and retail organizations.
3. Inventory is capitalized and labor and overhead costs are added to it until it is a finished good and is sold to a customer. This applies to a manufacturing organization.

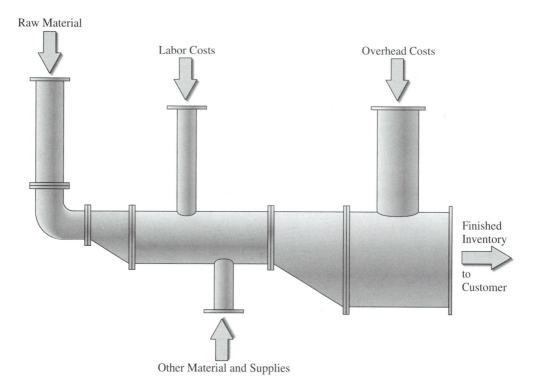

Figure 3–7   Manufacturing Inventory Cost Pipeline

The examples of depreciation, research and development, depletion of natural resources on land, amortization of intangible assets, prepayments, and inventory all apply the common accounting characteristic of capitalizing and then expensing the asset. Understanding this concept will help in understanding the impact of many technical decisions on the organization's financial statements. Managers, engineers, technicians, and other nonfinancial personnel can contribute their special or technical knowledge and combine it with accounting understanding to help improve the organization's financial results.

CASH FLOW AND THE ACCOUNTING EQUATION

For those in the organization who are analyzing and making decisions about capital expenditures, buildings, equipment, and other assets, understanding cash flow is vital to the decision process.

The term **cash flow** is often misunderstood. It has different meanings to different individuals. When it is a few days from payday, and a person's checking account is low, he or she may say, "I've a cash flow problem—all my funds are flowing out and none are flowing in." A process engineer who is analyzing the investment potential of a piece of equipment may say, "I need to summarize all of the cash flows of this equipment that are flowing out of the organization to pay for the product and process, and all of the cash flowing into the organization when the product is sold." The accountant or financial manager of the organization may say, "Our cash flow is going to be lower this quarter than in previous quarters." All of these comments, in their own way, are correct. Cash flow is the inflow and outflow from the cash account; it is directly related to the Accounting Equation. Cash flow is one of the many points where accounting and asset investment analysis meet.

If we could look at the check register of an organization or individual as deposits and withdrawals are made in the cash account, we would be able to see the cash flow in and out of the organization. Cash flow can also be determined indirectly from the Income Statement and Balance Sheet as well as directly from the cash account.

Often questions are asked when financial statements are studied:

"I see the organization made money last year, but where did all the profits go?"

"How can dividends be greater than profits?"

"What happened to the funds received from the bank loan?"

"We have many good potential investments to make in some new equipment that will save us money, but there are no funds available. What should we do?"

"Profits were low last quarter. How can we afford to invest in new equipment?"

"Are the profits and retained earnings on the Income Statement a measure of the funds available for investment into new equipment?"

The answers to these questions can be discovered with an understanding of cash flow and the Accounting Equation.

Recall the Accounting Equation:

$$\text{Assets} = \text{Liabilities} + \text{Owners Equity} + \text{Revenue} - \text{Costs}$$

Cash is an asset, and funds flow into and out of the cash account. Monitoring this flow and the residual cash amounts results in a Cash Flow Statement. This flow is shown in the expanded Accounting Equation in Figure 3–8.

Referring to Figure 3–8, the cash account can be increased or decreased from entries in different parts of the Accounting Equation. The changes in the Accounting Equation's five parts can affect the cash account and the cash flow.

Cash can be *increased* by:

1. Selling current or fixed assets.
2. Increasing current and term liabilities by borrowing funds.
3. Selling stock.
4. Retaining earnings from operations.
5. Selling services and products.

Cash is *decreased* by:

6. Investing in current assets such as inventory and marketable securities.
7. Paying current and term liabilities such as loans and accounts payable.
8. Investing in fixed assets such as buildings and equipment.
9. Repurchasing stock from shareholders.
10. Purchasing materials, labor, overhead, and administration expense items.
11. Paying taxes.
12. Paying dividends.

Assets = Liabilities + Owners Equity + Revenue − Costs

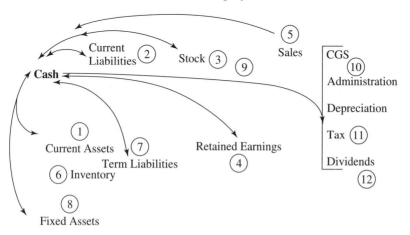

Figure 3–8   Cash Flows in the Accounting Equation

Notice that depreciation expense, amortization, depletion, and other noncash expenses do not increase or decrease the cash account.

Three Categories of Cash Flow

The above sources and uses of cash can be grouped into three categories:

1. **Cash Flow of Operations**—This includes cash received from the sale of products and cash used to pay suppliers, labor, and overhead costs. This is a measure of the organization's success for a period, resulting in profits and retained earnings from operations.
2. **Cash Flow of Investments**—This includes cash paid or received for assets, including equipment, buildings, land, or nonoperating assets such as patents and securities. For the growing organization, this is a net cash outflow into facilities, new and replacement of old equipment, development projects, and other profit-producing assets.
3. **Cash Flow of Financing**—This includes issuance and repurchase of company stock, borrowing, and repayment of loans and bonds.

Cash is exchanged between the three categories as the financial conditions of the organization require. For example, stock is sold and bank loans are obtained (financial cash flow) to provide funds for equipment, facilities, and other assets (investing cash flow). The investment of cash into equipment produces products that are sold and provides cash from sales (operating cash flow). The cash received from customers flows out of the organization to pay material suppliers and dividends to shareholders, and to reduce bank loans. Similar cycles are repeated continuously. The interrelationship of the three cash flows is shown in Figure 3–9.

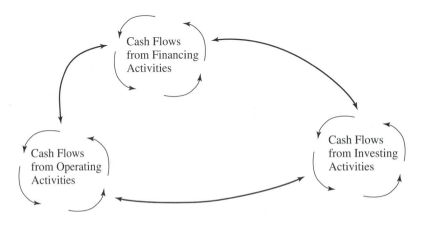

Figure 3–9   Three Categories of Cash Flow

> ## Total Cash Flow
>
> Cash Flow = Operating Cash Flow + Investing Cash Flow + Financing Cash Flow

Just because the organization is profitable (revenues exceed costs) on the Income Statement does not mean that all of the profits go into the cash account. Profits may be given to shareholders as dividends, inventories may be increased to support larger future sales, new equipment is purchased, or loans are repaid, leaving little cash in the checking account. The firm, as do individuals, may call this a "cash flow problem," meaning that there seems to be profits, but little cash is available.

> ## Land Poor
>
> Farmers that have purchased additional land to expand their operations may refer to themselves as "land poor" because they own the land, but have too little cash.

Organizationally, cash flow from operations and cash flow from investments are the responsibility of the design, manufacturing, operating management, and marketing parts of the organization. Typically they have the responsibility to design, manufacture and sell the product and invest in facilities. This usually includes decisions to invest in equipment, buildings, and inventory; hire personnel to produce the design and manufacture the product; and advertise and market the product successfully.

Senior management and financial and accounting personnel usually have the responsibility for the cash flow from financing. These activities include obtaining funds from banks and financial institutions, selling securities including common and preferred stock, selling bonds to provide funds for the operating part of the organization, and arranging for dividend payments to shareholders and payment of bond interest to bondholders.

Managers, engineers, manufacturing, operating personnel, and technicians who have the responsibility for determining the investments into equipment, buildings, R&D, inventories and other projects of the organization should have an understanding of the impact of cash flow on these potential investments. Even though an investment into equipment may have a high potential return and will save money compared to the present equipment, if the cash flow from investment is too low, it is possible that the good investment will go unfunded. For both individuals and organizations, there are almost always more good potential investments than there are available funds.

> Not all good investments can be funded from existing cash flows.

Sometimes internally generated profits do not provide enough cash, and external sources of funds, loans and security sales, must be obtained. This requires the technical, design, and manufacturing personnel who perform the investment analysis to work closely with the financial personnel who obtain the loans and sell the securities. When there is cooperation among financial, design, and manufacturing personnel, investment decisions are generally more successful.

Published annual reports usually include a Cash Flow Statement along with the Income Statement and Balance Sheet. The reader can see the details of inflows and outflows of cash between the three cash flow categories. For some, the Cash Flow Statement may give more information than the other two statements. Later in the text, we will discuss detailed examples of cash flow statements, cash flow and taxes, and the analysis of investments based on aftertax cash flow.

Measuring Cash Flow from Operations

Operating cash flow can be determined directly from the cash account, or indirectly from the financial statements. Cash, credit, and noncash accounts are shown on the financial statements. They can be separated to measure the flow of cash from and to operations. Depreciation, amortization, depletion, write-offs, and other noncash expenses are shown on the Income Statement. For credit sales, there may be a steady, but delayed, flow of cash from customers. So long as this flow is steady, it does not affect the residual cash in the cash account. Usually the focus is on the "bottom line." The question is asked, "Is there a profit for the month, quarter, or year?" Perhaps an additional, and sometimes more important question is, "Is there positive cash flow and enough funds for investment into assets, R&D, equipment, inventory, and buildings, to provide future profits?"

Warren Buffett and Cash Flow

Warren Buffett, one of the most successful investors in the U.S., states, "The current value of a business is determined by the net cash flows expected to occur over the future life of the business, discounted at an appropriate interest rate to the present. . . . So valued, all businesses from the manufacture of buggy whips to cellular telephones become economic equals."

—Berkshire Hathaway Co. 1989 Annual Report

Calculating Operating Cash Flow and the Income Statement

If the cash account entries are not available, the operating cash flow can be determined by working backward from the Income Statement. When the Income Statement is used to determine the operating cash flow, special attention must be given to the depreciation expense and other noncash deductions. Since the Income Statement shows both cash and noncash items, to determine the resulting cash flow for the period, the noncash expenses must be added back to the net profit after tax. The sum of profits after tax, net income, plus noncash expenses is the operating cash flow for the period.

Operating Cash Flow—A Working Approximation

Operating Cash Flow = Net Operating Profit After Tax + Depreciation
+ (Amortization + Depletion + Write-Offs + Deferred Taxes)

An organization's annual report often shows that depreciation, amortization, and other non-cash expenses are a significant part of the net income or profit after tax calculation. Referring to the abbreviated Income Statement for General Motors in 1994 and 1995 presented earlier in this chapter, we can see that the total of depreciation and amortization was more than twice the reported net income after taxes. Although this is not typical, it emphasizes the importance of considering cash flow as well as net income.

Determining Operating Cash Flow for Stollkeen Corp. Below is the Income Statement of Stollkeen Corp., a manufacturer of pumps and pressure vessels and a supplier to the chemical processing industry. Stollkeen has recently discovered that some of the inventory in its warehouse is obsolete and it has made the decision to reduce this inventory's value to zero and dispose of it. A special write-off will be made for the year's Income Statement for this inventory. Also, the organization purchased rights to some software two years ago that it intended to sell and distribute to its customers. Recently, competing software is being sold that makes Stollkeen's software obsolete. Stollkeen has decided to write off this software asset. Taxes of $10,000 have been deferred until next year and are included on the Income State-

---

Stollkeen Corp.
Income Statement for Year Ended 19xx

Revenue from Sales	$850,000
Cost of Goods Sold	400,000
Gross Profit from Operations	$450,000
Administration Expenses	200,000
Depreciation*	50,000
Write-off of Software*	45,000
Obsolete Inventory*	15,000
Profit Before Taxes	$140,000
Taxes*	60,000
Profit After Taxes = Net Income	$80,000
Dividends	35,000
Retained Earnings	$45,000

*Notes to the financial statements would normally explain these noncash deductions and special write-offs on the Income Statement.

ment as total taxes owed of $60,000. By using this information and the Income Statement, the operating cash flow can be determined.

The calculation of the operating cash flow for Stollkeen is shown below:

$$\text{Operating Cash Flow} = \text{Net Operating Profit After Tax} + \text{Depreciation} +$$
$$(\text{Amortization} + \text{Depletion} + \text{Write-Offs} + \text{Deferred Taxes})$$
$$\text{Operating Cash Flow} = \$80,000 + \$50,000 + (\$45,000 + \$15,000 + \$10,000) = \$200,000$$

If there are no other major adjustments, the $200,000 is the amount of funds in the cash account that are available for investment into equipment, buildings, research projects, and dividends distributed to the shareholders ($35,000 has been allocated to the shareholders). This cash can be used for replacement of equipment, purchase of new equipment, and other investments. Sometimes when an organization has large amounts of cash, it will even repurchase its stock in the open market as well as invest in assets and pay dividends. This repurchase has the effect of reducing the number of shares outstanding and increasing the value of the remaining outstanding stock.

If only the profit after tax, $80,000, is used as a benchmark of Stollkeen's financial success for the year, it would appear that they made $80,000 net income, and that the retained earnings of only $45,000 are available. When operating cash flow is considered, $200,000 is available for distribution to the shareholders or to be used for reinvestment. Since depreciation, software, and obsolete inventory have been on the balance sheet as assets, the write-off of these items is not a cash decrease from the cash account. These items are added back to the net income to determine available cash. As a result of this write-off, the assets shown on the Balance Sheet will, of course, decrease.

> Managers, process engineers, operating employees, salespeople, and shareholders can often learn more about the financial health and funds available for facilities investment by looking at the operating cash flow of the organization rather than the net income.

Investing Cash Flow

It is discouraging for managers, engineers, and operating personnel not to be able to invest in cost-reducing or quality-improving equipment for the organization. Even good investments with high returns must be unfunded if the organization has low cash flows. If, in addition, the availability of external funds through the securities markets or loans is too costly or unavailable, high-return projects may go unfunded.

The cash flow from investing category includes the investments into land, buildings, equipment, securities, and investments into other organizations including subsidiaries. This is the important responsibility of manufacturing engineers and technicians. Generally, for manufacturing organizations, the largest flow of cash is into operating equipment and buildings. The cash receipts from the investments into securities, land, and equipment are almost always less than the outflow of cash into investments. Typically, an organization has net cash invested in equipment and buildings, and has net cash flow from the sale of products and services.

Cash Flows from Financing Activities

This third category of cash flows includes borrowing from and repayments to banks, sale and repurchase of the firm's own stock, sales of bonds, dividends paid, and related financing activities. Financing cash flows are usually a function of the organization's need for cash. If operations can provide all the cash necessary for investing activities, borrowing and sale of securities may be low or zero. If the organization is growing, the operating profits are usually not adequate to provide funds for the investment into equipment, land, and buildings. Borrowing and/or sale of securities would have to be made.

Obtaining external funds for the internal expansion of the organization also depends on the prevailing interest rates for borrowing funds and the interest rates of bonds. When interest rates are high for loans and bonds, the organization is less likely to pursue outside financing. This may mean that some internal projects and investments, even those with high-potential returns, are not funded. Expansion is slowed, net income may be reduced, and other effects may result from the high interest rates. The opposite occurs when interest rates are low. External funds are obtained at low rates, projects and investments are funded. The organization expands, and hopefully profits increase. The internal funding of investments is often very sensitive to external interest rates and costs of capital.

Cash Flows

There are three categories of cash flow:
1. Cash flow from operations
2. Cash flow from investing
3. Cash flow from financing

Cash Flow and Economic Analysis

The funds available for investment into equipment, buildings, R&D, and other investments usually come primarily from operations and its profits, and from financing with the sale of securities and loans. The return on these investments provides the funds to pay shareholders and lenders, and to reinvest into other projects. The analysis and decisions to invest are the responsibility of process and design engineers, department managers, and others in the organization. Cash flow is at the heart of investment analysis and decision making. Investment decisions are not only based on the potential returns of the investment, but the availability of the cash with which the investments are made. Often, the energy required for seeking funds is equal to or even more than the effort to analyze the project.

Cash flows are the basis for the economic analysis that helps determine the financial success of the organization.

The accounting system records the history of the financial events including cash flows. Economic analysis and decisions are made based on this information mixed with technical, operating, customer, and other necessary information. Financial success is, in large part, based on economic analysis, which in turn depends on the organization's cash flows.

Last Thoughts on Financial Statements

The combination of income statements, balance sheets, and cash flow statements gives a detailed financial historical picture of the organization. They become the basis for future analysis of investments.

Although the details of the statements are important, it is also important to view the total financial picture. Understanding how individual investments and other financial decisions affect the organization's financial statements is an important skill for operating employees, engineers, and managers to have. As managers, engineers, technicians, and other operating personnel become more involved in team-based financial decisions, more financial knowledge is required.

A Caution

Along with the mathematics of depreciation and cash flow calculation, understanding the concepts and their financial impact on the organization is important.

SUMMARY

This chapter has extended the application of the Accounting Equation to the calculation of depreciation and other noncash expenditures, and to the three types of cash flows. Cash flows, depreciation, expensing and capitalization, and asset investment decisions are all interconnected in the Accounting Equation. These factors form the structure in which all the project, facility, and equipment investment decisions are made. Calculations and recommendations by engineers, managers, technicians, and financial personnel are based on these concepts.

- The Accounting Equation is the basis for matching the revenues and costs in the same period of time. Expensing and capitalizing of expenditures are methods that match the revenues and costs. Depreciation and other noncash expenses are the expensing asset values over time to permit the matching of the revenues and costs in the same period. Depreciation along with depletion, amortization, and write-offs are noncash expenses. They are entered on the Income Statement accounts as a result of decreasing the asset's values over time.

- Calculating depreciation is used rather than selling or appraising the asset. Mathematical methods are used to calculate the amount of value that the asset has lost in a period of time. These mathematical methods have evolved over time. They include straight line, sum-of-the-years-digits, declining balance, accelerated methods, unit-of-output methods, and most recently, the modified accelerated cost recovery system. The depreciation entries into the Accounting Equation as well as the selection and calculation of the depreciation methods have an effect on the organization's total expenses, profits, taxes, retained earnings, and cash flow.

- Other noncash and capitalization decisions, in addition to depreciation, include land, patents, trademark, goodwill, franchises, assets, and their amortization and depletion. These decisions have an effect on the cash flow, profits, and expenses of the organization. Inventories and materials can also be entered temporarily on the asset side of the Accounting Equation until they are matched with the products moving through the manufacturing system and to the customer, when they appear as an expense on the Income Statement.

- Intangible assets such as patents, copyrights, and leasehold improvements can be amortized over time. In-house research and development costs are usually expensed as they occur.

- When land is acquired, it is maintained at its purchased value until it is sold, unless it has natural resources that decrease in value over time, called depletion.

- Trends in depreciation methods change when Congress modifies them. The organization must choose between methods, and the decision has an effect on taxes and reported net income. Firms must consider many factors with respect to depreciation including the rate of inflation and the desire to report consistent earnings and optimize tax advantages.

- The three types of cash flows, operating, investing, and financing, are related to the Accounting Equation, capitalization, and noncash expenses. With net income, investing, and financing cash flows, funds available for investment by the organization are determined. Design engineers, process engineers, technicians, manufacturing managers, and financial personnel should have a thorough understanding of cash flows and their relationship to depreciation.

- Technical and financial reasons whether to expense or to capitalize expenditures are not always clear-cut. Judgment is required as well as interpretation of the Tax Code. An organization has a responsibility to employees, shareholders, and tax officials to represent its financial results using consistent and accepted accounting methods.

- As users of financial statements, engineers, managers, technicians, and other employees within the organization, and shareholders, customers, and suppliers outside the organization should understand the process and concepts of expensing, capitalizing, intangible assets, depreciation, and cash flows. They affect the organization's profits, taxes, and funds available for investment and distribution to owners and shareholders.

> When a team of engineers, technicians, managers, accountants, purchasing buyers, and operating personnel has the responsibility to recommend investments and improve the organization's financial results, it is important that they understand the impact of their decision on cash flows and the financial statements.

QUESTIONS

1. When a building is purchased, what are the entries into the Accounting Equation?
2. What is meant by cash accounting?
3. What is accrual accounting?

4. What are the entries into the Accounting Equation for capitalizing an asset compared to expensing?
5. When supplies that are to be used immediately are purchased, are they normally expensed or capital-

ized? What is a typical entry into the Accounting Equation?

6. What are the tax consequences of expensing versus capitalizing a cash outflow?

7. What criteria can be used to determine if an expenditure should be capitalized or expensed?

8. What is meant by matching?

9. Show the differences in entries into the Accounting Equation between purchasing a supply of inexpensive office materials and purchasing manufacturing equipment.

10. What is depreciation? How does it affect the accounting equation?

11. Why are buildings and equipment capitalized rather than expensed when they are purchased?

12. What are some common misunderstandings about depreciation?

13. What are the factors that affect whether an item is capitalized or expensed?

14. What are the different ways to determine the current value of a fixed asset such as equipment or a building?

15. What methods does the Tax Code currently permit for the depreciation of assets purchased since 1986?

16. Identify the following:
    a. MACRS
    b. Straight line
    c. SOYD
    d. DDB

17. With respect to mathematical depreciation methods, which one gives constant depreciation expense over the life of the asset?

18. What methods of depreciation calculation cause high depreciation expenses in the early years of the asset's life and low amounts in the later years?

19. Indicate the effect of straight line, MACRS, and SOYD depreciation calculations on the following:
    a. Asset value shown on the Balance Sheet.
    b. Taxes in the early years of the asset's life.
    c. Profits on the Income Statement in the later years of the asset's life.
    d. Cash flow in the early years of the asset's life.

20. What are the entries into the Accounting Equation when the asset is sold for a value greater than its current book value? For less than current book value? What is the effect on taxes owed?

21. What are the entries into the Accounting Equation if an asset is sold before its accounting life is completed?

22. Should repairs and maintenance costs be expensed or capitalized?

23. How should equipment installation and removal costs be handled?

24. Why is it important to estimate the life and salvage value of equipment?

25. How are buildings and structures depreciated under the MACRS method?

26. Are the lives specified by the MACRS method of depreciation the actual physical lives of the assets?

27. What are the depreciation lives specified by the MACRS method for the following:
    a. Small trucks.
    b. Computers.
    c. Manufacturing dies.
    d. Office furniture.

28. Considering large firms in the U.S., how have their methods of depreciation changed over the past 25 years? Why?

29. What are the advantages of using a depreciation method that relates the life of the asset to its output of units?

30. When would group depreciation be used?

31. How are research and development costs entered in the Accounting Equation?

32. What expenditures, in addition to buildings and equipment, are often capitalized by organizations?

33. What is the difference between the economic life and the accounting life of equipment?

34. What are the differences between depletion, depreciation, and amortization?

35. What are the types of assets that are amortized?

36. How can labor and overhead expenditures be added to inventory values? What are the entries in the Accounting Equation? What occurs over time to the inventory?

37. What are three different ways to enter inventory purchases in the accounting equation?

38. Why are the concepts of depreciation and other capitalizations important to employees, shareholders, suppliers, and customers?

39. What is meant by the "inventory pipeline"?

40. What is meant by the statement, "Cash flow may be more important than net profits"?

41. What are three types of cash flows?

42. How can operating cash flow be determined from the Income Statement?

43. What is the importance of cash flow to managers, engineers, or technicians?

44. How can the understanding of depreciation, capitalization, and cash flow apply to the economic analysis of equipment and buildings?
45. As the organization attempts to improve its financial results, why is it important for engineers, technicians, managers, and financial personnel to approach the effort as a team rather than individually?

PROBLEMS

1. Indicate which of the following expenditures would be expensed or capitalized and the reason for your decision.
   a. New conveyor system for the warehouse.
   b. Paper for the office copier.
   c. New roof for the office.
   d. Three used cash registers for a retail store.
   e. Hand tools that are used in the assembly department.
   f. Flowers for the reception area.
   g. Annual maintenance and repairs of elevator.
   h. Roof repairs for the warehouse.
2. Equipment is purchased that has an initial cost of $125,000. It has a 10-year life and its salvage value is estimated to be $12,000. Using a calculator or a spreadsheet, determine the following:
   a. Initial and annual entries into the Accounting Equation.
   b. Annual straight line depreciation expense.
   c. Book value of the asset after six years.
   d. A graph of book value versus time.
3. A building is purchased for $600,000 and is estimated to have a 30-year life with no salvage value. Determine the following using a calculator or a spreadsheet.
   a. Initial entries into the Accounting Equation.
   b. Annual straight line depreciation expense.
   c. Book value of the building at the end of the 14th and 25th years.
   d. A graph of book value versus time.
4. Office supplies are purchased for $600. Show the entries into the Accounting Equation.
5. Manufacturing equipment is purchased for $50,000 with a life of seven years and net salvage of $5,000 at that time. Straight line depreciation is used. After seven years, the equipment is sold for $10,000. Show the entries in the Accounting Equation when the equipment is purchased, annually during its life, and when it is sold.

6. A computer network system is purchased for a total cost of $200,000. It is estimated to have a five-year life and salvage value of $10,000 at that time. After five years, the system is disposed of for a price of $5,000. Straight line depreciation is used. Show the entries in the Accounting Equation when the purchase is made, annual entries during its life, and the final entries when the equipment is sold.
7. In Problem 5 above, what if the equipment is sold for $2,000 after five years? Show the initial, annual, and ending entries into the Accounting Equation.
8. In Problem 6 above, what if the system is sold after three years for $50,000? Show the initial, annual, and final entries into the accounting equation.
9. Using the information of Problem 2 above, determine the following using a calculator or spreadsheet:
   a. Sum-of-years-digits depreciation expense for each year of the equipment's life.
   b. Book value at the end of six years.
   c. Entries in the Accounting Equation if the equipment is sold after six years for $50,000.
10. Tooling for a press is purchased for $75,000. It is estimated to have a life of five years, and no salvage value at that time. Determine the following using spreadsheet or calculator.
    a. Annual depreciation expense using SOYD.
    b. Book value at the end of three years.
    c. Entries into the Accounting Equation if the tooling is disposed of at the end of three years with no salvage value at that time.
11. Referring to Problem 10, and using a DB rate of 1.5, calculate the depreciation expense for each of the five years. Assume the tooling is sold for $2,000 at the end of the fifth year. Show the annual entries into the Accounting Equation and the entries at the end of the fifth year.
12. A delivery truck is purchased for $35,000. Using spreadsheet software or calculator methods, determine the following:
    a. The annual depreciation expense using the MACRS method.
    b. The book value at the end of the second year.
    c. The depreciation using the straight line method. Plot the results of your calculations from parts a and c.
13. Office furniture is purchased for $75,000. Determine the following using spreadsheet software or calculator.

a. The annual depreciation expense using the MACRS schedule.

b. The annual depreciation expense using straight line.

c. The end of the year book value for parts a and b. Plot the results.

14. A press has a die-set that has an estimated life of 500,000 units. It has a first cost of $50,000 and a salvage amount of $3,000. If 60,000 units are produced in a year, determine the annual depreciation expense for the die.

15. Materials handling equipment is purchased for $250,000. Using a seven-year life and salvage value of 5% of the equipment's first cost, determine the following:

a. Straight line depreciation expense per year.

b. Book value of the equipment using straight line at the end of seven years.

c. Annual depreciation expense for each year using MACRS methods.

d. Book value of the equipment for the seven years using MACRS.

e. Plot the values obtained in parts b and d using spreadsheet software.

f. Show the entries into the Accounting Equation for straight line and MACRS if the equipment is sold at the end of the third year for $200,000. What are the tax consequences?

16. Earthmoving equipment is purchased by a construction company for $750,000. Using a five-year life and salvage value of 5% of the equipment's first cost, determine the following:

a. Straight line depreciation expense per year.

b. Book value of the equipment using straight line at the end of five years.

c. Annual depreciation expense for each year using MACRS methods.

d. Book value of the equipment for the five years using MACRS.

e. Plot the values obtained in parts b and d using spreadsheet software.

f. Show the entries into the Accounting Equation for straight line and MACRS if the equipment is sold at the end of the fourth year for $300,000. What are the tax consequences?

17. A company purchases five delivery vans for $12,000 each. They are estimated to have a life of 125,000 miles and they are depreciated using the unit-of-output method. At the end of the first year, each van has the following mileage:

Van	Mileage
1	10,000 miles
2	22,300
3	18,000
4	12,000
5	16,500

Determine the annual depreciation charge, assuming no salvage value, and the total depreciation for all the vans.

18. Indicate the accounting entries to be made for each of the following inventory systems:

a. Lawn mowers purchased and sold by a neighborhood hardware store.

b. Aluminum ingots purchased by a die casting company.

c. Labor costs at the hardware store in part a.

d. Equipment maintenance costs in the die casting company of part b.

e. Structural steel used in expanding an existing warehouse.

f. Diesel fuel for delivery trucks.

g. Potatoes for french fries purchased by a fast-food restaurant.

h. Crude oil purchased by a petroleum refinery.

19. Given the Income Statement below, determine the organization's operating cash flow for the period.

Income Statement for Year Ending 19xx

Sales	$4,500,000
Cost of Goods Sold	3,400,000
	$1,100,000
Expenses	200,000
Depreciation	400,000
Profit Before Tax	$500,000
Tax	$300,000
Profit After Tax	$200,000
Dividends	50,000
Retained Earnings	$150,000

How does the operating cash flow value affect the funds available for investment?

20. Given the Income Statement below, determine the organization's operating cash flow for the period.

Income Statement for Year Ending 19xx	
Sales	$800,000
Cost of Goods Sold	600,000
	$200,000
Expenses	30,000
Depreciation	40,000
Profit Before Tax	$130,000
Tax	$55,000
Profit After Tax	$75,000
Dividends	10,000
Retained Earnings	$65,000

How does the operating cash flow value affect the funds available for investment?

21. Inventory is purchased by a electronic components wholesale company from a manufacturer. The components have a value of $51.50 per unit and 100 units are purchased. Two weeks later a retail customer orders 25 of these components at a price of $67.00. Show the entries in the Accounting Equation when the components are purchased by the wholesaler and then when the units are sold to the retailer.

22. A manufacturer purchases $10,000 of raw material. During the manufacturing of the products, $40,000 of labor and $60,000 of overhead cost are spent. After completion, the products are stored in the warehouse, and three weeks later one-half of the finished products are sold for $65,000. Show the following entries in the Accounting Equation.
    a. When the raw material is purchased.
    b. When the labor and overhead expenses are incurred.
    c. When the finished product is sold.

23. Given the following Income Statement for the year, determine the operating cash flow for the company.

Electronics Corporation Income Statement for Year Ended 19xx	
Sales Revenue	$1,050,000
Cost of Goods Sold	750,000
Gross Profit From Operations	$300,000
Administration Expenses	150,000
Depreciation	40,000
Write-off of Obsolete Inventory	10,000
Amortization of Patents	5,000
Profit Before Taxes	$95,000
Taxes	40,000
Profit After Taxes = Net Income	$55,000
Dividends	20,000
Retained Earnings	$35,000

> They say that knowledge is power. I used to think so, but I now know that they mean money.
>
> —Lord Byron

DISCUSSION CASES

Hogan Inc.

Hogan is purchasing some manufacturing equipment that costs $550,000. It will cost $40,000 to revise the utilities and to install the equipment. The manufacturing engineer, Nancy Born, estimates that it will last eight years and have a salvage value at that time of $50,000. She estimates that there will be removal costs of $10,000 at that time.

She has been asked to recommend if the equipment should be depreciated using straight line or the modified accelerated cost recovery system. If MACRS is used, she would have to use a seven-year life. She isn't sure what impact the depreciation methods will have on Hogan's financial statements.

Nancy's boss has indicated that the product for which the new equipment will be used might be discontinued at the end of five years. If so, the equipment would have a salvage value of $80,000 and removal costs of $10,000 at that time.

1. Assuming that the equipment is kept for its full life, what are the effects on profits, taxes, asset value, and cash flow using MACRS and straight line?
2. Which method should Nancy recommend? Why?
3. What would be the effects of the product being discontinued?
4. What other factors should be considered?

Valley Services Inc.

Valley Services is a small maintenance company. It was started 10 years ago as a one-person business and it has grown rapidly. It provides repair services for commercial, industrial, and homes. One division provides lawn cutting, snow removal, leaf removal, and landscaping services to homeowners. Another division provides housekeeping and maintenance services for office and industrial customers, from routine daily office cleaning to large warehouse cleaning. A third division provides landscaping, repair, and maintenance services for apartments and condominium projects.

The organization is profitable. Jim Maxwell, the owner, works hard, sometimes 60 to 70 hours per week. He has the payroll, bookkeeping, and tax work done by a small accounting firm in the neighborhood, and he personally signs all the checks. The accountant, Bob Harrison, has been discussing with Jim the potential problems that Valley Services may be facing soon. Bob has said that growth is too fast to be supported by internal profits and Jim is going to have to consider borrowing or obtaining funds from outside sources if he wants to continue to grow. Jim has recently been approached by a competitor who wants to sell his business to Jim. Also, a small local industrial trash and waste company is for sale and Jim believes it would fit into his company's services.

Valley Services has some experienced and loyal employees. Some have been with Jim a number of years. Valley has a lot of equipment, including mowers, pickup trucks, vans, power cleaning equipment, tools, and related equipment. The company rents an old 20,000-square-foot garage that serves as storage for vehicles, repair facility, equipment storage, and offices for the company.

Although the company is very profitable, as the annual income statements show, Jim never seems to have enough cash in the checking account to pay the bills and purchase new equipment. Bob has talked to Jim about the organization's cash flow problems, but Jim does not understand the details of accounting. Jim knows that he may have to begin to borrow long-term funds from the bank in order to grow. He doesn't like the idea of the business taking on debt. Also, he has considered making the company's stock available to some of the more experienced and loyal employees. He feels that this would be a good motivator and reward for them, as well as a potential source of funds without borrowing. He has been approached by a relative who has offered to invest in the company in exchange for stock.

Late one night at the office, while thinking about all these problems, Jim jotted down the following list on the back of an envelope. He has been carrying the envelope with him during the last two weeks, but he does not have any conclusions or solutions.

Should the waste disposal company be purchased?

Is the purchase of the competitor a good idea?

Would the employees be interested in buying stock or receiving it as bonus?

How should the bank be approached for a loan?

Bob keeps talking about cash flow and net income—is this important?

How can we keep expanding and purchasing the equipment we need?

Disadvantages of having a relative as an owner?

Any other sources of money that I can use?

How can we be as successful as we are and *still not have any money*?

I need a plan—don't know how to put it together.

Our equipment is constantly needing repairs and it is wearing out. How can I have a continuous replacement program?

What should be done first? Second? Third?

1. What suggestions can you make to Jim?
2. Help Jim decide what to do first, second, and third.
3. What could Bob do to help Jim?
4. Why is Jim always short of cash if the company is profitable?
5. If you were Jim, what would you do?

Sommerville Software Inc.

Sommerville is a specialized software company. It produces custom software for its clients in the areas of design and manufacturing. It also has started to sell some of its proprietary software at trade shows and in technical publications. Sommerville is currently starting a new project, called Strawberry, which is expected to have wide applications.

The Strawberry Project will use approximately one-third of the organization's programming and related personnel for two years. A large part of the software, $500,000 worth, will be purchased from a subcontractor. The total cost of the Strawberry Project is estimated to be in the range of $800,000 to $1,200,000. The software, when completed, is expected to have a market life of between four and seven years. It will be unique and proprietary and the competition will have difficulty in creating similar products without being licensed by Sommerville, which Sommerville may not do if sales are as high as predicted. Somerville is not sure how to handle the accounting of the software that is generated internally or the software that is purchased from the subcontractor.

The organization has been very successful since its startup five years ago and has been growing at a rapid rate of doubling sales every 12 months. To provide funds for continued expansion, cash from the sale of securities is required. During the next 18 months Sommerville will publicly sell its stock. This requires registration with the federal Securities

Exchange Commission as well as state commissions. The process has begun and Sommerville has brokerage, law, and accounting firms working on the process. Everyone is very positive that the public sale will take place, and optimistic about the public's response to the offering. The sale of stock will provide much-needed capital for Sommerville to expand faster than it has in the past. It is expected that approximately $10,000,000 will be raised from the sale of the stock.

The organization is profitable, but management is concerned about the expenditures for the Strawberry Project. The directors, some of whom are employees, are divided over whether Strawberry Project expenses should be expensed as they occur or if they should be capitalized as assets and then expensed over the life of the sale of the software. Some of the members feel that if all the costs are expensed immediately, the financial statements that are sent to potential shareholders will show large costs and losses. These losses, they feel, would make the stock less attractive and it would not sell well. The price per share of the offered stock would be low. This group believes that the expenditures should be capitalized as an asset and then amortized over the next five years, matching the development costs with the expected sales.

A second group of directors believes that the potential shareholders are not concerned with current profits or losses. They are looking at the technical capabilities of the firm and its future profits, not its current financial statements. They feel that the project may not attain the forecasted sales. In that case, the accumulated costs as an asset would have to be written off in one large expense later, and then the shareholders would not understand why. This group wants to expense the costs of the Strawberry Project as they occur and to explain the details to the potential shareholders.

1. If you were considering purchasing this new security or were an employee, which position would you prefer? Why?

2. What are the advantages and disadvantages of each group's proposal?

3. What are the usual ways to handle purchased and internal development costs?

4. Which position should be adopted by Sommerville?

5. How does the fact that the firm is going public at this time affect the decision?

6. What other factors should be considered?

The Millionaires Club (B)

The student investment club has been formed and they are beginning to analyze companies with public-issued stock. They are searching for sources of financial information. Help them with the following tasks.

Working individually or in teams of 3–4 individuals per team, obtain the annual reports of 5–6 firms in the same industry. Each team should select an industry in which they are interested and that has a number of publicly held companies.

1. Annual reports are available on the World Wide Web or from the companies' investor or shareholders relations department. The names, addresses, phone numbers, and Web addresses are usually available from local brokerage offices. Most firms have Web home pages that can be used to request company information including annual reports. Some city and university libraries have annual reports on file. The research service Standard and Poors,

found in most libraries, gives information about publicly held firms. Some brokers' offices have annual reports that their clients can access.

2. After obtaining the annual reports, determine the different depreciation methods that the firms use to calculate their taxes and generate their income statements and balance sheets. Compare each organization's depreciation expense, profits before taxes, taxes, and fixed asset values over time. A spreadsheet may be useful with this analysis. Note if there are any trends in the data.

3. Determine the operating cash flow for each organization over the time periods of the annual reports. Compare this value with other organizations' operating cash flows.

4. All teams can report their results. See if there are significant differences in the way firms in different industries calculate depreciation. Determine if there are differences in depreciation and cash flows for established industries such as utilities or auto firms compared with newer industries such as computer manufacturers and software producers, and between service and manufacturing firms.

Harry and Joe's Repair Service

Harry and Joe were friends in high school. They were both interested in cars and were very skilled at repairing them. During the summer, they created summer jobs by taking care of neighbors' and friends' cars. They called their company Shade Tree Mechanics Co. They would pick up the car, bring it to Joe's backyard under a large shade tree, wash and wax it, do a tune-up, or perform other minor repairs as needed. After high school, they each went to work for large auto dealerships in the area. They then decided to create their own repair service. That was 20 years ago, and they have been very successful since then. They now have a large repair facility that specializes in cars and light trucks and employs 58 people, including the office personnel. They even have a popular talk show on the local radio station about car repairs.

They have purchased the latest computer test equipment, have many trucks and vans for pickup of cars and parts, extensive office computer systems, and a great deal of other equipment that is used in the repair shop. The total value of the shop, office, and trucks is approximately $400,000. The building that they own cost $200,000 eight years ago and they have since remodeled it and added on to the facilities.

During a recent lunch at the deli across the street, Joe and Harry were discussing the company's finances. Joe said, "The bookkeeper says that we made more money last year than ever!"

Harry said, "Yes, but where is it? All we ever do is buy new equipment for the shop or the office. We get a good salary from the business, but shouldn't we get some dividends as owners?"

Joe said, "We've got a cash flow problem."

Harry said, "Right. All the cash is flowing out and nothing to us!"

Joe agreed. "You're right, but the bookkeeper said that we are using the best depreciation methods we can, and that we are saving on taxes because of our depreciation methods."

"Depreciation, cash flow—accounting stuff! All I know is that for all the work we do and all the customers that are happy with our work, we sure don't get much in return! We're not getting any younger—what about our retirement?" Harry said.

Joe responded, "This afternoon, let's sit down with Margaret, the bookkeeper, and have her go over the numbers with us. She can explain things well. Maybe she can explain depreciation, cash flow, and the other accounting things. And maybe she can suggest some way we can get more cash out of the company and plan for our retirement."

"Good idea," Harry said. "Let's have dessert."

1. If you are Margaret, how do you respond?

2. Explain the concepts of depreciation and cash flow to Harry and Joe.

3. What financial measures could be used to show Harry and Joe their financial results?

4. How can Harry and Joe's plans for retirement affect the organization's finances?

FINANCIAL ANALYSIS AND TIME VALUE OF MONEY

Chapter 4   Return on Investment and Single Payment Calculations

> Remember, time is money.
>
> —Ben Franklin

KEY TERMS

Cash Flow The money moving into and out of the cash account on the Balance Sheet.

Compound Interest When money earns interest on the principal and previous interest.

Continuous Compounding Compounding interest when the compounding period is infinitely small.

Equivalence Money at one time period can be made equal to money at other time periods using TVM calculations.

F = Future Value Amount of money at a future point in time.

i = Generally, the effective interest rate per period.

Inflation The specific increase in price of a good or service over time, **g;** or general inflation in the economy, **f.**

Interest Applied generally, it is the payment or "rent" for the use of funds for some period of time such as a year.

$m =$ Number of periods into which a year is divided for the purpose of compounding calculations.

$n =$ **Time in Years.**

$P =$ **Present Value** Amount of money measured at the present time.

$r =$ Nominal interest rate.

Rate of Return (ROR) The interest to make the value of the investment's costs equal to the value of its revenues when both are measured at the same point in time. Also known as internal rate of return (IRR).

Return General term used to mean the amount of money coming back from an investment or project compared to the original investment. It may mean the revenue from the investment or the difference between the revenue and the original investment.

Return on Assets (ROA) The net income for a year from the organization's Income Statement divided by the total assets of the organization on the Balance Sheet.

Return on Equity Net profits divided by owners equity. From the shareholders' viewpoint, it is their return or return on investment.

Return on Investment (ROI) General term that may mean profits on an investment from its assets, interest, return, rent received by a property owner, return on assets, return on equity, rate of return, or other specific concepts.

Return on Net Assets (RONA) Net profits divided by total assets invested, less the annual depreciation for the organization.

Simple Interest When money earns interest on the principal only; no interest is earned or accumulated on interest.

Single Payment When a single initial amount earns interest over time to accumulate to a future amount.

Time Value of Money Diagrams Drawings representing the flows of money into and out of the project or investment.

Time Value of Money (TVM) Money has a greater value in the future than at present due to its earning capability.

LEARNING CONCEPTS

- Understanding and applying the concept of return on investment (ROI) analysis to problems.
- Relating returns and ROI to the accounting system, long- and short-run decision making, management goals, efficiency and productivity of the organization, personal investment decisions, products, taxes, inflation, and risk.
- Understanding single payment interest calculations.
- Using calculators, formulas, and tables to solve single payment calculations.
- Drawing TVM or cash flow diagrams.
- Applying simple, compound, effective, nominal, and continuous interest calculations.
- Understanding the concept of equivalence.
- Applying single payment calculations to solving multiple payment problems.

- Using the time value of money and interest concepts to calculate inflation, net ROI, and inflation's effect on project and investment returns.
- Understanding interest and value calculations of industrial and government bonds.

INTRODUCTION

The first step in the financial cycle, the recording of historical financial information, was discussed in previous chapters. Chapter 4 introduces the study of the time value of money and interest calculations. To measure the past financial success of investment and projects, and to analyze present and future projects and investments, interest or "rent on money" is considered. When using the Accounting Equation and historical accounting, no distinction is made between a current dollar and a dollar from last year's or next year's estimates. This concept assumes that all dollars in time are equal. By applying interest calculations, a dollar today can be extended into the future using earning capability or interest. The difference is due to interest or the time value of money.

In analyzing investments in equipment and projects, as well as an individual's investments, money at the present time can be expected to earn "rent," or interest. Returns on investments are measured in different ways. Savings accounts, mutual funds, certificates of deposit, returns on invested capital equipment, and related investments are generally expected to earn a return. Borrowed funds from banks, financial institutions, bondholders, and others usually require repayment of both principal and the interest on the loan. This is the first chapter of the section of chapters that presents the mathematical techniques for calculating interest on loans and returns on investments.

The topics of this chapter include different return on investment (ROI) concepts, simple and compound interest calculations for single payments and receipts, equivalence, inflation, and time value of money diagrams. The mathematics is presented using formulas, tables, calculators and spreadsheets.

Interest and Return on Investment Are Old Ideas

It's not certain when the idea of interest was introduced. There are records that show interest being charged in Babylon as far back as 1900 B.C. It probably existed before then. For example, when grain was borrowed it was repaid in a larger quantity. Banking existed in Babylon and the bankers' profits were in the form of interest on loans.

Historically, interest rates have ranged from 3% to over 30%. From 1200 to 1500 A.D. charging interest was generally illegal because of religious objections, but in the 1500s, with the rise of Protestantism, interest again became acceptable. Since then, interest has been a normal part of doing business in Western economies.

Return

A person reading financial statements is usually interested in net income or profits. A shareholder might ask, "I wonder how well the organization used all the money that was invested by the shareholders, past profits, bondholders, and the money borrowed from financial institutions?" A number of measures can help answer this question. Profits or net income from the year's operations is part of the answer. Two organizations might have similar net incomes for the year, but be quite different in earning capability. If one of the organizations had a large amount of capital invested and the other had a small amount invested, but each made approximately the same profits or net income, which is the more efficient? The one with the *smaller* investment would be the more efficient user of the capital since it had a higher return on investment.

Personal Savings and Investment Returns

If one bank will pay $5 interest after one year on a deposit or investment of $100, and a second bank will pay $5 interest after one year on an investment of only $75, assuming equal safety, the second bank would be the preferred place to save. The return to the depositor at the first bank is $5.00/$100.00 = 5% for the year, and the return at the second bank is $5.00/$75.00 = 6.67%. The second bank offers a higher return on investment. Note that both deposits were compared for the same time period, one year. The individual or organization that is able to earn a profit on a smaller amount of capital or invested money is the one that would receive the higher return.

Care must be used when discussing the concepts of returns, return on investment, and related terms. General usage may or may not agree with the specific definition of the terms. **Return** and **return on investment** may be used in a broad general sense to include all types of returns and investments by individuals and organizations. It includes the concept of a bank paying the depositor interest, of an organization earning a profit on invested capital, interest received on a certificate of deposit sold by a financial institution, a shareholder receiving dividends or profits from selling securities, a retail store investing in inventory and selling it to obtain a profit, or a bondholder receiving yearly, quarterly, or monthly interest payments. Time is an important part of the concept of all types of return. The specific meaning of the terms used here includes the use of flows of cash into and out of investments and the time when these occur. We will differentiate between the specific meaning of these investment terms and their general usage meanings.

A general financial goal is to achieve a return on investment. An employee, for example, who deposits part of a paycheck into the credit union wants a return, earnings or interest, on that deposit. A shareholder who purchases an organization's securities wants a return on the investment in the form or dividends or price appreciation of the stock's value. A family that is investing into a mutual fund retirement account wants to earn a return so that there will be more money available to the family at retirement than they originally deposited. An organization that purchases a new piece of equipment to manufacture a product wants to earn a return on that equipment investment.

> Time is the most valuable thing you can spend.
>
> —Theophrastus, 278 B.C.

The organization wants to invest its funds internally in many different activities to obtain an overall return for the organization that will meet its financial obligations. Investment into research and development will, hopefully, lead to new and profitable products. Purchasing new equipment will reduce manufacturing costs and increase the financial return from the purchase. Teaching employees new skills will improve their ability to serve customers or manufacture products better. Interest calculations and other decision-making techniques can be used to determine the relative advantages of one investment or project compared to another. The techniques can be used to estimate and measure the expected, current, and future financial performance of the investments. The following are some examples of different types of returns on investments.

Savings Account. An individual makes deposits into a savings account:

$$\text{Return on Savings Account} = \frac{\text{Interest Earned}}{\text{Deposit}}$$

The return on savings account investment is, of course, also called *interest*. The terms *return* and *return on investment* are often used generally to apply to many specific concepts.

Shareholder. A shareholder of a corporation is also interested in return. One type of return is the dividends that are received quarterly or annually compared to the amount invested in the stock. For example, if a security is purchased for $50 per share and the company pays the shareholder a dividend of $2 a year for every share that is held, the return on the investment in the security is:

$$\text{Return} = \frac{\text{Dividend per year}}{\text{Stock Price}} = \frac{\$2}{\$50} = 4\%$$

When return is calculated in this manner, it is also called the *yield* on the stock.

A shareholder may earn a return from the price appreciation of a security as well as from dividends. If the security is purchased for $50 per share and one year later the security is sold for $55 per share, the shareholder may say, "I made a 10% return [($55 − $50)/$50 = 10%] on that security." Stated in equation form, this is:

$$\text{Return on Price Appreciation} = \frac{\text{Selling Price} - \text{Purchase Price}}{\text{Investment}}$$

Real Estate Owner. The owner of a house that cost $100,000, who leases the property for $800 per month, which is $9,600 per year, may say, "I am earning a gross return of $9,600/$100,000, which is a return of 9.6%. However, I have property expenses of $3,000 per year, so my net return is ($9,600 − $3,000)/$100,000, which is a net return of 6.6%." The property owner would calculate net return as:

$$\text{Annual Return on Real Estate} = \frac{\text{Annual Gross Revenue} - \text{Annual Expenses}}{\text{Initial Investment}}$$

Corporation. An organization that invests $100,000 into a new piece of equipment that saves $15,000 in labor costs over the old equipment can also calculate its return on the equipment investment as:

$$\text{Return} = \frac{\text{Savings per year}}{\text{Investment into Equipment}} = \frac{\$15,000}{\$100,000} = 15\%$$

The production manager may say, "We made some improvements in that process by investing in some equipment and got a return on our investment of 15%."

The president of the corporation may say, "We had total assets of $1,000,000 at the start of the year and we had a net profit after tax of $50,000 for the year. Our **return on total assets (ROA)** was $50,000/$1,000,000 = 5% after tax!"

The chairman of the board of directors might say, "We had net income of $50,000 for the year and the value of our common shareholder equity (retained earnings plus common stock) on the Balance Sheet was $500,000, so our **return on common equity (ROE)** was $50,000/$500,000 = 10%!"

The production manager, president, and chairman are all calculating specific returns on investments that are important measures to each of them. Return on investment is a general measurement that is applied differently by each of them.

Small Business. A small business owner who has invested her savings into the business and borrowed $100,000 from the bank to invest into her company might say, "I invested $100,000, and after expenses for the year, the company had net income of $20,000 for the first year; my return was 20%. I am very satisfied with that return!"

$$\text{Return} = \frac{\text{Net Income}}{\text{Investment}} = \frac{\$20,000}{\$100,000}$$

This value is also called return on assets or ROA since the original total assets of the organization were $100,000. The net income is from the Income Statement and the total assets are from the Balance Sheet of the business.

Antique Collector. A collector of antiques purchases a piece of furniture for $500 and sells it one year later for $550. The profit was $50 on an investment of $500. The collector may say, "I made a return of 10% on that investment!"

$$\text{Return} = \frac{\text{Profit for the year}}{\text{Investment}} = \frac{\$50}{\$500} = 10\%$$

In the examples above, we can see that the general meaning of return is similar, but not exactly the same. The specific meaning may be return on assets, yield, interest, return on equity, or another term. Because the concept of return is important, we will be as specific as possible when using the terms. In the examples, the time period was one year. If the time periods were different, the calculation is a little more difficult and can be made using interest calculations. Savings accounts, stock dividends, profits on antiques, rental property investments, and savings made from purchasing new manufacturing equipment can be measured using interest calculations. The mathematics of money techniques introduced in this chapter can be applied to many different situations.

Annual Reports

For the organization's investments, the employees, shareholders, and managers would like to have the organization's assets earn the highest rate of return possible for a given risk level. When reading an annual report, ROA is an important ratio that a shareholder could calculate to measure the firm's financial success. For example, a firm that has $1,200,000 in total assets and earns $100,000 of net income or profit after tax in the current year has an ROA of $100,000/$1,200,000 = 8.3% for the year. This return could be compared to the organization's past return values as well as to the returns of other organizations. Return on equity (ROE) can also be calculated from the financial statements by dividing the net income for the year by the common shareholder's equity.

Not only are these returns a measurement of past financial results, they can be used as a decision-making tool when applied to future investments. An individual or organization might ask, "In the past, our return has been 10%. A new project that is being considered is estimated to have net income of $5,000 per year on an initial investment of $100,000. Should this investment be made?" Calculating the proposed return by dividing the annual return by the investment gives $5,000/$100,000 = 5%. If the estimates prove to be the actual returns of the project, it would have a return that is only half of the current historical returns. Perhaps this is not a project that should be financed, and the organization should search for others.

In terms of financial statements and the Accounting Equation, profits are the difference between revenues and costs, so ROA can be restated as:

$$\text{Investment Return} = \text{ROA} = \frac{\text{Revenue} - \text{Costs}}{\text{Total Assets}}$$

The corporation's directors and management can use this return as a measurement to compare with returns from other time periods and with other organizations' returns.

From the above, it can be seen that maximizing investment return for the organization can be done by increasing revenue and/or reducing costs, and/or reducing assets.

> The ROA percentage can be increased by not only increasing profits, but also by decreasing invested assets.

APPLICATIONS OF INVESTMENT RETURNS

The concept of returns or investment return can be applied to many different situations. Some of these are:

1. Project engineers and managers, process engineers, design engineers, technicians, and other technical decision makers not only have technical responsibilities, they also must design and implement their decisions about products, processes, and projects in an economical manner. A feasible technical decision must be made within financial boundaries.

2. The accounting system keeps the financial history of the organization. ROA is a decision technique using accounting data.

3. Decisions about the selling price of the product, the purchase of manufacturing equipment, and the selection of suppliers can be made using investment return concepts.

4. Return on assets may apply to both short- and long-run decisions of the organization.

5. Personal financial decisions can be made using ROA, yield, interest, ROE, or other calculations.

6. One of management's responsibilities is to maximize return on investment for shareholders by maximizing the returns of many individual investments and projects.

7. Financial return is one of many performance measurements made by the organization.

8. Individual products and departments can be analyzed using these techniques.

9. Investment returns are affected by taxes and inflation, which may be included in the analysis.

10. Many different financial measurement techniques are included under the general terms *returns, investment returns,* or *return on investments.*

11. Rate of return and other specific techniques can be calculated for future projects and investments or to measure the success of past investments.

> The French term *le roi* translates as *the king.* Coincidentally, ROI, or return on investment, is the king of financial measurements.

Before looking at the mathematical details of investment techniques and time value of money calculations, we will look briefly at the impact of returns on investments and the time value of money on the areas of engineering, production, marketing, accounting, product design, analysis, management strategy, individual products and processes, taxes, and inflation.

Past and Future Investment Returns

The accounting system deals with recording financial history, not future information or decisions. All of the procedures and techniques learned so far concerning the Accounting Equation are ways to record what has happened in the past. The financial records are the basis for the decisions, but the analysis, recommendations, and decisions are typically made by the technical personnel using time value of money, investment return calculations, historical data, and other information.

If we look at a typical Balance Sheet, we see that the inventory, buildings, equipment, and other fixed assets may represent as much as two-thirds or more of the total assets of the organization. The purchase, management, and maintenance of these assets are the responsibility of the engineering and manufacturing functions. Nonfinancial personnel make the decisions to purchase inventory, build, use, and maintain new production facilities, buy production tools, rework scrap, repair equipment, and so on.

In addition, the manufacturing departments hire the labor, use the material, and consume a large portion of the overhead expenses. These decisions are made by the manufacturing personnel, and the accounting function keeps track of these financial events after the decisions are made. It is important in decision making to be aware of the differences between the accounting function that records the financial event, and the actual making of the decision by the engineering and operating departments.

The marketing function in the organization also makes many financial decisions about product or service pricing, advertising expenditures, market research, and related costs. The sales function may also use accounting data to make their decisions. The financial accounting records are an important database for sales, engineering, and manufacturing decision making. This data combined with return on investment calculations and technical knowledge will result in successful decisions. The investment returns that the organization receives are based on how well engineering, management, and other personnel make these decisions about the organization's funds.

Xerox Corporation, for example, was able to earn a return on assets, as measured by net income divided by total assets, averaging almost 20% in the 1970s. In the early 1980s, however, this began to fall. ROI dropped over a five-year period from the 20% level to less than 5%. But with new leadership and corporatewide effort, the company was able to turn this around and increase the return by 1% per year during the late 1980s and early 1990s until ROI was back into the range of 14% to 18%. Return was a vital measurement tool for the employees and management of Xerox to use as a benchmark to see if the changes they were making had a positive impact on the companywide return on investment (R. Palermo and G. Watson, *A World of Quality* [Milwaukee: Quality Press, 1993], p. 15).

Return on net assets (RONA) is an important goal at National Semiconductor. By defining RONA as the net income for the year divided by the total assets less the year's depreciation expense, National Semiconductor has used the asset reduction concept to increase its RONA. Not only do managers, engineers, and others in the organization attempt to increase net income, they are also on the lookout for ways to reduce investments in assets as a way to increase the ratio.

Donald Macleod, chief financial officer at National Semiconductor, states, "One of the changes in our goals was to move away from profit and loss, bottom-line (only)–oriented measurement to one focused on return on net assets (total assets less depreciation). The investment base isn't just factories. It's inventories, it's receivables, it's all the conditions that go into the net assets on the balance sheet. . . . This encouraged the manager to get assets off his books. We closed some factories, but we also sold some factories" (W. Simon, *Beyond the Numbers* [New York: VNR, 1997], pp. 11-23).

Team-Based Decision Making

In many contemporary organizations, the decision process is team based. Organizations are finding that using individuals from different departments and giving the team the responsibility for analysis, decisions, and follow-up improvements can be successful. The composition of the team can be work unit–based with individuals from a common work group, or cross-functional with members from different functions. By combining engineering, financial, manufacturing, purchasing, marketing, and other personnel on a team, the analysis and decision making include more points of view, and decisions can be more successful.

The accounting function provides the past financial data to the other functions in the organization. The design personnel make decisions about a product's function, materials, and specifications. These design decisions are based on technical criteria. However, this decision process is made so that the product can be manufactured and delivered at a competitive cost, otherwise the customer will not purchase it. The design function must balance the economic criteria with the technical and customer criteria. These economic decisions require an understanding of such techniques as return on investment, breakeven analysis, make or buy decisions, and other financial analysis techniques. Successful managers, technicians, and engineers will be able to combine technical criteria and financial analysis to arrive at an optimum solution.

After the product or service is designed, the manufacturing and engineering staff must decide on the manufacturing and delivery process that will be used. These decisions include equipment purchase, material type, vendor selection, labor methods, assembly techniques, quality of the process, and related process decisions, plus the economic analysis of tooling selection, alternate methods of manufacture, and inventory availability. The analysis and decision making will likely include investment return monitoring.

Return on Investment and Management

ROI is one of the primary measures by which management's success is measured. ROI is usually a result of doing everything else right. If quality is continuously improving, markets are expanding, productivity is increasing, internal efficiency is improving, and other positive effects are occurring, then the ROI of the organization is healthy and likely to increase. Management has the responsibility of implementing a strategy that will improve ROI. This is important to the continuation of the firm and to the owner/shareholders. Improvement of ROI is a continuous process.

Stanley Mersman, CFO at Silicon Graphics Inc., states, "Once you're able to comprehend cash flow and communicate it inside your company, you can get your people to understand how the assets that are used by the company are just as important as the profits that drop to the bottom line. It relates over time to your stock price in the future and to building long-term shareholder value" (Simon, pp. 115-125).

ROI and Efficiency

The organization must be able to design and manufacture a product in terms of the returns that it will bring to the firm as well as the service it provides to customers. To do this successfully, the decision maker must analyze the present costs and also look at the long-range costs of production. ROI measures how efficiently assets are managed and deployed to give a return on those assets. ROI is analogous to efficiency.

$$\text{Efficiency} = \frac{\text{Output}}{\text{Input}}$$

Efficiency in physical terms is the relationship of output for a given input. Measuring the outputs and inputs of mechanical and electrical equipment to maximize efficiency is a primary test for the designer of products. Electrical, thermal, mechanical, and similar efficiencies are well-established measures of product design. Productivity, which is the output of products from a process compared to the inputs in labor, is an efficiency relationship also. ROI is similar to efficiency, relating profits or results (output) to assets or investments (inputs).

$$\text{Productivity} = \frac{\text{Output}}{\text{Input}}$$

$$\text{ROI} = \frac{\text{Profit}}{\text{Investment}}$$

For the individual who deposits $100 into a savings account that earns $5 interest a year, the investment return would be $5/$100 = 5% for the year. The return on investment is 5%, or a "productivity" of 5%. In terms of efficiency, the account was able to produce $5 for the year on an input of $100 at the beginning of the year.

Maximizing Return on Investments

Maximizing the return on investment can be achieved in two ways. One way is to increase the amount of profit earned on a given investment. An individual would prefer a 5% rather than a

4% credit union account, for example. A second way to maximize investment return is to invest a *smaller* amount. The idea is not to invest as much as possible for a given return, but to *invest as little as possible for a given return.* When we look at the total assets on the Balance Sheet of a large corporation and see billions of dollars, we sometimes conclude that the firm must be successful because it has all those assets. Not necessarily. The concept of maximizing ROI is to make as small as possible an investment (total assets) to obtain as large as possible profit. Talent and skill are not required to make a little money on a large investment. That's relatively easy. The difficult task is to make a lot of money with a small investment, or, "to invest a little to make a lot."

Product Analysis and ROI

Another use of ROI is to analyze the firm's individual products. Assume the firm manufactures and sells two products, A and B. Product A has a total investment of $1,000,000 and makes a profit of $200,000 per year before tax. Product B makes a pretax profit of $125,000 per year and requires an investment of $500,000 in total assets.

For Product A, ROI = $200,000/$1,000,000 = 20% for the year
For Product B, ROI = $125,000/$500,000 = 25% for the year

If we look only at the profit, it appears that A is the better product since it earns $75,000 more per year. However, it requires a higher investment. Product B is the better investment. It has a higher return based on a smaller investment. If the firm has a choice between expanding production of either product, Product B should be expanded rather than A. There are other considerations, of course, but maximizing ROI is an important criterion of the decision-making process.

The above example could be applied to an individual. If someone invests in different securities, the total profit from each security is not the only measure of the investment's success. The return on investment should be considered also.

Taxes, Inflation, and Return on Investment

Returns on investments can be calculated before or after taxes are paid. For both the individual and the corporation, taxes are a significant cost. ROI is often stated on an after-tax basis, and misunderstanding can result unless it is specified as before or after tax. The returns on projects and investments must be large enough to include inflation and taxes as well as the desired return.

Sometimes what appears to be a high return on investment may not be so high when inflation is taken into account. Generally, a 10% to 12% ROI after tax would be a good return for an individual or organization. However, in the 1970s, when inflation was in the mid-teen percentages, this "high ROI" was actually barely at breakeven or equal to the inflation rate. Fortunately, in the 1980s and 1990s, inflation has been low and has had a smaller impact on overall returns. When inflation is high, firms and individuals require a higher target ROI to offset the high inflation. This reduces the number of potential projects to invest in. By eliminating or postponing the low-return projects, fewer investments are made and fewer projects are funded. Products and investments that have an acceptable return in normal economic periods with low inflation may be discontinued if the inflation rate increases. When this occurs on a national scale, the general economy is weaker due to reduced economic activity. ROI has an impact on the general economy as well as individual projects.

When inflation is high, the interest rate charged on borrowed funds is also high. Inflation drives up the prices of necessities and other costs, which in turn reduces cash flows and funds available for investment. The combination of higher prices and higher interest rates has the effect of reducing the number of projects that get funded. Existing products and investments that are no longer meeting the desired ROIs may also be reduced or eliminated to focus on the higher return investments.

TIME VALUE OF MONEY CALCULATIONS

Now we are ready to apply the concepts of return on investment to specific problems. By using formulas, tables, and calculators or spreadsheets we can make the calculations that are necessary to solve interest and ROI problems.

To borrow a dollar today, the individual must be willing to pay back not only the dollar, but also the interest in the future. Thus, a dollar today has a different value now than a dollar in the future. This concept is the basis for the **time value of money (TVM)** or interest calculations.

Interest is the rent required when money is borrowed. Or, interest is the payment that is made to an investor for lending his or her money. To the borrower, interest is a cost. To the lender/investor, interest is a return on an investment. Interest is another way of stating ROI. One hundred dollars invested for one year that earns $5 interest has a 5% return. Or,

$$i = \text{interest} = \frac{\$5}{\$100} = \frac{\text{Profit}}{\text{Investment}} = 5\% = \text{ROI of savings account}$$

From the lender's viewpoint, the return on investment is 5%. To the borrower, the cost of borrowing is 5%. The borrower of the $100 can use the interest rate as a required minimum rate that they must earn on the borrowed funds to pay back the obligation. Interest to one person is ROI to the other.

Using Formulas, Tables, Calculators, and Spreadsheets

The solutions to most of the problems encountered in this book are found using the tables and formulas available in the book. Calculators and spreadsheets are also helpful in solving the problems.

Calculators with exponential functions are desirable. Some calculators are programmed with the time value of money formulas that are presented in this text. Programmable calculators can be programmed with the formulas used to make calculations. Refer to the calculator instruction manual for specific applications.

Popular computer spreadsheet software can be helpful with the time value of money calculations. Spreadsheets make some of the calculations faster, and they also permit the revision of values in the calculations to make "what-if" decisions faster than could be done with a hand-held calculator. What-if decisions let the user determine the impact of a change in one of the variables to the outcome of the calculation. For example, after analysis of equipment purchase seems to result in a favorable return on investment, then changes in material costs, maintenance costs, selling prices of the products, and related values can be made to quickly determine if the project is favorable under other conditions also. Rarely are investment or other financial decisions made on the basis of a single set of criteria. Also, some spreadsheet software has built-in functions of interest formulas used in this book. These functions are a substitute for the formulas or tables. Most examples presented assume that the reader has basic knowledge of calculators and spreadsheets.

Simple Interest

Although not often used, the concept of simple interest should be understood. **Simple interest** is the interest charged on the amount invested or loaned, not on the accumulated interest. Simple interest is different from compound interest, which includes the interest on interest. If Jim's Uncle Harry loans Jim $100 for three years at a simple interest rate of 8%, then the interest Jim owes at the end of the first year is:

$$\text{Interest owed at the end of the first year} = i = (.08)(100) = \$8.00$$

The interest owed at the end of the second year is $8.00, and it is also $8.00 at the end of the third year. The total that must be repaid to Uncle Harry is:

$$\text{Principal} + \text{Interest} = \text{Total Repayment}$$
$$\$100.00 + (3)(\$8.00) = \$124.00$$

Notice that the interest is the same for each year. This is simple interest. In equation form, simple interest is:

$$F = P(1 + ni_{simple})$$

$$\text{where } P = \text{Principal, \$}$$
$$F = \text{Future Value, \$}$$
$$n = \text{Time, years}$$
$$i_{simple} = \text{simple interest, \% per year}$$

Using the formula in the example,

$$F = \$100.00\,[1 + (3)(.08)] = \$100.00(1.24) = \$124.00$$

The $124.00 is the amount to be repaid to Uncle Harry at the end of the third year. Timing of receipts and payments is an important part of the time value of money calculations.

Generally, investments, loans, ROI calculations, and related financial calculations are not made on the basis of simple interest, but on compound interest. The reason is that, using the previous example, if Jim has used Uncle Harry's money for the second year, but does not pay him the first-year's interest, Jim is effectively borrowing $100 + $8 for the second year. Harry is loaning or renting out a total of $108 to Jim for the second year, not just the initial $100.

Compound Interest

Compound interest, unlike simple interest, is when the investor earns or the lender pays *interest on interest.* It is an exponential function of interest over time. Few if any lenders would consider simple interest rates for their loans. The period for compounding may be annually, quarterly, monthly, daily, continuously, or any other period of time. Compound interest, rather than simple interest, is the tool that is used for the calculations throughout this text. Unless specified otherwise, interest will always be compound interest. It is the basis for engineering economy analysis, calculation of returns on investments, personal and corporate loans, and many other applications.

Using the previous example of borrowing $100 for three years at 8% annual compound interest, the calculations are as follows:

First Year.

$$(\text{Principal Borrowed})(\text{Interest Rate}) = \text{Amount of interest}$$
$$(\$100.00)(.08) = \$8.00$$

The interest owed to Uncle Harry at the end of the first year is $8.00.

Second Year. Since the loan continues for the second year, Uncle Harry is, in effect, saying, "I will now loan you $108.00 for the second year." The second year's interest is calculated as:

$$(\$108.00)(.08) = \$8.64$$

This is the interest owed on the borrowed money for the second year. The total owed at the end of the second year is the original loan *plus* the first year's interest *plus* the second year's interest. Which is:

$$\$100.00 + \$8.00 + \$8.64 = \$116.64$$

The total amount owed to Jim's uncle at the end of the second year is $116.64.

Third Year. Since this is a three-year loan, Harry says, "I will now continue to loan you what you owe me at the end of the second year, $116.64, for one more year at 8% compound interest." The additional interest as a result of the third-year loan is:

$$(\$116.64)(.08) = \$9.33 \text{ (rounded)}$$

The interest owed for the third year is $9.33. The total that is to be repaid to Uncle Harry by Jim at the end of the third year is:

$$\text{Principal Borrowed} + \text{1st Year Interest} + \text{2nd Year Interest} + \text{3rd Year Interest} = \text{Total}$$
$$\text{owed at the end of the third year}$$

$$\$100.00 + \$8.00 + \$8.64 + \$9.33 = \$125.97$$

Notice the difference between this value, $125.97, and the value of $124.00 using simple interest. These calculations demonstrate the concept of compound interest, but the method is too cumbersome to apply to larger problems. The same results can be found more quickly using equations or a calculator.

Using the previous notation, and using i for compound interest per year, the formula for compound interest can be determined. The amount due at the end of the first year is:

$$P + Pi = P(1 + i).$$

The amount due at the end of the second year is the first year plus the second year amounts:

$$P(1 + i) + P(1 + i)i = P(1 + i)(1 + i) = P(1 + i)^2$$

By induction, for the amount due at the end of the third year:

$$P(1 + i)^3$$

And, generally,

$$F = P(1 + i)^n$$

Using this formula to solve the loan from Uncle Harry,

$$F = P(1 + i)^n = \$100(1 + .08)^3 = \$100\ (1.2597) = \$125.97$$

This is the same value found by using the long method.

Interest Equations for Single Payments

The term **single payment** covers situations when a single amount of money is invested into a project for some period of time at an interest rate to accumulate to an amount in the future. Other types of calculations where multiple payments are made into an investment over time will be covered in later sections.

The notation for single payment calculations is:

P = Principal, present worth, or present value at time zero or present time.
F = Future value or future sum in the future
n = Time in years
i = Compound interest rate per year, representing annual compounding unless specified
    otherwise

To find the value of 1.08 raised to the power of 3, use the exponential keys on your calculator. You may have to check the calculator's instruction manual to review the specific keystrokes. Some calculators may have the single payment compound interest formula hard-wired in. It can be accessed using special keys. The TVM formulas can also be programmed into programmable calculators. Spreadsheet software has this formula programmed into its standard functions. Table 4–1 shows compound interest and the growth of principal in tabular and graph format.

United Mutual Fund. What if $10,000 is invested into an account at United Mutual Fund for 23 years, earning 10% compounded annually? The solution using a calculator is shown below:

$$P = \$10,000, i = 10\%, n = 23 \text{ years}, F \text{ is unknown}$$
$$F = P(1 + i)^n$$
$$F = P\ (1 + i)^n = \$10,000(1.10)^{23} = 10,000(8.9543) = \$89,543$$

This is almost nine times the starting amount, and demonstrates the power of compounding. Of the total $89,543 future value: $89,543 − $10,000 = $79,543, the total interest earned over the period.

A College Fund. Frequently, rather than wanting to know the ending amount of an investment, we want to know the starting amount.

A family desires to have $50,000 available for a daughter's college expenses in 18 years. If 9% can be earned on the invested capital, what must be the starting amount? Using the basic single payment compound formula,

$$F = \$50,000, n = 18 \text{ years}, i = 9\%, P = ?$$
$$F = P(1 + i)^n$$
$$\$50,000 = P(1 + .09)^{18}$$
$$\$50,000 = P(4.7171)$$

Rearranging and solving for P,

$$P = \$50,000/4.7171 = \$10,600$$

Table 4–1   Interest Graphs

Future Value of $1.00 after 10 years at different interest rates.

n		1%	3%	5%	8%	10%	15%	20%	25%	30%	40%
1	1	1.01	1.03	1.05	1.08	1.10	1.15	1.20	1.25	1.30	1.40
2	1	1.02	1.06	1.10	1.17	1.21	1.32	1.44	1.56	1.69	1.96
3	1	1.03	1.09	1.16	1.26	1.33	1.52	1.73	1.95	2.20	2.74
4	1	1.04	1.13	1.22	1.36	1.46	1.75	2.07	2.44	2.86	3.84
5	1	1.05	1.16	1.28	1.47	1.61	2.01	2.49	3.05	3.71	5.38
6	1	1.06	1.19	1.34	1.59	1.77	2.31	2.99	3.81	4.83	7.53
7	1	1.07	1.23	1.41	1.71	1.95	2.66	3.58	4.77	6.27	10.54
8	1	1.08	1.27	1.48	1.85	2.14	3.06	4.30	5.96	8.16	14.76
9	1	1.09	1.30	1.55	2.00	2.36	3.52	5.16	7.45	10.60	20.66
10	1	1.10	1.34	1.63	2.16	2.59	4.05	6.19	9.31	13.79	28.93

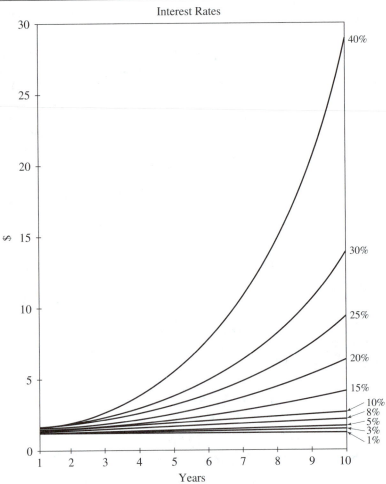

If \$10,600 is invested today at 9% compound interest, \$50,000 will be the amount at the end of the 18th year available for college.

The formula for the future value of $F = P(1 + i)^n$ can be rearranged to give the following:

$$P = F\left[\frac{1}{(1 + i)^n}\right]$$

This is the present worth version of the single payment compound formula rather than the previous future value version.

Capital Mutual Fund. If \$5,000 is deposited into a Capital mutual fund and it accumulates to \$14,980 in 11 years, what was the interest rate or return on investment earned?

$$P = \$5,000, \ F = \$14,980, \ n = 11 \text{ years}, \ i = ?$$

The basic single payment formula is,

$$F = P(1 + i)^n$$
$$\$14,980 = \$5,000(1 + i)^{11}$$

Rearranging the terms,

$$(1 + i)^{11} = \$14,980/\$5,000 = 2.9960$$

To solve, the 11th root of 2.9960 must be calculated:

$$1 + i = \sqrt[11]{2.9960}$$
$$1 + i = (2.9960)^{1/11}$$
$$1 + i = 1.1049$$

Subtracting 1 from both sides of the equation,

$$i = 1.1049 - 1 = .1049 = 10.49\%$$

The mutual fund has a return on investment of 10.49% per year for the 11 years.

Most calculators will solve for the unknown interest rate when the programmed functions are used.

Unknown Years for Single Payments. If \$3,987 accumulated to \$8,905 at a rate of 11.6%, what was the time span?

$$P = \$3,987, \ F = \$8,905, \ i = 11.6\%$$

The basic equation is $F = P(1 + i)^n$. Substituting into the equation,

$$\$8,905 = \$3,987(1.116)^n$$
$$8,905/3,987 = 2.2335 = (1.116)^n$$

To solve for an unknown exponent, use logarithms to the base 10 or natural logarithms. Using natural logs,

$$\ln (2.2335) = n \ [\ln (1.116)]$$
$$\ln (2.2335) = .8036$$
$$\ln (1.116) = .1098$$

Substituting into the equation,

$$.8036 = n(.1098)$$

Solving for n,

$$n = .8036/.1098 = 7.32 \text{ years}$$

Again, experiment with your calculator and spreadsheet to try these solutions. Function keys on many financial calculators will give an unknown number of years directly.

Summarizing to this point: Using only the one basic equation, we have seen examples of calculating a future value F, a present value P, an unknown interest/ROI rate, and finally solving for an unknown number of years, n.

The Rule of 72

There is a rule of thumb that can be used as an approximation, called the Rule of 72. This approximation says that for money to double in a single payment calculation, if time in years is multiplied times the interest rate in percent, the product will approximately equal 72. In equation form,

$$(n)(i) \simeq 72$$

For example, consider the problem of $1,000 invested for 10 years. What compound rate is necessary for the money to double itself?

$$(10)(i) \simeq 72$$
$$i \simeq 72/10 \simeq 7.2 \text{ or } 7.2\%$$

(Using the formula $F = P(1 + i)^n$, the exact answer is found to be 7.18%.)

This approximation can be checked by using the basic formula for compounding. How long does it take money to double at an interest rate of 12%?

$$n \simeq 72/12 \simeq = 6 \text{ years}$$

Remember that this approximation is only valid for F being *twice* the value of P.

INTEREST TABLES

Before there were calculators and computers, TVM calculations were made using interest tables. The Appendix of this text has these tables. These tables provide an alternate method of solving interest problems should you not have a calculator.

For example, if $850 is invested into a 10% account for 17 years, determine the value at that time.

$$P = \$850, \ i = 10\%, \ n = 17 \text{ years}$$
$$F = P(1 + i)^n$$

Using the basic equation and substituting,

$$F = \$850 \ (1.10)^{17}$$

At this point, rather than using the calculator to raise 1.10 to the 17th power, use the 10% tables in the Appendix. Find the 10% tables and go down the left column for an n value of 17. Then go to the right to the first column that has the heading *F/P* and find the value 5.054 in the body of the table. This *F/P* value is the value of $(1 + i)^n$, which is called the *F/P* value for $i = 10$, $n = 17$. A partial section from the tables is shown below:

10%	Table Values	
n	F/P	
15	4.177	
16	4.595	
17	5.054	This is the value for $(1.10)^{17}$
18	5.560	

Substituting this value into the formula and solving,

$$F = \$850(5.054) = \$4,296 \text{ (rounded to the nearest dollar)}$$

As a check, use your calculator to show that $(1.10)^{17} = 5.054$.

Which is better to use, the tables or the calculator? Both techniques have their advantages. The calculator gives values for all interest rates, while the table is limited to those that are given, and rarely are fractional interest tables available. Sometimes the tables do not contain the specific n value needed and you would have to use interpolation. Some find that the tables are faster for them and they make fewer errors using the tables than the calculator. Of course, a calculator with the functions built in is useful too. For now, you can use all of these methods of solution.

Using the Tables

If we want \$235,000 in 24 years and 12% can be earned, determine the starting amount.

$$F = \$235,000, i = 12\%, n = 24$$
$$F = P(1 + i)^n$$

Rearranging,

$$P = F/(1 + i)^n$$

From the tables,

		12% Table Values	
n	F/P	P/F	
23	13.552	.0738	
24	15.179	.0659	This is the value for $1/(1.12)^{24}$
25	17.000	.0588	

Go to the tables at 12% and $n = 24$, go to the second column, and find the *P/F* value: .0659. This is the reciprocal value from the first column. The first column in the tables is *F/P* and the second column is *P/F*. *Be careful not to confuse the two functions.*

The equation is,

$$F = P(1 + i)^n$$

and revising,

$$P = F/(1 + i)^n$$

Substituting in the equation,

$$P = \$235,000\ (.0659) = \$15,487$$

Using the Tables to Determine n

The tables can be used to determine an approximate value of n when P, F, and i are known.

Assume that we want to have \$2,000 grow to \$3,500 with a 10% estimated return. Determine the number of years required.

$$P = \$2,000,\ F = \$3,500,\ \text{and}\ i = 10\%$$

The basic equation is

$$F = P(1 + i)^n$$

Substituting,

$$\$3,500 = \$2,000(1.10)^n$$
$$F/P = 3500/2000 = 1.7500$$

Rather than using logarithms to solve for the unknown n, we can use the tables to give an approximate answer.

		10% Table Values	
n	F/P		
4	1.464		
5	1.611	The F/P value of 1.7500 is between	
6	1.772	years 5 and 6	
7	1.949		

Enter the 10% tables and find an F/P value nearest 1.75. It is between 1.611 and 1.772, which is a value of n between 5 and 6 years. For many engineering applications, this is a satisfactory approximation. If a more accurate value is required, interpolation can be used with the table, or logarithms or calculator functions can be used.

> A dollar today is not equal to a dollar tomorrow unless the desired ROI or interest is zero.

TIME VALUE OF MONEY DIAGRAMS

An important requirement for the solution of a problem is to state it accurately. To simplify a complex problem, a visual sketch or diagram is often helpful. A useful technique is to use a **time value of money diagram,** also called a cash flow diagram. To solve specific problems, we place both cash and noncash expenses, revenues, and profits on the diagram to describe the problem. So the diagram actually shows cash flow and noncash financial events. Let's see what a TVM diagram looks like.

If $4,500 is invested today for 12 years at 15% interest, determine the accumulated amount. Before solving the problem, draw a diagram, as shown in Figure 4–1.

There are different conventions used to draw these diagrams. The ones used here are:

- The horizontal line represents time with the present time, t_0, at the left end of the line.

- The horizontal line represents the investment or project.

- The investment, P, principal or present worth or present value, is shown as an investment into the project at the present time or time zero.

- The arrow coming from below represents dollars flowing into the investment at the present time.

- The future value, F, is shown at the right end of the line as an arrow upward on top of the horizontal line, meaning that the returns or profit of the investment are from or out of the project.

- The diagrams may or may not be drawn to scale. Ticks can be used to show the years or other periods. The life, n, of the project and the ROI or interest rate, i, are shown above or below the horizontal line.

- The diagrams and formulas for the single payments, unless stated otherwise, are based on the future value, F, at the end of the nth year, not the beginning of the last year.

- The initial payment, P, is at the *beginning* of the first year. This convention is used throughout the text, tables, formulas, and examples unless otherwise noted.

- A return, revenue, or withdrawal from the investment or project is shown by an upwards arrow coming out of the horizontal line.

- The arrows give direction to the flows into and out of the investment or project and are labeled with amounts if known.

- The unknowns can be solved using equations, tables, calculators, or spreadsheet.

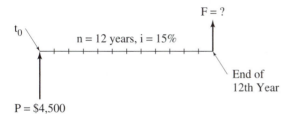

Figure 4–1   Time Value of Money Diagram

Drawing the diagram may seem like extra work. However, it can be a good habit to graphically describe a problem, especially more difficult problems. *Although the conventions listed above will be used for work in this book, individuals may make modifications for their own sketches.*

Solving the example problem,

$$F = P(1 + i)^n = \$4{,}500(1.15)^{12}$$

Using the tables or calculator,

$$F = \$4{,}500\ (5.3503) = \$24{,}076$$

Example of a TVM Diagram

Margo and Phil's Recreational Vehicle. Margo and Phil plan to retire in 15 years. They plan to purchase an RV and tour the country. They would like to have $20,000 available toward the purchase when they retire. What should they deposit today into a 7% account to have $20,000 in 15 years?

The first step is to draw the time value of money diagram, shown in Figure 4–2.

Using the diagram as a basis for the solution,

$$P = F\left[\frac{1}{(1 + i)^n}\right] = \$20{,}000\left[\frac{1}{(1.07)^{15}}\right] = \$20{,}000(.3625) = \$7{,}249$$

The value 0.3625 is found using calculator or tables.

By investing $7,249 now into an account that accumulates the interest of the 15 years at 7%, Margo and Phil would have $20,000 toward the purchase of their recreational vehicle at the end of the fifteenth year.

COMPOUNDING INTEREST MORE OFTEN THAN ANNUALLY

Two different types of interest have been introduced to this point, simple and compound. The interest that we used in the compound calculations was an annual rate. Actually, the notation of i is the annual rate of interest that takes into account the compounding during the year; more accurately, i is the *effective interest rate.* Another type of interest is *nominal interest.* It is the

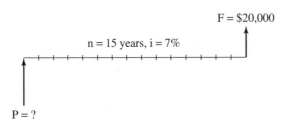

Figure 4–2   Margo and Phil's RV

interest rate without taking into account the effect of compounding during the year. The notation used for nominal interest is r.

Nominal interest is the annual rate that is not adjusted for the number of compounding periods. For example, if interest is stated to be a nominal rate of 10%, it is an annual rate of 10%. For our purposes, if no period is stated, the 10% is compounded annually. This is the type of compound interest that was used in the examples to this point in the chapter. The nominal rate may be stated to include a compounding period different from annual, such as 10% nominal interest compounded monthly.

Effective or actual interest is adjusted for the number of compounding periods. If the interest is stated to be 10%, but interest is compounded monthly, then the effective interest must be calculated to include the 12 periods per year.

Generally, the amount of interest earned on an investment over a period of time is $F - P$. The percentage earned is $(F - P)/P \times 100\%$. Dividing the year into m periods, and substituting r/m for i, and $P(1 + i)^n$ for F, gives:

$$\text{Effective interest} = i_{\text{eff}} = \left[\frac{F - P}{P}\right]100\% = \left[\frac{P\left(1 + \dfrac{r}{m}\right)^m - P}{P}\right]$$

Cancelling P above and below the line gives:

$$\text{Effective interest} = i_{\text{eff}} = \left(1 + \frac{r}{m}\right)^m - 1$$

where r is the nominal interest rate per year and m is the number of compounding periods per year.

If the nominal interest rate per year is 10%, and it is compounded monthly or 12 periods per year, the effective interest rate can be calculated.

$$i_{\text{eff}} = \left(1 + \frac{.10}{12}\right)^{12} - 1 = (1 + 0.0083)^{12} - 1 = 1.1047 - 1 = 10.47\% \text{ effective interest}$$

This effective rate per year can be compared to other annual rates.

> The rate of effective interest is equal to the nominal rate when compounding occurs annually and m is equal to 1 time per year.

Comparing Interest Rates

Circle City Credit Union versus Circle City Savings and Loan. Circle City Credit Union pays a nominal rate of 5% per year compounded quarterly, and Circle City Savings and Loan pays depositors a nominal rate of 4 3/4% per year compounded daily. Which has the better effective rate?

For the credit union,

$$i_{\text{eff}} = \left(1 + \frac{.05}{4}\right)^4 - 1 = 1.0509 - 1 = 5.09\%$$

For the savings and loan,

$$i_{\text{eff}} = \left(1 + \frac{.0475}{365}\right)^{365} - 1 = 1.0486 - 1 = 4.86\%$$

Throughout this text, interest is assumed to be effective interest unless stated otherwise.

Future Value and Effective Interest

If an initial deposit of $4,000 earns 20% nominal interest compounded monthly, what is the future amount of the account after 5 years?

The effective interest can be calculated as,

$$i_{\text{eff}} = \left(1 + \frac{.20}{12}\right)^{12} - 1 = (1.0167)^{12} - 1 = 1.2194 - 1 = 21.94\%$$

The diagram is shown in Figure 4–3.

The next step in finding the solution is to use this interest rate in the basic equation,

$$F = P(1 + i)^n = \$4,000(1.2194)^5 = 4,000(2.6961) = \$10,784$$

Another way to solve the problem directly is to combine the two steps into one, joining the effective interest equation and the basic single payment equation. Substituting into this equation gives,

$$F = P\left(1 + \frac{r}{m}\right)^{nm}$$
$$F = \$4,000(1 + .20/12)^{5 \times 12} = 4,000(1.0167)^{60}$$
$$F = 4,000(2.6960) = \$10,784$$

Note that this is the same answer that was found in the previous calculation.

To summarize, nominal interest, r, is an annual rate that does not consider compounding periods. Effective interest, i, is an annual rate that does consider compounding effects during the year.

Table 4–2 shows effective rates for the 10% nominal rate at different compounding periods.

Typically, when a bank or other financial institution makes a loan and the compounding on the loan amount occurs more frequently than one time per year, the lender is required to tell the borrower the effective rate for the loan. Some financial institutions may use 360 days per year

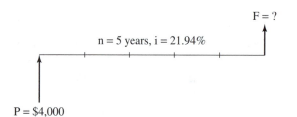

Figure 4–3   Future Amount and Effective Interest

Table 4–2 Effective Rates for 10% Nominal Rate

m *Periods per Year*	i *effective rate*
1*	10.000%
2	10.250%
4	10.380%
12	10.471%
52	10.515%
365	10.516%

*When $m = 1$, $r = i$

rather than 365 days per year. For example, even though a mortgage loan is quoted to the borrower at 9.5%, this is the nominal rate. The compounding is monthly. The effective rate would be 9.92% calculated using the above formula.

Of course, effective rates can be an advantage to a depositor into a credit union account that pays a nominal rate of 6% compounded daily. This would be an effective rate of 6.18%, which you can determine from the formula.

Credit Card Interest

Next, let's look at an example of credit card interest. Suppose a gasoline credit card charges you a rate of 1% per month if the balance is not paid. What are the nominal and effective rates?

The rate is quoted on a monthly basis, so the nominal rate is,

$$i = (12)(.01) = .12 = 12\% \text{ per year}$$

The effective rate is calculated using the formula and the interest is quoted at a monthly rate so that the nominal rate does not need to be divided by 12.

$$i = (1 + .01)^{12} - 1 = 1.1268 - 1$$
$$i = .1268 = 12.68\%$$

Sometimes the effective rate is called APR, which stands for Annual Percentage Rate. This is common in automobile financing and bank loans.

Continuous Compounding

Continuous compounding is not frequently used in engineering economy calculations, but it is encountered in savings accounts and other areas, so let's take a look at it now. The basic equation that was developed above can be used for continuous compounding except that the *m* value becomes infinite.

$$1 + i = (1 + r/m)$$

If we let $x = m/r$, then $r/m = 1/x$. The exponent becomes

$$m = (r)(m/r) = rx$$

and the formula becomes

$$1 + i = (1 + 1/x)^{rx}$$

The limit of $(1 + 1/x)^x$ as x becomes large is e, which has a value of 2.71828, so that,

$$1 + i = e^r \text{ or } i_{eff} = e^r - 1$$

Using the above equation, suppose $1,000 is invested into a continuously compounded account of 8% nominal interest for 5 years. What is the future amount?

$$i_{eff} = e^{.08} - 1 = (2.71828)^{.08} - 1 = 1.0833 - 1 = .0833 = 8.33\%$$
$$F = P(1 + i)^n = \$1,000(1.0833)^5$$
$$F = \$1,000(1.4919) = \$1,492$$

If only annual compounding, not continuous, is applied,

$$F = \$1,000(1.08)^5 = \$1,000(1.4693) = \$1,469$$

Example of Continuous Compounding

Susan's Loan. Susan has a quotation of a loan for $1,000 for one year at a rate of 9%, compounded monthly, and another quotation of 8.8%, compounded continuously. Both principle and interest on the loans would be paid on the one-year anniversary of the loan. Which is the better rate?

For the 9% compounded monthly rate,

$$i_{eff} = (1 + \frac{.09}{12})^{12} - 1 = (1.0075)^{12} - 1 = 1.0938 - 1 = 9.38\%$$

For the 8.8% continuously compounded rate,

$$i_{eff} = e^r - 1 = (2.71828)^{.088} - 1 = 1.0920 - 1 = 9.20\%$$

Polar Corporation's Multiple Payments and Time Value of Money

Suppose that Polar Corporation wants to make a number of investments into a project over time. The company needs to determine the future value of these investments. Initially $1,000 is invested into the project; at the beginning of the second year, $2,500 is invested; and at the beginning of the fourth year, $2,000. If the target ROI of the investments is to be 8% over the period, what is the accumulated amount at the end of year 7? What is the future value of the payments? What is the present value of the payments?

First, time value of money diagrams should be drawn to define the problem. These diagrams are shown in Figures 4–4 and 4–5.

For this problem, we can use the equation that relates P, present value, to F, future value. Rather than a single step to solve the problem, it will be necessary to make a number of calculations to take each of the three investments out to a future value, and then add these values together to obtain a total future value. Care must be taken to determine the correct number of years that each deposit accumulates.

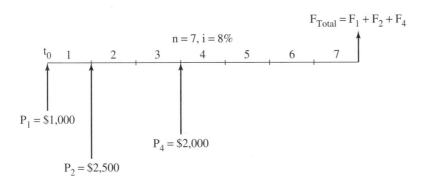

Figure 4–4   Polar Corp.: Future Amount of Multiple Payments

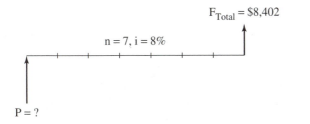

Figure 4–5   Polar Corp.: Present Value of Future Payments

The initial deposit of $1,000 is made at t_0 or the beginning of Year 1. It has a total of 7 years to accumulate to a future value.

The deposit of $2,500 is made at the beginning of Year 2, which is also the end of Year 1. It has a total of $7 - 1 = 6$ years to accumulate.

The deposit of $2,000 is made at the beginning of Year 4, which is also the end of Year 3. It has $7 - 3 = 4$ years to accumulate.

P_1, P_2, and P_3 refer to the beginning of the year when the deposit is made. The notation using the end of the year could also be used if preferred.

Using the basic formula of $F = P(1 + i)^n$ for each of the deposits, the future value of the $1,000 deposit is calculated as,

$$F_1 = \$1,000(1 + .08)^7 = \$1,000(1.7138) = \$1,714$$

Similarly, the future values of the other two deposits are calculated as,

$$F_2 = \$2,500(1 + .08)^6 = \$2,500(1.5869) = \$3,967$$
$$F_4 = \$2,000(1 + .08)^4 = \$2,000(1.3605) = \$2,721$$

To find the total future value of the three deposits, the future amounts are added:

$$F_{Total} = F_1 + F_2 + F_4 = \$1,714 + \$3,967 + \$2,721 = \$8,402$$

This problem can be taken further, and the present value of the deposits can be calculated. This could be done in three steps plus an addition, similar to the total future value calculation. Or the F_{Total} can be used to determine the present value at the beginning of the first year with the basic equation.

Rewriting the basic equation and solving,

$$P_{Total} = F/(1 + i)^n = \$8,402/(1.08)^7 = \$8,402/1.7138 = \$4,902$$

This problem demonstrates how the basic F/P equation can be used in multiple payment problems to determine future and then present values. This is a helpful technique to use when analyzing projects that have variable investments and receipts over time. The problem also demonstrates the concept of *equivalence,* which is the next topic.

Equivalence

Assume that as a lottery winner of $1,000,000 you were told that you would receive your money in payments of $50,000 each year for 20 years. Certainly you would be excited about the win, but after thinking about the terms of payment, would you really feel as though you had won $1,000,000? A normal reaction might be, "I didn't really win the million dollars since they are not going to pay me *now*."

In time value of money terminology, we would say that $50,000 each year for 20 years in the future is not equivalent to $1,000,000 today unless there is no time value of money and $i =$ 0. If the amounts are not equivalent, how could we make them equivalent? Should the future payments of $50,000 be smaller or larger to make them equivalent to the million dollars today? The payments would have to be more than $50,000 per year to be equivalent, we would say. How much more? In the next chapter, this type of calculation will be demonstrated. The term *equivalence* is used rather than *equal* because *equal* usually implies values and amounts measured at the same point in time, while **equivalence** implies values at different points in time.

Taking the Polar Corporation example of the three payments of $1,000, $2,500, and $2,000 invested into a project at 8%, we can say that those three payments are *equivalent* to the future value of $8,402 at the end of year 7 at 8%. Further, we can say that the present value amount, at t_0, of $4,902 is *equivalent* to the $8,402 amount at 7 years and 8%. And, finally, we can say that the $4,902 amount at t_0 is *equivalent* to the three deposits over time at 8% interest.

INFLATION AND TIME VALUE OF MONEY

Although inflation is not always considered in economic analysis for organizations and individuals, sometimes it must be part of the analysis. There are two general types of **inflation.** One type is the specific price increase of a good or service over time. An example is the change in price of movie theater tickets, postage stamps, a pound of hamburger, a gallon of gasoline, or fees for legal services. We have all experienced these price increases. For the organization, increases in wages, the price per pound of copper, the cost of an office copier, or an increase in the property tax rate are examples of this inflation.

A second type of inflation is national or an increase in the general price level. This is usually measured by one of two index measures. The Consumer Price Index (CPI) is a federal government index that measures the prices of a "market basket of items that individuals purchase." The price of this basket of the same goods and services is compared over time, and the difference is the general economy measure of inflation, or the CPI. A second indicator is the Producer Price Index (PPI). This is similar to the CPI except the items in the "basket of goods and services" are those that an organization would purchase. The change in price of this producer's basket is the PPI. The general inflation rate, whether CPI or PPI, is identified with the symbol f.

An individual may notice that the price of a gallon of gasoline increases in a year by 6%, but hears on the evening news that the CPI for the nation has only increased by 2%. The difference between an individual good or service's prices and the general price inflation is called *differential inflation*. In the gasoline case, the differential inflation is,

$$\text{inflation of a specific good} - \text{general inflation} = \text{differential inflation}$$
$$6\% - 2\% = 4\% \text{ differential inflation}$$

This differential can be negative when an item's price change is less than the general change in prices.

Example of Differential Inflation Rate

Abrahams Corporation. The price Abrahams has paid for copper ingots in the past year has increased from \$.80 to \$.89 per pound. The PPI for the year is reported to have increased by 2.5%. What is the inflation in the price of copper and what is the differential inflation rate?

The annual price change in copper is:

$$\text{Percent change} = \left[\frac{\text{Last Price} - \text{Original Price}}{\text{Original Price}} \right] 100\% = \left[\frac{\$.89 - \$.80}{\$.80} \right] 100\% = 11.25\%$$

The differential inflation rate for copper for the year is $11.25\% - 2.5\% = 8.75\%$.

Average Inflation Rate Over Time

For general inflation calculations, the rate of inflation, f, can be calculated using a formula similar to the F/P single payment formula considered earlier. The inflation rate, f, is substituted for the interest rate, i.

$$F = P(1 + f)^n$$

where F = last price of the goods or services
P = original price
n = time in years

When individual service and goods inflation rates are used, the symbol g is used for the rate.

Paramount Theater Ticket Prices. Alice found an old ticket stub from the Paramount Theater from five years ago, when the price was \$5.50. The current ticket prices are \$7.00. What was the average inflation rate of the tickets over the period?

$$F = P(1 + g)^n = \$7.00 = \$5.50(1 + g)^5$$
$$g = 1 - (\$7.00/\$5.50)^{1/5} = 1.0494 - 1 = 4.94\% \text{ per year}$$

From Alice's viewpoint, the price of the ticket has increased by almost 5% per year. If her income has also increased by 5%, she may not notice the change in prices. If the rate of change in her income is less than the inflation of the ticket prices, she is more likely to notice the change in prices. The general rate of inflation may or may not cause an individual's income to increase by the same amount. Some wage payment plans are adjusted to the general level of inflation. This is called a Cost of Living Adjustment (COLA). The adjustment may be based on the CPI or another inflation index.

Sometimes the differential inflation rate is negative as a result of decreasing prices of a product over time compared to the general inflation rate.

Personal Computers at Abrahams Corporation. Three years ago Abrahams Corporation purchased computers for $2,500. A computer with the same features costs $1,800 today. During the same period the PPI went from 123.0 to 130.6. What is the differential rate of inflation for the computers?

For the average general inflation rate,

$$f = 1 - (F/P)^{1/n} = (130.6/123.0)^{.33} - 1 = 1.020 - 1 = 2\% \text{ per year}$$

Since the price of the computers decreased, a deflation, the basic F/P formula can be used to determine the decrease in inflation.
For the computers,

$$F = P(1 + g)^n$$
$$g = (F/P)^{1/n} - 1 = (\$1,800/\$2,500)^{.33} - 1 = .8973 - 1 = -.1027$$
$$g = -10.27\% \text{ (note the minus sign indicating a deflation)}$$

The average differential inflation rate for the computers is $g - f = -10.27\% - 2\% = -12.27\%$. The computers have *deflated* in price an average of more than 12% compared to the general level of inflation, which was *increasing* at a 2% rate.

In economic analysis it is often assumed that the rate of inflation of specific goods and services such as raw materials, wages, power, and other costs is approximately equal to the general inflation rate and that the organization is able to raise its prices of the goods and services that it sells the same as the general inflation rate. When this is the case, inflation is often not taken into account. However, this can cause problems when the organization's prices do not increase at a rate similar to the prices it pays for the raw materials, labor, and overhead. The result is smaller profits on the Income Statement, since costs are increasing faster than revenues.

Another problem with inflation calculations is that they do not take into account changes in the quality of a good or service. For example, if the newer computers in the Abrahams Corporation example were faster, had more memory, or had better monitors, it would not be fair to compare the old and the new computers only on the basis of inflation. That would be an "apples to oranges" comparison. New equipment that is replacing old equipment may be more productive or produce less scrap. If this is the case, the inflation rate of the equipment would not be a fair comparison unless the different productivity and scrap rates are included in the analysis.

When inflation rates are low and the differential rates are small, neglecting the inflation/deflation rates is less important than when the rates are large.

Bill's Salary. Bill, an engineer for Botkins Corp., started 10 years ago at a salary of $28,000 per year. His current salary is $34,500. The average general inflation, f, for the past 10 years has been 2.3% per year. Has his salary increased in terms of inflation?

Using the basic equation of $F = P(1 + g)^n$, and solving for an unknown g,

$$\$34,500 = \$28,000(1 + g)^{10} - 1$$
$$g = (\$34,500/\$28,000)^{1/10} - 1 = (1.2321)^{.10} - 1$$
$$g = 2.11\%$$

Bill's wages have not quite kept up with the general inflation rate.

ROI and Inflation

Return on investment is a function of the return from the project and the rate of inflation. The higher the inflation rate for a given project's ROI, the lower the target or true ROI. For example, if a net 10% return on investment is desired and the inflation rate is 3% per year, how much does the investment have to earn to meet the target ROI *and* inflation? An *approximate* answer is that it must earn 10% + 3% = 13% to have a net ROI of 10%. In the 1970s, when inflation was greater than 10% for a number of years, the net return on corporate and personal investments was very low. Fewer investments are available when the inflation rate is high, since organizations and individuals consider the net return on investment after inflation, not just the project's ROI.

The combined inflation can be calculated using the following formula:

$$1 + i_{Combined} = (1 + f)(1 + i_{Actual})$$

where $i_{Combined}$ is the rate that includes the combination of inflation and actual desired rate without inflation, f is the general inflation rate, and i_{Actual} is the rate without inflation.

Using this formula, the combined rate in the example can be found:

$$1 + i_C = (1 + f)(1 + i_A)$$
$$1 + i_C = (1.03)(1.10) = 1.133$$
$$i_C = 13.3\%$$

It may seem that the difference of .3% between the approximate rate and the exact rate is small. However, in projects and investments involving large dollar values or long time periods, this small difference can become significant. Inflation may have to be considered in some situations.

Peggy's Certificate of Deposit. Peggy has invested in a CD that has a rate of 8.5%. The general rate of inflation during the holding period is 2.7%. What is the resulting return on her investment, considering inflation?

$$1 + i_C = (1 + f)(1 + i_A)$$
$$1 + .085 = (1 + .027)(1 + i_A)$$
$$1 + i_A = 1.085/1.027 = 1.0565$$
$$i_A = 5.65\%$$

INDUSTRIAL AND GOVERNMENT BONDS

Industrial, municipal, state, and U.S. government bonds pay quarterly, semiannual, or annual interest payments to the holders of the bond during the life of the bond. The bond is a loan to the organization from the bondholder. In this case, the initial amount is loaned to the borrower by the lender, interest is paid during the life of the bond, and the principal amount of the bond is paid to the lender at the end of the period.

For example, Margo and Phil decide to purchase a bond of Micro Corp. that is described as a $1,000, 8%, 2010 bond. Most industrial bonds have a face value of $1,000, the initial price. The interest rate is 8% on the $1,000 paid annually. And the principal is repaid to the lender in the year 2010. Each year, Margo and Phil receive an interest check for $80.00 ($1,000 × .08).

Zero Bonds

There are also bonds that pay zero interest to the holder during the life of the bond, but appreciate in value over their life. An example of this type of bond would be a bond that pays the purchaser (lender) $1,000 at the maturity date, but the bond is purchased for less than its potential value. The initial price of an 8% bond that is purchased 15 years before its maturity date can be calculated using the basic F/P formula,

$$F = P(1 + i)^n$$

$$where\ n = 15\ years$$
$$i = 8\%$$
$$F = \$1,000$$
$$P = unknown\ selling\ price$$

$P = F/(1 + i)^n = \$1,000/(1.08)^{15} = \$1,000/3.1722 = \$315.24$ is the initial purchase price of the bond.

No interest is paid to the holder during the 15 years, but the holder can redeem the bond that cost $315.24 at the end of the 15th year for $1,000.

In a future chapter, we will see what happens when the holder chooses to sell the bond prior to its maturity, and when the prevailing market interest rates are different from the 8%. A diagram of a zero bond is shown in Figure 4–6.

SUMMARY

Money has a time value. A dollar today is not equal to a dollar tomorrow unless the interest on that dollar is zero. Return on investment is a general concept that applies to savings accounts, returns from certificates of deposit, mutual fund returns, the earning capability of an organization, and the return from investing into assets, including buildings, equipment, projects, and inventories. Return on investments is a general term covering many different measurements. ROI

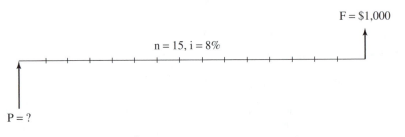

Figure 4–6   Zero Bond

is an important measure of a project's or investment's success. ROI is the tool that permits comparison between projects and investments. It is used by organizations and individuals to measure companywide results, individual projects, and personal investments. It looks not only at the net income or profits from an investment, but at the assets and costs that are required to obtain those profits. ROI is a "king" of the financial ratios and measures.

Topics presented in this chapter included:

- Return on investment concepts and applications.
- ROI as a measure of performance and a decision-making tool.
- Personal applications of ROI.
- Simple, compound, and continuous interest.
- The introduction of single payment and receipt interest calculations using formulas, tables, and other techniques to apply the formula $F = P(1 + i)^n$.
- Calculating unknown interest and unknown life using interest formulas.
- The Rule of 72.
- Time value of money or cash flow diagrams as an aid to solving TVM problems.
- Compounding more frequently than annually; effective and nominal interest.
- Solving multiple payment and receipt problems using the single payment formulas and tables.
- Equivalence and money's value at different points in time.
- The application of inflation to time value of money problems.
- General inflation, specific inflation, deferential inflation, and a project's net ROI based on inflation.

QUESTIONS

1. When people speak of a "return on their investment," what do they usually mean?
2. How is return on investment measured for an individual's savings account?
3. How could a shareholder measure the return on investment using the concept of yield?
4. How could a shareholder measure the return on investment using a securities purchase price and its current value?
5. How could an organization measure its return on the investment into new equipment?
6. How can a small business owner measure his or her return on investment?
7. What are some general applications of investment returns?
8. How do interest calculations and ROI relate to the financial statements? To interest calculations? To personal investments?
9. When did interest begin to be charged on loans?
10. What is the relationship of ROI to decision making?
11. Given a firm's annual report, describe how to determine the firm's historical ROA and its ROE.
12. How may a large organization such as Xerox or National Semiconductor use ROA or RONA to improve its financial performance?
13. How can organizational personnel such as design engineers, process engineers, and manufacturing managers use the concept of investment returns in their responsibilities?
14. What is the analogy between return on investment and efficiency and productivity?
15. What effect do taxes have on investment returns?
16. Why is simple interest rarely used?
17. How are the formulas for simple interest and compound interest different?

18. How can the interest tables be used to approximate unknown interest and time?
19. What is the Rule of 72?
20. How can a diagram of an interest problem help with its solution?
21. What is the difference between effective and nominal interest?
22. Why does "bigness" as measured by asset size of an organization not always imply its efficient use of capital?
23. The Small Business Administration gives "poor management" as a reason why most businesses fail. Explain this in terms of ROI.
24. If the profit from Product 1 is $10,000 and from Product 2 $50,000, the investment required for Product 1 is TA = $100,000 and for Product 2 TA = $800,000, determine the individual product returns and the combined returns for the two products.
25. In question 24, if an additional $100,000 can be invested in either Product 1 or 2 and a similar return can be obtained, which should be selected?

26. What organization personnel, in addition to the financial personnel, should understand the financial and accounting procedures of the organization?
27. Using newspapers, the World Wide Web, or other sources determine current interest rates of the following:
   a. Local savings and loan.
   b. Bank certificates of deposit for 5 years.
   c. Local credit union loan rate for a new car.
   d. Interest rate on 30-year Treasury bonds.
   e. Rate on 6-month Treasury bills.
   f. ROA and ROE from an organization's annual report.
28. What is an example of equivalence?
29. How does inflation affect the return on an investment?
30. How is general inflation measured? How is a specific product's inflation measured?
31. What is differential inflation?
32. What is the difference between zero bonds and other bonds?

PROBLEMS

For the following problems, make a diagram for each problem and then solve using any method including the interest tables, formulas, a calculator, or spreadsheet software. Check your answer using a different method of calculation. Use annual compounding unless stated otherwise.

1. An organization's financial statements show total assets of $1,230,000 and net income of $98,000 for one year. What is the ROA for the organization?
2. A stockholder is paid $4.10 in dividends on a stock that was purchased for $56.50. What is the return or yield on the investment?
3. If $1,400 is invested for 15 years and the rate is 8%, determine the accumulated amount if:
   a. Interest is compounded annually.
   b. Interest is simple.
4. If $500 is invested for 21 years at 10%, determine the accumulated amount if:
   a. Interest is compounded annually.
   b. Interest is simple.
5. An investment is made for 15 years at 8.6%. The starting amount is $4,000. Determine:
   a. The total accumulated amount at the end of 15 years.

b. In the answer from part a, what amount is interest?
6. A present value amount, $198, is invested for 35 years at 13% interest. Determine the future value.
7. A present value amount of $455 is invested for 40 years at 11%; Determine the future value.
8. If $13,784 will be received from an inheritance in 5 years, what is the current value if prevailing interest rates are 7.5%?
9. What is the value today of $1,345 that will be received in 18 years if the return on investment is 12%?
10. If $5,780 is to be received in 14 years, what is its value today if return on investment is 10%?
11. If $3,000 is borrowed now, and the loan is to be repaid in three years at an interest rate of 8%, determine the amount to be repaid then. Of this repayment amount, how much is principal and how much is interest?
12. If $4,500 is obtained from a loan at 10% interest, to be repaid with a single payment in four years, what is the total amount to be repaid? How much of this amount is principal and how much is interest?
13. Stuart Corporation must repay $150,000 in 5 years to an investor. How much must be deposited now

into an 8% account to accumulate to the required amount if taxes are neglected?

14. P and A Co. has to repay an investor loan of $400,000 in 8 years. How much must be deposited today into a 10% fund to obtain the required amount?

15. A student who has just graduated has decided to take the $10,000 graduation present that rich Uncle Harry has given her and put it into a mutual fund. She thinks the fund can earn a rate of 11%. Determine the future amounts, neglecting taxes on the earnings, in the fund at the end of years 10, 15, 18, 25, and 30.

16. Amy desires to have $10,000 in a tax-deferred retirement account in 10 years. She can invest $4,500 now. What must she earn on the investment?

17. Sam desires $15,000 for a down payment on a house in 5 years. If he has $8,600 to put into a mutual fund today, what return on investment must he earn?

18. If $10,500 is desired in 13 years and $3,780 is deposited today, what interest must be earned?

19. If $4,000 is invested today into a 7% account and it accumulates to $8,400, determine the number of years required, using an approximate method using the tables and an exact method using logarithms.

20. A van Gogh painting of "Sunflowers" was sold in 1987 at auction for $39,900,000. The original selling price in 1889 was the equivalent of $125. What is the return on the investment of this painting?

21. How long does it take money to double in an account if 12% interest is paid? Use the Rule of 72 and check your answer using the exact method.

22. What interest is required for money to double in 16 years?

23. The Technology Mutual Fund price increased from $16.78 per share on Oct. 8 and was at $20.17 on Aug. 15 of the following year. What was the annual compound rate of return?

24. A zero-type bond will repay $35,000 in 18 years and has a rate of 6.58%. What is the amount that must be paid for the bond today?

25. Karen is deciding in which bank or S&L to deposit $4,500. Determine the best of the following alternatives assuming the risk is the same for all.
   a. Bank X pays 6 3/4% compounded quarterly.
   b. S&L A pays 6 1/8% compounded daily (using 360 days per year).
   c. Bank Z pays 7% compounded annually.
   d. S&L C pays 6% compounded continuously.

26. If nominal interest is quoted at 11%, determine the effective rate when compounded
   a. Annually
   b. Semiannually
   c. Quarterly
   d. Monthly
   e. Weekly
   f. Daily (use 360 days per year)
   g. Continuously

27. A department store's credit application states that interest will be charged at a rate of 1 1/4% per month; what is the nominal and effective rate of this loan rate?

28. If Marcia deposits $7,000 into an account that pays 7.5% continuously compounded, and it is invested for 11 years, what is the accumulated amount?

29. Nominal interest is 11.38%. Compounding occurs monthly. What is the effective rate?

30. Obtain a sheet of 2- or 3-cycle semilogarithmic graph paper or use spreadsheet software. Recalling from algebra that a logarithmic function plotted on a semi-log graph will be a straight line, plot n on the rectangular scale x-axis, and the F/P factor on the vertical log scale. Select values of i from 1% to 100% and plot the family of interest lines. The exponential function will become a straight line on the simi-log graph, and the lines will have a common origin on the lower left corner and will radiate outward to the right. This graph can be used for interpolations of interest and time values. This graph can be plotted using paper and pencil or spreadsheet software.

31. Assuming that the Dutch really did purchase Manhattan Island for $24 worth of trinkets from the local Native Americans in 1626, what would the value of the island be today if a 6% per year appreciation rate is assumed? If monthly compounding is assumed?

32. An investor purchased 100 shares of a security on June 30 for $10.50 per share. Four years later, the security is worth $16.25. What is the ROI for the security? If the company paid $.35 per share dividends each year the security was owned, how does this affect the ROI?

33. A nominal interest of 8.75% is quoted by a bank for a mortgage. Using monthly payments, what is the effective interest rate?

34. A savings and loan has a nominal rate of 5% compounded quarterly, a credit union has a 4.5% nominal rate compounded monthly, and a bank has a nominal rate of 4.25% compounded daily. Which institution gives the best effective rate?

35. Payments are made of $1,500 into an 8% fund at the end of years 1, 2, and 4. Determine the accumulated amount at the end of the ninth year. What is the present worth of these amounts? Draw a TVM diagram first.

36. At the end of year 2, Spring Corp. invests $4,000, at the end of year 4, it invests $7,000, and at the end of year 8, $7,500. Determine the amount of money that the company would have to have now to make these future investments if it can typically earn a 10% return on its investments.

37. Jennifer, while looking through some old family papers, found a 1918 airmail stamp that her grandfather purchased. It is a 24-cent stamp that has the airplane printed upside down. She called the Stamp Shop, and they said that it would be worth approximately $125,000. What was the ROI on the "investment" if annual interest is assumed?

38. Summers Company buys gold for electronic component manufacture. At the beginning of the first year it cost $350 per ounce and the end of the fifth year they were paying $405 per ounce. What was the inflation rate?

39. Gasoline prices change from $1.45 per gallon to $1.29 over a two-year period. What is the rate of deflation?

40. The same size hamburger costs $3.50 today that cost $2.75 three years ago. What is the inflation rate?

41. Mary's hourly rate of pay has increased from $7.80 per hour to $10.50 per hour in 7 years. If the general rate of inflation in the country has averaged 2.9%, has her wage kept pace with the general inflation rate?

42. A certificate of deposit from a bank pays a rate of 9.05%. Over the same time period, the general inflation rate is 2.6%. What is the approximate net return and the exact net return?

43. An organization's financial statements show an average ROA of 7.5% per year after tax over 5 years. During the 5-year period, general inflation increased by 2.5% per year. What is the net ROA after tax for the organization?

DISCUSSION CASES

Midwest Products Inc.

Midwest has been in business producing plumbing fixtures for 65 years. It is well respected and has been profitable to the employees and the shareholders over the years. Customers have been loyal, but recently some customers have been purchasing from both domestic and foreign producers of the same products. The company has four plants with a total of 3,500 employees.

The company has three main product lines: bath and kitchen fixtures; do-it-yourself repair kits sold at hardware stores; and basic plumbing fixtures sold at plumbing supply companies. The sales for the three product lines are currently $4,000,000 for fixtures, $1,500,000 for do-it-yourself, and $6,000,000 for basic fixtures. Profits before tax for the three divisions are $300,000 for fixtures, $200,000 for do-it-yourself, and $300,000 for basic products. The investments in plant and equipment, inventory, and other assets for the three product lines are $3,500,000 for fixtures, $1,900,000 for do-it-yourself, and $5,100,000 for basic products.

Currently, sales have leveled off and for some products declined during the past two years. Profits are slowing and the sales department is reporting more and more difficulty getting orders. "Competition is fierce," the sales manager reports. Design and engineering departments are complaining that they have to come up with new products faster and faster. The manufacturing engineers and manufacturing managers complain that delivery requirements by customers are becoming more difficult to meet. Jim Morrison, the CEO of Midwest, had come up through the manufacturing and engineering departments and wants to turn the problems around. He feels that perhaps the company has become too established in doing things a certain way and these methods are holding the company back from solving the current problems. But he is not sure how to relate all of these problems, and how to describe the problem in order to formulate a solution.

1. What is the ROA for each of the product lines?

2. What is the percentage of profits on sales for each of the products?

3. What additional information would you like to know?

4. How can this analysis of ROA assist in formulating the problem and a solution?

5. What other approaches to the problem and solution would you recommend?

The Millionaires Club (C)

The students have been meeting to learn more about investments, the stock markets, and individual investing for retirement. They have placed some of their imaginary funds in money market accounts, some into individual stocks, and some into mutual funds. The investments have been successful so far. Some have increased, some decreased. They have learned a lot about reading annual reports, and how the financial markets function.

They are searching now for techniques that would be helpful in determining the relative success of the individual investments and how to monitor the overall portfolio. They are not sure how to decide if their decisions are good and how to determine when to sell the investment. Although they are interested in long-term results, they also want to be able to determine if and when an investment should be discontinued.

One of the members has suggested that they monitor the investments using ROI techniques incorporating time value of money and interest formulas. The initial price of the stock is P, the current value is F, n is the time held, and i, which is to be determined, equals ROI. By calculating this ROI, using the basic single payment formula and spreadsheets for each stock on a monthly basis, they would be able to monitor their investments with less emotion than they are currently doing.

1. How could the concepts of ROI apply to the evaluation of the group's overall performance?

2. How could ROI be applied to individual investments?

3. Can time value of money calculations be helpful to determine the success of an investment?

4. Can investments be quantitatively analyzed, or is investing an "art" based on intuition and experience only?

5. What other recommendations would you make to the Millionaires Club?

Uncle Harry's Loan

Carol's uncle has offered to loan her money for attending school. She estimates that it will cost about $10,000 per year for 4 years. She thinks that she can earn $3,000 per year with part-time work and summer employment. Uncle Harry has said that he is not in a hurry to get his money back after she graduates, so long as she is returning some of the funds on a regular basis. He wants the loan to be on a "businesslike" basis, and he wants to have some interest attached to the loan, but the rate can be as low as 4%. Harry has asked Carol to give him an idea of how much money she will need, when she will need it, and how and when the loan will be paid off.

1. Suggest the TVM diagram that Carol can use to show Uncle Harry the flows of money that she will require and her repayment proposal.

2. Is it necessary to keep this loan on a "businesslike" basis and have her sign a note?

3. What other factors should Carol consider when she gives Uncle Harry her proposal?

4. What will be owed when she graduates in four years? Five years after graduation?

5. What other things Carol should consider?

Carol's Retirement

Ten years after Carol has graduated and has paid Uncle Harry back the money she borrowed for school, he calls her and wants to discuss some ideas with her. At their discussion, he says that since she is his only niece, he wants to give her $10,000 as a gift. But there are conditions. He wants her to invest the $10,000 into an account that she can use for her retirement in approximately 25 years. She thinks it is a good idea, and believes that she can invest the $10,000 into a mutual fund that will average 10% return per year. She believes that the tax on the earnings of the account would be approximately 20%, and that inflation will average 2.5% per year over the 25 years.

1. Neglecting taxes and inflation, what is the future value of the account?

2. Including taxes and interest, what is the future value of the account?

3. If she could place the gift into a tax-deferred account, what would be the results?

4. What other factors should Carol consider?

Chapter 5   Annual Amount and Gradient Functions

KEY TERMS

Annual Amount (*A*) Investments into or returns from an investment that occur at the end of each year of a project's duration. Each amount is equal to other annual investments/returns. Also called Equal Annual Amount.

Deferred Annuity An annual amount investment received or paid with a starting time in the future.

Gradient (*G*) Annual series of investments or returns from investments in which each annual amount increases or decreases by a constant amount.

Perpetuity When the life of a project or investment is infinite, and annual returns or payments for the investment occur forever.

LEARNING CONCEPTS

- Calculating future and present values from annual amounts.
- Calculating future and present values from gradient amounts.
- Calculating present value of a future perpetual amount.
- Calculating annuities.
- Compounding annual amounts more frequently than once a year.

INTRODUCTION

The previous chapter covered the single payment functions for deposits and investments into projects at a point in time with returns occurring at another point in time. Chapter 5 presents calculations for multiple deposits and investments. Rather than a single payment into or from a project, annual payments occur throughout the life of the investment or project. They are usually assumed to be made at the end of each year.

Two general types of annual payments are introduced in this chapter. The first is the equal annual amount, *A*. This annual amount may be a deposit or investment into a project or it may represent the flow of money out of an investment on an annual basis. The second type, called a gradient, G, is also an investment or deposit into a project at the end of each year, but each annual investment is *larger or smaller* than the previous year's investment by a constant amount, G. The annual investments increase or decrease over time. The increase or decrease may be constant (a linear gradient) or may change (a nonlinear gradient). Only linear gradients are considered here.

Projects that have mixed flows of investments and returns are also covered in this chapter. Problems that can be described using both single payment and multiple annual payment calculations are presented. This type of analysis more closely represents the typical problems encountered frequently in product design and product and services production, as well as in an individual's financial decisions. The chapter also introduces the concepts of perpetual flow of funds and deferred annuities.

The concepts introduced have important application to product design, process engineering, purchasing, facilities management, equipment replacement, marketing, and other areas in the organization.

The types of calculations covered in the chapter are:

1. Converting an equal annual series of amounts to a future value.

2. Converting an equal annual series of amounts to a present value.

3. Converting an unequal series of amounts to a series of equal annual amounts.

4. Relating future perpetual amounts to a present value.

5. Converting a future series of amounts to a different future series of amounts, called deferred annuities.

These techniques will require the use of interest tables, calculators, TVM formulas, or spreadsheets. We will continue to use the time value of money diagrams introduced previously to visually describe the problem and to help in the solution of the calculations. These diagrams will be expanded to include multiple payments. Let's begin.

EQUAL ANNUAL AMOUNT AND FUTURE VALUE

The notation for equal **annual amount** is A. It occurs at the end of the year. All the A values must occur annually and be equal. Most interest tables, including these in this book, show A values that are at the end of the time period. However, some interest tables show values at the beginning or middle of the period. Be sure to check to see the time used in your tables. The formulas introduced here are for equal annual amounts that occur at the end of the year. If other times are required, the formulas should be modified to apply to other than end of year payments. Some calculators have the TVM functions programmed in with function keys representing A. See your specific calculator instruction manual for details.

Annual Amount to Future Sum

The basic function, diagrammed in Figure 5–1, is:

$$F = A \left[\frac{(1 + i)^n - 1}{i} \right]$$

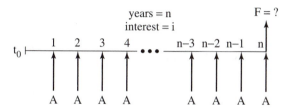

Figure 5–1   Annual Amount Formula and Diagram

Figure 5–1 shows a diagram of annual payments into an investment that has a compound interest rate of i. The ending value, F, is to be determined. Rather than make a series of calculations using the F/P function for each year, a general formula can be determined.

Generally, each A value can be taken to its future value, F, for its respective number of years,

$$F = A(1 + i)^{n-1} + A(1 + i)^{n-2} + A(1 + i)^{n-3} + \ldots + A(1 + i)^2 + A(1 + i)^1 + A$$

Multiplying each term in the above equation by $(1 + i)$ gives,

$$F(1 + i) = A(1 + i)^n + A(1 + i)^{n-1} + A(1 + i)^{n-2} + \ldots + A(1 + i)^3 + A(1 + i)^2 + A(1 + i)$$

Subtracting the first equation from the second equation and factoring out A and F gives,

$$F(1 + i) - F = A(1 + i)^n - A$$

$$F = A\left[\frac{(1 + i)^n - 1}{i}\right]$$

The term in the brackets is referred to here as the *F/A factor* with the notation $(F/A, n, i)$.

This function converts receipts, deposits, or investments made for n years of A amount at the end of each year, with the project earning an interest rate or ROI of $i\%$ per year. For example, if annual end of the year deposits of \$1,000 are made into an 8% account for 10 years, the future value can be calculated as (see Figure 5–2):

$$F = A\left[\frac{(1 + i)^n - 1}{i}\right]$$

$$F = \$1,000\left[\frac{(1.08)^{10} - 1}{.08}\right] = \$1,000\left[\frac{(2.1589 - 1)}{.08}\right] = \$1,000(14.4866) = \$14,487$$

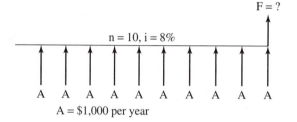

A = \$1,000 per year

Figure 5–2   Future Value Example

The answer is rounded to the nearest dollar. It is important to remember that the investments into the project are made at the *end of the year*. The last annual payment is deposited at the end of the tenth year, which is also the time when F is calculated. The value in the brackets, F/A factor of 14.4866, can also be found in the interest table in the Appendix under the heading of 8%, in the F/A column, at an n of 10 years. Be careful not to use the reciprocal A/F factor. Some calculators may have this formula programmed into their function keys, or this formula can be put into programmable calculators. See your calculator instruction manual for specific details. Either the formulas or the tables can be used to make the calculations. Some spreadsheet software has these functions as well.

Looking at the answer, the total investments into the project are

$$(10 \text{ years})(\$1,000/\text{year}) = \$10,000$$

The total interest earned is $\$14,487 - \$10,000 = \$4,487$ for the 10 years.

This type of calculation would be common when an organization invests an equal amount into a project at the end of each year over a period of time. Also, this calculation can be applied to the repayment of a loan that the firm obtains and can be used to calculate a repayment schedule. It also applies to an individual who deposits an equal amount into a retirement account over a period of time.

For another example of this type of retirement calculation, assume that a quality control engineer deposits $\$2,000$ per year into her retirement account, say a mutual fund, that earns 10% annually over a period of 30 years. What would be the accumulated amount in the retirement account? The time value of money diagram is shown in Figure 5–3. In addition to the diagram showing all 30 arrows, two alternative TVM diagrams are shown, a block diagram and a single arrow diagram.

Diagram (a) in Figure 5–3 shows each of the annual deposits into the account as arrows into the horizontal line with the $A = \$2,000$ notation below the arrows. An alternative way to draw the diagram, shown in part (b) of Figure 5–3, is to use a dashed line to represent the 30 individual arrows with the notation $A = \$2,000$ in the box area. It is understood that the box represents the end of year annual investments of A. The third method, shown in part (c) of the figure, uses a single arrow with the notation $A = \$2,000$. It is understood that an A value occurs at the end of each of the 30 years. Throughout this book, the diagram in part (a) will be used, but the other diagrams are options that may be useful.

The tables or the formula can be used to solve this problem. An abbreviation of the required function is often written as:

$$F = A(F/A, n, i) = \$2,000(F/A, n = 30, i = 10\%)$$

Using the tables, the F/A value for $n = 30$ years and $i = 10\%$ gives $F/A = 164.4940$. Substituting,

$$F = \$2,000(164.4940) = \$328,988 \text{ at retirement}$$

This amount is a significant retirement amount for most individuals. Prove to yourself that the same answer can be found using the equation.

The total deposits by the engineer are,

$$(30 \text{ years})(\$2,000/\text{year}) = \$60,000$$

deposits over her working career, and the total interest earned is

$$\$328,988 - \$60,000 = \$268,988 \text{ over the 30 years}$$

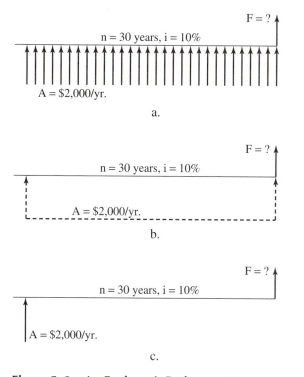

$F = ?$

$n = 30$ years, $i = 10\%$

$A = \$2,000/\text{yr.}$

a.

$F = ?$

$n = 30$ years, $i = 10\%$

$A = \$2,000/\text{yr.}$

b.

$F = ?$

$n = 30$ years, $i = 10\%$

$A = \$2,000/\text{yr.}$

c.

Figure 5–3   An Engineer's Retirement Investment

Retirement Investments

Although tax laws may change, the current rules for Individual Retirement Accounts do not require taxes to be paid on the earnings of the account until the withdrawals are made after retirement. In some cases, depending on an individual's income, tax status, and other employee
benefits, part or all of the initial investments can also be deducted from the annual income when
the deposits are made. Each individual's tax situation will be different.

There are other types of retirement accounts in addition to IRAs. One popular type is a
401k plan (403b for nonprofit organizations). In this type of account, the organization or individual deposits an amount from the employee's pay into a fund on a pretax basis. This differs
from ordinary company retirement funds because the 401k deposits stay with the individual
even after he or she leaves the organization. The employee may also have some voice in the
type of fund into which the money is deposited.

Ben's IRA and Unknown Annual Amount. As another example, assume that a recent
college graduate, Ben, has just started working as a process engineer, and decides to start a retirement account. He decides that he would like to retire at age 62; he is now 23 years old. He
wants to have one million dollars in his account when he retires. If he thinks he can earn an ROI
of 11% in the account, what must be the annual deposits?

For the problem, $F = \$1,000,000$; $n = 62 - 23 = 39$ years = number of end of year annual
deposits; $i = 11\%$; $A =$ unknown. The diagram for this problem is shown in Figure 5–4.

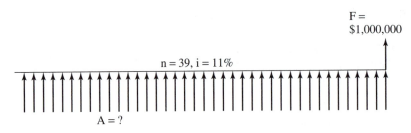

Figure 5–4   Ben's IRA

Solving the retirement problem using the formula, the basic *F/A* equation is:

$$F = A\left[\frac{(1+i)^n - 1}{i}\right]$$

This can be rewritten as:

$$A = F\left[\frac{i}{(1+i)^n - 1}\right]$$

Substituting into the equation,

$$A = \$1,000,000\left[\frac{.11}{(1.11)^{39} - 1}\right] = \$1,000,000\left[\frac{.11}{57.5593}\right] = \$1,911 \text{ per year}$$

If interest tables for 11% and *n* of 39 years are available, this problem could be solved using the tables.

Highways and Bridges Inc.—Present Worth of Annual Payments. A civil engineering consulting firm, Highways and Bridges Inc., wants to move in 10 years from rented space into their own building. The company needs $500,000 for this facility in 10 years and believes it can earn 15% ROI. How much must be deposited annually? What amount must be deposited today to accumulate to $500,000? Taxes and inflation are neglected. Figure 5–5 shows the diagram.

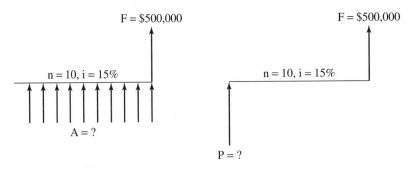

Figure 5–5   **Highways and Bridges Inc.**

Either the tables or the formulas can be used to solve for the annual payments. Using the tables, $A = F(A/F, n = 10$ years, $i = 15\%)$. Substituting,

$$A = \$500{,}000(0.0493) = \$24{,}650/\text{year}$$

The second part of the problem, present worth, can be found using the single payment function developed in the previous section. Using the single payment tables, $P = F \ (P/F, n = 10$ yr, $i = 15\%)$.

$$P = \$500{,}000 \ (0.2472) = \$123{,}600$$

In terms of equivalence, \$123,600 today is equivalent to \$24,650 annually for 10 years, which is equivalent to \$500,000 in 10 years at 15% interest.

The direct relationship between A and P can be developed by combining the F/P and F/A relationships.

$$F = P(1 + i)^n \text{ and } F = A \left[\frac{(1 + i)^n - 1}{i} \right]$$

Since F is common to both equations, the right-hand sides of the equations may be set equal to each other.

$$F = P(1 + i)^n = A \left[\frac{(1 + i)^n - 1}{i} \right]$$

Solving for P,

$$P = A \left[\frac{(1 + i)^n - 1}{i(1 + i)^n} \right]$$

Now we have direct relationships for F and P, A and F, and P and A.

Applying this equation to the previous calculation and solving directly,

$$P = \$24{,}650 \left[\frac{(1.15)^{10} - 1}{.15(1.15)^{10}} \right] = \$123{,}600$$

Problem Definition and Solution

Now that the relationships have been presented for the values of P, A, and F, these relationships can be combined to solve many types of financial problems. The general solution approach used is:

1. State the known and unknown factors in the situation.
2. Sketch a TVM diagram of the deposits, investments, withdrawals, and returns over time. Show information on the diagram.
3. Select the appropriate functions and equations that relate the known and unknown values.
4. Solve the mathematics using formulas, calculator, tables, approximations such as the Rule of 72, or spreadsheets.
5. Check the answer for logic and math.

By using this procedure, a complex problem with many different flows of money over time can be simplified, reduced, and then solved. Rounding to the nearest whole dollar amount is normally done.

Annual Amount to Present Worth

The previous *P/A* function may be rewritten in terms of *A/P* as shown below:

$$A = P\left[\frac{i(1+i)^n}{(1+i)^n - 1}\right]$$

Loan Repayment for Lex Corp.—An Unknown Annual Amount. If Lex Corp. borrows $8,000 from a local bank at 8.75% for 5 years, to purchase equipment, what amount must be repaid annually at the end of each year?

$$P = \$8,000, n = 5 \text{ years}, i = 8.75\%, \text{ and } A \text{ is unknown}$$

The diagram for the flow of funds is shown in Figure 5–6.

$$A = \$8,000\left[\frac{.0875(1.0875)^5}{(1.0875)^5 - 1}\right] = \$8,000(0.2554) = \$2,043 \text{ per year}$$

If $2,043 is paid to the lender at the end of the first year and at the end of each of the next four years, the loan will be paid off at the end of the fifth year. Sometimes a lender calls this process amortizing the loan. Lex Corp. is paying a total of (5 years)($2,043) = $10,215 over the next five years to repay the use of $8,000 received now. Or, in time value of money terminology, we can say that $8,000 today is *equivalent* to $2,043 of annual payments over the next five years at 8.75%. The terms of the loan could have been negotiated to a monthly or quarterly basis. The terms of loans, especially commercial loans, are unique to each borrower and lender.

Determining Maximum Loan Amount. Bill Saran, a student who works part-time, needs to purchase a car. He estimates that he can pay a maximum of $2,800 per year for the car loan over the next 4 years. If the current loan rate for used cars is 10%, what is the maximum price that he can pay, assuming the loan can be set up to be repaid annually?

$$A = \$2,800, n = 4 \text{ years}, i = 10\%, P \text{ is unknown}$$

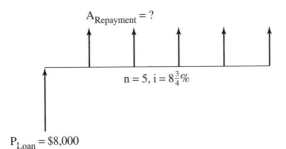

Figure 5–6   Lex Corp.'s Loan

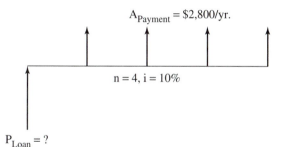

Figure 5–7   Bill's Loan

The time value of money diagram is shown in Figure 5–7. The tables can be used to solve this problem.

$$P = A(P/A, n = 4, i = 10\%)$$

From the tables, the P/A factor is 3.1699.

$$P = \$2,800(3.1699) = \$8,876, \text{ the maximum loan amount}$$

Generally, a personal loan of this type would be repaid on a monthly basis. Although many different terms may be agreed upon by the lender and borrower for individual loans, the usual payment basis in the United States is a monthly schedule.

Savings from Investing in New Equipment at Hollings Co. At Hollings Co. an improvement team composed of a department supervisor, process engineer, equipment maintenance manager, purchasing buyer, tool designer, and operator has estimated that the redesigning of an assembly process will save approximately \$4,500 per year. The savings will come from increased productivity and reduced scrap. The improvement will require some new tools and equipment, however. The process is estimated to have 7 more years of working life and the desired ROI is 12%. What is the maximum that can be spent on the new tools and equipment? Sometimes time value of money is not understood or used in cases such as this. Before looking at the TVM application, consider the case if TVM is not used.

With no interest considered (interest rate is zero), the solution is to multiply the number of years times the savings per year.

$$\text{Total Savings} = (\text{No. of Years})(\text{Savings per Year}) = (7)(\$4,500) = \$31,500$$

The conclusion, without considering the time value of money, is that if the tooling and equipment costs less than \$31,500, the investment should be made. Caution: It is difficult to reduce a process to money without considering product quality, worker safety, and many other nonfinancial factors. But for our purposes here, we will consider the costs and savings only. Now, considering interest, the maximum amount to invest in the tooling and equipment can be calculated.

The capital that might be invested into the process has alternative uses. It could be placed in financial alternatives such as CDs, savings, or other investments, or the money could be invested into some other improvement in the organization. The value of the invested money has

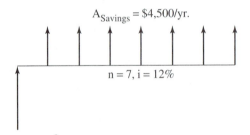

Figure 5–8   Savings at Hollings Co.

been targeted as 12% ROI by the company. Any decision to invest these funds must take the ROI into account.

The solution using time value of money is:

$$A_{Savings} = \$45,000/year, n = 7 \text{ years}, i = 12\%, P \text{ is unknown}$$

The diagram is shown in Figure 5–8. Using tables for the *P/A* values,

$$P = A(P/A, n = 7, i = 12\%) = \$4,500(4.5638) = \$20,537$$

The maximum amount to spend on the improvement, to earn 12%, is $20,537. Comparing this to the previous calculation using 0% interest, we have $31,500 vs. $20,537. If an error was made and more than $20,537 was invested, an ROI of less than 12% would be realized. And if the organization wants to gain a return of 12%, then the maximum to invest is $20,537, not $31,500.

Finding Unknown Years or ROI Using the *A/F* or *A/P* Factor

Sometimes it is necessary to solve for an unknown ROI or *n*. For example, a project's first cost, annual costs, and the desired ROI are known, but the life of the investment is not known. Or the annual costs, future value, and life are known, but the resulting ROI is unknown. In these and other cases, the solution can be made quickly if a calculator that solves for unknown *n* or *i* is used. Interest tables can be interpolated for an exact answer. A third and quick method is to use the interest tables for an approximate answer. Often, if a calculator is not available, the interest tables are used to find an approximate solution of ROI or *n*, which is useful in investment decision making.

Jane's Investment. Jane is considering investing into a mutual fund that has had a 12% annual return over the past few years. If she made deposits of $3,000 each year, how long would it take for her account to be worth $40,000, if the fund continues to earn at the 12% rate and she neglects taxes? Figure 5–9 shows the diagram.

Although the exact method of solution is done by interpolation of the interest tables or by using a calculator that has the function programmed in, there is an approximate method that is also useful. This is done using the interest tables. Calculating the *F/A* factor for Jane's investment,

$$F/A = \$40,000/\$3,000 = 13.3333$$

Going to the 12% tables and looking under the *F/A* column, the value is found to be between 8 and 9 years. A partial section from the 12% interest table is shown:

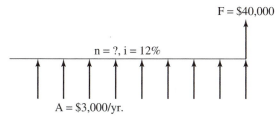

Figure 5–9   Jane's Mutual Fund

12% Interest Table

n	F/A	
8	12.300	(The value of 13.3333 is between 8 and 9 years)
9	14.776	

This quick approximation indicates to Jane that during the eighth year, her account would accumulate to her desired amount of $40,000. The end of year annual deposits of $3,000 would be made by Jane for 8 years. By calculator, the exact answer is 8.43 years. This can be checked using the formula too.

Harold's Unknown ROI. If Harold wants to have $800,000 in his retirement account, in 30 years, and he deposits $1,900 annually, what interest rate must he earn, neglecting taxes, over the life of the investment?

$$F = \$800,000, A = \$1,900, n = 30, i \text{ is unknown}$$

The TVM diagram is shown in Figure 5–10. In this situation, the quick solution is to use the interest tables for an approximate answer. This time, however, the approximate answer will be an ROI, interest rate, rather than n.

The F/A factor is $800,000/$1,900 = 421.0526. Next, go to the tables and search for a value of F/A of 421.0526 at an n of 30 years. A partial section of the table is shown below.

12% Interest Table			15% Interest Table	
n	F/A Factor		n	F/A Factor
30	241.333	(Value of 421.0526 between these factors)	30	434.745

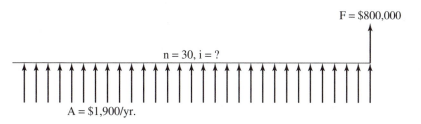

Figure 5–10   Harold's ROI

From the tables we see that the value 421.0526 is between 12% and 15% interest and is nearer 15%. This is an approximate answer. Other interest tables can be used to determine a closer approximation. An exact answer, ROI = 14.8%, is found using a calculator.

Hollings Co. Process Improvements. Techniques of solving for unknown n or i values have application to investments in equipment that save annual costs, as in the Hollings Co. example. If the target ROI, initial cost of the equipment, and annual savings are known, n can be determined. This n can then be compared to the estimated life of the equipment.

If a process improvement team at Hollings considers process improvements that have an initial cost of $50,000 and annual savings of $8,000, and a target ROI of 12% is desired, how long must the improvements be used for the investment to meet the target ROI? The P/A factor is $50,000/$8,000 = 6.2500. In the 12% interest tables, this P/A value is found between an n of 12 to 13 years. (The exact answer using a calculator is 12.23 years.)

What if the team believes the process will be in service for only 8 years? Should they reconsider the investment? Using 8 years, P of $50,000, and annual savings, A, of $8,000, ROI can be found using previously demonstrated methods. The resulting ROI is only 5.8% if the life is 8 years. You can check this percentage using a calculator or tables. At this shorter life, the project does not meet the target of 12%. See Figure 5–11 for the diagrams.

Justifying Product and Process Improvements at Morgan Co. A common use of time value of money concepts is to financially justify different types of improvements. Often quality improvements, productivity improvements, safety considerations, inventory reduction, and other process improvements cause an *increase* in initial costs and investments. These additional costs must be offset and justified with the benefit of the improvements.

Morgan Co. uses improvement teams to improve processes and products. Currently, a Morgan team is looking at a department that has some quality problems of high scrap rates. The team is trying to change some production methods to improve productivity while reducing scrap. The following costs and savings apply to the process. Scrap reduction is estimated to be $430 savings per month and increased output would be valued at $200 per month, but to obtain these savings, a higher quality raw material would be purchased that would cost an additional $1,500 per year. New equipment and tooling would cost $20,000, revised utilities would be a one-time cost of $1,200, and additional employee training would be $500. If the process is expected to be used for 7 more years and the desired ROI is 12%, determine if the improvements are justified.

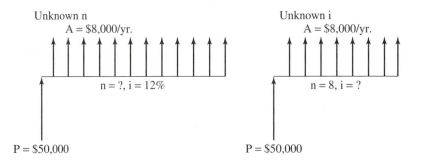

Figure 5–11   Savings at Hollings Co. and Unknown *n* and *i*

Summarizing the savings:

Scrap Reduction = (12 mo)($430/mo)	= $5,160/yr
Productivity Improvement = (12 mo)($200/mo)	= $2,400/yr
Total Annual Savings	$7,560/yr
Additional annual cost of the material	−$1,500/yr
Net savings after increased material cost	$6,060/yr

Summarizing the investment's first costs:

Equipment and tooling	$20,000
Utilities revision	1,200
Training	500
	$21,700

The TVM diagram is shown in Figure 5–12.

A number of approaches to this type of problem can be used by the improvement team.

Annual amount solution—One approach is to find the annual value of the initial investment using the target desired ROI of 12% and $n = 7$ years.

$$A = P(A/P, n, i) = \$21,700(.2191) = \$4,755/\text{year}$$

Since this amount is less than the actual annual savings of $6,060 per year, the investment is justified at 12%.

Present value solution—Another method can be used to solve this problem. The future net savings per year can be brought back to the present and then compared to the actual initial cost.

$$P = A(P/A, n, i) = \$6,060(4.5638) = \$27,656$$

Since the present value of the future savings is greater than the actual initial cost of $21,700, the investment is justified because it meets the 12% ROI target.

ROI solution—Finally, a third method is to find the actual ROI that the investment would earn.

$$P/A = \$21,700/\$6,060 = 3.5809$$

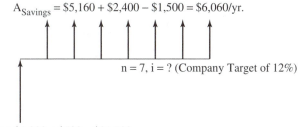

$A_{\text{Savings}} = \$5,160 + \$2,400 - \$1,500 = \$6,060/\text{yr}.$

$n = 7, i = ?$ (Company Target of 12%)

$P_{\text{First Cost}} = \$20,000 + \$1,200 + \$500 = \$21,700$

Figure 5–12   Morgan Co.'s Scrap Reduction and Productivity Improvement

Searching the tables for this value at 7 years, we find the nearest interest amount with this P/A value is near 20%. Since this is greater than the target ROI of 12%, the investment is justified. Using a programmable calculator, an exact answer of 20.24% ROI is found.

Interpolation of Interest Tables

Although finding unknown interest or n values is faster using a programmed calculator, i or n can be solved by interpolating the interest tables. The interpolated answer will be very close to the one found using the calculator.

Solving the above unknown interest problem by interpolation, using the 20% and 25% tables with the actual $P/A = 3.5809$, $n = 7$, gives the following:

P/A Values from Interest Tables

	20%	25%
$n = 7$	3.6046	3.1611

Using differences,

$$20\% \text{ value} - \text{actual } P/A \text{ value} = 3.6046 - 3.5809 = .0237$$
$$\text{Difference of 20\% and 25\% } P/A \text{ values} = 3.6046 - 3.1611 = .4435$$

Using proportions,

$$\frac{25\% - 20\%}{0.4435} = \frac{x\%}{0.0237}$$

$$x = \frac{(.05)(0.0237)}{0.4435} = 0.0027 = .27\%$$

$$\text{Value} = 20\% + 0.27\% = 20.27\% \text{ by linear interpolation}$$

This value can be compared to the value obtained using the calculator, 20.24%. (The difference is due to the linear interpolation of the nonlinear function.) The calculator, using the programmed formula, is faster if it is available. The approximate ROI answer obtained, near 20%, may be acceptable in the analysis by the improvement team.

The previous discussion presented examples of problems that use present value, future values, annual amounts, project life, and ROI. By using diagrams to describe the problems, complex problems can be simplified. Then, by using calculators, interest tables, formulas, and approximate methods, the problems can be solved.

The annual amounts we have been considering have been equal annual amounts. Next, we can look at annual values that are not equal to each other: gradient functions.

GRADIENT FUNCTION

The annual amount calculations are applicable only when the annual amounts are equal. Sometimes we want to solve problems that have increasing or decreasing annual amounts, called **gradient** amounts.

Example of Gradient Function Calculations

Maintenance Cost of a Welding Robot The cost of maintenance of a welding robot for the first year of use is $2,000. If the maintenance cost is estimated to increase by $500 each year over the 6-year life of the robot and ROI is 12%, what are the present value and the future value of the maintenance cost?

Until now we have not seen a function that will convert an unequal amount to a present or future amount. Now, consider the function that relates an unequal annual amount to an equal annual amount.

A' = The first year starting amount for the series
G = The amount that the series increases or decreases each year—the gradient amount

For the robot,

$$A' = \$2,000, G = +\$500/\text{year}, n = 6, i = 12\%, A \text{ is unknown}$$

The diagram of the problem in Figure 5–13 shows the first year's cost of $A' = \$2,000$. This amount increases at a constant rate of $500 per year. End of the year payments are used. To find F, the future value of the annual maintenance cost, a six-step solution using the F/P relationship could be used. This is time consuming. A more direct method is the gradient function.

Referring to the diagram in Figure 5–13, the end of the first year shows the $2,000 cost plus no increase. The end of the second year shows $2,000 + $500 or $A' + G$. The end of the third year: $A' + 2G$ or $2,000 + $1,000. This progression continues to the end of the sixth year, when the cost is $A' + 5G$.

The function for the relationship between A and A' is:

$$A = A' \pm G \left[\frac{1}{i} - \frac{n}{(1+i)^n - 1} \right]$$

The derivation and TVM diagram for this calculation are shown in Figure 5–14.

The diagram in Figure 5–14 shows A' amounts for each year. In addition, there is an increasing G amount starting in the second year of the series. To determine F, the future value of the A'

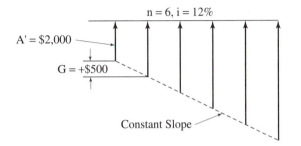

Figure 5–13   Welding Robot's Maintenance Cost

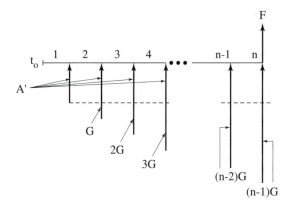

Figure 5–14   Gradient Formula and Diagram

and the G values are taken to the end of the nth year and added together. Generally, this function is as shown in the equation below:

$$F = G(F/P, n-2, i) + 2G(F/P, n-3, i) + \ldots + (n-2)G(F/P, n=1, i) + (n-1)G$$

Multiply this equation by $(F/P, n = 1, i)$:

$$F(F/P, 1, i) = G(F/P, n-1, i) + 2G(F/P, n-2, i) + \ldots + (n-2)G(F/P, n=2, i)$$
$$+ (n-1)G(F/P, n=1, i)$$

Subtracting the second equation from the first equation:

$$F - F(F/P, n=1, i) = -G(F/P, n-1, i) - G(F/P, n-2, i) - \ldots -G(F/P, n=2, i)$$
$$- G(F/P, n=1, i) + (n-1)G$$

Factoring,

$$F(1+i) - F = G[(F/P, n-1, i) + (F/P, n-2, i) + \ldots + (F/P, n=2, i) + (F/P, n=1, i) + 1] - nG$$

The bracketed amount equals $(F/A, n, i)$ from the development of the F/A function previously, and

$$Fi = G(F/A, n, i) - nG$$

Multiply both sides of the equation by $(A/F, n, i)$, which is the reciprocal of the F/A factor:

$$Fi(A/F, n, i) = G - nG(A/F, n, i) \text{ and } A = F(A/F, n, i)$$

$$A = \frac{G}{i} - \frac{nG}{i} (A/F, n, i)$$

which equals,

$$A = G\left[\frac{1}{i} - \frac{n}{i} (A/F, n, i)\right] = G\left[\frac{1}{i} - \frac{n}{i}\left(\frac{i}{(1+i)^n - 1}\right)\right]$$

The i above and below the line cancel, and the resulting bracketed value is the A/G factor. This value, when multiplied by the gradient amount, G, converts the increasing G over n years to an

A. This value can be *added* to A' to give the equivalent total A for the gradient series. If it is a decreasing gradient, the A/G factor times G is *subtracted* from the A' value to convert the series to an annual amount.

Returning to the maintenance costs of the robot, which are $2,000 for the first year and increase by $500 each year for 6 years, and solving, we have

$$A = A' \pm G\left[\frac{1}{i} - \frac{n}{(1+i)^n - 1}\right]$$

This function can be abbreviated as: $A = A' \pm G [A/G, n, i]$.

$$A = \$2,000 + \$500\left[\frac{1}{.12} - \frac{6}{(1.12)^6 - 1}\right] = \$2,000 + \$500(2.1721) = \$2,000 + \$1,086 = \$3,086$$

This annual amount calculation can be used to find the present value and future value of the maintenance costs. From the tables,

$$F = A(F/A, n = 6, i = 12\%) = \$3,086 (8.1152) = \$25,043$$
$$P = A(P/A, n = 6, i = 12\%) = \$3,086 (4.1114) = \$12,688$$

This solution of the present worth and future values can be stated in terms of *equivalence.* The starting amount of $2,000 that increases by $500 each year for 6 years is equivalent to a present value of $12,688 at 12%, to a future value of $25,043, and to an annual amount of $3,086.

With the gradient function, our introduction of the basic interest formulas is complete. The TVM concepts and mathematics can be applied to many types of problems that are encountered in product design, manufacturing, marketing, purchasing, process engineering, facilities management, project management, and many other types of personal applications.

Using the Tables for Gradient Calculations

Determine the present value of an investment that has a first-year value of $1,000 but that decreases by $40 each year for 10 years. The ROI is 20%.

$$A' = \$1,000, G = -\$40, n = 10, i = 20\%, \text{P is unknown}$$

Using the tables and referring to Figure 5–15,

$$A = A' - G (A/G, n, i)$$
$$A = \$1,000 - \$40 (A/G, n = 10 \text{ yr}, i = 20\%)$$
$$= 1,000 - 40 (3.0739)$$
$$= 1,000 - 123 = \$877 \text{ per year}$$

Solving for P,

$$P = A (P/A, n = 10, i = 20\%) = \$877 (P/A, n = 10, i = 20) = \$877 (4.1925)$$
$$P = \$3,677$$

The Wilsons' Retirement Account. The Wilson family invests annually into their IRA over the next 25 years. They are able to deposit $1,000 the first year and increase this amount by $50 each year over the life of the investment. They earn an ROI of 10%. What is the accumulated amount at the end of the period?

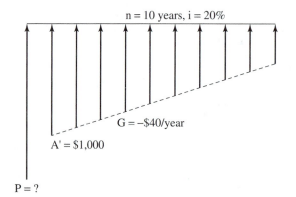

Figure 5–15   Gradient and Interest Tables

The known factors are:

$$A' = \$1,000, \ G = +\$50, \ n = 25 \text{ years}, \ i = 10\%, \ F \text{ is unknown}$$

The TVM diagram is shown in Figure 5–16.

The tables can be used for this problem. The first step is to convert the unequal investment to an equal annual investment.

$$A = A' + G(A/G, n = 25, i = 10) = \$1,000 + \$50(A/G, n = 25 \text{ yr}, i = 10\%)$$
$$A = 1,000 + 50(7.4580) = 1,000 + 373 = \$1,373/\text{year}$$

Next, convert the equal annual amount to a future amount:

$$F = A(F/A, n, i) = \$1,373(F/A, n = 25, i = 10\%) = 1,373 \ (98.3471)$$
$$F = \$135,031, \text{ the amount accumulated in the retirement account}$$

Usually the problems encountered in financial decisions require the conversion of unequal annual payments to an A, P, or F. In the case of P and F, there is a two-step solution. First solve

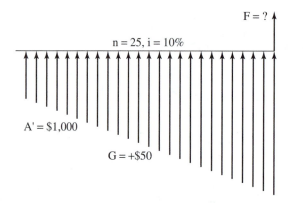

Figure 5–16   Wilson Family's Retirement Account

for A and then solve for P or F. The equations of A/G and A/P or A/F can be combined to give a direct, single-step solution. For our purposes, the longer two-step solution is easier to use to describe the problem. If these gradient functions are used frequently, the combination of functions can be developed and used.

Some financial decisions involve nonlinear gradient functions. There are equations not covered here that can be developed for nonlinear applications. If there are dollar flows that do not fit into the functions developed here, a step-by-step solution can be used by taking each individual value to a point in time and summing those values.

PERPETUITY

There is a special case application of the P/A relationship. When n is very large or assumed to be infinite, annual amounts into the future are called payments in **perpetuity.** This has applications for long-term investments, endowments, and engineering applications when the life of the project is long.

Examples of Perpetual Calculations

Perpetual Calculations at Pharmaceutical Inc. A large pharmaceutical company, Pharmaceutical Inc., desires to set up an endowment fund that would provide $100,000 per year for scholarships to students majoring in chemistry. If an ROI of 8% can be earned, how much must be invested so that these scholarships will be available forever, in perpetuity?

The formula for this calculation is a special case of the following:

$$P = A\left[\frac{(1+i)^n - 1}{i(1+i)^n}\right]$$

Using limits and letting n approach an infinitely large value, it can be shown that the P/A relationship becomes:

$$P = \frac{A}{i}$$

This can also be shown intuitively. If P is invested into an account that earns i interest each year, then $P \times i = A$ is available for withdrawal each year. If this amount is withdrawn from the account annually, leaving P to earn $P \times i = A$ for the next year, this may continue indefinitely, and

$$P \times i = A; \ P = \frac{A}{i}$$

For the Pharmaceutical Inc. scholarships,

$$P = \frac{A}{i} = \frac{\$100,000}{.08} = \$1,250,000$$

This amount, $1,250,000, earning 8%, will yield $100,000 annually for the scholarships.

Value of a Rented Facility. The concept of perpetuity can be used to establish the current value of a product, equipment, or facility that is earning an annual return. For example, if a facility is rented to a tenant for $5,000 per month, and the operating expenses, maintenance, and taxes are $15,000 per year, what is the facility's current value if the landlord desires a 12% ROI?

$$P = \frac{A}{i} = \frac{\text{Net Annual Income}}{\text{ROI}} = \frac{(12 \text{ mo.})(\$5,000/\text{mo.}) - \$15,000}{.12} = \frac{\$45,000}{.12} = \$375,000$$

This answer, $375,000, is not exact, but an estimate of the facility's value. This method is used frequently when trying to establish "ballpark estimates" of the value of industrial and commercial property. Other concepts, such as the cost of building the current property and the value of comparable nearby property, also influence the final estimate of value.

The above perpetuity calculation is also known as the *income method* of estimating property or investment value. An assumption of this technique is that the investment will provide the income perpetually into the future.

Value of a Security. This same technique could be applied to a security that is paying dividends to shareholders. One approach (not the only one) would be to divide the dividend amount paid annually by the desired ROI. For example, if a security is paying $8.50 per share dividend and the desired return on investment is 10%, then the current value is estimated as:

$$\text{Present Value of Investment} = P = \frac{A}{i} = \frac{\text{Annual Dividend}}{\text{Desired ROI}} = \frac{\$8.50}{.10} = \$85.00$$

Of course, the anticipated change in the future dividend payments and the price increase or decrease can also be considered to establish the security value.

Buffett's Method of Determining Security Value

Warren Buffett, successful investor and CEO of Berkshire Hathaway, an Omaha-based investment firm, is often asked how to determine the value of a security. To Buffett, it is like a bond. A bond's value is equal to the cash flow from future interest payments, discounted back to the present. A stock's value is figured the same way; it equals the anticipated earnings per share brought to the present using a conservative interest rate such as the rate on 30-year government bonds. This price, along with other information, is used to determine the fair and current intrinsic value of the security. (Mary Buffett and David Clark, *Buffettology* [New York: Rawson-Scribner Assoc., 1997], p. 82)

Value of a Proprietary Product or Process For another application of this type of perpetual calculation, consider a product that is a proprietary or secret process. It is providing a net profit of $150,000 per year, calculated by subtracting manufacturing and marketing costs from sales. If the organization desires a return on investment of 15%, then the current value of the product to the firm is equal to:

$$P = \frac{\$150,000}{.15} = \$1,000,000$$

Again, this is an approximation method that would only apply in a perpetual situation.

DEFERRED ANNUITIES

An annuity is the same as what we have called an annual amount. A **deferred annuity** is an annual payment or investment that will occur in the future, rather than starting immediately. An example would be a firm's investment into a fund that would accumulate for some period of time, and then the firm would make annual withdrawals from this fund.

For example, assume that Micro Corp. places $100,000 into an investment that has a return of 10%. The investment is placed there for four years, and then at the end of the fifth year, annual withdrawals begin and continue for six years. During this period, the funds continue to earn 10%. What is the annual amount that may be withdrawn during the six-year period? The solution of this type of problem requires the use of more than one of the time value of money functions. The first step is to draw the TVM diagram, as shown in Figure 5–17.

$$P = \$100{,}000, \, n_1 = 4 \text{ years}, \, i = 10\%, \, n_2 = 6, \, A \text{ is unknown}$$

From the diagram in Figure 5–17, we see that the initial investment P of $100,000 will accumulate to a future value F. The first step is to solve this part of the problem.

$$F = P \, (F/P, \, n_1 = 4 \text{ years}, \, i = 10\%)$$

Using the tables and substituting,

$$F = \$100{,}000 \, (1.4641) = \$146{,}410$$

This is the accumulated amount in the fund at the end of the fourth year. This is also the amount that will remain in the fund and from which A amount will be withdrawn each year for the next six years. This amount can now be thought of as P which will be available four years from now. This amount will earn 10% and there will be six withdrawals of A amount.

$$F = P' = \$146{,}410$$

Solving for A given P' can be done using the A/P function and the tables,

$$A = P'(A/P, \, n_2 = 6, \, i = 10\%) = \$146{,}410 \, (.2296) = \$33{,}616/\text{year}$$

This amount of $33,617 is paid out of the fund at the end of each year for six years and then the fund is "empty." This annual payout, known as a deferred annuity, is equivalent to the initial investment of $100,000 now.

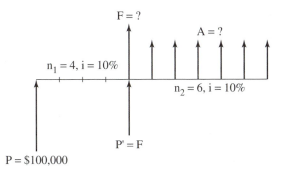

Figure 5–17   Deferred Annuity at Micro Corp.

Example of a Deferred Annuity Calculation

Susan and Stuart Sims' College Fund. Susan and Stuart Sims deposit $1,000 per year into a college fund account for their son, Stuart Jr., starting on his first birthday and continuing through his eighteenth birthday. Then there are four withdrawals from the account to pay for college expenses. If the fund earns 8%, what is the value of the withdrawal amounts? Figure 5–18 shows the diagram.

The known elements are

$$A_1 = \$1,000, \, n_1 = 18 \text{ years}, \, i = 8\%, \, n_2 = 4 \text{ years}$$

The unknown values are A_2, F_1, and P_2.

Figure 5–18 shows the annual deposits of $1,000 accumulating to the future amount, F_1. To solve for this amount, the tables are used.

$$F_1 = A_1 \, (F/A, \, n = 18 \text{ yr}, \, i = 8\%) = \$1,000 \, (37.4502) = \$37,450$$

This amount, $37,450, is available at the end of the eighteenth year for *reinvestment* for the next four years. This reinvestment, P_2, is equal to F_1, $37,450. There will be four withdrawals from this fund starting at the end of the nineteenth year through the twenty-second year with no additional annual deposits.

The calculation of the value A_2 can be made using the tables:

$$F_1 = P_2 = \$37,450$$
$$A_2 = P_2 \, (A/P, \, n = 4 \text{ yr}, \, i = 8\%) = \$37,450 \, (.3019) = \$11,307/\text{year}$$

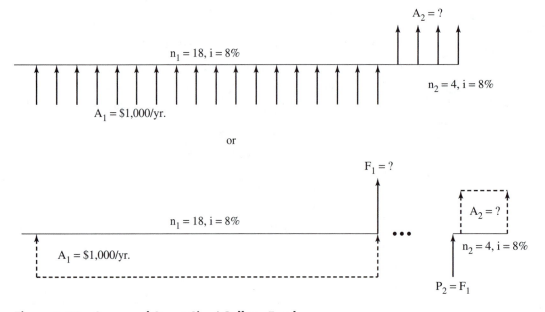

Figure 5–18   Susan and Stuart Sims' College Fund

This amount, $11,307, is withdrawn at the end of each of the four years for college expenses. The residual earns 8%. At the end of the fourth year, the account is empty.

The total of the investments made by the parents is (18 yrs)($1,000/yr) = $18,000. The total amount available for college expenses is (4 yrs)($11,307) = $45,228.

This type of deferred annuity is the basis for many retirement funds. The investor and/or employee deposits an annual amount into the fund over his or her working career. Upon retirement, withdrawals are made annually (or monthly). It also applies to an R&D activity when an organization invests funds for a number of years before realizing a salable product from which returns can be obtained.

Compounding More Frequently Than Annually

In the discussion of single payment calculations, the problem of compounding periods more frequent than annually was presented. Now, the concept of more frequent compounding can be applied to the annual payment functions.

Tech Corp. has borrowed $100,000 for investment into a new material handling system. The loan is for five years at an annual rate of 9% compounded quarterly, with quarterly repayments required. What are the quarterly payments? The diagram for this problem is shown in Figure 5–19.

The known values are

$P = \$100,000$, $n = 5$ years, $i = 9\%$ compounded quarterly, $m = 4$, A/m is unknown

The basic formula for A/P is:

$$A = P\left[\frac{i(1 + i)^n}{(1 + i)^n - 1}\right]$$

If i is replaced with r/m and n is multiplied by m to give the exponent nm, substituted into the basic equation, we have:

$$\frac{A}{m} = P\left[\frac{\frac{r}{m}\left(1 + \frac{r}{m}\right)^{nm}}{\left(1 + \frac{r}{m}\right)^{nm} - 1}\right]$$

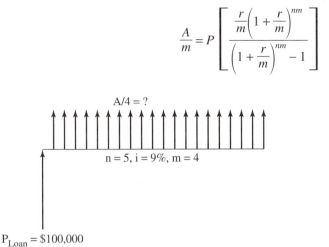

Figure 5–19   Tech Corp.'s Loan for a Material Handling System

The value of P is \$100,000; $A/m = A/4$; the value of $i/m = .09/4 = .0225$; and $nm = (5)(4) = 20$. Substituting these values into the modified equation,

$$\frac{A}{4} = P \left[\frac{\frac{.09}{4}\left(1 + \frac{.09}{4}\right)^{20}}{\left(1 + \frac{.09}{4}\right)^{20} - 1} \right]$$

$$\frac{A}{4} = \$100,000 \left[\frac{.0225(1.0225)^{20}}{(1.0225)^{20} - 1} \right] = \$6,264 \text{ per quarter}$$

Generally, substituting r/m for i, and nm for n, into the basic equations will convert them to functions for applications requiring compounding more frequently than once per year.

SUMMARY

This chapter presented single payment values as related to multiple and annual payments. Both equal and gradient annual payments were discussed. The following concepts and calculations were covered:

- Future value from annual amount.
- Annual amount from future value.
- Present value from annual amount.
- Annual amount from present value.
- Unknown ROI and unknown number of periods.
- Annual amount, present value, and future value from unequal gradient amounts, with increasing and decreasing gradient functions.
- Present value from future perpetual amounts.
- Deferred annuities.
- Annual amounts based on compounding periods more frequent than annually.

QUESTIONS

1. What is the definition of the notation A?
2. What is the advantage of drawing a time value of money diagram?
3. Why is it an advantage for an individual to start a retirement account at an early age compared to waiting until later in his or her working career?
4. Is there an advantage to using the tables compared to using formulas and a calculator to solve problems?
5. What function would be used to calculate the payment amount of a loan?
6. What formula is used to calculate the value of a retirement program?

7. How can an approximate unknown ROI or number of years be calculated using the F/A relationship from the interest tables?
8. What is the difference between a gradient and an annual amount?
9. What types of problems are typically solved using gradient functions?
10. What is the equation that permits the calculation of a present value from a perpetual annual amount? What occurs when the ROI value is reduced? Is this logical?
11. Why is it important to draw a diagram to solve deferred annuities problems?

PROBLEMS

For the following problems, draw a time value of money diagram and solve using tables, a calculator, formulas, or spreadsheets. As an optional exercise, if your calculator can be programmed, program it for the functions introduced in this chapter. If you have access to spreadsheet software, determine the time value of money functions of the software and the protocol to access them.

1. $A = \$1,000$, $n = 25$, $i = 15\%$. Determine F.
2. $A = \$400$, $n = 130$ years, $i = 12\%$. Find P.
3. $F = \$1,000,000$, $n = 141$ years, $i = 12\%$. Find A.
4. If $n = 23.75$ years, $i = 13.65\%$, $A = \$435.89$ per year, determine the future value.
5. If A is $\$3,600$, $i = 12\%$, $n = 34.8$ years, determine the future value and the present value of this amount.
6. Deposits of $\$345.79$ are made annually for seven years into an account paying 10.67%. Determine the future value at the end of the period.
7. A no-load mutual fund has grown at a rate of 18.4% in the past and this is expected to continue. If deposits of $\$1,600$ are made for 10 years, what is the accumulated amount?
8. Jane Morrow purchased 100 shares of a security on March 1 for $\$24$ per share and sold the stock on October 8 of the next year. What ROI did she earn?
9. Jane Morrow is establishing an Individual Retirement Account. She plans to deposit $\$2,000$ per year into the fund until she retires at age 62. She is now 25 years old. Jane expects her IRA to be invested in mutual funds and to average the same rate of return as the Dow Jones Industrial Average has grown over the past 50 years, which is 11%. What will be the amount available at retirement? What if she postpones her retirement until age 65, and continues to make deposits? What if Jane is fortunate enough to earn a return of 13% over the period?
10. A company desires to accumulate $\$150,000$ in four years to expand its office building. If the company can earn 9%, what equal amount must it deposit annually?
11. A quality control technician is participating in a 401k retirement program at his job. He deposits $\$175$ per month in the fund. If the fund is anticipated to earn 10% over the 27 years that he will participate, what is the accumulated amount? What would be the

accumulated amount if the company held the twelve payments until the end of the year in a noninterest-bearing account and then sent the accumulated amount into the fund at the end of each year?
12. A recent college graduate has decided on a goal of having $\$1,000,000$ at retirement, and believes that she can earn a return of 12% by depositing into mutual funds for her IRA. She is 25 years old and would like to retire at age 60.
    a. What annual amount must she deposit?
    b. What annual amount would be required if she waits until she is 35 to start the deposits?
    c. From part a, what is the accumulated amount if at age 60 she decides to continue the deposits and work three more years?
13. What rate of interest is required to have annual payments of $\$45$ accumulate to $\$900$
    a. in 15 years?
    b. in 12 years?
14. If monthly payments of $\$125$ are made into an account that pays 12% nominal interest, compounded monthly for 15 years, what will be the accumulated amount at the end of the fifteenth year? (Hint: Use the basic A/F equation $m = 12$, divide all interest values by m, then multiply all n values by m, and divide A by m.)
15. The warehouse manager at Johnson Corp. is considering installing additional insulation in the warehouse. He estimates that the annual savings would be a reduction in heating cost from the current $\$16,700$ per year to approximately $\$9,500$ per year. Johnson Corp. desires to earn 15% ROI, and the cost of installing the insulation is expected to be $\$21,000$. The warehouse is expected to be used for another 10 years.
    a. Is the insulation project a good investment?
    b. What if Johnson Corp. only used the facility for seven more years?
    c. What effect would the insulation have, if any, on the resale of the facility in 10 years?
16. A lottery winner wins the One Million Dollar Prize. She finds out, after winning, that the payments are actually made annually over the next 20 years in installments of $\$50,000$ each. What is the value of the prize today if ROI is 8%?
17. Electronics Inc. has purchased a machine for $\$140,000$. They borrow $\$100,000$ from the bank at

a 9% nominal rate. The loan agreement requires monthly repayments over the next five years.

a. Determine the repayment amount.

b. What is the effective interest rate?

18. A monthly technical publication, *Technology Journal for Electronics Manufacturing,* offers a three-year subscription for $375. If the subscriber wants to earn 12% ROI, what is the cost of each monthly copy?

19. A machine can be purchased for $150,000 with a life of 10 years and no salvage value. It is anticipated that the machine will save $25,000 per year over its life. What is the ROI on the machine?

20. At Tech Corp., Rob Hendricks is studying the vector valve department and finds that a new tool could be designed that would save $2,600 in material and $5,000 in labor per year. The current process will be used for five more years. If Tech Corp. normally desires a 15% ROI on its investments, what is the maximum that can be spent on the revised tooling and equipment to improve this process? What if Rob *did not consider* the time value of money? What would be the maximum to spend on the revision?

21. Deposits of $800 are made the first year into a 7% account. This amount increases by $30 for the second year and continues to increase over a total of 10 years.

a. What is the equivalent annual cost of these future deposits?

b. What is the present value of these deposits?

22. If an initial deposit of $3,000 decreases by $100 each year for 15 years, what is the equal annual value of the deposits if i = 12%? What is the future value at the end of the 15th year?

23. Maintenance cost on equipment is expected to be $1,000 for the first year and will increase by $200 each year over the 8-year life of the equipment. If interest rates are 14%, determine the equivalent annual amount of the costs. Determine the present value of the maintenance costs.

24. An investor plans to deposit $1,000 the first year into an 8% account and to increase the deposit by $25 each year for 30 years. Find the future value of the investment.

25. Rob Hendricks, a process engineer at Tech Corp., is establishing a retirement fund. He currently earns $30,000 per year. He wants to take $1,500 of his gross wages and put it into the fund. His salary will increase over the years, and he will be able to add additional deposits of $60 per year over his working career. Specifically, the deposits will be $1,560 for the second year and $1,620 for the third year. This rate of increase will continue for a total of 28 years. What is the value of the account if Rob earns 10% on his investments? What is the value of the account if at the end of the 18th year of the fund, Rob's Uncle Harry leaves him $25,000 in his will and Rob invests all of this amount into the fund? Neglecting taxes, what is the total amount in the fund at the end of the 28 years?

26. If dividends of $5.50 per year are expected to be paid on a security for the unlimited future, and the security is selling for $62 per share, what is the anticipated return on investment?

27. A large consulting engineering firm is renting a parking lot for their employees to park their cars. The rent is $1,000 per month. The consulting organization desires to earn 10% return on investment. If the firm could purchase the lot, what is the current value based on perpetual use of the lot?

28. An endowment fund is established by an art museum. The fund earns 8% per year. The museum would like to use only the proceeds from the fund—none of the principal—for annual operating funds. The goal is to have the fund provide $40,000 per year. What is the required size of the fund?

29. Nancy and Tom own a small house that they rent to tenants for $650 per month. The annual taxes, insurance, maintenance, and other expenses are $2,500 per year. They are considering selling the property. The appraiser they hire says that he will determine the value of the house in three different ways. He will first compare the house to similar homes in the area; second, consider the value based on current construction costs; and finally, consider the income potential of the home. For the latter method he will use a rate of return of 8% and divide the net annual income by this rate. What is the value using the latter method? Is this a valid approach?

30. A school loan is made to Joe Moss. He will receive $5,000 each year for four years. He will repay the loan with annual payments over a ten-year period starting on the anniversary date of the loan in the fifth year. His loan rate is 7.5%. Determine the annual payments Joe must make.

31. Max and Helen have decided to establish a college fund for their daughter's college expenses. They an-

ticipate that there will be deposits into the account beginning on her second birthday through her eighteenth birthday. There will be four withdrawals of $10,000 each starting on her nineteenth birthday. No deposits will be made during this four-year period. If the fund can earn an average of 8%, what is the value of the payments that Max and Helen must make into the fund?

32. Twin sisters, Maggie and Molly, are 22 years old. They have $2,000 per year to invest or spend as they choose. Maggie decides to start a retirement fund and she opens an IRA account at a discount brokerage. She will put $2,000 into the fund for six years and then make no more deposits for the next 37 years, when she will retire and go on a world cruise with her sister. Molly decides to spend her $2,000 each year for the next six years. Then she will begin to make deposits of $2,000 into her account for the remaining 37 years until she retires and goes on the cruise with Maggie. How much does each sister have at retirement? Interest earned is 10%.

33. Assume that the average college graduate earns $275,000 more than the nongraduate over their respective working careers. This amount is spread equally over 35 working years. The graduate invests $10,000 per year for four years in education. The individual wants to earn at least 10% return on investment.
    a. Does the investment meet the ROI criterion?
    b. What is the actual ROI?
    c. What other factors could be considered in this problem?

34. What single amount at the end of year eight is equivalent to equal annual payments over fifteen years of $3,000/year if interest is 10%?

35. The maintenance costs of a new piece of assembly equipment and its computer control system are estimated to be $1,000 the first year and will increase by $200 each year for the life of the equipment, which is 8 years. The desired ROI is 10%.
    a. Determine the equal annual amount of the maintenance costs.
    b. What is the present value of the maintenance costs?
    c. What is the present value of all costs if there is an additional major overhaul of the equipment costing $10,000 that occurs at the end of the fourth year?
    d. What is the total equivalent equal annual cost of the maintenance over the life of the system including both the annual and the fourth year overhaul?

36. The town of Centerville is building a new city hall. It plans on selling bonds to finance the construction. The initial cost of construction is $2,400,000. The maintenance costs for the structure are expected to be $100,000 the first year and to increase by $6,000 each year over the life of 50 years. There will also be major roofing, HVAC, painting, and structural revisions occurring at the twentieth and fortieth years that will cost $300,000. Interest is 8%. What is the total amount of the bond offering that will include the initial cost, annual maintenance, and special major repairs?

37. A county park is considering the construction of a small lake by building an earth dam on Slick Rock Creek. The cost of the dam construction would be $1,150,000, and other improvements would cost $220,000. The county is anticipating that the lake would be used for swimming, boating, fishing, and picnicking and that fees could be charged for these services. If ROI of 6% is desired for the investment of taxpayer funds, and the usable life of the park is 25 years, determine the annual net cost of the park. If it is assumed that the dam life is perpetual, and the other improvements have a 25-year life, what is the annual cost?

DISCUSSION CASES

Kendrick Electronics

A small manufacturing company, Kendrick Electronics, is currently renting warehouse space for $10,000 per month to a large manufacturer. The manufacturer has approached Kendrick asking if they are interested in selling the warehouse. It is located some distance from the current Kendrick location and Kendrick has ample space to expand at their current location as their sales increase.

The property taxes on the warehouse are $2,600 semi-annually. Insurance costs are approximately $4,400 per year for property and liability coverage. Maintenance on the exterior of the structure, including roof, painting, and general repairs, is approximately $8,000 per year. The tenant is responsible for parking lot repairs, maintenance, and snow removal. The tenant is responsible for all other repairs and replacement of the internal additions to the structure including the HVAC system.

Kendrick currently is growing at a rate of approximately 15% per year. They are planning to add on to their existing facility sometime within the next two to four years. Their growth rate means they need additional working capital from borrowing as well as from profits. The idea of selling the old warehouse is interesting, and engineer George Morris, assistant to the production manager, has been asked to come up with a price range for discussion purposes with the tenant.

1. How can George use the concept of perpetuity investments to come up with a price range?

2. Are income taxes a factor in this sale?

3. What other sources of information could George use to get "ballpark" price ranges for the property?

4. What other factors should be considered in this situation?

Roberts Technical Supply Co.

Joe Roberts, the owner and president of Roberts Technical Supply Co., is considering the application of time value of money concepts to his financial decisions. RTS is a manufacturing firm specializing in small plastic components. Most of its customers are in the electronics or automobile business. The present economic analysis is done by the accounting department. Joe feels that he would like to have the product designers, process engineers, and heads of departments do more justification of their equipment requests, process improvement decisions, and product design decisions using time value of money techniques.

He plans on having some training sessions to explain some of the concepts to key managers and then hold additional sessions for department managers as well as design and process engineers. The decision that he has to make before the training begins is how to establish a target interest rate to be used in future calculations.

Joe knows that the current interest rate on short-term deposits, where the company invests its cash when not required in the checking account, is 5.15%. The rate at which the company can borrow money from the National Bank is usually the current prime interest rate plus 1%. Prime is currently 7.5%. He remembers that his personal investment portfolio has a mutual fund that has earned an average of 12.45% per year over the past five years. He looked at the RTS financial statements for the past three years and notes the following information.

	This Year	Last Year	2 Years Ago
Profit After Tax	$150,000	$110,000	$120,000
Total Assets	$1,750,000	$1,570,000	$1,450,000

The current effective tax rate for RTS is 42%, which includes some state and local taxes as well as the average federal rate.

1. What are the alternative uses for RTS's cash and what is their ROI?

2. What has been the annual ROI/ROA for the total company over the past three years?

3. Does Joe's ROI on his mutual fund have anything to do with the company's ROI?

4. Should the internal ROI used by RTS be based on before or after tax? Why?

5. What are the effects on the decisions that are made in the company by not using time value of money/ROI analysis?

6. Are there any other factors that Joe should consider?

Tech Corp.

Rob Hendricks, a process engineer at Tech Corp., is responsible for the initial manufacturing engineering as well as the improvement of the process during its life. Products at Tech Corp. have a long manufacturing life since they are producing new components as well as replacement parts for in-service repairs. He is concentrating on the improvements that he would like to make in different processes. Tech Corp. uses a cross-functional team approach for continuous improvement. Rob is the team leader for different teams. A team is usually comprised of the process engineer, department supervisor, one or more operators, designer, setup supervisor, and quality technician. Sometimes middle and upper management leads or participates in the teams, and sometimes marketing and finance are represented.

Rob's current problem is that he has a number of improvements that are being finalized and the request for funds to implement the improvements is the next step. Rarely does upper management turn down these requests unless there is a cash flow problem, and then they are only postponed for a month or two. Management feels that the team usually finds the best solution to problems, and a team's recommendation is usually supported. Rob is trying to improve the analysis of these recommendations by combining quality improvement, methods techniques, productivity improvement, cost reduction, and time value of money into one document.

For the flow adjustment regulator department, Rob and the team have completed their process analysis and improvement study. They have looked at the recent quality problems that have resulted in both defective products caused in the process as well as field service reports of failures at the customer's application site. A pilot study indicates the internal scrap reduction as a result of this effort should amount to $1,100 per month. The field failure costs due to warranty and replacement should reduce those costs by $650 per month. To reduce these costs, the methods are being changed and a different raw material will be used that will add an annual cost of $14,500 per year.

The team has also made some method improvement suggestions that will improve the safety of the operation and improve productivity. Currently, 350 units per shift are produced, and the revised method will permit 390 units per shift to be produced. The value of the additional 40 units of production, considering direct labor, fringe benefits, and reduced down time, will be an approximate savings of $125 per week. The team has recommended that new tooling be designed and built to reduce equipment changeover time from one model to another. The cost to design and build it is estimated to be $35,600. The revised equipment is also necessary for the improved output per shift as well as the changeover reduction. To implement these improvements some utilities will have to be changed at a one-time cost of $1,200; operator training costs of $900 will be required; and removal of the old equipment will cost $2,100. In addition to the other savings, the team feels that due to the changeover improvement, less in-process and finished goods will need to be stored. This will free approximately 450 square feet of storage space. The cost accounting department says that a square foot of warehouse space is worth about $4 per year. This space is already being considered for expansion by an adjoining department.

The team feels that they have done a good job with this problem. Quality and safety will be improved, setup time for changeovers will be reduced, output per shift will be improved, methods will be improved, and inventory and floor space will be reduced. It is Rob's responsibility to summarize all of this work and do a cost analysis; then the team will make its presentation to management. Rob wants to use spreadsheet rather than manual analysis for this summary so that he can do some "what-if" calculations. One problem is that he is not sure if this product and process will be used for four, five, or six more years. And, although the company usually uses 15% as a desired target ROI, he would like to try other values and determine the actual ROI on this project.

1. Set up a spreadsheet showing all the costs and then all of the revenue summarized.

2. Apply time value of money calculations to this problem and make any reasonable assumptions.

3. Set up the spreadsheet so that different n values may be tried since the life of the product is not fixed.

4. What is the impact of the life of the product not being fixed on the recommendations? What if the product is discontinued in three years?

5. What other factors should Rob and the team consider?

6. What recommendations should the team make to management?

Chapter 6   Annual Amount Applications

KEY TERMS

Annual Operating Costs The costs associated with a process other than initial and residual values.

Annual Total Cost, A_T Determined by adding the equivalent annual cost of the initial investment, A_P, to the annual operating costs, A_O, and subtracting the annual equivalent value of the salvage value, A_S.

Interest-Only Loan A loan that requires only interest to be paid during the life of the loan. The principal is paid at the end of the loan period.

Lease Purchasing the use of the equipment without owning the equipment.

Residual Salvage Value The estimated value of the asset, project, or investment at the end of its use.

Risk The probability of a project or investment not meeting the desired or targeted outcomes.

Risk/Return The relationship that establishes a proportional return on investment compared to the investment's risk.

Sensitivity The amount of change in the output or results in an investment compared to the amount of change in the inputs.

What-If Calculations Changing the values of the inputs to see the effect on the output conditions.

LEARNING CONCEPTS

- A seven-step approach for solving TVM applications.
- Total annual cost calculations using the TVM formulas, $A_{Total} = A_{Operating} + A_{First\ Cost} - A_{Salvage}$
- Annual cost calculations using $A_{Total} = (P - S)(A/P) + iS + A_{Operating}$
- Lease versus purchase decisions.
- Estimating salvage value.
- Determining target or desired ROI values.
- Calculating loan repayment amounts, total interest, future balance using TVM and equivalence techniques.
- Types of industrial and commercial loans.
- "What-if" calculations, sensitivity, and what-if tables.

- Calculating unit costs.
- TVM calculations applied to personal financial decisions including home mortgage loans, retirement accounts, and valuation of returns on securities.

INTRODUCTION

Of the many calculations using time value of money and equivalence techniques, the annual cost calculation is one of the most frequently used. The determination of product unit costs, comparing purchase with lease options, the economic justification of equipment, loan calculations, and personal loan and retirement account calculations are applications of the annual amount determination.

This chapter considers the analysis of the organization's and the individual's investment analysis and decisions using a seven-step approach to solving time value of money problems. Different types of problems, when reduced to their diagrams, are solved with similar techniques. The chapter demonstrates the application of the time value of money tools developed in the previous two chapters to typical investment projects.

A seven-step procedure relates the initial cost of equipment to its annual cost, annual operating cost, and salvage value. Combining the seven-step procedure, annual cost calculation, and TVM diagrams provides a basis for analyzing complex projects and investments.

There is sometimes confusion concerning the use of depreciation calculations. Here, depreciation as an annual cost is compared to the annual cost based on time value of money. The chapter gives a simple approach to compare benefits and costs of product and process improvements. Determining and selecting rates of ROI, and estimating residual salvage value of projects and equipment are discussed.

The concept of "what-if" calculations and sensitivity analysis is applied to examples that demonstrate how changing the input costs and revenues can affect the output variables and project outcome. Risk and ROI are discussed to relate their interdependence. What-if tables are presented.

The discussion of lease and purchase alternatives uses equivalence techniques that permit comparison of their relative costs.

The chapter discusses types of industrial loans, loan terminology, and loan calculations. Leases are presented as a technique of obtaining funds for equipment, facilities expansion, and product/process improvements. Basic examples of relating loans and sources of funds to the justification of a project or investment are presented.

Finally, the chapter presents a discussion of financial problems encountered by individuals such as home mortgages, retirement accounts, and security valuation. These investments can be analyzed using similar techniques as used in the analysis of an organization's projects and investments.

AN APPROACH TO SOLVING TVM RELATIONSHIPS

The basic mathematical relationships using the interest functions and tables were presented in the previous chapters. As more complex problems are encountered, a seven-step solution can be applied and the calculations can be simplified. The steps to this approach are shown below.

Steps for Solving Time Value of Money Applications

1. Describe the situation using a TVM diagram.
2. Summarize the known and unknown factors using the diagram.
3. Mathematically determine the unknown using equivalence and TVM calculations.
4. Make "what-if" calculations to determine sensitivity.
5. Look outside the problem to determine the impact on other projects and sources of funds.
6. Consider the effect of noneconomic factors on the solution.
7. Combine economic and noneconomic factors to make a recommendation or decision.

Drawing a diagram reduces the investment or project to its critical factors and describes the project visually. The drawing conventions can be modified to describe the problem accurately.

Stating the known and unknown factors indicates the direction that the calculations must take to arrive at a solution. Frequently there are a number of steps to the solution, which these can be shown on intermediate diagrams.

The tools for the solution and the calculations may be formulas, programmable calculators, interest tables, spreadsheets, or any other equipment or software available.

The what-if step is important to see what impact changing one or more of the variables has on the financial outcome of the project. This step tests the sensitivity of the solution with respect to its inputs. This may also permit the decision to be viewed in terms of its risk or probability of failure to meet the investment project's financial targets. Estimates are generally used for salvage, ROI, and life values. These are predictions and estimates, not known amounts, which can be combined with the historical accounting values of costs and revenues. The amount of variability can be tested using the what-if calculations.

The first step to solving a problem is defining it.

Since a single financial decision or investment rarely stands alone, the impact of outside factors on the investment should be determined. The proposed investment usually has an impact on other products, processes, facilities, and investments. These relationships can be identified. The source of funds can also be considered in the analysis.

It is rare for a project or investment decision to be made exclusively on financial criteria. Usually, noneconomic factors must also be considered. Employees, shareholders, the community, customers, suppliers, and others may be affected by the investment. The organization's structure and culture, customer perceptions, and legal implications of the investment must be considered. This is the most difficult step because the variables cannot be reduced to numbers.

The nonquantitative nature of the problem makes it difficult to assess the risk of the project. In addition, this step requires the skills, technical abilities, and experience of team members and management as well as their economic analysis abilities.

Selecting a Value for ROI

In the previous TVM examples and problems, i or ROI was given. In practice, the determination of the ROI values to be used in calculations may be accomplished by the following methods:

1. The owners of the firm or the directors of the organization can state the target or minimum return that they want the organization to earn. For example, the board of directors could say, "We are responsible to the shareholders, customers, and employees of the organization and want to earn 15% before tax return on our investments."

2. Alternative investments such as treasury bills or bonds (both government and corporate), mutual funds, or other investments could establish the minimum or target return values. For example, if the current average interest rate on corporate loans is 9.5%, this could become the minimum ROI that the firm would desire on its projects.

3. Another method of determining the target or minimum desired ROI is to look at previous projects to see what their actual ROI has been. If the average ROI on recent investment has been 12%, this value can become a goal for future projects. Higher or lower ROI goals could be made when the risk of the particular investment is considered.

The target or minimum ROI that an individual or organization uses in TVM calculations could be based on a combination of the above methods. The target ROI is different under different economic environments. In the early 1970s, when inflation was in the 15% to 18% range, firms would avoid any project or investment that was lower than the inflation value. When cash can be invested in government securities at high rates, it is difficult to find internal projects in the organization that have ROIs that meet or beat inflation or low-risk government investments.

The target ROI will also vary depending on the firm's profits and cash flow. When profits are high and cash is available, projects with relatively low ROIs may be funded. Likewise, when profits are low and cash is tight, few investments are made, even if they have a high potential return. When market interest rates are low, firms and individuals borrow money to make investments, and many projects may be created. When interest rates are high, investment is generally lower. The Federal Reserve considers these factors when making adjustments to the financial system's interest rates. They try to keep interest rates low enough to encourage investment, but not so low that the economy is too active and reaches maximum capacity, causing inflation to develop.

What-If Calculations

Frequently, the desired ROI is a *range* of returns. For example, the organization may say that a pretax ROI of between 10% and 15% is the goal. For calculations, different percentages can be used. In some calculations, a small change in ROI can cause a large change in outcome. Spreadsheets are an ideal technique for this type of **what-if calculation.** Once the spreadsheet is set up for one percentage, it is relatively easy to try other percentages to see what occurs.

> What-if questions look for risks, rewards, the sensitivity of outcomes, and best solutions.

In addition to selecting different values of ROI, project life, operating costs, and salvage value can be changed to see the impact on the outcome. For example, an organization interested in purchasing a new piece of equipment could ask, "If the operating costs are twice what we have estimated, what is the impact on the cost of the product this equipment produces?" Or, "If this equipment is used for three years less than anticipated, what is the impact on the ROI of the investment?"

The what-if questions are unlimited, but the revised calculations are relatively easy once the initial solution has been established.

Sensitivity of Outcomes to Inputs

The what-if analysis considers the changes that happen as the variables are changed. The degree of this change is called **sensitivity.** If factors can be changed in the investment with small impact on the result, the outcome has a *low sensitivity*. If the outcomes change a large amount for small changes in inputs, the investment has *high sensitivity*.

The case of annual cost, as will be seen, is often affected very little by a large change in estimated residual value, for example. This indicates that the annual cost has a low sensitivity to the salvage value. Or, if operating costs are large compared to the other components of total annual costs, the project has a high sensitivity to operating costs. For a large, complex financial analysis, both what-if and sensitivity analysis are important in the decision-making process.

> Sensitivity is the degree of change in an output for a given change in input.

By looking at the possible what-if situations, the risks of the situation can be identified. The risk of incorrectly estimating the residual salvage value of a item of equipment is often small, for example.

RISK

Risk is the likelihood or probability of a project or investment not meeting the targeted outcomes.

> Higher risk often requires a higher target ROI to compensate for the risk.

Someone who purchases a lottery ticket generally knows that the desired outcome of winning the prize has a small chance of occurring. The investor who purchases a bond from the U.S. Treasury believes that there is a high likelihood of receiving the specified interest amount and the return of the principal, since the Treasury has a long history of paying bond interests and principal. The chance or risk of not receiving the funds is very low.

Projects also carry risks. Some projects, for which similar investments have been made in the past, may have well-known costs and outcomes. The risk of not meeting the desired outcome is small. Other projects, such as new research ventures, may have a high likelihood of not meeting the desired outcome. This risk of this latter type of project is higher.

When a project or investment has a perceived higher risk, the desired or target ROI is often increased in an attempt to compensate for the risk. By investing in a number of similar high-risk investments, if some projects fail to meet the desired outcome, the higher returns of the others will compensate for the failures. Conversely, often a lower ROI is accepted for a lower risk investment since it is perceived that a number of investments of this type have high chances of success and failure is rare.

Some risks can be controlled within the organization. Minimizing risk is part of the responsibility of the engineers, technicians, and managers of the projects. For risks outside the organization, the best management strategy is to avoid the risk since it can't be controlled. For example, mountain climbers can take care of their equipment to minimize the risk of failure, but they cannot control the weather. Equipment failure is an internally controllable risk; the weather is an external risk. Climbers check the weather reports carefully, and don't climb in threatening weather. Avoiding the risks that cannot be controlled is a strategy to minimize investment risks.

> Risk should be controlled, minimized, or avoided.

Market factors such as customer demand are normally thought of as outside the firm's control. But rather than avoid the market, good organizations learn all they can about the customers' wants and needs, and then incorporate this information in the design and production of the good or service. Quantifying customer-related information and taking "the voice of the customer" inside to the design and manufacturing functions is called Quality Function Deployment. This technique, perfected by some Japanese firms, doesn't modify the risk, but changes the services or product to conform to the market requirements. If the customer wants a red product, determine that before the product is designed so that when the customer asks for it, it is available. In this way the risk of not selling an otherwise well-designed product is reduced.

Diversification is also a part of minimizing risk. Determine the risk potential before committing funds. Control the internal risks as much as possible, and then attempt to avoid, diversify, or prepare for the uncontrollable external risks.

OPERATING COSTS

The annual costs related to the project or equipment are operating costs. These costs include materials, labor, insurance, heat, utilities, salaries, maintenance, repairs, property taxes, advertising, marketing expenses, and any other costs associated with the investment. These costs are usually stated on an annual basis and have no time value of money associated with them. The assumption in including these costs in the TVM calculations is that they occur at the end of the year. If this assumption is not satisfactory for some situations, the costs can be converted to monthly or quarterly amounts using the appropriate formulas. The notation is $A_{Operating}$ or A_O.

Salvage Value

Equipment that has been in service for some time is likely to have decreased in value. This is the basis for the concept of depreciation used in the Accounting Equation. Salvage value is the market value amount that can be realized when the item is sold in the future. In the analysis of investments, the revenue received from the sale of the used equipment at the end of the project's life is a factor. An estimate of this future value is usually included in the analysis. There are a number of approaches to determine this estimated value. Below is a summary of ways to make this estimate.

Estimating Residual Salvage Value

Base the estimate on similar equipment that has been disposed of in the past.

Look at used equipment values in the open market.

Estimate the salvage value of the components of the equipment based on "scrap value."

Base the residual value on a percentage of the first cost of the equipment. The percentage can be derived from previous equipment sales.

Neglect the residual salvage value.

Use a combination of the above techniques.

The accounting record-keeping function can provide the value of past equipment sold for its **residual salvage value.** This data can be used to estimate the residual value of the current purchased equipment. Consider such characteristics as similar use, equipment type, and manufacturer.

For many types of equipment, a used equipment market exists. Comparing the prevailing values of used equipment is a basis for estimating the future value of new equipment. Individuals do this frequently with automobiles and other personal assets. By checking current selling prices, the individual can estimate the life and future value of the automobile or other assets.

Many fixed asset items that are used until they are completely worn out may still have some salvageable materials or components that may have some value. Scrap value such as for steel or copper can be calculated if the equipment contains large quantities of those materials. Electronic components may have precious metals or contain components that can be sold.

Some organizations that have large quantities of a particular equipment item may have accurate data that can be expressed as a percentage of the first cost for the salvage value for a particular life of the item. Examples of this calculation would be a warehouse facility that has many different forklift trucks of different ages. The firm retires and replaces some of the trucks each year, and is able to collect accurate data on the residual value compared to the first costs.

Residual salvage value is usually small when it is annualized.

Neglecting the residual salvage value at first may seem like a weak method. However, if the value is small compared to the first cost and the other operating costs, this may be acceptable. In the application of time value of money, the factor to bring the future salvage value to an annual amount, the *F/A* factor, is small. When the small *F/A* factor is multiplied times the estimated salvage value, the product, annual value of the salvage, is very small. This small value can sometimes be neglected without having a major impact on the investment decision.

> Because of disposal costs, sometimes residual salvage value is a cost rather than a revenue.

Sometimes at the end of an asset's life the net residual salvage value is less than zero. That is, the removal costs are greater than the remaining value of the asset. In this case, money is paid to remove the asset. Rather than being a revenue to the firm at the end of the asset's life, there is a cost of removal. In terms of time value of money, this cost can be annualized and included with the other annual costs rather than being a revenue and offsetting some of the costs.

Depreciation as Part of Annual Cost

When a fixed asset such as equipment or buildings is purchased, the accounting department writes a check to the supplier, which causes the cash account to decrease in the firm's checkbook. At the same time, an asset account of equipment is created and is increased by the value of the new equipment. Viewing the Balance Sheet at this point in time, cash would be lower because of the check to the supplier.

From previous discussion, this depreciation expense is determined using various schedules and formulas. Sometimes this annual depreciation expense is thought to be the annual cost of the equipment and this amount is used in economic calculations. This is an incorrect technique. Only if the return on investment desired is zero is the annual depreciation the same as the annual cost of the equipment that is used for economic analysis. If there is any time value of money included in the analysis, then the depreciation expense is only the accounting cost, not the equipment value used for the economic analysis.

The TVM method of determining the annual cost of an asset includes not only the value of the asset spread over its life, but also the desired return on that investment. Using time value of money, the first cost of the equipment is annualized. This annualized cost is then added to the **annual operating costs.** Finally, the estimated residual salvage value of the equipment is brought back to an annual basis. This annualized revenue is then subtracted from the other annual costs to give a net **annual total cost** that includes the equipment's first cost, its operating cost, and its residual value.

Davis Corp.'s New Equipment. Jim Peterson is a manufacturing engineer for Davis Corp., a small metalworking shop. A new press is being purchased and Jim is trying to calculate the annual cost of the equipment so he can do some cost estimating of the new products the equipment will produce. The press costs $130,000. He estimates that it will last 10 years and have a salvage value at that time of $15,000. He knows that when the equipment is purchased, the accounting department will write a check to the supplier, decrease cash in the checkbook by

$130,000, and increase a fixed asset account, equipment, by $130,000. If straight-line depreciation is used, the equipment is assumed to decrease in value by the following formula:

$$\text{Dep} = \frac{P-S}{n} = \frac{\$130,000 - \$15,000}{10 \text{ years}} = \$11,500/\text{yr}$$

Each year, the accounting department will reduce the value of the asset by $11,500 and increase costs, depreciation expense, by $11,500. On the Income Statement, there is an annual expense of $11,500/yr. Is this the "cost" of the equipment? From the accounting and Income Statement point of view, the annual cost is equal to the assumed $11,500. If Jim assumed that the cost of the equipment is the annual depreciation (*and incorrectly not include time value of money*) he would use the value of $11,500 per year as the "annual equipment cost."

If the estimated annual operating costs of the equipment including power, material, supplies, overhead, space costs, labor, and related costs are $5,000 per year, the total annual cost using depreciation would be:

$$\text{Total Cost per year} = \text{Depreciation/yr} + \text{Operating Costs/yr}$$
$$= \$11,500/\text{yr} + \$5,000/\text{yr}$$
$$= \$16,500/\text{yr}$$

However, if the organization did not purchase the equipment for $130,000, it could put the money into an investment that pays a return. If the firm invests the funds into a 10% account, it would earn ($130,000)(10%) = $13,000 per year. If the funds are invested into a 12% fund, $15,600 could be earned. Also, the organization includes the cost of the interest that the firm would pay to a lender, such as a bank, if it borrowed the funds to buy the equipment. Whether the money is borrowed from a bank or if the firm "borrows" the funds from its own cash drawer, the annual cost of the equipment should include the equipment value and time value of money, not just the annual depreciation.

> If depreciation is used as annual cost, ROI = 0%.

If the use of depreciation calculation is incorrect, since it does not include the time value of money, how should comparisons be made? Using the concept of equivalence, measuring dollars at the same point in time, is the better approach. The key to financial decisions is being able to compare the costs of one decision with the costs of other decisions. In order to make a comparison, the dollars must be compared *at the same point in time* (equivalence).

Analyzing projects and investments requires both the determination of how to obtain funds for the project and the determination that the project is economically justified. Time value of money analysis is applied to the following alternatives by Jim Peterson at Davis Corp.

Davis Corp. Alternatives for Source of Funds

1. Purchase the equipment using our own funds.
2. Lease the equipment.

3. Purchase the equipment using a bank loan.

4. Combinations of the above and other alternatives.

CALCULATING ANNUAL COST INCLUDING ROI

If Davis Corp. desires to earn 15% ROI on the invested capital of $130,000, a time value of money calculation must be made. Jim can draw the diagram shown in Figure 6–1.

Refer to the figure for this discussion. The arrow below and into the horizontal line on the left end represents the investment by Davis into the equipment valued at $130,000. The arrow coming out of the line at the right end represents salvage revenue coming back to Davis when the equipment is sold after 10 years. The salvage is normally a revenue. This annualized revenue is subtracted from the other costs, which reduces the net annual cost of the project. Thus it is shown as a revenue arrow above the line on the TVM diagram. If the net value of the equipment at the end of its life is less than zero, and the removal value is greater than the salvage value, the amount is a cost. In this case, the arrow is shown as a cost below the line. Davis wants to earn 15% return on its investments. If it does not invest $130,000 in the new press, Davis will invest the money in other investments that will earn a target of 15% ROI.

To calculate the annual cost, the first cost is converted to annual cost using the *A/P* factor for *n* of 10 years and *i* of 15%. The value of the future residual salvage value is brought back to the annual revenue. The net annual cost is the difference between this revenue and cost.

First cost: $P = \$130,000$, $S = \$15,000$, $n = 10$ yr, $i = 15\%$, A_{Total} = unknown

$$A_{Total} = A_P - A_S = P\,(A/P, n = 10, i = 15) - S\,(A/F, n = 10, i = 15)$$
$$A_{Total} = \$130,000\,(.1993) - \$15,000\,(.0493)$$
$$A_{Total} = \$25,903/yr - \$739/yr = \$25,164/yr$$

The calculation above shows that in order to earn 15% return on investment, Davis must receive or "charge itself" $25,903 per year on the investment of $130,000. The equivalent an-

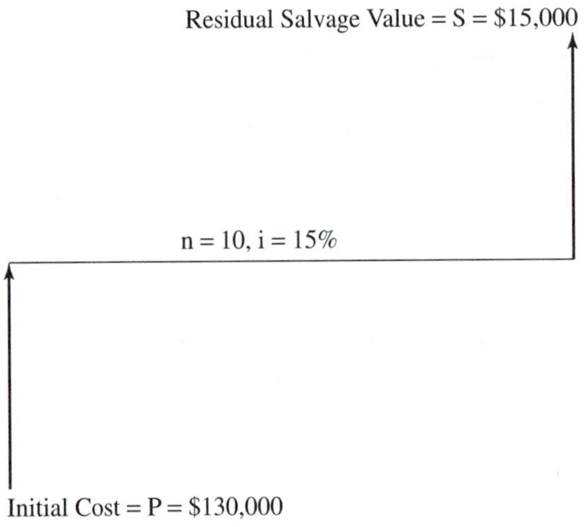

Residual Salvage Value = S = $15,000

n = 10, i = 15%

Initial Cost = P = $130,000

Figure 6–1   Annual Cost of New Press at Davis Corp.

Annual Value of Salvage $= A_{Salvage} = \$739/year$

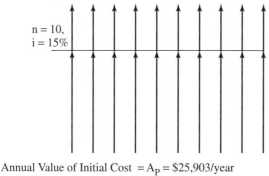

n = 10,
i = 15%

Annual Value of Initial Cost $= A_P = \$25,903/year$
$A_{Total} = A_P - A_S = \$25,903 - \$739 = \$25,164$ per year

Figure 6–2   Revised Diagram of Cost of Press at Davis Corp.

nual benefit of the future residual salvage value is $739/year for a net annual cost of
$25,164/year. The revised TVM diagram in Figure 6–2 shows the annual cost of the equipment
below the line and annual revenue of the sale of the equipment. Compare this amount, $25,164,
with the depreciation amount of $11,500. It is different by more than a factor of two times!

It is important to recognize that an equipment investment is like any other investment and
must have its own return, as would a savings account, certificate of deposit, bond, or other investment. When making comparisons between alternative pieces of equipment, make or buy decisions, and lease or buy decisions, the TVM approach is used to calculate the annual cost.

Modified Annual Amount Formula

The formula $A_{Total} = A_P - A_S$ can be modified for calculation purposes to the following form:

$$A_{Total} = (P - S)\,(A/P, n, i) + (i)(S)$$

The identity, $A/F = A/P - i$, can be shown to be true by the following:

$$A/F + i = A/P$$

$$\frac{i}{(1 + i)^n - 1} + i = \frac{i(1 + i)^n}{(1 + i)^n - 1}$$

$$\frac{i}{(1 + i)^n - 1} + \frac{i[(1 + i)^n - 1]}{(1 + i)^n - 1} = \frac{i(1 + i)^n}{(1 + i)^n - 1}$$

$$\frac{i + i(1 + i)^n - i}{(1 + i)^n - 1} = \frac{i(1 + i)^n}{(1 + i)^n - 1}$$

The i cancels in the numerator of the left part of the equation. The identity remains and
demonstrates that $A/F = A/P - i$:

$$A = A_P - A_S$$
$$A = P(A/P) - S(A/F)$$

Substituting,

$$A = P \, (A/P) - S \, (A/P - i) = P(A/P) - S(A/P) + Si$$
$$A = (P - S)(A/P) + Si$$

This formula can be used as an alternate equation form to calculate A_{Total}. Applying this equation to the previous problem,

$$A_{Total} = (\$130,000 - \$15,000) \, (.1993) + (.15)(\$15,000)$$
$$= \$22,914 + \$2,250 = \$25,164$$

This is the same answer found previously. This formula requires using only one value from the table. Some users may prefer it.

Adding Operating Costs to the Annual Equipment Cost

An additional calculation can be considered. If the equipment has annual operating expenses, which include maintenance, power, operation, indirect materials, direct labor, and materials, use the notation $A_{Operating}$. It can be added to the annual cost found from the present value and salvage value.

The revised and complete equation for analyzing the initial, annual, and future costs would be:

$$A_{Total} = A_{Operating} + A_{Pres} - A_{Salvage}$$

or

$$A_T = A_O + A_P - A_S$$

If Jim Peterson at Davis Corp. estimates that the cost of power, maintenance, supplies, overhead, space, and other operating costs are \$5,000 per year, the annual cost of the equipment is calculated using the formula for A_T:

$$A_{Total} = A_{Operating} + A_{Pres} - A_{Salvage}$$
$$A_T = A_O + A_P - A_S = \$5,000/yr + \$25,903/yr - \$739$$
$$A_T = \$30,164/year \text{ including operating costs}$$

This approach is used often in the calculation of annual costs. Comparing the cost of purchasing the equipment to leasing, comparing make or buy decisions, and other cost comparisons can only be annualized using TVM calculations.

The alternative formula that includes the operating costs can be used also. The revised formula becomes:

$$A_{Total} = (P - S)(A/P, n, i) + (i)(S) + A_{Operating}$$

Some may prefer to use this formula to calculate the total annual cost. Most examples in this text will use the annual cost formula showing annualized first costs, annual operating cost, and annualized residual value.

Sensitivity of Estimating Residual Salvage Value

For the previous example, $n = 10$, $i = 15\%$ gives an A/F value $= .0493$. For most ranges of life and interest rates, the resulting A/F value is small. Therefore, even if the estimate of the residual salvage value is *incorrectly* estimated, the impact of the error is reduced by multiplying the salvage value times the small A/F factor. A_S has a low sensitivity to the salvage value change.

In the previous Davis Corp. example, $A_{Salvage}$ is \$740/year based on the estimate of salvage value of \$15,000. What if the residual value is actually \$25,000? Would this have a significant impact on the purchase decision? Using this revised value, $(S)(A/F, n = 10, i = 15\%) = (\$25,000)(.0493) = \$1,233$. The revised total annual cost would be

$$A_T = \$25,903 + \$5,000 - \$1,233$$
$$A_T = \$29,670/year$$

Compare this revised annual cost of \$29,670 with the original annual cost of \$30,164. A \$10,000 difference in salvage value estimate results in only:

$$\$30,169 - \$29,670 = \$499/year \text{ difference in annual cost}$$

An "error" in the estimate of residual value of 66% (\$25,000 − \$15,000)/\$15,000 causes a difference of less than 2%, \$499/\$30,169, in the annualized cost. The conclusion is that a large error in the estimate of residual salvage value results in only a small difference in the estimated annual cost due to the A/F factor value. And the larger the value of n and/or i, the smaller the A/F factor and the smaller the resulting $A_{Salvage}$ value. The A_T has low sensitivity to a change in salvage value.

Gradient Operating Costs

Frequently when estimating operating costs of equipment, we find that maintenance and repair costs are not equal throughout the life of the equipment. A gradient is often used to represent these costs. Assume that the operating costs for the new press can be divided into two types. The first group of operating costs is for power, space, supplies, taxes, labor, and related costs and these are \$3,000 per year. The maintenance and repair costs for the first year are \$1,800. They increase by \$300 per year over the life of the equipment. Using the same equipment first cost and salvage, what are the total annual costs under these conditions? See Figure 6–3.

$$P = \$130,000, S = \$15,000, A_{Oper} = \$3,000 \text{ per year}, A'_{Maint} = \$1,800,$$
$$G_{Maint} = \$300/yr, n = 10 \text{ yrs}, i = 15\%, A_{Total} = \text{unknown}$$

First, solve the gradient function and convert it to an annual cost amount:

$$A_{Maint} = A'_{Maint} + G_{Maint} (A/G, n = 10 \text{ yr}, i = 15\%) = \$1,800 + \$300 (3.3832)$$
$$= 1,800 + 1,015 = \$2,815/yr$$

The next step is to solve for the total annual costs:

$$A_T = A_P + A_O + A_M - A_S$$
$$= \$130,000 (A/P, n = 10 \text{ yr}, i = 15\%) + \$3,000$$
$$+ \$2,815 - \$15,000(A/F, n = 10 \text{ yr}, i = 15\%)$$
$$A_T = \$25,903 + \$3,000 + \$2,815 - 739 = \$30,979/year$$

This is the total annual cost of the equipment if the gradient maintenance cost is included.

If frequent calculations will be made similar to those above, using computer spreadsheets will be helpful in solving the repetitive calculations.

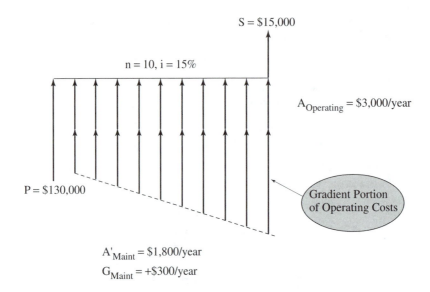

$S = \$15,000$

$n = 10, i = 15\%$

$A_{Operating} = \$3,000/year$

$P = \$130,000$

Gradient Portion of Operating Costs

$A'_{Maint} = \$1,800/year$

$G_{Maint} = +\$300/year$

Figure 6–3   Gradient Press Operating Costs at Davis

Economic Justification of a New Press at Davis Corp. Jim Peterson is continuing to consider the purchase of the new press. The old press is working at its maximum capacity. But recently it has required more and more maintenance and repairs. The new equipment is anticipated to require less down-time for maintenance and it has some features that permit faster die changeover than the existing equipment.

Overall, Jim estimates that an additional output due to reduced down-time of 10 hours per month will be available from the new equipment. Also, the current maintenance and repair costs will decrease. Jim estimates that increased production from reduced down-time will be worth $200 per hour. And he believes that other maintenance/repair costs will be reduced by $700 per month. The annual savings can be calculated as follows:

Annual savings of up-time = (10 hr/mo)($200/hr)(12 mo/yr) =	$24,000/yr
Annual savings of maint/repair = ($700/mo)(12 mo/yr) =	+8,400/yr
Total annual savings =	$32,400/yr

Comparing the above amount of $32,400/year to the annual cost of the equipment:

Annual cost of proposed equipment =	$30,979
Annual savings from reduced maintenance and higher productivity =	−32,400
Difference =	$1,421/yr

Recall that the annual costs were calculated with a 15% ROI. The annual savings are *greater* than the annual cost. The equipment purchase seems justified based on the savings. Now, Jim Peterson and Davis will have to decide if the purchase of the new equipment is possible with the funds available to the organization. Sources of funds include internal cash, borrowed funds, or perhaps a lease.

If the bank is charging 8% on loans, this would reduce the net return to Davis. If the new equipment is important for Davis to accept a lower return on the investment, it can be considered. If Davis believes that funds can be invested in other projects that have higher returns, it may postpone this acquisition. The complete analysis of existing equipment versus proposed new equipment, including tax considerations, is deferred until a later chapter. However, the above analysis demonstrates the basic considerations.

Lease Alternative. Jim Peterson is considering **leasing** the new press rather than purchasing it. The leasing firm has given Jim a quotation. They require a $5,000 initial payment and quarterly payments of $6,500 over the 10-year period. The operating costs would be the responsibility of Davis and would be the same as if the equipment were owned.

There are a number of approaches to this calculation. The most important consideration is to compare the lease costs with the purchase costs at the same point in time. One of the most common problems in making financial decisions similar to these is that the dollars are viewed at different times. This is a comparison of "apples to oranges." Measuring dollars at the same point in time is, of course, the concept of equivalence.

Since the purchase alternative costs have been converted to an annual basis in the previous calculation, one approach is to bring the lease alternative costs to an annual basis for comparison with the buy alternative. The difficulty with this calculation is that the lease costs are quarterly, and there is no formula or table that will directly convert the quarterly payments to annual payments. The solution is to bring the quarterly payments to the present worth, add the down payment of $5,000 to the present worth, and then move the dollars to an annual basis. Then compare the purchase costs that have been annualized to the annual lease costs. See Figure 6–4.

For this calculation, quarterly payment = $6,500, $n = 10$ years, $m = 4$, $i = 15\%$, P_{Total} is unknown.

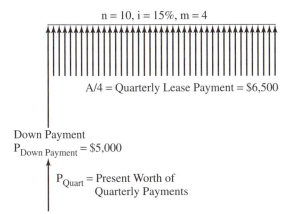

Figure 6–4   Lease Alternative for Cost of Press at Davis

The modified formula for the quarterly payments brought back to the present is shown below:

$$P_{\text{Quart}} = \frac{A}{m} \left[\frac{\left(1 + \frac{i}{m}\right)^{mn} - 1}{\frac{i}{m}\left(1 + \frac{i}{m}\right)^{mn}} \right] = 6{,}500 \left[\frac{\left(1 + \frac{.15}{4}\right)^{40} - 1}{\frac{.15}{4}\left(1 + \frac{.15}{4}\right)^{40}} \right]$$

Solving

$$P_{\text{Quarterly Payments}} = \$133{,}581$$

The next step is to add the present value of the future payments to the initial down payment and then take this amount out to an annual amount:

$$P_{\text{Total}} = P_{\text{Quarterly Payments}} + P_{\text{Down Payment}}$$
$$= \$133{,}581 + \$5{,}000 = \$138{,}581$$

This amount, \$138,581, is the equivalent value of the future quarterly payments with an ROI of 15%. The next step is to take the total present value out to an annual basis. This annual amount can be added to the operating costs to arrive at a total annual cost for the lease alternative. See Figure 6–5.

$$A = P_{\text{Total}} \,(A/P, n = 10 \text{ yr}, i = 15\%) = \$138{,}581(.1993) = \$27{,}613/\text{yr}$$

Adding this annual amount to the annual operating and maintenance costs,

$$A_{\text{Total Cost}} = A_{\text{Lease}} + A_{\text{Operating}} + A_{\text{Maintenance}} = \$27{,}613 + \$3{,}000 + \$2{,}815 = \$33{,}428/\text{year}$$

Comparing Purchase and Lease Alternatives

Comparing the annualized purchase amount of \$30,979/year with the lease cost of \$33,428 gives a difference of \$2,449/year higher for the lease than the purchase. No salvage value is considered because the equipment belongs to the leasing company.

There could be, however, an advantage if Davis chooses the more costly lease alternative. The advantage is that they do not have to pay the total of \$130,000 now. If they do not have the money

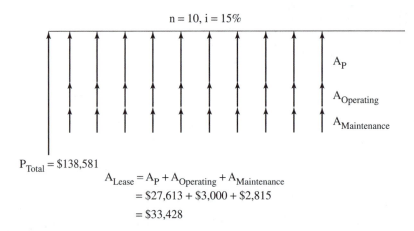

Figure 6–5   Annual Cost of Lease of Press for Davis Corp.

to purchase the equipment and choose not to arrange a bank loan for the funds, then the lease may be a good solution for them. Frequently, the final decision of any analysis is based on many factors. The lowest cost alternative may not be selected if other factors are more important. In this example, the annual costs of the purchase alternative versus the lease alternative are actually very close. Cash flow and sources of funds are often important components of investment decisions.

> A lease is usually more expensive, but it has other advantages, such as reducing initial costs and cash flow.

This example demonstrates the importance of comparing *all costs at the same point in time.* This is the concept of equivalence, which was demonstrated in previous chapters. Equivalence is the basis for almost all engineering financial decisions.

A Loan for Davis Corporation. Another alternative that Davis management is considering is to obtain a loan from a local bank for the purchase of the new press. The terms that the bank quoted are for a 10-year loan with annual repayments at a rate of 8%. Jim wants to calculate the loan repayment amount. (See Figure 6–6.)

For this calculation,

$$P = \$130,000, n = 10 \text{ years}, i = 8\%, A = \text{unknown}$$
$$A = P \, (A/P, n = 10, i = 8\%) = \$130,000 \, (.1490) = \$19,374/\text{year}$$

Jim is considering the advantages and disadvantages of a loan. He calculates the total interest that would be paid on the loan.

Total interest paid = Total payments − Principal = (10 years)($19,374/yr) − $130,000
$193,740 − $130,000 = $63,740 Total interest paid to bank

If funds are limited at Davis, the company can use the loan to purchase the equipment now and pay for it over the next 10 years.

Suppose that Davis has the cash to buy the equipment outright, but desires to use the cash for some other purpose. The loan alternative makes this possible. The disadvantage, of course, is the interest cost paid to the bank. If the lower net return is acceptable to Davis, then this loan alternative is a possibility. If Davis has a target of 15% ROI, and the bank loan has an interest rate of 8%, then the net approximate ROI to Davis is only 7%.

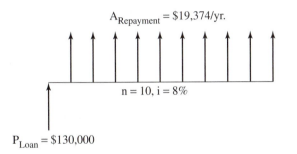

$A_{\text{Repayment}} = \$19,374/\text{yr.}$

$n = 10, i = 8\%$

$P_{\text{Loan}} = \$130,000$

Figure 6–6   Bank Loan Alternative for Davis Corp.

Another type of loan that Davis is considering is an **interest-only** type of loan. This type of loan is on a year-to-year basis. The only requirements are that the loan be paid in full within 10 years and that annual interest be paid. Davis could borrow the $130,000 today and next year would owe 8% interest: (.08)($130,000) = $10,400. Davis can normally choose to pay part or all of the principal balance on the loan at that time. If some of the principal is paid, the principal balance is reduced and the next year's interest is lower. Or, if Davis chooses not to reduce the principal, the interest would be the same as for the first year. This type of loan has the flexibility to allow Davis, based on its profits and cash flow position, to choose to pay none, some, or all of the principal. If only interest is paid during the 10 years, the final principal payment, called a "balloon payment," would be due at the end of the tenth year. Jim calculates the total interest paid for this last option below:

Total interest paid with no principal reduction = (10 years)(interest per year)
= (10)($10,400) = $104,000 total interest paid over the life
of the loan

As another option, assume that the loan with interest only for 10 years, at a rate of 8%, is paid as follows. The first three years, Davis pays only the interest of $10,400 each year. For years 4, 5, and 6, Davis pays $20,000 in principal in addition to the interest on the loan balance. In year 7, Davis pays $30,000, and then pays the balance at the end of the eighth year. The schedule of loan balance and payments is shown in Table 6–1.

Table 6–1 can be calculated manually or on a spreadsheet and the anticipated payments can be entered to update the table. The spreadsheet table can be used to answer what-if questions. The advantage of the interest-only loan is the flexibility that it gives Davis in setting a repayment schedule that can be tailored to fit the company's cash, cash flow, and profit situation over the 10-year period.

The costs of a project such as the one at Davis can be separated into material, labor, and overhead, including financing costs. Deciding whether and how to fund such a project may be a team-based decision. Upper management can coordinate the team decision process so that

Table 6–1   Interest-Only Loan Calculations—Davis Corp.

Year	Balance at End of Year	Principal Paid at End of Year	Interest Paid 8% Rate	Balance at Beginning of Next year
1	130,000	0	10,400	130,000
2	130,000	0	10,400	130,000
3	130,000	0	10,400	130,000
4	130,000	20,000	10,400	110,000
5	110,000	20,000	8,800	90,000
6	90,000	20,000	7,200	70,000
7	70,000	30,000	5,600	40,000
8	40,000	40,000	3,200	0
9	0	0	0	0
10	0	0	0	0
Total Interest Paid			66,400	

maximum investment benefits are gained with an optimum cost of funding. Of course, noneconomic factors must be considered, such as training employees to use the new equipment, changing the process methods, making sure the new process is safe, allocating space, improving product quality, and so on.

> Selecting investments and funding investments is like the chicken and the egg. Which comes first?

Expanding facilities, replacing equipment, and investing in new products are all decisions that must be considered when deciding how to provide funding. Some organizations look at all the potential uses of funds in terms of their respective ROI, and then look at the funds available from profits, securities sales, loans and leases separately. Of course, when profits are high and interest rates are low, more projects can be funded.

The Davis example demonstrated four ways to finance the purchase of a new piece of equipment. These included:

1. Purchase from cash reserves without a loan.

2. Lease the equipment.

3. Obtain a loan with annual repayments over the life of the equipment.

4. Obtain an interest-only loan with flexible repayment of principal.

Many other alternatives could be considered, using different time periods, or variable or adjustable interest rates. Most lenders will set up loan terms that suit the borrower. Generally, there is no such thing as a "standard loan." This is especially true for commercial loans as opposed to consumer loans.

Production, marketing, and other nonfinancial personnel may not be directly involved in the negotiation of loans or sale of securities, but it is important that they have an understanding of the process. The contemporary approach to making financial decisions is to optimize the benefits to the customer and minimize the costs to the total organization. As organization structures become flatter, team approaches are applied to cost reduction, financial decisions, and profit improvement; more knowledge of the financial process is required by technical, production, and marketing personnel.

These types of decisions do not take place in isolation. The organization is likely considering more than one project or investment. Each investment must be considered in terms of the other project requirements of the organization, the savings and ROI they will earn, and the firm's current profits, available cash, and borrowing capacity.

Improving a Product at Sims Medical Microelectronics Corp. In this example, a product is being redesigned. The manufacturing process is being revised to improve the product's quality and to increase capacity utilization. This is an example of process improvement that results in lower costs.

Kate Bower is the design engineer responsible for the A-43 assembly. She has been working with the process engineer on design improvements that would make the product easier to manufacture and that would also reduce scrap. Kate has been analyzing the field failure reports and meeting with major customers. She has improved the design to reduce the complaints and field failures. The revised design meets customer requirements and manufacturing improvement standards. Currently, 2,000 units are produced each month.

Manufacturing Benefits

Scrap Reduction/Month = 30 units @ $150 per unit
Increased Output/Month = 100 units @ savings of $15 per unit in labor costs

Dollar Value of Redesign to Manufacturing

Scrap Reduction = (30 units/mo)($150/unit)(12 mo/yr) =	$54,000/yr
Output Increase = (100 units/mo)($15/unit)(12 mo/yr) =	+$18,000/yr
	$72,000/yr

Value of Redesign in Reduced Field Failures

Field Failure Reduction/Month = 6 units @ average cost of $250 per unit = $1,500/month
Reduced Installation Time by Customer = 30 minutes per unit @ $10 per hour = $5/unit

Dollar Value of Field Failure Reduction

Failure Reduction = (6 units/mo)($250/unit)(12 mo/yr) =	$18,000/yr

Total Dollar Value Manufacturing and Field Improvements

$72,000 + $18,000 = $90,000

Since the reduced installation time reduces the customer's costs and not Sims', this is not included in the project's benefits. However, the improved quality and response to a customer need helps with marketing. In cases like this, the marketing department sometimes estimates the added customers that can be expected from an improved product. In some cases, the price of the redesigned product is increased. Additional customers, repeat customers, and/or additional profits, if included, would contribute to the financial benefits of the redesign.

Design and Manufacturing Costs of the Redesigned Product

250 hours of Design Time @ $90 per hour =	$22,500
Prototype Testing 135 hours @ $140 per hour =	+$18,900
Revised Equipment Tooling =	+$15,000
Operator Training Costs =	+$ 1,300
Marketing Information Revision =	+$ 2,000
Total Initial Costs for Redesign	$59,700

Increased Supplies for improvements = $1.90 per unit

($1.90/unit)(2,000 + 100 increase units/yr)(12 mo/yr) = $47,880/yr

Revised Materials for improvements = $1.00 per unit

($1.00/unit)(2,000 + 100 increase units/mo)(12 mo/yr) = $25,200

Total Additional Annual Costs	$73,080/yr

Summary of Redesign Costs and Revenues

Savings/benefits from the redesign =	$90,000/yr
Initial cost for redesign =	$59,700
Additional annual cost =	$73,080/yr

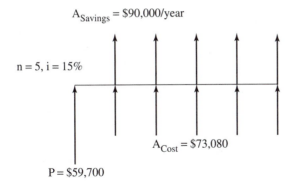

A_Savings = $90,000/year

n = 5, i = 15%

A_Cost = $73,080

P = $59,700

Figure 6–7   Kate's Redesign of Component A-43

Sims has recently earned 12% ROI on similar new projects. The board of directors want a 15% ROI on all projects. Kate has discussed the improvement with the A-43 product marketing manager and he indicated that the product has five to six years of remaining life. The TVM diagram is shown in Figure 6–7.

Using the initial costs of $59,700, $n = 5$, $i = 15\%$, the annual amount of this cost can be calculated:

$$A_P = P(A/P, n = 5, i = 15\%) = \$59,700(.2983) = \$17,809$$

We add this annual amount to the annual cost: $73,080 + $17,809 = $90,889. This total annual cost is greater than the annual benefits of $90,000. The what-if comparison is shown in Table 6–2. It is based on $n = 4$, 5, and 6 years and $i = 12\%$ and 15%.

By calculating other lives and returns, Kate can determine the sensitivity of the annual costs to the changes in n and i. Kate observes that at an ROI of 15%, the board of directors' target, only the 6-year life costs less than the annual savings of $90,000 (the 5-year alternative is very close, however). For the 12% ROI alternatives, the costs of both the 5- and 6-year alternatives are less than the annual benefits of $90,000. If the product life is 4 years or less, Kate observes, neither the 12% nor 15% ROI alternatives are less than the benefits. This kind of what-if information is helpful to Kate, her boss and team, management, and others that look at the project costs and revenues to determine if the project is warranted.

The redesigned product has a reduced installation time. This is a benefit to the customer. Although this cost is not calculated as a benefit to Sims, it could be considered. This improvement would help retain existing customers and perhaps obtain new ones. Ultimately it would likely increase market share and profits for Sims.

Table 6–2   What-If Comparison of Annual Total Costs for Different n and ROI for Design Component A-43

Life of Design = n Years	ROI = 12%	ROI = 15%
4	$92,735	$93,990
5	$89,641	$90,889
6	$87,601	$88,755

Nonfinancial benefits including the shorter installation time and the improved quality are difficult to quantify, but they are part of the decision process. What-if analysis can also show the sensitivity of the potential outcomes to the inputs such as life and ROI. Other questions could be asked, such as, "What if the scrap units do not decrease by the proposed 30 units per month?" or "What if the actual productivity increase from the redesign is only 80 units per month?" These are the types of questions that Kate can expect to be asked when her proposal is reviewed by her boss and the senior managers. She can prepare for these questions in advance with her what-if analysis.

Calculating Product Unit Cost at Sims Medical Microelectronics. Another component made by Sims is the RT-106 assembly. This equipment costs $150,000 with a life of 8 years and an estimated residual salvage value of 10% of the first cost. The material costs are estimated to be $2.60 per unit, and labor time estimated costs are 1.6 hours at an average cost of $10.50 per hour. The general overhead, which includes salaries, taxes, marketing expenses, office costs, and related costs, is calculated at 200% of the direct labor costs. Specific operating expenses for the equipment include power, maintenance, repairs, space costs, and related expenses of $20,000 per year. Marketing estimates that approximately 4,000 units per year can be sold. The project manager is calculating the product's unit cost (see Figure 6–8). An ROI of 15% is desired by Sims.

For the equipment,

$$P = \$150,000, \; S = (.10)(\$150,000) = \$15,000, \; n = 8 \text{ years}, \; i = 15\%$$
$$\text{Material} = \$2.60/\text{unit}; \text{ labor} = (1.6 \text{ hours/unit})(\$10.50/\text{hour}) = \$16.80/\text{unit}$$
$$\text{Overhead} = (2.00)(\$16.80) = \$33.60/\text{unit}; \text{ annual operating costs} = \$20,000/\text{year}$$
$$\text{Output} = 4,000 \text{ units per year; unit cost} = \text{unknown}$$

$$
\begin{aligned}
A_{\text{Total}} &= A_{\text{M+L+OH}} + A_{\text{Oper}} + A_{\text{P}} - A_{\text{S}} \\
&= \$212,000 + \$20,000 + \$150,000(A/P, n = 8, i = 15\%) \\
&\quad - \$15,000(A/F, n = 8, i = 15\%) \\
&= \$212,000 + \$20,000 + \$33,428 - \$1,093 = \$264,335/\text{year}
\end{aligned}
$$

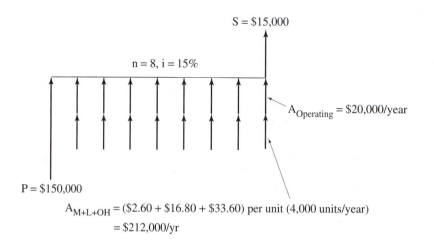

$S = \$15,000$

$n = 8, i = 15\%$

$A_{\text{Operating}} = \$20,000/\text{year}$

$P = \$150,000$

$A_{\text{M+L+OH}} = (\$2.60 + \$16.80 + \$33.60) \text{ per unit } (4,000 \text{ units/year})$
$= \$212,000/\text{yr}$

Figure 6–8   Calculating Unit Costs of the RT-106 Assembly

To find the unit cost, divide A_{Total} by the annual volume of 4,000 units per year:

$$\$264,335 \text{ per year}/4,000 \text{ units per year} = \$66.08 \text{ per unit}$$

This unit cost can be used to determine the selling price to the customer, or it can be used to compare the cost of buying the component from an outside vendor with this cost to manufacture.

Jim Peterson's New Home and What-If Questions. Just as the time value of money techniques are applied to an organization's investments, they can also be applied to an individual's financial decisions. The following example shows the application of TVM analysis to the purchase of a home, one of the largest purchases many families will make. The amounts are usually larger for the corporation than the family, but the calculations are similar.

Jim Peterson of Davis Corp. wants to purchase a house for $120,000 with a $32,000 down payment from the sale of his current home. He plans to have a 30-year mortgage with an interest rate of 10%.

Two solutions to the calculation of the loan repayment schedule are:

1. An approximate solution of the payment amount for quick answers.
2. An exact solution that takes into account the monthly payment schedule.

Jim's loan repayment, exact solution—For the exact solution, the A/P formula must be revised using $m = 12$ months per year. For the calculations, $P_{Loan} = \$120,000 - \$32,000 = \$88,000$, $n = 30$ years, $m = 12$, $i = 10\%$. See Figure 6–9.

The modified formula becomes:

$$\frac{A}{m} = P\left[\frac{\frac{i}{m}\left(1 + \frac{i}{m}\right)^{nm}}{\left(1 + \frac{i}{m}\right)^{nm} - 1}\right]$$

Substituting into the formula,

$$\frac{A}{12} = 88,000\left[\frac{\frac{.10}{12}\left(1 + \frac{.10}{12}\right)^{360}}{\left(1 + \frac{.10}{12}\right)^{360} - 1}\right]$$

Solved, the monthly loan repayment is $772.26 per month.

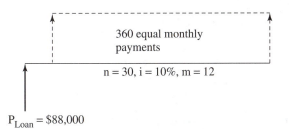

360 equal monthly payments

$n = 30, i = 10\%, m = 12$

$P_{Loan} = \$88,000$

Figure 6–9   Jim's Loan Repayment—Exact Method

Jim's loan repayment, approximate solution—An approximate solution is quick to calculate with the tables or formulas. It gives a close approximation to the exact answer. Using the A/P formula and solving for A, the annual amount, and then dividing the annual amount by 12 months gives the approximate monthly payment.

$$P = \$120,000 - \$32,000 = \$88,000, n = 30 \text{ years}, i = 10\%, A = \text{unknown}$$
$$A = P\ (A/P, n = 30 \text{ yr}, i = 10\%) = \$88,000\ (.1061) = \$9,334.97/\text{yr}$$

Dividing the annual amount by 12 months,

$$A/12 = \$9,334/12 \text{ months} = A/12 = \$777.91 \text{ per month}$$

This method gives an answer that is approximately $6 more than the exact solution value. This approximate method is useful if only annual interest tables are available and a quick answer is needed.

Calculating total interest paid—Jim wants to know how much interest he will have paid to the bank over the life of the loan. The interest paid can be found by calculating the total payments to the bank over 30 years and then subtracting the original amount borrowed.

The total interest paid over the life of the 30-year loan is calculated using the exact monthly payment:

$$\text{Total Interest} = \text{Total Payments} - \text{Original Loan Amount}$$
$$\text{Total Interest} = (12 \text{ mo per yr})(30 \text{ yr})(\$772.26 \text{ per mo}) - \$88,000$$
$$= \$278,031.60 - \$88,000$$
$$\text{Total Interest} = \$190,013.60 \text{ over the 30-year loan period}$$

Jim is surprised to find that the total interest paid is more than twice the amount of the loan! This, of course, assumes that the loan continues for 30 years. Today, most families move before their loan is completely repaid.

Lenders hope that borrowers don't calculate the total interest paid on a loan. They may not borrow, if they do!

A 15-year loan—Jim is considering reducing the time period of the loan to 15 years. He wonders how much more the monthly payments would be, and what the savings in interest would be, compared to the 30-year loan.

Using the same principal and interest, and solving for the exact monthly payment amount, with $n = 15$,

$$\text{Monthly payment} = P\ (A/P, n = 15 \text{ yr}, i = 10\%, m = 12) = \$945.65/\text{mo}$$
$$\text{Total interest paid} = (12 \text{ mo per yr})(15 \text{ yr})(\$945.65) - \$88,000$$
$$= \$82,217 \text{ total interest paid over 15 years}$$

This shorter loan period increases the monthly payment from $772.26 to $945.65, an increase of $173.39 per month. The total interest paid decreases from $190,013 to $82,217, a decrease of $107,796. Jim must decide if he can afford the additional monthly payment in order to reduce the interest paid. Perhaps a compromise of 20 years would be better for Jim.

Table 6–3   What-If Table of Jim's Loan Calculations

Number of Years for Loan	Monthly Payment	Total Interest Paid
15	$945.65/mo	$82,217
20	$849.22/mo	$115,813
30	$772.26/mo	$190,013

If the 20- or 15-year loan is attractive to Jim, and if the monthly payments are too high, there is a possible compromise. He could obtain the 30-year loan with the lower monthly payments, and if his financial position improves, pay *additional* principal on the loan as he makes the monthly payments. This is a flexible solution. It has the advantage of shortening the loan period and reducing the total interest paid without the higher monthly payments. The loan agreement should provide for the option to make the additional payments if Jim desires.

Other loan terms—The above calculations assumed a constant interest rate, but a lower interest rate might be negotiated with the 20- and 15-year time periods. A shorter period of time on the loan may mean a rate of perhaps $\frac{1}{8}$ to $\frac{1}{4}$ percent lower than for the longer loan term.

Other loan types include flexible or variable rate loans, where the interest rate changes over the life of the loan. An advantage of this type of loan is that if market interest rates decline during the life of the loan, the loan rate also decreases. The disadvantage is that if market rates increase, the loan rate and monthly payments also increase.

Paying the loan completely before the 30th year—Most loans can be paid off before the end of the loan period without a penalty payment. The borrower should make sure that this "early payoff" clause is in the loan contract. Jim is thinking that he might want to move from this house before the end of the 30-year period, and he wonders what he would owe on the loan at the end of 10 years.

Using the above 30-year example, if Jim pays off the loan at the end of 10 years, what is the balance due? The initial loan = P = $88,000, monthly payment = $772.26/month, n = 10 years, m = 12. The balance of the loan at the end of the tenth year is unknown.

The TVM diagram in Figure 6–10 shows the initial loan amount of $88,000. To find the balance owed at the end of the tenth year, a two-step calculation is made.

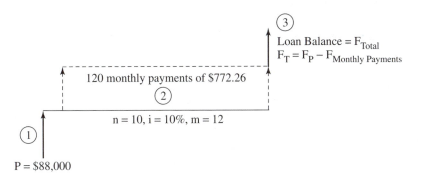

Loan Balance = F_{Total}
$F_T = F_P - F_{Monthly\ Payments}$

120 monthly payments of $772.26

$n = 10$, $i = 10\%$, $m = 12$

$P = \$88,000$

Figure 6–10   Jim's Loan Balance at the End of Ten Years

1. The first step assumes that *no payments* are made on the loan. The resulting amount owed at the end of 10 years is calculated. To calculate this value, $88,000 is taken to 10 years in the future at 10% compounded monthly.

$$P = \$88,000, n = 10 \text{ yr}, m = 12, i = 10\%, F_P = \text{unknown}$$

Using a calculator,

$$F_P = P(F/P, n = 10 \text{ yr}, m = 12, i = 10\%) = \$238,220$$

This amount would be the balance due if no payments had been made.

2. The second step is to calculate the future value of the payments that Jim *actually* made during the 10-year period. This amount is the future amount of $772.26/month for 10 years at 10%.

$$\text{Monthly payment} = \$772.26, n = 10 \text{ yr}, m = 12, i = 10\%, F_{A/12} = \text{unknown}$$

Using a calculator,

$$F_{A/12} = \$772.26 \ (F/A, n = 10 \text{ yr}, m = 12, i = 10\%) = \$158,194$$

This is the future value of the monthly payments at the end of 10 years. To find the balance that is owed by Jim at the end of the tenth year, the future amount that would be owed if no payments were made is reduced by the future amount actually paid.

$$F_{\text{Total}} = F_P - F_{A/12} = \$238,220 - \$158,194 = \$80,026, \text{ the balance of the loan at the end of 10 years}$$

Jim is surprised at this large amount still owed. His total payments to the lender have been $(10 \text{ yr})(12 \text{ mo/yr})(\$772.26/\text{mo}) = \$92,671$. Unfortunately for Jim, most of his payments went toward interest rather than principal. For example, the first year's payments of (12 months)($772.26/mo) were $9,267. The *approximate* (not calculating the monthly compound value) first-year interest of 10% on $88,000 is $(.10)(88,000) = \$8,800$. This means that of the $9,267 paid during the first year, $8,800 was interest and $9,267 - \$8,800 = \$467$ was the approximate reduction of principal for the first year. In the early years of a loan such as this, the largest portion of the payments are applied to interest. Toward the end of the loan, the largest portion of the payments apply to the principal.

At the end of the tenth year, Jim will owe $80,026 in order to pay off the loan. *If* there has been inflation during the 10 years and *if* the value of the home has increased to be more than the original price, it is likely that Jim will be able to sell the home for an amount greater than he paid for it, and have a profit. This is possible, but not guaranteed.

Kate's Individual Retirement Accounts. Kate, a design engineer at Sims, is considering opening an Individual Retirement Account (IRA). IRA deposits are made annually into an approved fund managed by banks, mutual funds, and other financial institutions. Currently the interest and appreciation of assets in the account are not taxed until withdrawals are made during retirement. Because interest income, dividends, and profits from sale of investments in the account are not taxed until withdrawn, they compound faster and can be reinvested into other investments. The annual deposits may or may not have tax advantages, depending on the individual's income. There may be restrictions and penalties on withdrawal before the retirement period begins. Of course, the tax laws change frequently, so a tax professional should be consulted for up-to-date information.

Because of increased lifespan, company retirement plans being reduced, increased costs of health care, inflation, revisions in Social Security, and taxes, it is more important than ever for individuals to prepare for their own retirement. Employer and federal government retirement funds may not be enough to sustain one's standard of living at retirement.

Kate is aware of Social Security benefits and the retirement benefits that Sims Corporation offers, but she thinks she should have additional retirement investments. She knows that Social Security benefits could change by the time she retires. Sims could go out of business or be purchased by another firm, which might change her benefits.

Kate thinks that she can deposit $1,500 per year into the account. She is now 26 years old and plans to retire in 34 years. One mutual fund that she looked at has had an average ROI of 11% over the past few years, and, although there are no guarantees about the future, she thinks this rate of growth will continue. To determine the accumulated amount in her retirement account, she makes the following calculations (see Figure 6–11):

$$A = \$1,500/\text{year}, \ n = 34 \text{ years}, \ i = 11\%, \ F = \text{unknown}$$

Using the F/A relationship and solving,

$$F = A \ (F/A, \ n = 34 \text{ yr}, \ i = 11\%) = \$1,500 \ (306.8374) = \$460,256 \text{ in the account at the end}$$
$$\text{of 34 years}$$

Kate's total deposits would be (34 yr)($1,500/yr) = $51,000. The total earnings in the account are:

$$\text{Total Earnings} = \text{Accumulated Amount} - \text{Deposits}$$
$$= \$460,256 - \$51,000 = \$409,256$$

Kate is surprised at the amount that can be accumulated. This is because compounding occurs without taxes being deducted from the earnings. If the earnings were taxed, the accumulated amount would be significantly less.

> Usually we work for money. Compounding lets money work for us.

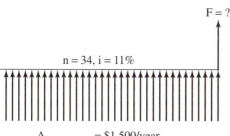

$A_{\text{IRA Deposit}} = \$1,500/\text{year}$

Figure 6–11   Kate's IRA

Changing the ROI—What if Kate is fortunate and the account actually earns an average return of 12% over its life? What would be the effect of the additional 1% on the accumulated amount? The original TVM diagram applies except the ROI is changed. The calculations are shown below:

$$F = A \ (F/A, \ n = 34, \ i = 12\%) = \$1,500(384.5210) = \$576,781$$

The amount is *significantly* greater with the additional 1%. Are these example rates realistic? The Standard and Poor's average of stocks, for example, has increased at a rate of over 13% per year for the period 1980 to 1995. There are more than 3,000 mutual funds, and their ROI performance varies greatly. Since 1920, the Standard and Poor's average has increased by approximately 11% annually. However, there are never any guarantees as to future returns.

Personal Retirement Plans

- Start as soon as possible. Compounding is maximized when *n* is large.
- Select the best investments for your individual situation, considering risk and return.
- Monitor the investment over its life.

Earning $1 million for retirement—Kate has decided that she would like to have a total of $1 million in her retirement account. She thinks she can average an ROI of 12% for 37 years by working three years longer. The what-if calculations for the annual amount are shown in Figure 6–12.

Using the *A/F* relationship, solve for the unknown annual amount.

$$F = \$1,000,000, \ n = 37 \text{ years}, \ i = 12\%, \ A = \text{unknown}$$
$$A = F \ (A/F, \ n = 37 \text{ years}, \ i = 12\%) = \$1,840/\text{year}$$

Is it really possible for an individual to accumulate $1 million for retirement? Everyone's situation is different, of course, but given a large enough *n* and a reasonable ROI, and the current tax laws, a large amount can be accumulated. It is important to start early. Waiting until re-

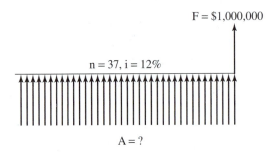

Figure 6–12   Kate's Million-Dollar IRA

tirement is a few years away does not permit the exponential function to compound long enough.

> A large n is the exponent that makes an IRA successful.

Other Types of Retirement Accounts. There are other alternatives besides IRAs that individuals can consider for retirement accounts. One popular account is called a 401(k) or 403(b) for not-for-profit organizations. These accounts allow the employer to withhold an amount from the employee's pay before tax is calculated. The withheld amount is sent to a financial institution outside the organization and under the employee's management. If the employee leaves the organization, the account stays with him or her. A company-managed retirement fund is usually not transportable if the employee leaves (there are some exceptions to this).

There are other types of retirement options for self-employed individuals. A tax advisor can give helpful advice.

> Q. If someone does not have a retirement plan, when is the best time to start one?
> A. Today!

Kate's Investment in Sims Common Stock. Kate has been purchasing Sims common stock, using the employee stock purchase plan, since she began working there. The purchases are made by deductions from her paycheck. She first purchased Sims stock at $55 per share three years ago. She has received dividends of $1.80 per share each year of her ownership. The value of the security is currently $71 per share. Although she doesn't plan on selling the securities, she wants to calculate the return on her investment. (See Figure 6–13.)

There are two primary ways to gain from common stock, preferred stock, and bond investments. There may be a gain from *dividends or interest,* or gain from the *price appreciation.*

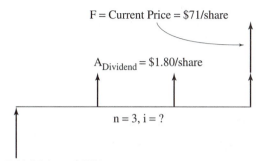

F = Current Price = $71/share

$A_{Dividend}$ = $1.80/share

n = 3, i = ?

P = Initial Price = $55/share

Figure 6–13   ROI on Kate's Investment in Sims Stock

The dividend can be assumed to continue into the infinite future, and the return can be calculated as a perpetuity.

$$A_{Div} = (P)(i_{Div})$$
$$i_{Div} = A_{Div}/P = \$1.80/\$55 = 3.27\%$$

The ROI due to the price appreciation is calculated using the F/P relationship and solving for an unknown i.

$$F = P(1 + i)^n$$

$$\frac{F}{P} = (1 + i)^n$$

$$\left[\frac{F}{P}\right]^{1/n} - 1 = i$$

Substituting into the last equation,

$$\left[\frac{\$71}{\$55}\right]^{.333} - 1 = i = 8.88\%$$

The total ROI can be found as,

$$ROI_{Total} = ROI_{Dividend} + ROI_{Price} = 3.27\% + 8.88\% = 12.15\%$$

The above calculation can be set up on a spreadsheet for monitoring investments. ROI is one of many measures of the performance of investments. The advantage of using TVM calculations compared to other monitoring techniques is that time is considered. Some techniques look only at the profit or percentage increase or decrease in the security's value. It makes a difference, as we know, in terms of ROI if an investment increases from $10 a share to $15 a share in one year or five years.

SUMMARY

This chapter has covered applications of time value of money and equivalence for examples of financial decisions for both organizations and individuals.

- Using the seven-step approach to TVM problems, including the cost of purchases, leases, and bank loans.
- Selection of a value for ROI.
- What-if, risk, and sensitivity of investment outcomes based on the investment inputs and variables.
- Calculation of A_{Total} using interest calculations, and time value on money analysis.
- Techniques of estimating equipment salvage values.
- Comparison of equipment annual costs based on purchase using internal funds, leasing, and bank loan.
- The justification of new equipment based on savings due to increased productivity, reduced down-time, and improved quality using time value of money analysis.

- Unit cost calculation of products based on the time value of money.
- Using what-if determination to assist with investment analysis and decisions.
- The application of equivalence and time value of money to an individual's home purchase loans, including what-if analysis of loan duration, total interest paid, and calculation of loan balance.
- Calculation of retirement accounts, and security returns for individuals.

QUESTIONS

1. What are the seven steps to solving time value of money problems?
2. State three approaches to establishing a target ROI percentage for an organization.
3. Why is it important to do what-if calculations when doing financial analysis?
4. When doing what-if calculations, what is meant by sensitivity?
5. If the risk of a project or investment is high, why is the desired return usually also high?
6. What is risk?
7. What are some examples of low-risk and low-return investments? What are some examples of high-risk and high-return projects?
8. What are some ways to estimate residual salvage value of equipment?
9. How is salvage value incorporated into time value of money calculations?
10. Why can equipment salvage value sometimes be neglected when calculating annual costs?
11. Can the removal costs of an asset be higher than the residual value?
12. How is annual cost calculated using the depreciation method? Why is this not a valid technique for determining equipment's annual costs?
13. What is the formula for calculating the annual cost given the first, salvage, and operating costs?
14. How does the concept of equivalence apply to calculating annual equipment costs?

15. What is an alternative formula for calculating annual costs?
16. What are the advantages and disadvantages of leasing equipment compared to purchasing?
17. How is equivalence used to compare lease costs to owning costs?
18. How can the formula for calculating annual amount from present value be modified to include monthly rather than annual amounts?
19. Why is the calculation of a product's unit cost based on time value of money rather than depreciation?
20. What is the difference between a loan that is repaid in regular payments over the loan's life compared to an interest-only loan?
21. Why would an organization or individual prefer an interest-only loan over other types of loans?
22. Is there a significant difference between the total interest paid on a 30-year compared to a 15-year home mortgage loan? Why?
23. Why does the balance owed on a home loan not decline very much in the early years of the loan?
24. What is an IRA? What is a 401(k) plan? What are the tax advantages?
25. Why should an individual consider other retirement plans in addition to Social Security and company-sponsored retirement programs?
26. How can time value of money be applied to determining the ROI of a security?

PROBLEMS

1. Some equipment costs $230,000. The salvage value is estimated as $15,000; the equipment's life is 9 years; and the desired ROI is 15%. Determine the annual cost of the equipment.
2. Calculate the annual depreciation for the equipment in problem 1. Why is the difference between the depreciation amount and the annual cost important? Use the straight-line method.
3. Some equipment costs $60,000 with salvage value estimated as 5% of its first cost. Operating expenses are $4,500 per year; desired ROI is 10%; and the life of the equipment is 7 years. Calculate the an-

nual costs based on equivalence and time value of money.

4. New furniture for the reception lobby for Figens Co. is being purchased: five couches at $1,200 each, 14 chairs at $500 each, three tables at $400 each, and a receptionist's desk for $2,000. The furniture is expected to last 8 years and have a residual salvage value of 10% of its original cost at that time. Calculate the following:
   a. The annual cost using straight-line depreciation.
   b. The annual cost using an ROI of 12%.

5. An individual is considering the purchase of a new lawn mower. The cost of the mower is $350, the estimated residual value after 5 years is $25, and maintenance, fuel, and related costs are expected to be $35 per year. Determine the following:
   a. What is the annual cost based on depreciation?
   b. If the owner is able to earn 10% on investments, calculate the annual cost of the mower using time value of money. Draw the TVM diagram.
   c. Compare the difference between the answers in parts a and b.
   d. What is the effect on annual costs if the residual value is $5?
   e. What is the effect on annual costs if the buyer uses a credit card to purchase the mower and the interest on the credit card is 15%?
   f. What is the effect on annual costs if the mower is purchased on sale at a 20% lower price?
   g. Considering the above answers, on what factors is the annual cost most sensitive?

6. Canbert Corp. is purchasing some new material handling equipment for their warehouse. The initial cost is expected to be $100,000 with a salvage value of $10,000 after 6 years. Annual operating costs including maintenance, power, taxes, labor, and related costs are estimated to be $13,000 per year. The firm typically targets 12% for all of its investments. Determine the following:
   a. Draw the TVM diagram.
   b. Calculate the annual cost using time value of money.
   c. Calculate the annual cost based on the depreciation.
   d. Compare the answers of parts b and c.
   e. What is the effect on the annual cost if the salvage is only $1,000?
   f. What is the effect on annual cost if the equipment only lasts 4 years?

g. What is the effect on annual cost if the target ROI increases to 20%?
   h. Considering the above answers, on which factors is the annual cost most sensitive?

7. A small family-owned flower shop is considering purchasing a new delivery van. The price of the van is $14,000. The residual value is estimated to be $2,000 at the end of 5 years. Operating costs of fuel, repairs, taxes, insurance, and related costs are estimated to be $2,500 per year. The business has a goal of 10% return on all of its investments. Calculate the following:
   a. Draw the TVM diagram, and calculate the annual cost of the van using the ROI of 10%.
   b. What is the annual cost if the van is kept for 6 years and the value at the end of that time is $100?
   c. Some additional options could be added to the van that would cost $1,800. Calculate the revised annual cost based on the life of 5 years and salvage of $2,000.
   d. What if the van lasts only 4 years? What if the operating costs are $3,000?

8. Software Inc. is considering the purchase of new computer equipment totaling $150,000 with an estimated salvage value of $25,000 after 3 years. Maintenance, repairs, supplies, and other operating costs are estimated to be $13,000 per year. Determine the following:
   a. The annual cost based on the depreciation.
   b. The annual cost using a desired ROI of 15%.
   c. If the funds to purchase the equipment are borrowed at a rate of 8% for 3 years, what is the effect on the annual costs?
   d. What is the annual cost if the equipment is purchased from internal funds, and it is sold for $40,000 after 2 years?

9. New equipment is purchased for $40,000. It has a residual salvage value at the end of 10 years of $2,500. The desired ROI is 12%. Operating expenses are expected to be $2,000 the first year and increase by $500 each year during the life of the equipment. Draw the TVM diagram. Calculate the equivalent annual operating costs and the total annual costs.

10. Many Channels Inc., a cable TV firm, is considering purchasing three new installation/repair trucks. The trucks would cost $32,000 each. The estimated life is 4 years with a residual salvage value of

$5,000 each. The annual operating expenses of taxes, repairs, fuel, insurance, and related costs are estimated to be $8,000 per year for each truck. The firm targets 15% ROI on its investments. As an alternative, Many Channels has received a quotation from Cheap Lease Corp. for leasing the trucks for a total cost of $35,000 per year for all three trucks. Many Channels would be responsible for the same operating costs. Determine the best alternative.

11. Mary Tellas, the production manager at Rogers Corp., is considering a new computer numerical control machine for a cost of $250,000, with annual operating expenses of $20,000 per year. The equipment is expected to last 7 years and have a residual value of $20,000 at that time. The firm requires a 12% return on its investments. A lease is also being considered. The lease would require a down payment of $20,000 and monthly payments of $4,300 for the 7-year period. Compare the purchase and lease alternatives on an annual cost basis.

12. Canbert Corp. is considering a bank loan in order to purchase $100,000 of new materials handling equipment. The loan rate would be 8.5% for 6 years. Determine the repayment amount if:
    a. Annual end of year repayments of principal and interest are made.
    b. Monthly repayments of principal and interest are made.
    c. Determine the total interest paid over the life of the loan in part a.

13. Dugan Co. is considering a new piece of equipment that costs $20,000. It is estimated to have a life of 10 years with no salvage value at that time. General operating expenses are expected to be $2,000 per year, and repairs to the equipment are expected to be $300 the first year and increase by $50 per year over the life of the equipment. An ROI of 12% is desired. Calculate the annual total costs of the equipment.

14. Calculate the unit cost of a product if the material cost is $3 per unit, labor is $5 per unit, and overhead is calculated at a rate of 150% of labor. The first cost of the equipment needed for production is $50,000, with a life of 12 years, and a salvage value of $4000. Five thousand units per year are required. ROI is 15%. Annual operating costs of $1,000 per year are also required. What is the unit cost?

15. Software Inc. is purchasing new computer equipment that costs $150,000. Software is a young company with variable sales and profits. They are considering an interest-only loan that would have a 10% annual rate. At the annual anniversary date, Software could pay off the loan, pay part of the principal and interest, or pay only the past year's interest. The only requirements are that the interest be paid annually and the loan be finalized within 5 years. The first year they pay interest only; the second year they pay interest and $75,000 toward the principal. At the end of the third year, they pay the interest and the remaining principal. What is the total interest paid?

16. Jones and Jones, a brother and sister partnership, have a commercial cleaning service with 35 employees. They are purchasing new vacuum cleaners, floor waxing equipment, tools, and related equipment for $45,000. They estimate the equipment will last 4 years and have no salvage value at that time. They estimate repairs and maintenance of the equipment will be $800 per year. Additional supplies will be $300 per year. Calculate the annual cost of the equipment using an ROI of 15%.

17. Equipment costing $80,000, having a salvage value of $5,000, a life of 5 years, and operating costs of $20,000 per year is expected to produce 5,000 units of product each year. The material and labor costs are estimated to be $6.73 per unit. Additional general overhead is estimated to be $5,000 per year. If a desired ROI is 15%, determine the total annual cost of the equipment and the unit cost of the product. If the life of the equipment is 7 years, what is the product unit cost?

18. Mark Patrick, a process engineer, is considering whether to purchase new equipment. He estimates that the material for the process would cost $14.70 per unit, with a labor cost of $11.00 per hour, and the product would require 1.15 hours per unit. Equipment would cost $56,500 and have a salvage value of 10% of the first cost after 10 years and operating costs of $4,000 per year for maintenance and repairs, space costs, and power costs. General overhead would be estimated at 185% of the direct labor costs. The firm desires an ROI of 12%, but Mark wants to base his calculations on 10% and 15% as well as the desired 12%. He would also like to consider the case if the equipment only lasts 8 years rather than 10 years. Calculate the unit costs of the product based on the given information, and set up a what-if table for the unit costs for the different ROI

rates and the two estimated lives of the equipment. Quantity required is 10,000 units per year.

19. Henson Co. is considering buying a new delivery truck. The current truck has recently had a large number of repairs, maintenance costs, and downtime. When the current truck is down for repairs, either customer deliveries are postponed or sometimes another truck is rented for a few days. Julie is the delivery schedule manager in the sales department and is trying to analyze the cost of purchasing a new truck. She estimates that the savings, if the new truck is purchased, would be 19 days per year at $75/day, and the costs of repairs would be reduced by $4,100 per year. The benefits to customers would be faster and more reliable deliveries. This has some monetary value to the customers, and should help Henson retain and perhaps increase their customer base. The truck that Julie is considering costs $23,000 and would have a salvage value of $2,000 after 5 years. Julie has received recent memos from upper management concerning an ROI target of 10% for investment projects. She knows of some projects that have been accepted at an ROI of 8%. Also, she thinks the truck might last only 4 years, or perhaps as long as 6 years. Calculate the following:

a. Benefits per year from purchasing a new truck.
b. The annual cost of purchasing the new truck, and a what-if table for $n = 4$, 5, and 6 years versus ROI = 8% and 10%.
c. How should the benefits to the customer be evaluated?

20. A product is being redesigned along with the manufacturing process in order to reduce field failures and to reduce the process scrap. The unit costs $4 per unit and the annual output rate is 3,500 per year. The current field failures are costing $450 per month and the revised design and process would reduce the field failure cost to $200 per month. The redesign costs are anticipated to be $500 and the improved process would cost approximately $1,500 for tooling, operator training, and revised software. The desired ROI is 12% and the product is anticipated to be used 4 more years. Determine if the investments in the improved design and process are warranted. What if the product is produced only 3 more years?

21. Mary Tellas, the production manager at Rogers Corp., is purchasing a new home. She plans on using the money she receives from the sale of her existing home, $23,000, as a down payment in the new home. The new home costs $105,000. The mortgage company has quoted a rate of 10% for a 30-year mortgage, and 9.5% for a 15-year loan. Determine the following:

a. The monthly payment for the 30-year loan using the approximate method.
b. The monthly payment for the 30-year loan using the exact method.
c. The monthly payment using the exact method for the 15-year loan.
d. The total interest paid for the 30-year loan.
e. The total interest paid for the 15-year loan.

22. Nancy Brock is buying her first home. It costs $85,000, and she will make a down payment of $15,000. She is planning on a 30-year loan at 8% interest.

a. What is the exact monthly payment she will pay to the mortgage company?
b. If Nancy is transferred to a new job in another city after owning the home for 6 years, what is the balance due the loan company when she sells the house?

23. Software Inc. is considering a bank loan for purchasing new computer equipment costing $150,000. The loan would be at a rate of 8% for 3 years.

a. What is the quarterly repayment amount of principal and interest?
b. What is the total interest paid over the 3 years?
c. What is the balance owed by Software Inc. if, after 2 years of payments, they pay off the remainder of the loan? What is the payoff amount?

24. A borrower can afford to pay $350 per month for a car. If the current car loan rate is 10% for a 4-year loan, what is the maximum price of the car?

25. An IRA can be established with High Growth Mutual Fund. The deposits of $1,800 per year would be made into the fund for 23 years. The anticipated return is 10%. What will be the accumulated amount? What if 11% can be earned?

26. John wants to have $500,000 when he retires. He plans to retire at age 64 and his current age is 28. He thinks he can earn 10%. What is the annual payment that he must make?

27. Maria has decided to have $1,200 each year deducted from her pay and placed into her firm's 401(k) plan. If this continues until she retires in 21 years, and the fund earns 10%, what is the accumu-

lated amount in the account? What is the total amount deposited by Maria?

28. An individual begins depositing $400 per year for the first year and increases the amount by $50 each year for a 30-year period. The account earns 12% ROI. What is the accumulated amount in the retirement account?

29. An individual puts $4,000 into a mutual fund. The value increases to $7,100 in 7 years. What is the ROI of the account?

30. A security is purchased for $12 per share. It pays $1.00 per year in dividends for 4 years, at which time the security price is $17.50 per share. What is the ROI of the dividend, the ROI of the price appreciation, and the total ROI?

31. A security increases from $49 to $68 per share in 8 years. During that time it paid a dividend of $1.50 per share. What is the total ROI?

DISCUSSION CASES

Beaker Tool and Die Co.

George Beaker has a medium-size tool and die shop. He does work for some large firms in the automotive and construction equipment industries. His firm is well known for its quality and speed of delivering custom dies. He has 65 employees that work two shifts five days a week with third shift and weekend overtime as the orders require. His production facilities are state of the art. He prides himself on having the latest equipment that provides the best quality available and the most efficient output. He has approximately 40 tools that range in value from $25,000 to over $300,000. George spends a significant portion of his time looking for and analyzing new equipment as well as updating current equipment. He sells his older equipment at machine tool auctions or through trade publications.

George also works closely with the sales people to price and quote the jobs accurately so that they can stay competitive in the market and make money. The accountant gives the sales people and George the cost information. Estimating the material and labor costs is not difficult. George feels that the difficult part of calculating the costs is determining the cost of the equipment. Currently, the cost accountant has historical records of the expenses other than material and labor. These other expenses, overhead costs, are totaled and then divided by the total hours the plant is open to arrive at an overhead cost on a per hour basis. The overhead costs include power, depreciation of the building, sales commissions, office expenses, equipment depreciation, taxes, quality costs, maintenance, and all the other nonmaterial and nonlabor costs.

The sales people estimate the material, the labor hours required, and the labor cost, and then estimate the total machine hours required to do the job and multiply the machine hours times the machine hourly overhead costs. The estimate is given to George and he checks the figures and adds on a profit amount based on what he thinks the market will bear. The formal quotation is then sent to the customer. The company uses straight-line depreciation for its equipment accounting. George uses this annual depreciation amount as the equipment's cost when he determines the prices for customers. He is aware that time value of money can be used, but he isn't sure if that applies to his business or not.

George wants to make sure that the prices are actually reflecting the true costs of the production of the work. He wonders that maybe some of the more specialized and costly equipment is not covered in some of the quotes in the overhead allocation method that they are currently using. Conversely, he thinks that the current method of overhead costs for the older, less costly equipment may be overstated with the overhead method. He would like a better method, but he fears that another method would be too complex, requiring computers and difficult calculations. He is thinking that a two-level system of overhead costs would be better. One level of overhead costs would include the office expenses, taxes, sales commissions, and other general costs. The second level of overhead costs would include the specific equipment costs for the particular job being quoted. The cost accountant thinks this is too much extra effort and would not improve the estimates. George is not sure what to do.

1. Is the current system acceptable? What problems might the current system be causing?
2. Would the two levels of overhead expenses be too difficult to administer? Would it improve the cost estimates?
3. How could the concepts of time value of money apply to Beaker's problems?
4. How could a two-level system of overhead costs be implemented?

Sally and Hal Winters' Retirement

The Winters are both employed. Sally works for a large Fortune 500 company as a sales representative. Hal is a quality control manager at a medium-sized firm. Sally is 29 and Hal is 32. They have one child. They often discuss the family finances. It seems that even with their two incomes, money is a problem. Sally feels that they should consider the various retirement plans that are available. Hal says that they are both employed at firms that have excellent benefit plans, including medical and retirement plans. And, he says, "We don't have any extra money to put into a retirement plan. The mortgage, living expenses, and college fund seem to take all our income."

Hal feels that the small amount they could put into a retirement account would not be worth the effort. With Social Security and their company plans, he feels that they are adequately covered. Sally's company has a 401(k) plan and she wonders if she should have some money deducted from her check for that account. The bank they use for checking accounts has IRA accounts that are advertised in the monthly statement. They both agree that it is difficult to save any money. Their credit card payments are $200 per month. The college fund is $50 a month and they aren't always able to make regular deposits to that account. They aren't sure what to do.

1. What should Hal and Sally do about retirement accounts?
2. Do you agree with Hal that they have good coverage?
3. Is the bank a good source for information on retirement accounts?
4. What else should they consider?

Unit Costs at Matrix Corp.

Linda Henderson, a project engineer for Matrix Corp., is responsible for a new product. She has been working with the designers, process engineers, manufacturing supervisors, quality technicians, and marketing personnel who are involved with the new product. Equipment required for the process is estimated to cost $456,000. Suppliers and others estimate the life of the equipment between 8 and 10 years with a salvage value of between 5% and 10% of the first cost of the equipment. Annual operating costs of the equipment are expected to be $5,000 for the first year and increase by $1,000 per year over the life of the equipment.

The firm uses straight-line depreciation for all of their equipment. The marketing department has indicated that 25,000 units will be sold annually. Material costs are $5.00 per unit, labor costs are $8.00 per hour, and each unit is expected to take .75 hours per unit. General overhead is calculated as 110% of the labor costs.

Upper management has stated that they expect all projects and investments to have a return of at least 12%. Some recent projects have been producing an ROI of as low as 5% and some as high as 15%. Linda isn't sure about the ROI for this project. She knows that the initial advertising is being prepared and that the sales department needs to be able to state the price of the new product to potential customers.

1. What should Linda do to assist the sales department to obtain a price? Can the seven steps be used to help the sales department?

2. Should she decide on the ROI for the project, or should she prepare alternate costs at different ROI values, present this information to upper management, and let them decide?

3. How can she present the calculations in a useful form?

4. What should be her response when she is asked why she did not use depreciation as an annual cost?

5. What other things should Linda consider?

In-Line Skate Corp. (A)

A rapidly growing, small manufacturer of in-line roller skates is attempting to solve some marketing and production problems. The sales are growing so rapidly that no one has any time to try to improve the product or process. Hans Ford is an engineering technology student working at the company on an internship, and he has been given the task of gathering some information and doing some initial calculations for the most popular product line, the Go-For-It Skate. This product is very popular with younger skaters and skateboarders who are changing to in-line skating. The company has a good reputation with this group of customers. A number of winners of national skating contests have won using In-Line's Go-For-It (GFI) model.

There are some complaints from dealers about products that are being returned by customers as defective. The company has an on-the-spot replacement policy; there are no long delays for warranty returns to the factory. The GFI model costs the buyer $450 at a retail store. Dealers make a 50% mark-up in price from their cost of $225. This good margin for retailers permits them to put the skates on sale and still make a profit. The skates have been in such demand and short supply, however, that reduced-price sales are infrequent. The manufacturing cost to In-Line of the GFI model is $155 per unit. A defective product is returned to the factory and good parts of the defective unit are recycled. Usually this amounts to about $35 worth of material and labor per defective pair.

Sales have been running at 2,500 units per month. They could be greater, but the production area is at maximum capacity. Customer returns are approximately 4% of the delivered product. Hans has determined that the primary cause of the product field failures is the wheel bearing and mounting assembly. He has met with product designers, process engineers, and quality personnel and has collected some information. The existing equipment can be revised at a cost of $150,000 to improve the bearing and mounting assembly. The component parts and the process are expected to be used for three more years with no salvage at that time. Additional material and labor costs would be $.90 per unit. Additional operating costs would be $1,000 per month. The quality department estimates that if the improvements are implemented, the field failures would drop to 1% or 2% of output.

Hans is expected to present a written report at a management meeting next week. He isn't sure how to proceed. He has talked with the accounting personnel and understands that they use straight-line depreciation. Hans is not able to find any specific advice about a target ROI. All that he has been told is that the company "wants to earn a good return on all of its investments."

1. What calculations should Hans make in preparation for the meeting?

2. Since the company has no specific ROI goal, should he just use depreciation as the annual cost of the equipment?

3. What are some noneconomic benefits of the redesign of the product and process?

4. Can the seven steps help Hans with his calculations and presentation?

The Millionaires Club (D)

A group of students have formed a club to learn about investment in securities and setting up retirement accounts. They have no money, but have been investing "on paper" in securities and mutual funds. Each does some research on a company: reading and analyzing the annual reports, searching for recent articles about the company in the financial press, obtaining company catalogs, talking with employees and management, and talking to customers. After the analysis and discussion, the group makes a decision "on paper." The group would like to use time value of money to determine how well they have done so far, and perhaps use TVM as a way to eliminate some poor investments. They are not sure how to proceed.

Current Portfolio of the Millionaires Club

	Buy Price	Current Price	Time Held
100 shares Henderson	$40⅜/sh	$44¼/sh	10 months
1000 shares Bomax	$5/sh	$5.50/sh	18 months
Money market fund	$1,200	$1,200	12 months
Hi Tech mutual fund	$6,500	$7,200	16 months

1. Calculate the ROI of the individual investments and the overall portfolio ROI.

2. Is ROI a valid measurement of the investment performance?

3. What other information should the group consider?

4. Should the group try to diversify the portfolio more?

In-Line Skate Corp. (B)

The company's sales are growing rapidly. The company is looking for additional storage space for raw, in-process, and finished goods, and some additional office space. The company has been considering leasing some additional warehouse space that could also serve as an office area. A 5-year lease would be $8,000 per month. Initial remodeling costs would be $25,000.

The firm is also looking at buying an older warehouse nearby. The down payment would be $50,000; the mortgage would be for $550,000 for 15 years at an interest rate of 10%. The company has the cash to purchase the warehouse outright, but this could reduce the speed at which they are able to expand into other products as well as slow the growth of current product expansion.

The additional space would improve efficiency a great deal. The owners are not sure if they want to purchase since the loan is for 15 years. The lease seems expensive, but it is only for 5 years. They are not sure what to do. (Refer to In-Line Skate Corp. Case A.)

1. What are the relative costs of the alternatives?

2. What noneconomic factors should be considered?

3. What is the best recommendation to the owners you would make?

4. If the firm wants to earn 15% ROI on its investments, how would this affect the decision?

5. What other factors should be considered?

Chapter 7 Analyzing, Selecting, Monitoring, and Evaluating Projects and Investments

KEY TERMS

Annual Amount Method Converting all project and investment costs and revenues to annual basis for comparison with other annual costs.

Benefit/Cost Analysis The summation of all revenues divided by the summation of all costs to determine a ratio that is compared to other projects' B/C ratios. This analysis may or may not include TVM.

Discounted Cash Flow (DCF) Similar to the present worth method. The comparison of present worth of costs and revenues.

External Rate of Return Desired rate, target rate of return, or alternative rate available at financial institutions, used to analyze projects and investments.

Four Basic TVM Methods Present worth, annual amount, future worth, and internal rate of return.

Future Worth (Future Value) Method Converting all costs and revenues into a future worth for comparison with other future worth.

Internal Rate of Return (IRR) The interest rate at which the revenues and costs of an investment or project are equivalent.

Payback The ratio of a project's initial cost to its annual benefits. It may or may not include TVM.

Present Worth (Present Value) Method Summation of project's or investment's present worth of future costs and present worth of revenues for comparison with other present worth amounts.

Project Management Components Analysis, selection, monitoring, and performance measurement of the project's schedule, costs and revenues, quality, and related factors.

Ranking Techniques Selecting projects that have the best IRR, present worth, annual amount, life, quality contribution, schedule, initial costs, and related factors.

Sources of Funds The three major sources of funds for the corporation are profits, borrowing, and sale of securities. Individuals can borrow and use profits as sources.

LEARNING CONCEPTS

- Application of four basic TVM techniques of analysis, selection, monitoring, and evaluation to projects and investments.

- IRR and other factors as project ranking tools.
- Calculating the required project life to meet a required IRR.
- Sensitivity and what-if analysis in project selection.
- Other popular analysis techniques such as discounted cash flow, payback, benefit/cost analysis, MAPI, and life cycle analysis.
- Applying project analysis, selection, and monitoring concepts to personal investments.
- Using the four basic techniques as a performance measurement tool.

INTRODUCTION

> **Components of Project Management**
>
> Analysis of Project
> Selection of Project
> Monitoring of Project
> Performance Measurement of Project

The management of projects and investments has four primary components: analysis, selection, monitoring, and performance measurement.

The analysis of investments and projects is done using one or more of the four basic time value of money methods: present worth*, annual amount, internal rate of return, or future value* calculations.

Selection is limited to the funds available and the economic desirability of the projects. What-if analysis can be applied to the alternatives. Both quantitative and qualitative factors can be considered. Ranking of investments using ROI, initial cost, risk, and other factors can help in selecting projects.

Other popular analysis and selection techniques include benefit/cost analysis, payback, discounted cash flow, life cycle costs, MAPI, and proprietary techniques that have been developed by many organizations. When applying these techniques, remember to use TVM too.

Monitoring projects and investments can be done using the same basic techniques used for analyzing and selecting. Evaluations can be made using time value of money techniques so that future projects can be selected more successfully.

This chapter concludes with a discussion of performance evaluation of investment and projects using the four basic techniques.

What-If and Sensitivity Analysis

The method of asking what-if questions and extending the analysis to include other possible situations was introduced in the previous chapter. In practice, the analysis and selection of

---

*The terms *worth* and *value* are used interchangeably in this book.

alternatives would normally include what-if and sensitivity analysis in order to arrive at a complete solution. Because of space limitations, the examples in this chapter do not include what-if analysis as complete and thorough as might be applied in practice.

THE FOUR ANALYTICAL METHODS

Four general techniques are used to compare alternatives. These techniques are annual amount, present worth, future worth, and internal rate of return.

Four Techniques of Analysis

Present Worth
Annual Amount
Future Worth
ROI or Internal Rate of Return

All four methods are based on the principle that different alternatives have costs and revenues occurring at different points in time. It is not possible to compare alternatives directly unless TVM techniques and equivalence are used, since dollars at different points in time cannot be compared. To make a comparison without applying TVM would be like comparing "apples to oranges." With time value of money techniques, dollars at different points in time can be made equivalent and comparisons can be made.

Annual Amount Method

This method, presented in the previous chapter, is often the preferred technique. It requires that all of the costs and revenues be brought to an **annual amount** using time value of money, and then comparing alternatives on an annual basis. The lowest value for the annual amount is the most economical alternative. This method can handle alternatives with different lives, one of the reasons that makes this a preferred method. This technique, as do the others, uses the organization's or individual's target or desired ROI to move the dollars to an annual amount for each alternative.

Present Worth Method

In the **present worth** method, all costs and revenue dollars of the project are brought to the present for comparison. Using a desired or target ROI for the calculations, the present worth of each alternative can be compared and the projects with the largest difference between the present worth of the revenues and the present worth of the costs are selected. Care must be taken to use this method *only if the lives of the alternatives are equal.* If the lives of the alternatives are different, this technique may give an incorrect solution. This technique is also known as the *discounted cash flow (DCF).*

Future Value Method

In the future value method we sum the future value of the project's costs and compare them with the sum of the future value of the revenues. Like the present worth method, the **future value** method is not valid for comparing alternatives with different lives. If the service lives are equal, the costs and revenues of the alternatives are taken to a future sum and then they can be compared.

Return on Investment or Internal Rate of Return Method

This method considers all revenues and costs of the project or investment and determines the internal ROI or **internal rate of return (IRR)** of the investment. Rather than using a desired or target ROI in the calculations, this method requires solving for an *unknown interest rate*.

Example of Lease or Buy Decision: Jennings Co. As an example of the four analysis techniques, consider a firm that wants to purchase some manufacturing equipment. The investment is analyzed using the four basic methods. The goal is not just to determine the present worth, annual amount, future worth, or IRR, but also to compare the costs and benefits of the leasing vs. purchasing. There are two steps: first, to perform the analysis; second, to make the selection. The analysis step is mostly mathematical calculations. The second step includes both calculations and nonquantitative judgments.

Normally, only one of the methods would be used to make this decision. We will apply all the methods to demonstrate the differences in the techniques and to show that all the techniques indicate that the same alternative is the best.

Carol Taylor, an industrial engineer technician with Jennings Co., is analyzing new equipment and trying to determine if the equipment should be purchased or leased. Equipment can be purchased for an initial cost of $50,000. It has a life of 6 years and salvage value of 10% of its first cost. The desired ROI for investments at Jennings is 15%. Or the equipment can be leased for $13,200 per year. If operating costs are the same under both alternatives, determine the best alternative. The TVM diagram is shown in Figure 7–1.

Equivalence and TVM techniques can be used to move all of the dollars to the same point in time so they can be compared. Carol would normally use only one of the four methods to make her analysis. In this case, as a demonstration, all four methods are used.

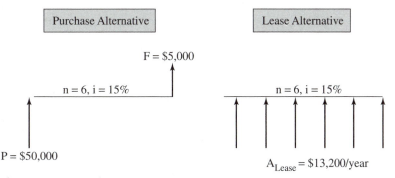

Figure 7–1   Jennings Co.

Present worth method—The following box shows the steps of the present worth method:

Steps of Present Worth Method

1. Use only for alternatives that have identical service lives.
2. Select desired ROI.
3. Using *P/F*, *P/A*, or other TVM relationship, bring all revenues and costs to present worth.
4. Compare the present worth of the alternatives.
5. Select the lowest net present worth alternative.

Using this approach, all the costs of each alternative are brought to t_0 using a target ROI of 15% and then the costs are compared at that point in time. The first cost of $50,000 is already at the present time. The residual salvage value is brought to the present and subtracted from the present worth of the first cost to obtain the net present worth.

$$P_{buy} = P_{50,000} - P_{5,000}$$
$$P_{buy} = \$50,000 - \$5,000(P/F, n = 6 \text{ years}, i = 15\%) = \$50,000 - \$5,000(.4323)$$
$$P_{buy} = \$50,000 - \$2,162 = \$47,838$$

The lease is on an annual basis and can be converted to a present worth using the TVM functions:

$$P_{lease} = P_{13,200} = \$13,200(P/A, n = 6 \text{ years}, i = 15\%) = \$13,200(3.7845) = \$49,955$$

When we compare the present worth of the two alternatives, we see that the buy alternative has the smallest present worth, and this would indicate that the buy alternative is preferred at the given ROI. This conclusion is only valid if the two alternatives are for the same service life.

Present worth analysis for alternatives is only applicable when the alternatives have the same service lives.

Annual amount method—The following box shows the steps of the annual amount method:

Steps of Annual Amount Method

1. Select desired ROI.
2. Using *A/P* and *A/F* calculations, bring all revenues and costs to annual amount.
3. Determine the net annual cost by subtracting costs from revenues.
4. Select the alternative with the lowest annual cost.

We use the technique of bringing the first cost to an annual amount and the salvage to an annual amount, then subtracting the annual salvage from the annual value of the first cost.

$$A_{buy} = A_P - A_S$$
$$= P(A/P, n = 6 \text{ years}, i = 15\%) - S(A/F, n = 6 \text{ years}, i = 15\%)$$
$$= \$50,000(.2642) - \$5,000(.1142)$$
$$= \$13,212 - \$571 = \$12,641 \text{ per year}$$

The lease alternative is already an annual amount of $13,200 per year. Comparing the two annual amounts, we see that the buy alternative is the least expensive for the given ROI. This buy alternative selection is the same as determined by the present worth method.

Future worth method—The following box shows the steps of the future worth method:

Steps of Future Worth Method

1. Select desired ROI.
2. Using *F/A* and *F/P*, bring all revenues to a future amount.
3. Determine the net future worth by subtracting the future costs and the future revenue.
4. Select the alternative with the lowest net future worth by subtracting the sum of the future worth of the project's costs from the future worth of the revenues.

The application of the future worth method is similar to that of the present worth method. Move the first cost of the equipment to the end of the sixth year, and then, since the salvage is already at a future worth, it can be subtracted for the future worth of the first cost.

$$F_{buy} = F_P - F_S$$
$$= P(F/P, n = 6 \text{ years}, i = 15\%) - \text{salvage value}$$
$$= \$50,000(2.313) - \$5,000 = \$115,653 - \$5,000 = \$110,653$$

The annual lease amount is taken to a future worth:

$$F_{lease} = A_{Lease}(A/F, n = 6 \text{ years}, i = 15\%)$$
$$F_{13,200} = \$13,200(8.7537) = \$115,549$$

Using this method, the future worth of the buy alternative is the best alternative.

Internal rate of return (IRR) or ROI method—The IRR is calculated by finding the interest rate at which the revenues equal the costs. This equivalence must, of course, occur at the same point in time. In some situations, as in this example, there are no known revenues. The choice is between two or more projects that have known costs, but unknown revenues. In addition, it would be difficult to determine the revenue or profit values, if any. Replacing worn-out or obsolete equipment that is part of a profitable process cannot be evaluated based on the revenue or profit that single item contributes. Without the equipment, the total process would not be able to produce any products. The approach in these cases is to compare alternatives that are capable of delivering the required service to the product, process, or service, and to select the best of many alternatives. "Best" may be determined by economic analysis or by a combination of economic and nonfinancial judgment.

The modification of the IRR technique when only the costs of alternatives are known is to determine the rate of return at which two alternatives have equal costs. This rate can be compared to the target or desired return to determine which alternative is most desirable at a rate above or below the equivalent IRR.

If revenues are unknown or cannot be determined, a method of equivalent IRR can be used. Equivalent IRR determines the interest rate at which the two alternatives' costs are equivalent. This rate is compared with the desired rate of return.

To continue the solution of the Jennings Co. example, the equivalent IRR will be determined. The given interest of 15% is initially neglected. The approach is to set the two alternatives equal to each other at an unknown interest value. This requires several trials, selecting different ROI values until a solution is found. Here, different interest values are selected to bring the purchase alternative cost to its annual amount. At low rates of interest, the buy alternative is found to be best and at high interest rates, the lease alternative is found to be the lowest cost. By selecting different interest values, the final equivalent ROI is found. By comparing this calculated equivalent IRR value with the desired ROI value, the best alternative can be determined.

Steps of IRR Method

1. Assume an ROI value. Neglect desired or target ROI.
2. Using TVM equivalence functions, move all dollar amounts to the same point in time. This can be present worth, annual amount, or future worth.
3. If the cost of each alternative *equals* the value of the other alternative at the same point in time, then the assumed ROI is the equivalent ROI.
4. If the alternatives' dollar values are *not equal,* assume a different ROI. Solve for the equivalent dollar amounts at the same point in time. Compare the dollar values. If the dollar values are not equal, repeat the process by changing the ROI until the equivalent values are equal.
5. When the equivalent ROI is found,
   a. Compare the value to a target or desired ROI.
   b. Determine which alternative is best at the equivalent ROI and target ROI.

The first step is to select an interest value to try, say 12%. Then, the two alternatives are compared by taking the amount of the purchase and the amount of lease to the same point in time in order to establish an equivalent comparison. This point in time may be based on present worth, annual amount, or future worth.

Assume an ROI value of 12%. Using the annual amount approach, take the first cost of the buy alternative to an annual amount. Bring the salvage to an annual amount. Find the net annual cost at 12%.

$$A_{buy} = A_P - A_S$$
$$= \$50,000(A/P, n = 6, i = 12) - \$5,000(A/F, n = 6, i = 12)$$
$$= \$50,000(.2432) - \$5,000(.1232)$$
$$= \$12,161 - \$616 = \$11,545/\text{year}$$

The lease cost is $13,200 per year. The two annual amounts are not equal, therefore 12% is not the equivalent interest rate. Next, try an ROI value of 20%.

Try a 20% ROI.

$$A_{Buy} = A_P - A_S$$
$$= P(A/P, n = 6 \text{ years}, i = 20\%) - S(A/F, n = 6 \text{ years}, i = 20\%)$$
$$= \$50,000(.3007) - \$5,000(.1007)$$
$$= \$15,035 - \$504 = \$14,531 \text{ per year}$$

The lease amount of $13,200 per year is on an annual basis already and can be compared to the $14,531 per year.

The purchase annual amount is *larger* compared to the annual lease amount. Since the 12% calculation gave an annual amount for the buy alternative that was *less than* the annual lease amount of $13,200, the interest value that would make the two annual amounts equal to each other must be between 12% and 20%.

Next try 15%.

$$A_{buy} = A_P - A_S$$
$$= P(A/P, n = 6 \text{ years}, i = 15\%) - S(A/F, n = 6 \text{ years}, i = 15\%)$$
$$= \$50,000(.2642) - \$5,000(.1142) = \$13,212 - \$571 = \$12,641 \text{ per year}$$

Since the lease cost of $13,200 is between $12,641 and $14,531, the equivalent ROI is between 15% and 20%.

The process can be continued to find the exact ROI at which the two alternatives are equal. This is found to be near 16%. Some calculators and spreadsheets have the capability to solve directly for an unknown interest rate. The function key on the calculator or the term in the software menu may be IRR for Internal Rate of Return. Check the specific documentation for your software or calculator to determine the correct procedure.

Looking at Table 7–1, note that the exact answer is between 16% and 17%. That is where the annual lease cost equals the annual cost of purchase. As the desired ROI increases, the buy alternative has a higher annual cost, making the lease more attractive.

This example of comparing an equipment lease with a purchase demonstrates the four methods of comparing alternatives. For this example, all four methods indicated that the purchase alternative is preferred. Any of the four methods would generally give the same conclusion about the best alternative. In the actual solution of this type of problem, only one method needs to be used to determine the best alternative.

Example of Choosing One of Two Systems: Tigan Corp. This organization is making a decision between two systems that control toxic byproducts from a process. Installing the

Table 7–1   Comparison of Lease and Buy Alternatives for Different ROI Values

ROI Rate	Annual Lease Cost	Annual Cost of Purchase
		$A_T = A_P - A_S$
10%	$13,200	$10,832
12%	$13,200	$11,545
15%	$13,200	$12,641
16%	$13,200	$13,012
17%	$13,200	$13,388
20%	$13,200	$14,531

equipment must be done because of environmental regulations. The critical question is which system will provide the service at the lowest cost.

Alternative	A	B
First cost	$50,000	$65,000
Annual operating cost	$5,000/year	$2,000 per year
Cost to remove system after 8 years	$10,000	$7,000
Desired ROI	12%	12%

This problem could be analyzed by any of the four methods. Present worth or future worth could be used because the lives of the alternatives are equal. Annual amounts of the two alternatives could be calculated and compared. IRR cannot be compared directly because there are no known revenues for the installation. The modified approach to IRR, equivalent internal rate of return, can be used by determining the value at which the costs are equal to each other at the same point in time and then comparing this value with the desired ROI to determine the best option.

For this example, the annual amount and the equivalent IRR for the alternatives will be calculated. The TVM diagram is shown in Figure 7–2.

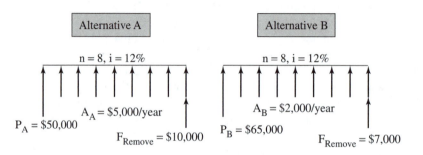

Figure 7–2   Tigan Corp.'s Pollution Control Equipment

Note in Figure 7–2 that the removal costs at the end of the equipment's life are considered a cost, and unlike residual salvage value, are shown below the horizontal line.

Annual amounts for the equipment—For alternative A,

$$A_{\text{Total}} = A_P + A_{\text{Operating}} + A_{\text{Removal}}$$
$$= \$50,000(A/P, n = 8 \text{ yr}, i = 12\%) + \$5,000 + \$10,000(A/F, n = 8 \text{ yr}, i = 12\%)$$
$$= \$50,000(.2013) + \$5,000 + \$10,000(.0813)$$
$$= \$10,565 + \$5,000 + \$813 = \$16,378/\text{year}$$

For alternative B,

$$A_{\text{Total}} = A_P + A_{\text{Operating}} + A_{\text{Removal}}$$
$$= \$65,000(.2013) + \$2,000 + \$7,000(.0813)$$
$$= \$13,085 + \$2,000 + \$569 = \$15,654/\text{year}$$

Since the life and ROI are the same for alternative B and alternative A, the same TVM factors apply. The annual amount of the costs for alternative B are less than for alternative A. Therefore, based on an ROI of 12%, alternative B would be selected.

Equivalent IRR—For this method, different ROI values are selected until the costs of the two alternatives are equal to each other at the same point in time (equivalent). The point in time for this solution can be annual, present, or the future. Since the annual amount for 12% has been calculated, the annual amount can be continued as the time to which to bring the costs of the alternatives. At 12%, the annual costs of A and B are not equal, therefore 12% is not the equivalent IRR. Other values must be tried.

Try 15%. For alternative A,

$$A_{\text{Total}} = A_P + A_{\text{Operating}} + A_{\text{Removal}}$$
$$= \$50,000(A/P, n = 8 \text{ yr}, i = 15\%) + \$5,000 + \$10,000(A/F, n = 8 \text{ yr}, i = 15\%)$$
$$= \$50,000(.2229) + \$5,000 + \$10,000(.0729)$$
$$= \$11,143 + \$5,000 + \$729 = \$16,872/\text{year}$$

For alternative B,

$$A_{\text{Total}} = A_P + A_{\text{Operating}} + A_{\text{Removal}}$$
$$= \$65,000(.2229) + \$2,000 + \$7,000(.0729)$$
$$= \$14,489 + \$2,000 + \$510 = \$16,999/\text{year}$$

The two values are very close at 15%. The actual value is slightly less than 15%. This is the equivalent IRR for the two alternatives. Below 15%, alternative B is the lower cost alternative. If Tigan Corp. desired an ROI greater than 15% (the equivalent IRR), then alternative A would be preferred.

This example demonstrates the application of the annual amount and the modified IRR, equivalent IRR, techniques for selecting the best project or investment. Present worth or future worth methods could have been used since the service lives are equal.

Examples of Choosing One of Three Systems: Morgan City Hospital. Three different radiographic (X-ray) systems are being evaluated by the medical staff of Morgan City Hospital.

Any of the three will meet the needs of the radiology department. The desired ROI is 10%. The costs of the systems are shown below:

X-Ray System	A	B	C
First cost	$250,000	$300,000	$375,000
Annual operating cost	$75,000	$60,000	$80,000
Residual salvage value	none	$36,000	$45,000
Life	5 years	7 years	9 years

The first question to answer is which of the four methods can be used? Since the lives are different for the three systems, to use the present worth or future worth methods would require the common denominator calculation. This is a lengthy calculation and will give the same conclusion as the shorter annual amount method. Because there are no revenues or profits, the internal rate of return method cannot be directly applied. However, the equivalent IRR techniques can be used, except that the solution to this problem requires three parts. Since there are three alternatives, A would be compared to B, then A to C, and then B to C. By elimination, the annual amount method is the most direct solution for this analysis.

For alternative A,

$$
\begin{aligned}
A_{\text{Total}} &= A_{\text{First Cost}} + A_{\text{Operating}} - A_{\text{Residual Value}} \\
&= \$250,000(A/P, n = 5 \text{ yr}, i = 10\%) + \$75,000 - \$0 \\
&= \$250,000(.2638) + \$75,000 \\
&= \$65,950 + \$75,000 = \$140,950/\text{year}
\end{aligned}
$$

For alternative B,

$$
\begin{aligned}
A_{\text{Total}} &= A_{\text{First Cost}} + A_{\text{Operating}} - A_{\text{Residual Value}} \\
&= \$300,000(A/P, n = 7 \text{ yr}, i = 10\%) + \$60,000 - \$36,000(A/F, n = 7, i = 10\%) \\
&= \$300,000(.2054) + \$60,000 - \$36,000(.1054) \\
&= \$61,620 + \$60,000 - \$3,794 = \$117,826/\text{year}
\end{aligned}
$$

For alternative C,

$$
\begin{aligned}
A_{\text{Total}} &= A_{\text{First Cost}} + A_{\text{Operating}} - A_{\text{Residual Value}} \\
&= \$375,000(A/P, n = 9 \text{ yr}, i = 10\%) + \$80,000 - \$45,000(A/F, n = 9 \text{ yr}, i = 10\%) \\
&= \$375,000(.1736) + \$80,000 - \$45,000(.0736) \\
&= \$65,100 + \$80,000 - \$3,312 = \$141,788/\text{year}
\end{aligned}
$$

In this example, alternative B is the lowest cost choice. This assumes, since the service lives are different, that the annual costs of the *future units* have the same future A_T as the current units.

Summary of the Four Basic Techniques

Present Worth

- The major disadvantage is that it cannot be used if the alternative lives are different.
- It is relatively easy to calculate manually.

- This method requires using a desired or target ROI value.
- It is easy for computer applications because most spreadsheets have present worth functions.
- It can be used as a step in the process to determine an equivalent ROI.

Annual Amount

- Annual amounts can be compared with other annual financial data.
- A desired or target ROI value is required for calculations.
- A major advantage is that alternatives with different lives can be compared.
- It can be used as a step in determining equivalent ROI.
- It is one of the most often used of the four techniques in decision making.

Future Worth

- It cannot be used if the alternatives have different lives.
- It requires a desired or target ROI value.
- It is easy to calculate manually.
- Resulting answers are difficult to relate to other types of financial data.
- It can be used as an intermediate step to IRR calculations.

IRR

- It does not require a target or desired ROI estimate.
- It may give multiple solutions in some cash flow methods.
- ROI between alternatives is a breakeven solution.
- This method can be used with alternatives having different lives.
- Projects can be ranked based on their ROIs.
- A project's ROI can be compared to other internal and external ROIs of the firm.
- It is difficult to calculate manually without special calculator or spreadsheet functions.
- It is difficult to interpret equivalent IRR if there are more than two alternatives.
- IRR must be modified to equivalent IRR if revenues are not known.

> Decision makers need a combination of analytical and quantitative skills plus judgment, common sense, and experience.

Some Cautions About the Four Techniques

The usual situation is for competing alternatives to have different service lives. In these cases, the ROI or annual amount methods would be used since the present worth or future worth methods are generally not valid when lives are different.

If the desired ROI is higher or lower, then it is possible that different conclusions could have been made.

Sometimes, what-if calculations are made by trying different ROI values, salvage, life, or operating costs to see the sensitivity of the results for the given changes.

The equivalent internal rate of return solution of the two alternatives must be compared to the desired or target ROI to determine which alternative is the best.

> Almost anyone can make a little money by investing a lot of money. The goal is to make a lot of money from investing a little money.
>
> —Anon.

With equivalent IRR calculations, the returns must be compared to the desired or target ROI values. It is not always obvious which alternative is the best above or below the equivalent IRR value. Additional analysis may be required.

Another problem may exist with IRR calculations. Depending on the arrangement of the flow of funds, it is possible to obtain multiple, and mathematically correct, IRR values for the same problem. This usually occurs when the project has many changes from net costs to net revenues over its life. These multiple changes may yield a solution of multiple values. Therefore the annual amount method is even better to use. However, IRR and equivalent return methods still are important, especially when used as a project ranking tool.

If the service lives are equal, any of the four techniques will give valid answers under most conditions. The present worth method is frequently applied when analysis is done by financial personnel. The annual amount method takes different lives into account and since many costs are annual operating costs, they are already in annual terms. Many budgets and decisions are made based on annual data so that the annual amount method is "comfortable" for many individuals. The ROI or equivalent IRR method has the advantage of permitting the ranking of alternatives based on their rates of return.

OTHER ANALYSIS METHODS

Although the methods presented here are useful for most economic analysis, other techniques may be encountered. These others are usually derivatives of the four basic methods. Note that some of the other analysis methods, however, do not include time value of money. These other methods include:

Benefit/cost analysis

Payback

Discounted cash flow

Life cycle analysis

MAPI

Proprietary techniques

Benefit/Cost Analysis

Benefit/cost (B/C) analysis is a common technique that is often used for analysis of alternatives. The benefits may be expressed as present worth of revenues, profits, or savings. The present worth of cost and expenses are divided into the benefits to give the B/C ratio.

Example of B/C Analysis: Barrows Construction Co. Barrows is working with one of its clients to design and construct an addition to the client's facilities. Patty Smithson is the project engineer and project manager. Patty has estimated $80,000 for the cost of the construction. The client has estimated the net future savings per year of the project as $30,000 for each of 6 years, and desires a 12% return on investment.

Patty says that time value of money can be applied to the B/C calculation (see Figure 7–3).

$$\text{Benefits/Costs} = \frac{\Sigma \text{ Present Worth of Benefits}}{\Sigma \text{ Present Worth of Costs}}$$

Bringing the savings benefits to the present is calculated:

$$P_{\text{Save}} = A(P/A, n = 6, i = 12\%) = \$30,000(4.1114) = \$123,342$$

Dividing this by the present worth of the costs,

$$B/C = \$123,342/\$80,000 = 1.5$$

The client tells Patty that in the past, their calculation has been to add the revenues and divide by the sum of the costs without any interest calculation. In this case $B/C = (6)(\$30,000)/\$80,000 = 2.2$. And they have a rule of thumb that all projects should have a B/C of at least 2.0 in order to be funded. Patty explains the details of time value of money. If they desire an ROI of 12%, the present worth of the future savings is used rather than the arithmetic sum of the future savings. She indicates that with a ratio of 1.5 at 12%, the client is exceeding the 12% target return. She calculates the IRR of the project to be almost 30%. (Check this using your calculator.) The client says that he is very happy with that return.

Payback

During a discussion with Patty, the client indicates that his organization sometimes uses the **payback** method to determine the acceptability of investments. It is calculated as the number

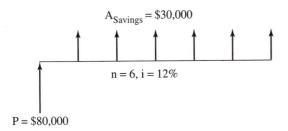

Figure 7–3   Barrows Construction's Client

of years required to pay back the original cost of the investment. For the current project it would be calculated as:

$$n_{Payback} = \text{Investment/Annual Savings} = \$80{,}000/\$30{,}000/\text{year} = 2.7 \text{ years}$$

The client says he likes the simplicity of the calculation and normally likes to have a payback of no more than four years in order for the project to be funded. Patty explains that the calculation doesn't consider the savings that occur for the project past 2.7 years. She explains, "Another project that has only three years of service. The same initial costs and revenues per year would have the same $n_{Payback}$ of 2.7 years. The two projects would have different IRRs and different A_{Total}. Using time value of money and equivalence calculations to determine the project's return is a preferred method to measure the project's acceptability."

Managing the Shareholders' Investment?

At a shareholders' meeting of a large firm, the vice president of finance was discussing how, for the current plant expansion, they were doing "payback calculations" to justify all of the new equipment purchase decisions. After the meeting, the manager was asked to explain what he meant by payback. He was glad that a shareholder was interested in something besides the free refreshments, and he explained, "The first cost of the equipment is obtained from a supplier's quotation, and this figure is divided by the amount of money we will save per year. The result is the number of years required for it to pay for itself." He went on to say that if the payback was not three years or less, the equipment would not be considered. He even took out an envelope from his pocket and wrote "$100,000 computer divided by $40,000 savings per year equals payback of 2.5 years. Okay to buy."

He was asked about the operating costs. He said that they were considered in the savings. He was asked about the equipment life. He said that it didn't matter so long as the equipment lasted longer than the payback period. He was asked about the salvage value, about alternative projects, about the TVM, and other factors, and responded that they were *not* important!

The above discussion between the client and Patty may seem unlikely with our knowledge of TVM. However, some organizations and individuals use the non-TVM-based B/C and payback methods to make project and investment decisions today. With our understanding of time value of money, equivalence, and interest calculations it also may be difficult to imagine that shareholders may still hear that their funds are being managed using techniques like those described by the vice president of finance (in the accompanying box).

Discounted Cash Flow

Another technique sometimes encountered is **discounted cash flow (DCF).** This technique is the same technique, with variations, as the present worth method presented here.

There is an important caution that should be considered, however. The DCF techniques that are used by some organizations *do not* differentiate between projects' lives. Typically, DCF uses the interest value to reduce each alternative's future revenues and costs to a present worth. Using a desired ROI, the present worth of the project's costs is subtracted from the present worth of the project's revenues and savings. The larger the difference, the better the investment, it is concluded. If a number of alternatives are analyzed, the ones with the largest difference are typically selected.

Caution should be used when using the DCF method of investment selection. If the alternatives have different service lives and are compared on a DCF basis, an incorrect conclusion may be made. IRR or annual amount techniques are more applicable.

Life Cycle Costs

Some organizations, including government agencies, use life cycle costs or life cycle analysis. This name implies that "life cycle" includes the initial costs, operating costs, residual salvage values, revenues, and profits of the project—all costs and revenues over the life of the project. The term life cycle usually includes a return on investment and time value of money calculation. Although there are many variations of the basic concept, the technique is usually based on sound methods. There are some applications of life cycle analysis, however, that do not include TVM. Care should be taken when encountering the various techniques labeled life cycle costs to ensure that time value of money and equivalence are applied.

MAPI

Although not as popular now, at one time this technique was used a great deal. It was made popular by the Machinery and Allied Products Institute, a trade association of equipment manufacturers. A handbook was published with forms, tables, and graphs to be used in the analysis of purchasing new equipment. It was primarily applied to replacement analysis and used the terms "Challenger" and "Defender" for the identification of new and existing equipment. The technique uses TVM. The only difficulty with it is that it is a very "cookbook" approach. That is, it is a "fill in the form, look up the numbers in tables and graphs" approach that is typically applied without understanding the concepts or principles of the calculations. Even though it uses time value of money, the shortcoming is that the technique is often applied without consideration for exceptions and special circumstances. It is still in use by some organizations.

Proprietary Analysis Techniques

There are many variations of analysis and selection techniques. Some have different names but actually are similar to the basic four methods of analysis. Many organizations customize the techniques that have been presented here with their own terminology and special considerations. These are generally proprietary, and like some financial analysis and manufacturing techniques are kept private within the organization. Each of these different techniques can be examined for its use of the correct concepts and methods to ensure that it is a truly effective tool.

Example: Valuation of Bonds

Many industrial, local, state and federal bonds are issued at a face value of $1,000. The interest rate at which they are offered is determined by a number of factors, the primary of which is the prevailing market interest rates. If the offered bond interest rate is *lower* than the prevailing rate, either the bonds would not sell or they would sell at a price lower than their face value. If bonds are issued at *higher* than prevailing rates, the issuer is obligated to pay a higher than necessary rate to the holders. If, for example, a 10-year bond is issued for 7% interest, the purchaser pays $1,000 initially, will receive an annual interest check for $70, and will be repaid the $1,000 at the end of the tenth year.

What if the purchaser sells the bond during the 10-year period? How is the selling price determined? If during the holding period the prevailing market interest rates increase to 9%, the original bond's value will be adjusted to reflect this higher rate. A buyer of the 7% bond who will only receive $70 each year (this amount is fixed) will not be willing to pay $1,000 for the bond now even though the issuer will pay $1,000 at the maturity date. The price of the bond will decrease as the market interest rates increase. Likewise, if the market rates of other bonds decrease to 6%, for example, the price of the bond will increase since buyers are willing to pay a higher price in order to receive the $70 per year payment.

Bob Harris wants to determine the value of a bond that he has inherited from his uncle's estate. The bond has 5 years of life remaining before it is redeemed and pays 7% on the face value of $1,000 annually. The holder will receive the $1,000 at the end of the period. The prevailing market rate is 9%. The TVM diagram for the calculations is shown in Figure 7–4.

The present worth of the future interest payments plus the present worth redemption amount of $1,000 can be calculated.

$$P_{\text{Bond Value}} = P_{\text{Interest}} + P_{\text{Redemption}} = \$70(P/A, n = 5, i = 9\%) + \$1,000 \ (P/F, n = 5, i = 9\%)$$
$$P_{\text{Bond Value}} = \$272.28 + \$649.93 = \$922.21$$

If the prevailing interest rates had decreased and were now 6%, the current value of the bond would be:

$$P_{\text{Bond Value}} = P_{\text{Interest}} + P_{\text{Redemption}} = \$70(P/A, n = 5, i = 6\%) + \$1,000 \ (P/F, n = 5, i = 6\%)$$
$$P_{\text{Bond Value}} = \$294.87 + \$747.26 = \$1,042.13$$

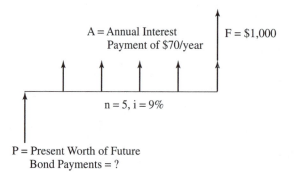

Figure 7–4 Bob Harris's Bond

If Bob chooses to sell the bond now and interest rates are 9%, he would receive the $922.21. If he holds on to the bond until maturity, he would receive $70 annually for 5 years and $1,000 when the bond is redeemed in 5 years.

Example: Unknown Equipment Life

Two different telephone systems are being considered for purchase by Merrit Corp. One system has an initial cost of $85,000 and an operating cost of $15,000 per year. The second system has a first cost of $110,000 and an operating cost of $2,000 per year. If the organization has a desired ROI of 15%, how long must the more expensive system be used for it to be economical?

Although similar to the payback method, this method uses TVM while the traditional payback method does not. The solution is presented below (and see Figure 7–5).

This is a trial type of solution, selecting different n values and trying to find one that makes the cost of the two alternatives equal at that n. All the values can be taken to a present worth, future amount, or annual amount using TVM equivalence methods. The solution below demonstrates the use of the annual amount; the other methods would yield a similar solution.

Try 3 years first. Alternative 1:

$$A_{\text{Total}} = A_P + A_{\text{Operating}}$$
$$= \$85,000(A/P, n = 3 \text{ yr}, i = 15\%) + \$15,000$$
$$= \$85,000(.4380) + \$15,000$$
$$= \$37,230 + 15,000 = \$52,230/\text{year}$$

Alternative 2:

$$A_{\text{Total}} = A_P + A_{\text{Operating}}$$
$$= \$110,000(P/A, n = 3 \text{ yr}, i = 15\%) + \$2,000$$
$$= \$110,000(.4380) + \$2,000$$
$$= \$48,180 + \$2,000 = \$50,180/\text{year}$$

Since the two annual amounts are not equal, the correct life value is not 3 years.
Try 5 years. Alternative 1:

$$A_{\text{Total}} = A_P + A_{\text{Operating}}$$
$$= \$85,000(A/P, n = 5 \text{ yr}, i = 15\%) + \$15,000$$
$$= \$85,000(.2983) + \$15,000$$
$$= \$25,357 + \$15,000 = \$40,357/\text{year}$$

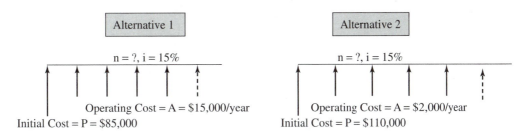

Figure 7–5   Merrit Corp.'s Telephone System

Alternative 2:

$$A_{Total} = A_P + A_{Operating}$$
$$= \$110,000(.2983) + \$2,000$$
$$= \$32,813 + \$2,000 = \$34,813$$

The difference between the two annual amounts using $n = 5$ is $\$40,357 - \$34,813 = \$5,544$. This value is greater than the difference when $n = 3$ years was calculated. Next, try a value lower than 3 years, such as 2 years.

Try $n = 2$. Alternative 1:

$$A_{Total} = A_P + A_{Operating}$$
$$= \$85,000(A/P, n = 2, i = 15\%) + \$15,000$$
$$= \$85,000(.6151) + \$15,000$$
$$= \$52,285 + \$15,000 = \$67,285/year$$

Alternative 2:

$$A_{Total} = A_P + A_{Operating}$$
$$= \$110,000(.6151) + \$2,000$$
$$= \$67,663 + \$2,000 = \$69,663/year$$

What if Table 7–2 shows the annual amount, the differences in annual amounts, and n?

From the calculations and Table 7–2, it can be seen that the difference increases above 3 years. If other values greater than 5 years are calculated, the difference is even greater than at 5 years. The annual amounts are equal between 2 and 3 years. From the calculations, the annual amount of alternative 2 is lower than for alternative 1 when n is greater than 3 years. Therefore it is concluded that the second phone system is the most economical *if it lasts longer than 3 years*. Both telephone systems would be expected to have a life greater than 3 years. The longer the second alternative lasts, the lower its annual amount. Another way of considering this calculation is to think of the break-even point between 2 and 3 years. The longer the period over which the higher initial costs and the lower operating costs are spread, the lower the annual amount. Or, it can be said that the organization is investing its funds at 15% and the longer the life of the systems, the better the return on the second system.

Unequal Service Lives

Comparison and analysis of alternatives with different service lives requires some special comments. Assume that a new robot is being considered by Young Corp. One robot has a life of 6

Table 7–2   Net Annual Amount vs. *n*

Years	Annual Amount of Alternative 1	Annual Amount of Alternative 2	Annual Amount Difference, $A_1 - A_2$
2	$67,285	$69,663	−$2,378
3	$52,230	$50,180	+$2,050
5	$40,357	$34,813	+$5,544

years and another has a life of 8 years. Each has different first costs and salvage value. The four basic methods—present worth, annual cost, internal rate of return, and future value—are not equally applicable in this case of different service lives. The present worth or future value of the two alternatives cannot be compared, for example, because the present worth and future values are based on different service lives or *n* values. There are different first costs, operating costs, salvage values, and *n* values. This may cause an incorrect selection. One alternative may appear to have a lower present worth or future value that is due to the life rather than the costs.

Using the two other basic methods, IRR and the annual amount, is preferable for alternatives with different lives. These two methods give their respective answers in percent per year for IRR and dollars per year for the annual amount method. Thus, even though the alternatives are different service lives, the solutions are comparable since they are comparing annual percent to annual percent or annual dollars to annual dollars.

There is an implicit assumption in applying these two acceptable methods. The 6-year alternative will be replaced at the end of its sixth year, but in the initial analysis, it is assumed that the IRR or the A_{Total} of the future seventh and eighth years will be the *same* as in the first 6 years. In practice, of course, the percent return and annual cost may be higher or lower due to changes in costs, revenues, and other future events. This assumption is generally more acceptable than basing the selection of an alternative from present worth or future value calculations of alternatives with different service lives. There is another valid method that can be used: the common denominator method.

Common Denominator Technique

In the Young Corp. example, the two robots can be compared using present worth or future value when the lives of the projects are equal. Where one project is 6 years and the other is 8 years, the common denominator of the two lives is 24 years. The assumption using this method is that if a 6-year alternative is selected, when it wears out, it is replaced with a second, third, and fourth alternative with the same costs as the first 6 years for a total of 24 years of service. If the 8-year alternative is selected, it is replaced with a second and a third alternative for a total of 24 years of service. The cost of the replacement units are the same as for the initial robot for both alternatives. Now, the two 24-year service lives can be compared using present worth or future value. All of the costs associated with each alternative over a 24-year period can be brought to the present or taken to the future and compared. A diagram of the two alternatives using the common denominator method is shown in Figure 7–6.

The diagrams show the two alternatives of Robot A for 6 years and Robot B for 8 years. The 6-year alternative is assumed to be replaced at the end of 6 years, again at the end of the twelfth year, and at the end of the eighteenth year for a total of 4 different robots to give a service life of 24 years. The 8-year alternative, Robot B, is assumed to be replaced after the eighth year, and again after the sixteenth year for a total of 24 years of service. Since the service lives are now equal, the present worth or future value of each may be calculated and compared.

Some assumptions are part of the common denominator method used in this example:

1. The costs, revenues, and other characteristics of the future purchases are the same as for the first purchase.
2. The organization requires a robot's service for 24 years.

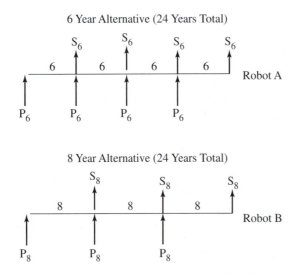

Figure 7–6   Robots at Young Corp.—Different Lives, Common Denominator Method

If the common denominator method is used, and then the annual costs for the alternatives are calculated from these values, the resulting annual amount is the same as the annual amount that would be calculated if only one period of the alternative is used to calculated the A_{Total}. Prove this for yourself using data that you choose.

This brings the analysis back to the use of the annual amount calculation for each of the two alternatives. Although the common denominator method is used by some organizations, the annual amount method will ensure the same selection with less calculation.

> When comparing alternatives with different service lives, present worth can only be used with the common denominator technique. A better solution is to compare the alternatives using annual amount or internal rate of return.

Example: An Earth Dam at Bald Mountain State Park. The average annual cost to residents and farmers of a small creek's annual flooding is $40,000. An earth dam is constructed that controls and limits the flooding. What is the present worth of the dam? (See Figure 7–7.)

If an ROI of 8% is the prevailing or desired rate of return, the limit of the P/A function is:

$$P = \frac{A}{i}$$

$$A = iP$$

For the dam, the present worth of all of the perpetual future savings per year at 8% ROI is:

$$P = A(P/A, n = \inf, i = 8\%) = \$40,000/.08 = \$500,000$$

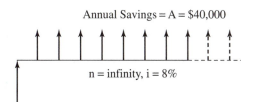

Figure 7–7   **Earth Dam**

The present worth of the dam today is the maximum to spend for the construction of the dam, assuming that it lasts forever.

Warren Buffett's Valuation of the *Washington Post*. Warren Buffett, one of the most successful investors in America, made his wealth by buying firms and by investing in securities. Beginning with $105,000 in 1957, his company, Berkshire Hathaway, is currently valued at more than $31 billion. As CEO of Berkshire's investments, he has purchased candy companies, insurance firms, shoe companies, furniture stores, newspapers, 7% of the outstanding stock of Coca-Cola Co., and other investments. To determine a fair price to pay for a company or its stock, he has applied TVM techniques similar to the ones presented here (R. Hagstrom, Jr., *The Warren Buffett Way* [New York: J. Wiley, 1994]; Berkshire Hathaway, *1997 Annual Report*).

Buffet uses the concept of a firm's "owner's earnings" to estimate the future annual funds that are available to the owners. Owner's earnings are related to cash flow. Buffett defines owner's earnings as:

Owner's Earnings = Net Income + Depreciation − Capital Reinvestment

The above values are annual values. If the firm is investing an amount back into capital equipment equal to the annual depreciation charge, only the net income is available to the firm and the shareholders. If the capital reinvestment is larger or smaller than the depreciation, this would decrease or increase the amount of funds left for the firm and the shareholders.

By assuming that the flow of these earnings continues into the future, the present worth of the future earnings can be calculated by dividing the annual earnings by the target interest rate (present worth of a perpetual future amount). Buffett typically uses the prevailing U.S. government bond rate as a target rate of return.

For the *Washington Post* Co., for example, the owner's earnings per year in 1973 were $10.4 million. This was determined from the financial statements:

Net Income + Depreciation and Amortization − Capital Expenditures = Owner's Earnings
$13.3 million + 3.7 million − $6.6 million = $10.4 million

Dividing the owner's earnings by the prevailing government bond rate of 6.8% yields:

$P = A/i = \$10.4$ million$/.068 = \$153$ million (approximately)

The total market value (market price × number of shares outstanding) of all the shares of the *Post* were $80 million in 1973. Thus, Buffett concluded that the true value of the company, $153 million, was greater than the price if he purchased all of the company's stock. He also believed that the owner's earnings were going to increase, rather than stay constant, over the next few years so that the present value would be even greater. (See Figure 7–8.)

Whether 100 shares of stock or the total company is purchased, this valuation technique can be applied. These techniques not only apply to the valuation of securities, but also to the valuation of divisions, patents, products, services, and other investments or projects.

Most individual investors do not use this technique of determining if an investment into a security or mutual fund should be made. They tend to follow the crowd and buy what others are buying. Buffett does his analysis using TVM techniques and avoids the crowd. He is able to remove himself from the day-to-day market fluctuations by looking at his investments from Omaha using his own valuation methods.

> Risk comes from not knowing what you're doing.
> —Warren Buffett

Example: Bergson Co.'s New Warehouse. Bergson Co. wants to build a new warehouse. One alternative would cost $750,000, with operating costs of $50,000 per year, and no salvage after 20 years. A second alternative would have a first cost of $500,000, operating costs of $60,000 per year, and no residual salvage value after 15 years. If the firm's desired ROI is 15%, what is the best alternative?

With different lives, the present worth or future amount methods would not be appropriate unless the common denominator method is used, which is a cumbersome calculation. The ROI method could be used to find the break-even or equivalent ROI at which the two methods are equal and then that value could be compared with the target ROI of 15%. A more straightforward method is to use the annual amount method. (See Figure 7–9.)

For alternative 1:

$$A_{Total} = A_P + A_O = \$750,000(A/P, n = 20, i = 15\%) + \$50,000$$
$$\$750,000(.1598) + \$50,000 = \$119,850 + \$50,000 = \$169,850/year$$

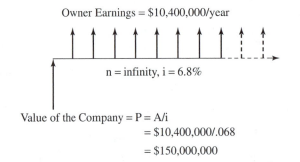

Owner Earnings = $10,400,000/year

n = infinity, i = 6.8%

Value of the Company = P = A/i
= $10,400,000/.068
= $150,000,000

Figure 7–8   Buffett's Valuation of the *Washington Post*

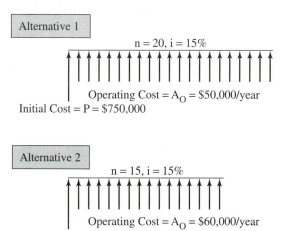

Figure 7–9 A Warehouse for Bergson Co.

For alternative 2:

$$A_{Total} = A_P + A_O = \$500,000(A/P, n = 15, i = 15\%) + \$60,000$$
$$= \$500,000(.1710) + \$60,000 = \$145,500/year$$

On an annual cost basis, alternative 2 is the least expensive. Since the lives are different, it is assumed that if the 15-year alternative is selected, the additional years would cost the same as the first 15 years. This may require purchasing a new structure or extensive repair and remodel of the old one. Also, as always, there are noneconomic factors that should be considered before a final decision is made.

Example of IRR: Williams Corp. As an example of solving for an unknown internal rate of return, Williams Co. is considering a project with the following revenues and costs. This example uses a trial solution. If calculators and spreadsheets are available that can solve directly for the unknown IRR, they may be used.

Assume that the flows of costs and revenues from the project are as shown in Figure 7–10. The original diagram is shown on the left. This diagram can be simplified by taking the net value of the revenues and costs that occur in the same year. Either diagram can be used for the solution. The one with fewer variables is easier to solve manually.

The problem is to determine the return on investment for the project. Even after the revision and simplification, the problem doesn't fit into the interest equations or tables. The only method to use to solve the problem for the unknown i is to select an i, and try a solution to see if the \$3,000 investment is equivalent to the revenues of \$2,000 at the end of years 2 and 3. If it isn't equivalent, try a different i and continue until a solution is found.

The usual solution method (not the only one, however) is to select an interest rate, bring all of the revenues back to a present worth at time zero, then bring all of the costs back to a present worth at time zero. If the two present worths are equal, the interest rate selected is correct; if they are not, the assumed i value is not correct. An alternative solution would be to bring the

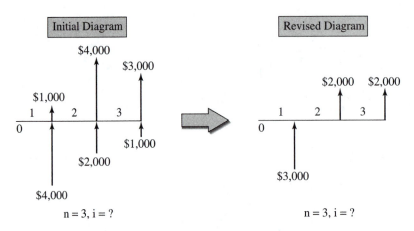

Figure 7–10   Williams Co. Project—IRR Calculations

dollars to either an annual amount or a future amount. When the costs and revenues are the same at the same point in time (equivalence), then the return used to make the calculations is the IRR of the project.

For an initial trial, select 10%. Bring the costs back to t_0 at 10%:

$$P_{3,000} = \$3,000(P/F, n = 1, i = 10) = \$3,000(.9091) = \$2,727$$

Bring the revenues back to t_0 at 10%:

$$P_{2,000} = \$2,000(P/F, n = 2, i = 10) = \$2,000(.8264) = \$1,652$$
$$P_{2,000} = \$2,000(P/F, n = 3, i = 10) = \$2,000(.7513) = \$1,503$$

The total present worth of the revenues is:

$$\$1,652 + \$1,503 = \$3,155$$

This is greater than the present worth of the costs, so the selected i value is not 10%. **Try 15%.** Present worth of the cost:

$$P_{3,000} = \$3,000(.8696) = \$2,609$$

Present worth of the revenues:

at year 2,	$P_{2,000} = \$2,000(.7561)$	= \$1,512
at year 3,	$P_{2,000} = \$2,000(.6575)$	= \$1,315
	Sum of present worth	= \$2,827

The sum of the present worth of the revenues is still larger than the present worth of the costs. But the difference between them is smaller than at 10%, so the solution is going in the right direction.

Try 20%. The present worth of the cost:

$$P_{3,000} = \$3,000(.8333) = \$2,500$$

The present worth of the revenues:

$$P_{2,000} = \$2,000(.6944) \qquad = \$1,389$$
$$P_{2,000} = \$2,000(.5787) \qquad = \$1,157$$
$$\text{Sum of the present worth} \qquad = \$2,546$$

Table 7–3 shows the project's IRR and the present worth of the costs and the revenues.

The table shows that the net present worth for the project (present worth of revenues minus present worth of costs) is close to zero at an IRR of 20%. If Williams Co. invests in this project and if the revenues and costs actually occur as estimated, they will make an ROI of approximately 20% on their investment. If that return is satisfactory to Williams and other nonfinancial factors are acceptable, the investment can be made.

The four methods of project and investment analysis are applicable to many different types of financial problems for both individuals and organizations. These methods can be amplified by including taxes and the concepts of replacement analysis, topics that are presented in later chapters.

PROJECT MANAGEMENT—PROJECT ANALYSIS, SELECTION, MONITORING, AND PERFORMANCE MEASUREMENT

Good project management is important for the success of the contemporary organization as it moves into the next century. Traditional line and staff management is applicable for products and services that have a high volume and long lifespan, but as customer response time and innovations have shorter cycle times, project management becomes increasingly more important. New product development, process improvements, advertising and promotion activities, construction of new facilities, creation of software, and custom-designed and built systems all require short, intensive project management and organization.

> Time value of money can be used to analyze, select, monitor, and evaluate projects, and it can be integrated with scheduling, cycle time reduction, productivity improvement, quality improvement, and team management of projects.

Time value of money techniques are vital to project analysis, selection, monitoring, and measurement. The management of the project's quality, schedule, and costs can all be integrated with TVM techniques. (See Figure 7–11.)

Table 7–3 IRR vs. Present Worth for Williams Co. Project

IRR of Project	Present Worth of Revenues	Present Worth of Costs	Net Present Worth $= P_{Revenue} - P_{Costs}$
10%	$3,155	$2,727	+$428
15%	$2,827	$2,609	+$218
20%	$2,546	$2,500	+$46

Project Management

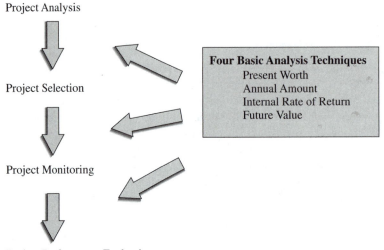

Figure 7–11   **Project Management and Time Value of Money**

> Good analysis, selection, monitoring, and performance measurement of
> projects and investments is often part science and part art.

Capital Investment for Projects

Financial planning and budgets for ongoing products, processes, and services is different from analysis, selection, and planning for one-time investments. The financial plan or budget can be divided into two major components.

1. The *operating budget* is the estimation of the next period's sales income and the respective costs of those sales. The costs are typically divided into materials, labor, and overhead expenses. This is an estimate of what the future Income Statement will look like for the next accounting period, usually a year.

2. The *capital budget* is the decision to fund projects, equipment, and investments, and improve or expand the organization's fixed asset base. This includes investment in development, new products, research, and related projects. This part of the budget process normally requires the use of TVM techniques.

Capital investments can be analyzed and selected using TVM techniques. One selection technique is to rank the projects in order of their potential ROI, initial cost, life, present worth, or other criteria. There are also nonmathematical considerations that may apply: worker safety, customer satisfaction, quality, and related factors. Parts of these factors can be quantified; experience and judgment, in addition to TVM, are required. The calculations and applications of

TVM may sometimes be the easiest and most direct part of the analysis and selection process. Judgment ability often makes the difference between average and excellent success. When Warren Buffett, chairman of Berkshire Hathaway, analyzes a firm's management, he places a great deal of importance on the manager's ability to allocate the firm's capital over time. Deciding what to do with a company's funds is critical to the organization's success.

> I have not learned how to solve difficult business problems, but what I have learned is how to avoid them. We are successful because we know how to identify one-foot hurdles that we could step over rather than because we acquired any ability to clear seven footers.
>
> —Warren Buffett

Analysis of Projects and Investments

As we have seen from the material presented in this chapter, analysis of projects and investments is primarily based on the four methods of present worth, future worth, annual amount, and internal rate of return. This analysis step, the most quantitative of the components of project management, is the beginning of the project management process. The four methods can be used individually or together to provide a financial basis for selecting projects.

Project Selection

After analysis of potential projects and investments comes the selection of the projects. Projects can be ranked by their net present worth if they are of equal length; by annual costs and benefits; or by the estimated return on investment. A strict application of this last approach would be to rank the projects from highest to lowest IRR and invest into the highest return projects first. In practice, some of the high IRR projects are also high risk, so risk analysis is part of the funds allocation process. The IRR ranking method must also take into account the initial investment and diversification of the funds into a number of projects as well as overall optimization of the organization's plan.

> There are two rules to investing. The first rule is, don't lose money. The second rule is, don't forget rule number one.
>
> —Benjamin Graham

For example, one project may require a majority of the organization's funds, but management may not want to "put all of the investment funds into one basket." Diversification of funds can be considered. The ROI/IRR ranking is only one approach for project selection. It is a powerful one, however. The IRR ranking and the annual amount can be combined. The initial cost, the project's risk, and life can be included to give a more complete picture of the possibilities. What-if analysis can be applied to present a picture of outcomes based on variation of the project's inputs. As more information is added to the selection process, the analysis becomes more complex. A balance between simple and complex selection techniques is often needed.

Some Cautions About Project Analysis and Selection

Project engineers, managers, and technicians should note these cautions:

1. The consideration of cost and revenue estimates in the analysis stage may be overly optimistic. Senior management may set a desired ROI. Everyone connected with the potential project can become excited, take ownership of the ideas, and want the project to become a reality. Too high estimates of revenues and too low estimates of potential costs are made in order to meet the ROI targets. Equipment lives may be overestimated. Overall, the project looks good and it is approved. Later, the actual costs are a little higher than estimated, revenues are a little lower, and equipment life is shorter than estimated. What-if and sensitivity analysis can help identify the outcomes from this overoptimism.

2. Project and investment monitoring can use TVM techniques during the project's life. This information can be compared to the original estimates to determine the cause of any variance. This feedback helps the project management team make some mid-course corrections if they are necessary. Also, the project team can apply this new information to future projects. TVM techniques are important in the original analysis and in the monitoring step as well.

3. A third caution is to make a final evaluation of the project using time value of money applications. Upon completion of the project the actual results can be compared to the original planned performance. By observing, recording, and distributing information about the differences between the planned and the actual performance, future projects' results can be improved. Too often the project team disbands after the project is completed without a thorough analysis of the results.

4. A fourth consideration is for the project team to be aware of the project's funding source. Are the funds coming from internal or external sources? For internally financed projects, the target or desired ROI is used with the analysis, and if the project goes as planned, the ROI is realized. For projects that are funded using outside money such as from banks, the sale of bonds, or other financial sources, the cost of the capital and the interest paid becomes a factor.

Overoptimism applies to the individual as well as the organization. It is easy to be optimistic about potential investments such as mutual funds, individual securities, and other investments. As the investment results become a reality, they may or may not meet the original estimates of returns. Too often the individual will live with a low return, "hoping and praying that things improve" rather than facing the reality of a loss.

Individuals, like organizations, can use time value of money methods to analyze, select, monitor, and evaluate projects and investments. Continuous monitoring and final analysis using TVM techniques will help keep the focus on the actual results, not just the original estimates. This feedback can be used to make future decisions more successful.

Allocating Funds for Projects

For organizations and for individuals, wants and needs usually exceed available resources. More projects are available than funds. Part of the process of analyzing and selecting alternatives is allocating and rationing limited funds among a number of possible projects. The heart of this rationing process is the use of TVM techniques. Examples of allocating funds include:

1. A firm deciding which equipment to improve or replace.
2. Individuals deciding where to place their retirement funds.
3. Deciding which new products and services to develop.
4. Quality improvement teams allocating their time and money to different quality improvement projects.
5. A research group assigning personnel and funds to projects during the coming year.
6. A family allocating its funds to home improvement projects.
7. A utility company deciding on those projects that will improve service to customers during the next year.
8. An organization allocating funds that will improve response time to customers and delivery of the firm's products.
9. Making safety improvement projects.

These examples cannot be analyzed on a quantitative basis only. Many nonfinancial factors are involved such as customer needs, quality of products and services, safety of employees, morale of personnel, training of personnel, and so on.

Friendly Financial Skies

The application of TVM analysis can be very useful in the capital decision-making process. Rarely is it used exclusively, however. For example, a medium-sized firm wanted to buy a small corporate plane. The firm had narrowed the selection process down to two different models based on time value of money analysis. The marketing and technical personnel of the aircraft firm had prepared an extensive financial analysis of the firm's past air travel costs, the equipment costs, and a history of the executives' travel requirements. The executives were invited to come to a nearby airport to take a short flight to a nearby town, have lunch, and return on the other aircraft so they could compare the two models. Later, the firm received an order for one of the aircraft, and the executives were asked what factors they thought were most significant in making their selection. Was it the operating costs, the speed factors, initial costs, seating capacity? The response was that the interior design of one model was preferred. Although either aircraft would have been a sound economic choice, sometimes the response to visual or emotional factors makes the decision.

The selection process is usually a combination of quantitative and qualitative factors. Often the financial analysis will eliminate the undesirable alternatives using quantitative methods. As the alternatives are narrowed, more qualitative factors may be introduced into the decision process.

W. Edwards Deming, management consultant and quality expert, strongly encouraged managers to use data and factual information in arriving at decisions rather than using simply intuition or "gut feelings." He believed that too many decisions are made incorrectly because readily available data and factual statistical information are ignored. If U.S. management has a bias, it may be, as Deming suggests, toward intuitive decisions. The best decision methods combine data, statistics, and analysis with experience, judgment, common sense, and other qualitative factors. Frequently, when good decisions and alternative selections are analyzed after the fact, it is evident that there was a balance between quantitative analysis and qualitative judgment.

Sources of Funds for Projects and Investments

Sources of funds are not constant over time. Some years, profits are high. Sometimes the interest rates are low and borrowed funds are readily available. Other years are good for the sale of securities to bondholders and shareholders. The financial climate is a dynamic and continually changing environment. The process is one of matching the selected projects with the funds that are available internally and externally to the organization. (See Figure 7–12.)

Available funds are rarely equal to the number of potential projects. There must be some selection process to ration the funds. The best investments must be selected and the least needed projects must be eliminated.

Project Ranking Example: Olden Corp. Different projects can be analyzed using the TVM techniques that have been described. A typical result is shown in Table 7–4.

Table 7–4 of potential projects for Olden Corp. can be analyzed from a number of different viewpoints. Ideally, an investment with a high return and low risk is desired. In addi-

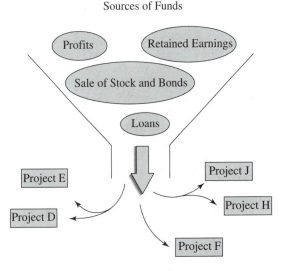

Figure 7–12   Funding and Selecting Projects

Table 7–4   Project Ranking and Internal Rate of Return for Olden Corp.

Project	Risk	Initial Cost	Internal Rate of Return
A	High	$190,000	13%
B	Moderate	$58,000	4%
C	Moderate	$500,000	22%
D	Very high	$45,000	11%
E	Low	$150,000	7%
F	Moderate	$405,000	6%
G	High	$156,000	15%
H	Low	$75,000	12%
I	Low	$180,000	8%

tion, it would be desirable to have diversified investments so that "all the investment eggs are not in one basket." Ranking of projects by risk, initial cost, or return can be made (see Table 7–5).

As might be expected, the IRR rank is not consistent with the risk rank. Some compromises must be made. Various strategies can be implemented at this point to aid in the selection of the projects. Note that Project C has the highest IRR. Although the risk is moderate, the initial investment is the highest of all the projects, $500,000. This project may consume most of the organization's funds for the period. This limits diversification. If the project is successful, a high return is realized, but if the project fails to meet its target of 22% ROI, there are no other projects to offset its failure.

Project G has the next highest return of 15%. Its risk is high, and it has a medium initial investment of $156,000. Is this a good project to invest in?

Table 7–5   Ranking of Olden's Projects by IRR

Project	IRR	Risk	Initial cost
C	22%	Moderate	$500,000
G	15%	High	$156,000
A	13%	High	$190,000
H	12%	Low	$75,000
D	11%	Very high	$45,000
I	8%	Low	$150,000
E	7%	Low	$150,000
F	6%	Moderate	$405,000
B	4%	Moderate	$58,000

Should a large number of projects be funded to spread the risk of failure, or should a few projects be considered that have the highest potential return to maximize potential return?

From the ranking of projects and investments by potential ROI, the organization could fund those projects with the highest returns until the funds are gone. In the example, if a total of $1 million is available and IRR is used, projects C, G, A, H, and D would be funded. These five projects would have a total initial cost of $966,000. If Project I were also funded, the initial costs would exceed the amount of funds available. Perhaps the firm could consider additional financing or revise the order of the projects in terms of lower risk or another characteristic. There are, of course, many different combinations of projects and many different criteria that can be used to select the investments. IRR is an important, but not the only, criterion of the selection decision.

Ranking Methods for Project Selection

1. Internal rate of return.
2. Initial cost of project.
3. Risk of project.
4. Present worth (if lives are the same).
5. Nonfinancial factors.
6. Combination of the above.

Monitoring Projects and Investments

After the project is under way, it can be monitored using time value of money techniques and the four basic methods of analysis. Periodic monitoring of the project can be made with regards to costs, budget, scheduling, productivity, profits, and also to see if the project is meeting the financial goals that were estimated at the beginning of the investment. Corrective action can be taken during the duration of the project if it is not meeting the original goals. For long-life projects and investments such as new product development and investment into manufacturing processes and equipment, the monitoring phase may be the longest phase. Analysis and selection, assuming funding is available, may occur within days or weeks. Project monitoring may occur over many months or years.

IRR as a Project Evaluation and Performance Measurement Tool

After completion of the project or investment, final evaluation is made of the project's performance. In many organizations, a great deal of effort and time is spent on the analysis and selection of projects and investments, but a small, if any, amount of time on monitoring and evaluating performance. Individuals, too, spend a great deal of time selecting investments. Bookshelves are full of popular "Selecting Investments" titles yet there are very few "Monitoring and Measuring Performance" titles.

Time value of money techniques are as applicable to the monitoring and performances measurement of projects and investments as to the original analysis and selection.

Estimating the future IRR of a potential project is an important technique to apply in the analysis and selection stage, but it can also be used as a historical performance measurement tool. If an organization wants to achieve continuous financial improvement, the ROI of past projects should be measured, not to point a finger at failures, but so the project managers and project teams can improve their cost and revenue estimating abilities. When the project is new, everyone is optimistic and it is easy to overestimate the IRR of an investment. Comparing actual costs, revenues, and rates of return with those that were originally estimated in the analysis phase of the project along with the reasons for the differences is an important part of project management. Customer satisfaction, quality, customer response times, scheduling, efficiency, productivity, profitability, and other factors can also be combined with the ROI and other TVM techniques to measure performance. By monitoring current and past projects, the organization can become more effective at future project analysis.

It is also possible to use the ROI of a project as part of the performance measurement and reward system of a project team, department, or work group. Before introducing any rewards based on IRR, however, it is important to have complete understanding of the technique by the participants.

Example of Project Performance Results: Jones Co. The Jones Co. has been working on a project for three years. The actual costs and revenues from the project are shown in Figure 7–13. The costs at the start of the project were $3,000. The revenue at the end of the first year was $2,000; at the end of the second year, $5,000; and at the end of the third year, $8,000. The project's actual costs were $3,000 at the end of the first year; $2,000 at the end of the second year; and $2,000 at the end of the third year. Originally, the project was selected because it had a high estimated return of 32%. Did it actually produce this return? See Figure 7–14.

If the project actually returned the estimated 32% ROI, then the present worth of the actual revenues at 32% would equal the present worth of the actual costs at 32%. Checking at 32%,

$$P_{Revenues} = \$3,000(P/F, n = 2, i = 32\%) + \$6,000(P/F, n = 3, i = 32\%)$$
$$= \$3,000(.5739) + \$6,000(.4348)$$
$$= \$1,722 + \$2,609 = \$4,331$$
$$P_{Costs} = \$3,000 + \$1,000(.7576) = \$3,000 + \$758 = \$3,758$$

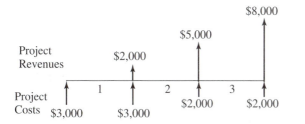

Figure 7–13   Determining Actual ROI for a Project at Jones Co.

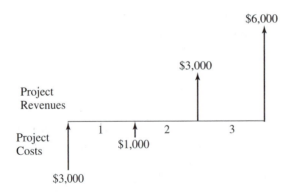

Figure 7–14   Determining Actual ROI for a Project at Jones Co.—Revised Diagram

The two present worths are *not* equal at 32%. The present worth of the revenues is greater than the present worth of costs. The *actual* IRR is very near 40%. (Check this using your own calculator.) The project manager can go back to the original estimates and compare them with the actual results to discover the root causes of the differences. This feedback can improve future project estimates as well as serve as a performance measurement of the current project and the project's team.

Optimizing Project Selection

If ROI, risk, initial cost or other ranking techniques are used to determine the selection of projects, care should be taken to ensure that the total organization's ROI is optimized, not just a single project's return. Placing the organization's funds into one project or investment decreases the funds available for other projects. The overall organizational strategy should be optimized. One investment into a project may be successful, but it may also reduce the effectiveness of other projects.

For example, if one piece of equipment is improved in quality and speed of output, this may cause other equipment in the same work area to become overloaded and have to work overtime to keep up with the increased productivity of the revised equipment. This could actually increase overall costs and reduce the work area's total productivity. Another example is investment into increased marketing and advertising for a product that has a high potential return on investment. This may cause increased costs of inventory holding, order handling, shipping, and customer delays of other products.

> I would rather be vaguely right about investments than precisely wrong.
> —John M. Keynes

Projects are not independent. Most have components that are related to and somewhat dependent on other parts of the organization. Because of this interrelationship, the decision to improve the productivity of one component may cause an imbalance in the other components.

These relationships are often subtle and difficult to predict in advance. As one manager or engineer tries to improve the efficiency, productivity, costs, and quality successfully in one area, other work areas may even become less efficient due to the relationship with the improved process.

Therefore, projects and investments should be compared to each other as well as to the goals of the organization to determine the best allocation of funds. This analysis of interrelationships can be a final part of the project selection process.

The mix of projects must take into account the risks of each project and the overall impact on the organization; internal funding versus external sources of money; and the overall impact of the individual project as well as the individual project returns. Optimizing and balancing the mix of projects and investments are as important as individual project selection.

> Project and investment analysis, selection, monitoring, and performance measurement is a balance between science and art; mathematics and judgment; experience and estimates.

SUMMARY

This chapter has presented the financial techniques for project management—analysis, selection, monitoring, and performance measurement of projects and investments. The topics of this chapter include the following:

- The four methods of project analysis: present worth, annual amount, internal rate of return/ROI, and future worth.

- Examples using the four techniques included lease versus buy analysis, alternative projects, industrial bond valuation, analysis of alternatives with unknown service lives, and projects having different service lives.

- Application of the four techniques to different service life projects using IRR, annual amount, and the common denominator approach.

- Examples of the four techniques included analysis of perpetual life projects, using present worth techniques for the valuation of securities, warehouse construction, IRR application to a project that has unequal costs and revenues over time, and what-if analysis using the four methods.

- Other analysis methods discussed are benefit/cost analysis, payback, discounted cash flow, life cycle analysis, MAPI, and some other proprietary analysis techniques.

- Project management and TVM techniques using analysis, selection, monitoring, and performance measurement were discussed.

- Project management topics included capital investment budgets, project analysis, selection, allocating funds for projects and investments, sources of funds for projects, and project ranking as a selection technique.

- Project and investment monitoring, performance measurement, and optimizing project selection were presented with emphasis on IRR/ROI as a useful tool.

QUESTIONS

1. What are the four basic techniques for analyzing investments and projects?
2. Which of the four methods is applicable to unequal service lives?
3. What is IRR? How is it determined?
4. How can lease versus buy alternatives be analyzed using TVM techniques?
5. State the steps for solving a flow of dollars using each of the techniques:
   a. Present worth
   b. Annual amount
   c. Future worth
   d. Internal rate of return
6. What is meant by net present worth?
7. What is meant by the statement, "Project and investment selection is part science and part art"?
8. How can individuals apply the four basic techniques to their personal financial decisions?
9. What are the four components of project management?
10. How can sensitivity and what-if analysis be used with project analysis?
11. What is discounted cash flow?
12. Why does the payback method often give invalid conclusions?
13. What is the benefit/cost method? How can it be modified to give valid conclusions?
14. What is life cycle analysis? Is it different from the application of the basic TVM technique?
15. What is the MAPI method? Is it valid? What is a shortcoming of this technique?
16. Budgets are usually made up of two parts. What are the parts?
17. What is meant by "rationing funds"?
18. What are some typical nonfinancial factors that may influence the selection of an investment or project?
19. What are common sources of funds the organization can use for projects or investments?
20. What are project and investment ranking criteria other than IRR?
21. What are the disadvantages of using ranking as the *only* approach to selecting investment or alternatives using time value of money techniques?
22. Why is it important to monitor and evaluate the past performance of projects and investments?
23. How can individuals use IRR to monitor and measure the performance of their own investments?
24. What are the advantages and disadvantages of selecting a few high-return projects for investing the organization's funds rather than many projects?
25. What are the problems of relying exclusively on quantitative techniques? On qualitative techniques?
26. What are the potential problems when the initial project or investment's costs and revenues are estimated?
27. When funding projects and investments, what are the differences between using internal funds versus borrowed funds?
28. Why is diversification in project investments important?

PROBLEMS

1. A student accepts a job at a summer camp for 3 months. The pay includes room and board and an end-of-the-job bonus of $1,000 if she stays the full 3 months. If she has a value of 10% for her personal ROI, what is the value of the bonus at the beginning of the work period?
2. A coach of a minor league baseball team has a $125,000 per year contract for the remaining 4 years. If management wants to buy out his contract, what is the minimum offer that he should accept now if his desired ROI is 8%?
3. Boggs Co. has been asked by another firm if they would like to purchase the firm's rights and production equipment for Product 22-L. Boggs is interested because it would round out their product line. The owner of the product has asked Boggs to submit an offer. Boggs has estimated that the revenue per year for the 22-L would be $75,000 per year for 8 years. The annual manufacturing and marketing costs would be $41,000 per year. The initial startup costs for Boggs would be $25,000. If Boggs has a desired ROI of 15%, what is the maximum they should offer for the 22-L product line?
4. A new office building is constructed at a cost of $3 million. It is estimated to have a life of 50 years with a value at that time of $200,000. It will have maintenance costs of $30,000 per year. It will also have major repairs costing $80,000 that occur at

years 10, 20, 30, and 40. And it will have additional repairs at the end of year 25 costing $400,000. Determine the value of the annual amount if the target ROI of the firm is 15%. (Hint—first find the present worth of the funds, then the annual amount.)

5. The Sudz yer Duds chain of do-it-yourself laundry and dry cleaning is anticipating the purchase of 5 new washers at a cost of $1,000 each, which would have $500 per year operating costs and no residual salvage value. If the increased speed of the washers and reduced operating costs will earn $1,200 total per year for the company, and the organization's desired ROI is 12%, what must be the minimum life of the new equipment in order to meet the return?

6. A security that can be purchased for $75 per share is expected to pay a dividend of $6.50 per year. A second security can be purchased for $48 per share that pays a dividend of $4.75 per year. Using the present worth and IRR methods, determine the best investment if any price appreciation of the securities is neglected. Assume the dividend will continue indefinitely into the future.

7. A Midwestern state with a number of state parks is considering increasing the fee charged to tourists that visit the parks. This would provide an additional revenue of $25,000 per year. Determine the current value of this project if the fee increase is considered to last forever and the state has a desired ROI of 10%.

8. A small city is purchasing land to create a city park. The land costs $50,000. There will be construction costs of $20,000 to create a small pond and picnic area. Funds are valued at 11% for the city. The annual maintenance is expected to be $3,000 per year forever. What is the annual cost of the park if it is assumed to be perpetual?

9. Jim can pay his dues to an organization of $50 per year, or he can pay $1,000 now for a "lifetime membership." If Jim plans to live for a long time, what is the IRR of the membership? Assume a large or infinite time span.

10. In order to meet some new environmental standards, engineers and technicians at a chemical firm have designed a new waste control system for one of the processes. The cost of the installed design is expected to be $350,000. It will have little or no residual value at the end of 12 years. The savings over the current disposal costs are expected to be $65,000 per year. The firm typically desires a return of 12% in projects such as these.
a. Determine the IRR for the project.
b. Compare the answer in part a to the target value of 12%.
c. What nonfinancial benefits are associated with this type of project?

11. A new computer network system is being considered for an organization. The initial cost of the system is $400,000. Annual maintenance and operating costs would be $25,000 per year. After 6 years the system is expected to be worth $65,000. The organization has a desired ROI of 15% for its projects. Determine the equivalent annual cost of the system.

12. Some new material handling equipment is being considered for purchase or lease. The initial cost of the equipment would be $150,000, with a life of 8 years, salvage of 5% of the equipment's first cost, and operating costs of $20,000 per year. The firm desires a 12% ROI. A leasing firm has quoted $35,000 per year for the same system. The firm would still have the same operating costs to pay, but would not have the salvage value with the lease. Determine the following:
a. The annual cost of the equipment at 12%.
b. The advantages and disadvantages of the purchase and lease alternatives.

13. A new heating and air conditioning system is being purchased to replace the existing system for a small insurance office. The new installation would cost $6,500 initially, would have a life of 10 years, and is expected to save $420 per year in energy costs. If the organization desires a 12% ROI, determine the net annual cost of the new system including the energy savings. Salvage value is negligible. A more efficient system can be purchased for an additional $1,800 that would save a total of $490 per year in energy. Should the organization consider the more efficient system?

14. Manufacturing equipment is designed and built at a cost of $150,000. It is estimated that it will save $18,000 per year in production costs. The life of the process is expected to be 9 years, with a salvage value of 8% of the first cost of the equipment. Determine the IRR for the investment. What if the actual savings from using the equipment is only $12,000 per year? What if the equipment only lasts 6 years?

15. A product is expected to be produced and sold for a 5-year period. The annual sales are expected to be $45,000 per year. The cost of the equipment is $100,000 initially, with operating costs including labor, material, and overhead of $10,000 per year, and a residual value of $10,000 at the end of 5 years. If the firm has a desired ROI of 10%, determine the present worth and the annual amount of the investment.

16. Two alternatives are being considered as investments. Alternative A has a first cost of $100,000, annual costs of $5,000, and no salvage value at the end of 6 years. Alternative B has a first cost of $125,000 and salvage of $10,000 after 8 years. The operating costs for the equipment are the same for each alternative. If the firm desires an ROI of 8%, determine the following:
   a. The present worth of each. Is this a valid approach?
   b. The annual cost of each alternative.

17. Susie can buy a bond for $890. It has a redemption value of $1,000 in 4 years that pays an interest of 7% or $70 per year to the holder. As an alternative, Susie could buy 100 shares of common stock selling at $8.50 per share. She thinks it will double in 5 years but it does not pay any dividends to the holder. If commissions are neglected, what are the IRR values for each of the investments? Are the risks the same for both investments? What do you recommend to Susie?

18. Mike Rodriguez has an investment fund for his son's education. He has been recommended a bond that pays 6.7% on its face value of $1,000. The bond has been on the market for a few years and its selling price has fluctuated slightly over the years. Bond prices fluctuate as prevailing interest rates change in the marketplace. It currently has 11 more years until its due date, when $1,000 will be paid to the holder. The bond's current price is $965. What is the actual rate of return that the bond is paying over its life when the appreciation and the annual interest are considered?

19. A subscription to a technical journal costs $75 paid now for the next year's issues, or the subscriber can pay $125 for two years rather than pay for two single years' subscriptions. What is the return on the investment of the longer subscription period?

20. For a new product, an initial investment of $500,000 is required. The revenues and costs for each of the five years of the product's life are:

Year	Sales Revenue	Costs
1	$100,000	$100,000
2	$200,000	$150,000
3	$300,000	$100,000
4	$500,000	$150,000

The expected residual value of the equipment is 10% of its first cost of $500,000 at the end of the five years. Determine the IRR of the investment.

21. A firm is considering purchasing some new equipment at an initial cost of $120,000, a life of 8 years, and a residual salvage value at the end of 8 years of $14,000. Or the firm can lease the same equipment for $29,000 per year. The costs of operation such as labor, repairs, maintenance, power, supplies, and related costs are the same for both alternatives. If the firm desires an ROI of 12% on its investments, determine the following:
   a. The best alternative using present worth calculations.
   b. The best selection using the future worth method.
   c. The best selection using annual amount methods.
   d. Apply equivalent internal rate of return calculations and determine the rate at which the net costs of the purchase alternative equals the lease alternative. Compare this rate to the firm's desired ROI. Does this method give the same conclusion as in parts a, b, and c?
   e. What are the noneconomic advantages and disadvantages of the lease versus the purchase?

22. An individual is considering leasing a new car. The initial cost of the car is $18,000 with an estimated life of 5 years and residual salvage value of $4,000 at that time. The car dealer has suggested that the car can be leased for $375 per month for the same period of time. If the operating expenses are the same for the purchaser under both alternatives, what is the best alternative? The individual wants to earn 10% on the investment. Compare the alternatives on an annual amount basis.

23. Three alternatives are being considered for a revised computer network system in a small engineering design office. The organization currently has an ROI objective of 15% on its investments. Analyze the al-

ternatives using present worth and annual amount techniques and determine the best alternative.

Project	1	2	3
Initial hardware cost	$50,000	$70,000	$90,000
Annual operating costs of supplies and repairs	$5,000	$4,000	$2,500
Initial software cost	$10,000	$8,000	$8,000
Salvage after 5 years as a percent of first cost of hardware	5%	7%	8%
Misc. taxes, power, and space costs per year	$4,000	$4,000	$3,000

24. Dodson Inc. is considering purchasing some new processing equipment. The carbon steel alternative has been quoted at $40,000 initially. It would have an 8-year life with residual salvage of $2,000. The operating costs would be $3,000 per year with a major overhaul at the end of the third and sixth years of $4,000. A more corrosion-resistant alternative of stainless steel is offered at $60,000 with a life of 12 years, an operating expense of $1,500 per year, and no major overhauls.
    a. If Dodson desires an ROI of 15%, which is the best alternative?
    b. Should a present worth or annual amount calculation be made?
    c. What is the equivalent IRR at which the two alternatives are equal?
    d. Present the annual amount and equivalent IRR in table form from part c.

25. Four designs for new tooling are proposed. The organization's target return on investment is 15%. No residual salvage value would exist for any of the alternatives. They would each have a 6-year life.

	A	B	C	D
First Cost	$60,000	$80,000	$100,000	$40,000
Savings/year	$15,000	$25,000	$31,000	$12,000

Determine each alternative's IRR and rank them.

26. A family is considering having a new furnace installed in their home. They have received quotations from two companies. They desire to earn 8% on their investments. One furnace is expected to cost $1,800 initially, have operating costs of $1,000 per year, and last 10 years. The second alternative is advertised as a "high efficiency" model and would cost $2,600 initially, have annual costs of $800 per year, and last 12 years. Determine the best alternative. Which of the four analysis methods should be used?

27. A power utility is considering treating the wood utility poles with a preservative and anti-termite solution. The cost for an untreated pole is an average of $450. To treat these poles would cost an additional $40 per pole in material costs and $15 in labor. It is estimated that the treatment would extend the life of the pole from an average of 35 years to 50 years. If the utility desires an 8% return on its investments, and there are 350 poles expected to be set within the next year, should these poles be treated?

28. Two baggage handling systems are being considered for a small airport. The first system has an initial cost of $800,000, operating costs of $50,000 per year, and no salvage value. The second system that is being considered has an initial cost of $1.2 million and operating costs of $25,000 per year. If the airport management considers 12% ROI a target, how long must the more expensive system be used in order for it to be the most economical? Is this a reasonable expectation? Solve this problem on an annual amount basis. What other factors should be considered?

29. A quality improvement project is being considered by an organization. The initial cost of the revised system is expected to be $30,000. It is expected to save $13,000 per year in scrap costs. The organization has a desired ROI of 15% for all of its investments. Calculate the payback using the incorrect method *without* TVM to determine the number of years before the improvements will pay for themselves. Then, using TVM techniques, determine the *correct* number of years for the improvements to pay for themselves. Which method gives the longer time period? Why?

30. A "power saving" air conditioning system is being considered. It is estimated to cost $5,000 more than the inefficient system and expected to save $1,300 per year in power costs. If the firm has a target ROI value

of 12%, determine the payback using the *incorrect* method that does not consider TVM. Next, calculate the correct payback period using TVM techniques. Is this a reasonable expectation? If, under actual use, the system only saves $300 per year in power costs, how long will it take for the system to pay for itself? Is this a reasonable time period?

31. For the coming year, an organization is considering the following projects for investments. Rank the projects using three methods: IRR, risk, and initial cost. State any conclusions after the ranking is completed.

Project Ranking and Internal Rate of Return

Project	Risk	Initial Cost	Internal Rate of Return
A	Moderate	$10,000	10%
B	High	$8,000	15%
C	Low	$20,000	6%
D	Very high	$5,000	16%
E	Moderate	$15,000	12%
F	Low	$12,000	7%
G	High	$18,000	25%
H	Very low	$7,000	10%
I	Moderate	$25,000	13%

What are the advantages and disadvantages of using the three ranking systems?

32. An engineering technician is considering the following investments. Rank the investments by IRR, risk, and minimum investment amount. Should some of the lower potential return investments be considered because of diversification or should only the best returns be considered?

Investment	Estimated IRR	Risk	Minimum Investment Amount
Treasury bills	5.5%	Very low	$10,000
Mutual fund G	12%	Moderate	$5,000
Mutual fund H	15%	High	None
Savings and loan	5%	Low	None
Certificate of deposit	8.3%	Moderate	$5,000
Company 401(k) plan	8%	Moderate	$200 per month (maximum of $600 per month)

33. A small software consulting firm is considering its budget for next year. It believes that the maximum available funds for the coming year would be $50,000 from current profits and retained earnings, and an additional amount of $30,000 could be borrowed from the bank at 8.5% for 5 years. They are trying to determine the projects to fund for the coming year. Some of the projects are for the development of new software to sell and some are cost-reduction projects. The accompanying table shows the IRR, initial cost, potential risk, and the importance as defined by the marketing department. Rank the alternatives and make recommendations.

Project	Importance to Marketing Dept.	Risk	Internal Rate of Return	Initial Investment Required
13-R	Minor	Low	10%	$10,000
126-CC	Low	Moderate	15%	$23,000
30-I	High	High	20%	$30,000
85-F	Moderate	Moderate	12%	$14,000
46-MM	Very high	High	23%	$27.000
443-H	Low	Moderate	5%	$4,000
3-T	Moderate	Low	13%	$45,000
36-YT	High	High	18%	$39,000

34. Circle City is building a new bridge over a small river. The city council has decided to fund the bridge by selling bonds, and to pay the bondholders from future taxes. The cost of the bridge is as follows:

Right-of-way purchase	$50,000
Design	$40,000
Construction	$2,200,000
Maintenance	First year costs of $10,000 increase by $4,000 each year for 50 years

The bonds would be sold for $1,000 each. The city council wants to sell enough bonds to pay for the right-of-way, design, construction, and future maintenance. The total value of the bond issue must be determined. The rate on the bonds is 4.5%.

DISCUSSION CASES

Speck Corp.

As a producer of quality wood products, Speck has become successful in its field. Profits and sales are expanding. Customer satisfaction is high. James Speck is considering the methods they use to evaluate investments in new products and equipment. Currently, the management estimates the profits available for the coming year. From this value, they look at the equipment that is wearing out and needs to be replaced. This equipment replacement is the first priority for funds. Then they look at the new products that the design and marketing team is suggesting for the coming year; the costs, investments, and revenues are estimated; and management makes a decision about funding the new products. This is usually a lively discussion with many differing opinions since the new products are risky and usually only 50% of the new products that are introduced in a given year are successful.

After the old equipment decisions and the new products decision are made, if there are any funds remaining, the management group considers the possible dividends to be paid to the owners. The dividends have been very good in recent years. Jim Speck is not sure if the techniques the organization is using for the investment of funds could be improved. He thinks that as long as business is good and profits are high, the decision process is easy, but he thinks that if the firm should run into a period of low profits, the decision process would break down.

The management group is experienced and works well together. Jim doesn't want to introduce anything into the decision process that would decrease the group's enthusiasm or efficiency such as too much accounting analysis of the decisions. He does, however, feel that most of the decision making is based on intuition. This seems to have worked well, but he wonders if there is anything else they could do to improve the decisions.

1. Is this situation typical in firms or is it unusual?
2. What could Jim do to supplement the decision-making abilities of the group?
3. Should the existing decision-making process be completely changed to a more quantitative technique?
4. What are the differences in making equipment decisions when profits and sales are high compared to when they are low?
5. What other things should Jim consider?

Computer Chip Co.

Computer Chip (CC Co.) has been discussing the possibility of purchasing a small supplier, James Components, that has been a vendor for Computer Chip for a number of years. The supplier also sells its products to others in the computer industry. CC Co. has been very successful in the past few years, and has funds available to purchase the supplier. CC Co. has been considering creating its own in-house production capability to produce the same product that James has been selling to CC Co. But if the supplier could be purchased for a good price, it would eliminate all the startup costs for CC Co. to purchase the business outright.

James has current sales of $3,500,000 per year. The profit after tax is currently $370,000 for the year. James is expected to have sales and earnings of at least the current amount well into the future. The management at James would remain in place if CC Co. purchases James.

CC Co. has a desired ROI of 10% for most of its projects, but sometimes it will target a lower rate if the project is critical to the firm's competitive position in the industry. CC Co. is a publicly held company, but the stock of James is all held by the firm's management.

CC Co. is trying to arrive at a value that is a fair price to pay for James. They are aware that there are a number of different methods to determine the value of a firm. They are not sure what to do.

1. What is the price that CC Co. should pay for James?
2. What other considerations should CC Co. make?

Toby Corp.—Buying a Patent

Jim Toby has developed a small organization that purchases patents from inventors. Some patents are created and owned by individuals who cannot afford to develop the patent into a product or choose not to apply their creative talents to marketing and manufacturing but to pursue other inventions. Jim Toby buys the rights to the patents so that he can either manufacture and market them or find some other firm to produce the product. An inventor has just contacted Jim and they want to negotiate a price on the use of the patent by Jim and his company. Jim has done his analysis on the development costs, production costs, and revenue costs.

Jim's estimates of the new product's cash flows are that an initial cost to get the product into design, manufacturing, and marketing is $25,000; cost at the end of the first year is $25,000; at the end of the second year, $20,000; and at the end of the fourth year, $35,000. The revenues are $20,000 for each of the first and second years, $80,000 at the end of the third year, and $55,000 at the end of the fourth year.

Now that the evaluation of the patent and its product revenues and costs have been estimated, Jim is trying to estimate how much he should offer to the patent holder, and also if the patent has the potential to be a good investment for Jim and his company. Most of Jim's projects have earned 15% ROI.

1. What should Jim do?
2. What nonfinancial factors should Jim consider?
3. What other factors should be considered?

Meyers Co.

At a meeting of the Meyers Co. board of directors finance committee the topic of budgets and capital equipment expenditures was discussed. The finance committee is made up of three members of the board and the treasurer/VP—finance. The treasurer presents the budget for the coming year, which is in the form of a projected Income Statement. The treasurer also presents the proposed capital equipment expenditures for approval. The final item is the request for outside funding such as bank loans, common and preferred stock sale of securities already authorized by the shareholders, and any large lease commitments that the firm will require within the coming year. All of these items require board approval, but the committee is the first reviewer and is responsible for knowing the details of the request when it is presented to the board.

Jim Kai is the treasurer of Meyers. He has been there since he graduated from business school with his MBA in finance 22 years ago. He has worked his way up to his present posi-

Proposed Capital Investments for Meyers Co.

Investment	DCF Amount in $	Payback in Years
Warehouse addition	+$200,000	3.6
CNC machine	+$40,000	2.8
Computer network revision	+$50,000	1.7
Delivery vans	+$120,000	2.6
Remodeled lobby	+$200,000	3.8
Painting facility	+$400,500	2.4
Revised utilities in plant	+$25,000	4.6

tion. He has presented the request for some capital equipment items to the committee in the usual form of both discounted cash flow and payback ranking, as shown in the accompanying table.

The discounted cash flow is calculated by bring the future revenues and savings of the project back to the present using the firm's desired ROI of 12% and subtracting from this value the present value of the costs brought back to the present. Jim eliminates any negative DCF projects since a negative indicates that the DCF of the costs is greater than the DCF of the revenues. He proposes that only the highest DCF projects and the shortest payback alternatives be funded.

Payback is found by dividing the initial investment by the annual savings. This gives an answer in the number of years that is required to recover the first cost of the investment. Meyers has had a rule of thumb that a project should pay for itself in 3.5 years. There have been some exceptions to this in the past, however. The firm has been using both payback and DCF to evaluate investments for the past 10 years. The current profits are ample to pay for these projects, and there would be some cash left over after funding all of these projects.

Stuart Michaels, an engineer, has been brought on the board because of his knowledge of quality systems, Total Quality Management, and related quality concerns. Meyers is undergoing a complete overhaul of its quality system and needed some assistance at the board level to advise the internal quality functions on major quality changes. Stuart has experience running his own companies over the past 20 years. He feels that the analysis of the new projects using discounted cash flow and payback are incorrect, and the firm may be making serious errors in their analysis.

Jim, the treasurer, defends the analysis, saying that all of the cost data comes from the production and design personnel based on actual quotes and sound accounting data. Although there have been two or three projects over the past 10 years that haven't worked out, Jim believes that these analysis methods have helped the company invest its funds profitably. The other directors don't really have any experience in the analysis, although those on the committee have a little experience using DCF in their respective organizations. The committee doesn't know what to do. The board meets next week to consider these and other budget items, so something has to be done.

1. Who is correct, Jim or Stuart? Are they both correct?

2. What are the difficulties of dealing with decision makers who may not be familiar with mathematical techniques of analysis?

3. Could there be some other projects that are not being considered that should be considered?

4. What is the general legal and ethical responsibility of the directors with respect to investing company funds?

The President's Memo

Bill Abbot, of Process Engineering, arrives at his office Monday morning and finds the following memo from the company president.

To: Bill Abbot, Process Engineering
From: Harry Mills, President
Ref: Next Quarter's Capital Investment

Bill:

I know you have some experience with the analysis of capital investments from your previous assignments. I need your help on this one. We present our requests to the board each quarter for capital investments. In the past, the financial people have made the presentation for us and are able to explain the financial analysis to the board, and I usually discuss the technical details of the investment. The chairman of the board told me at lunch the other day that there have been a number of lawsuits against directors in the U.S. lately for not actively managing their firms' money. We have never had any trouble in this area, but we seem to be getting more questions at the shareholders' meetings about how we invest the company's money. The finance people have been using payback, DCF, and benefit/cost analysis, which do not use interest calculations, to analyze our investments. I feel that we should have some analysis using interest calculations. I know you have recommended to me in the past that we should tighten up our analysis, selection, and monitoring of our projects and investments. I would like your analysis of the following projects, and may want your help at the next directors' meeting to answer questions. I need your analysis this week.

Thanks for your help,
Harry Mills

Projects

Product 18-4 Development. The new product group feels that we should be in this market immediately. They estimate an initial startup cost of $400,000, which would put us right there as one of the big three in this market. This initial cost would cover R&D, prototypes, initial advertising, market research, and related costs. Annual equipment maintenance would be $5,000 per year, material and labor costs would be $12.00 per unit, revenue would be estimated at $25 per unit, and the initial equipment would be an additional $100,000 with an 8-year life and salvage of $15,000 at that time. Annual marketing costs would be $10,000 per year. The sales are estimated to be 2,000 units for the first year, 4,000 units for the second year, 6,000 units per year for years 3, 4, 5, 6, and 5,000 units per year for years 7, 8, 9, and 10.

Dept. 14 Equipment Upgrade. This department is asking for $100,000. They estimate that the new equipment would replace two workers at $8.58 per hour including fringes. They work a standard single shift in that department, as you know. The equipment would last 5 years and have a salvage of $5,000 then. Maintenance and annual operating expenses would be $8,000 per year.

Computer Department Request. The computer department is requesting another high-speed printer and additions to the network. The equipment can be leased for $35,000 per year. The savings on this investment are estimated to be 2 hours per shift. The department works 2 shifts per day, 5 days per week. The accounting department estimates that the hourly rate in the computer area is $110 per hour. I personally feel that we have given too much to the computer department in the past 3 years. They always want new equipment and it never lasts as long as the manufacturing equipment.

Apex Corp. Our subsidiary, Apex, is requesting $400,000 for plant expansion, equipment, inventory, and related investments. Last year, as you may know, their pretax earnings were $100,000 on an asset base of $1 million and this year they are expected to earn $200,000 pretax on an asset base of $1.4 million. I wish we had more details on this one.

Plant Engineering. The plant department needs $250,000 for new equipment, machine controls, and revised utilities. The average equipment life is 12 years, and they estimate that they will save $40,000 per year as a result of this investment. I think they do a great job and we should give them all the help we can.

Assembly Requests. The assembly department wants to lease new equipment for $25,000 per year. Maintenance costs for the new equipment are approximately $5,000 per year. Equipment could be purchased for $250,000 that would have an 8-year life and the same operating expenses. I don't know the expected salvage value.

As you may know, the board of directors wants to earn 15% pretax return on our investments. Also, for your information, it looks like current earnings, borrowing, and stock sale would total $500,000 to $600,000 in new funds available for investment this coming quarter. I'll be visiting our South American plants this week, and I'll see you when I return on Friday.

1. Analyze the budget requests.
2. What additional things would you like to know if you were Bill?
3. What are the responsibilities of directors in the management of the organization's funds?
4. What should the organization know about future investment analysis and selection?

Millionaires Club (E)

Amy Cook, the treasurer of the student investment club, is trying to prepare a report for the next club meeting. She wants to summarize all of the investments to date so that everyone can see how the club is doing. She has the information from the records, but she isn't sure how to analyze and present the information.

Stock A: 125 shares purchased Sept. 25 and sold March 15 next year; cost was $12.50 per share and commissions were $50; the selling price was $18.90 and commissions were $62.00.

Stock K: Purchased as a new issue on July 10 of last year for $15 per share; 200 shares were purchased; still own the security today, June 5, and the quote is $16.50 per share.

Stock D: Bought 50 shares for $45 per share and commission of $95, 18 months and 5 days ago; the quote today is $55 per share and a $1.10 per share dividend was paid last December.

Stock R: Purchased 58 shares 5 months ago for $34 per share and commission of $58; it is currently quoted at $29 per share, and it paid an annual divided of 50 cents per share last month.

Stock M: 1,000 shares purchased at $2.25 per share 4 months ago, including commissions. Current quote is $1.80 bid and $2.05 asked.

Stock C: Purchased 300 shares 13 months ago at $67.50, and the current price is $81.50 per share. Commissions on the purchase were $150.

Stock Y: Purchased 55 shares for $29 per share 20 months ago and commissions of $60; 25 shares were sold 6 months ago for $34 and commissions of $50. The current value is $31 per share.

Amy is going to use her spreadsheet software to make her analysis and report.

1. Set up a spreadsheet to make the calculations and prepare the report to the group.

2. How can IRR/ROI analysis help the club make better decisions?

3. How can the calculations be combined with the initial decisions to improve the decision making of the group?

III FINANCIAL DECISION MAKING

Chapter 8   Breakeven Analysis

KEY TERMS

Area Graph Plot of unit costs, revenue, and profit versus a specific quantity of output.

Asset-Intensive Firm When the organization has very high fixed costs with respect to variable costs.

Breakeven (BE) The point at which total costs equal total revenue.

Capacity A measure of the maximum output an organization, department, or equipment can produce in a time period. It is usually fixed in the short term.

Capacity Imbalance Also called bottlenecks or constraints. This occurs when components of a process produce at different rates and inventory accumulates because of the imbalance.

Cost-Volume-Profit Relationship Same as breakeven.

Fixed Cost Costs that are not dependent on the volume of output, such as rent, depreciation, salaries, advertising, and other overhead costs.

Isocost Quantity of output at which two processes or products have equal costs.

Labor-Intensive Firm When the organization has very low fixed costs with respect to variable costs.

Linear Breakeven The breakeven case when the cost and revenue components are linear.

Make or Buy The financial decision to either produce the product or service or to purchase it from an outside vendor.

Nonlinear Breakeven When the breakeven components, variable and fixed costs and revenues, are dependent on output and are nonlinear.

Semifixed Costs Costs that are a hybrid of variable and fixed costs. They are sometimes independent of output and sometimes dependent on output volume. Examples include taxes, salaries, and power expenses.

Total Cost and Revenue Graph Plot of variable costs, fixed costs, total costs, and total revenue versus total quantity of output.

Total Costs The sum of variable costs and fixed costs of a product or process.

Total Revenue The product of the quantity sold times the selling price, determined by the demand curve.

Unit Cost Graph Plot of unit costs (variable, fixed, and total) versus total quantity produced, which shows declining costs as quantity of output increases.

Variable Cost The material and labor cost of producing a good or service.

LEARNING CONCEPTS

- Calculating breakeven and isocost points at which revenue equals costs or costs of two processes or products are equal.
- Calculating profits from known costs, revenues, and outputs of processes or organizations.
- Drawing and plotting total graphs, area graphs, and unit cost graphs showing revenue, costs, profits, quantity, and capacity of firms, processes, and departments, also capacity imbalances and bottlenecks that produce higher costs and lower efficiency.
- Calculating isocosts between two or more processes.
- Time value of money and breakeven calculations.
- Make or buy decisions.
- Asset- and labor-intensive firms, and monopoly organizations justified on breakeven concepts.

INTRODUCTION

This chapter builds on the topics of financial statements and time value of money. Breakeven (BE) calculation is a useful and powerful tool that is used frequently to determine the volume of product or service that must be produced to maximize profitability. It is used to choose between two or more alternatives and to determine make or buy decisions. When time value of money analysis is combined with breakeven classifications an even more powerful decision and analysis tool is created. Process engineers, designers, purchasing agents, sales managers, and department managers can apply breakeven analysis to their decisions. Breakeven analysis is sometimes also referred to as "cost-volume-profit (CVP)" analysis.

The topics presented that utilize BE analysis are linear and nonlinear BE; BE graphs, including area and unit cost graphs; isocosts between two processes or products; impact of changes in the components of BE on the profit of the activity; BE's relationship to department and organization capacity; TVM and BE; make or buy analysis and decisions; managing asset- and labor-intensive organizations; and BE analysis of monopolies.

LINEAR BREAKEVEN

The simplest breakeven concept is the **linear breakeven** case. It is a good first approximation and may apply to many situations.

Figure 8–1 shows a linear breakeven graph. Quantity of output per year is shown on the horizontal axis and total costs and revenues in dollars are shown on the vertical axis.

Fixed Costs

Fixed costs (FC) per year are shown as a horizontal line. They represent those costs that *do not vary* with output. Examples of these costs would be rent, depreciation, advertising, building maintenance, management salaries, property taxes, and other fixed or overhead costs. Although it is true that over the long run all costs can vary, these costs are fixed with respect to output in the short run. With fixed costs, there is the expense of rent or lease costs, depreciation, taxes, management salaries, and other nonvariable costs even at zero output. This is the opposite of variable costs, which are typically zero at zero output.

Variable Costs

The **variable costs** (VC) are drawn on the graph starting from the origin and sloping upward to the right. The slope is the cost per unit, which includes material and labor costs. The line starts at the origin since if no output is produced, no variable costs are expended. Variable costs are those costs that *do vary with output,* such as direct and indirect labor and direct and indirect material costs.

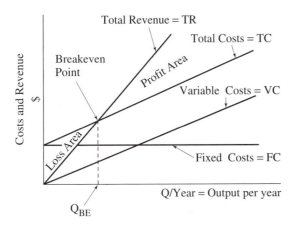

Figure 8–1   Linear Breakeven Graph

Total Costs

Total costs (TC) are found by adding fixed to variable costs. In this linear case, the total cost line is parallel to the variable cost line and intersects the vertical axis at the same point as the fixed cost line.

Total Revenue

Total revenue (TR) is the selling price multiplied by the quantity sold. The TR line on the graph starts at the origin because if no product is sold, no revenue is received. The line slopes upward to the right and intersects the TC line. The point of intersection of TC and TR is the breakeven point.

Total revenue and variable and fixed costs are assumed to be linear. Generally, to sell more units, the price must be reduced (decreasing the slope of the revenue curve). The total revenue for any level of sales is the quantity sold times the selling price. To understand the concepts of BE, we will first assume that the cost and revenue functions are linear. Later in the chapter, we will look at the more realistic nonlinear case of BE calculations.

Total Revenue from the Demand Curve

Figure 8–2 shows the linear demand curve that determines the relationship between the quantity of units sold and the selling price. It shows that few units are sold at a high price and that larger quantities are sold as the price decreases.

Quantity Sold/Year	Price/Unit	Total Revenue
10	$2.00	$20.00
15	$1.80	$27.00
20	$1.60	$32.00
25	$1.40	$35.00
30	$1.20	$36.00
35	$1.00	$35.00
40	$.80	$32.00
45	$.60	$27.00
50	$.40	$20.00

Here, the demand curve is shown as a linear relationship for simplicity. Generally, the curve is downward sloping and decreases in slope further away from the origin, which indicates that to sell a larger quantity, more than a linear decrease in price must occur. The relationship between selling prices and quantity determines the total revenue function. This information is typically a responsibility of the marketing and sales departments.

Breakeven and Profit

The point of intersection of total revenue and total costs is the **breakeven point.** This comes from the definition of the Income Statement: when revenue equals costs, profit is zero, the BE point. Directly below this point on the horizontal axis is the breakeven quantity (Q_{BE}). This is

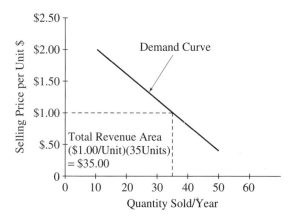

Figure 8–2   Typical Demand Curve to Determine Selling Price vs. Quantity Sold

the quantity that, if produced and sold, would yield zero profit. To the left of this point, the costs are greater than the revenue; thus there would be a loss or negative profit. An output and sales to the right of the Q_{BE} is the area of profit, since revenues are greater than costs in this area. (See Figure 8–1.)

The triangular area bounded by *TR* and *TC* to the right of the breakeven point represents the profit area. This area expands as *Q* increases to the right. At a higher output, profits are greater. This is due to the increasing variable costs, but constant fixed costs. To the left of the breakeven point is the triangular loss area where costs are greater than revenues. From the graph, maximum profit would occur at an infinite output. This is unrealistic, of course, and is one reason why the linear case of breakeven is an imperfect representation of reality. For simplified analysis, however, costs and revenues can be assumed to be linear over short ranges of output.

Breakeven

At breakeven, the total revenues equal the total costs.

Example of Breakeven Calculations: Agricultural Products Corp. Assume that material and labor costs are $5 per unit, fixed cost is $15,000 per year, and the selling price for the product is $8 per unit. Determine the BE point for Agricultural Products. What is the profit if 5,000 units per year are produced and sold? What is the profit if 6,000 units per year are sold? If 3,000 units per year are sold?

At the BE point,

$$TR = TC$$
$$TR = \text{(selling price/unit)(quantity sold/year)}$$
$$TC = FC + VC = FC + \text{(cost/unit)(quantity/year)}$$

Substituting into the equation,

$$(SP)(Q) = FC + (VC)(Q)$$
$$(\$8/\text{unit})(Q \text{ units/year}) = \$15,000/\text{year} + (\$5/\text{unit})(Q \text{ units/year})$$
$$8Q = 15,000 + 5Q$$
$$3Q = 15,000$$
$$Q = 5,000 \text{ units/year for breakeven}$$

If a quantity greater than 5,000 units per year is produced and sold, there would be a profit. If a quantity less than 5,000 units per year is produced and sold, a loss would result. The graph of the BE, costs, and revenues is shown in Figure 8–3.

At an output and sales of 5,000 units per year, no profit is earned because $TR = TC$ and profit is zero. At 6,000 units per year (to the right of the BE point) the revenue is:

$$(6,000 \text{ units/year})(\$8/\text{unit}) = \$48,000/\text{year}$$

The cost at 6,000 units is:

$$TC = FC + VC$$
$$= \$15,000/\text{year} + (\$5/\text{unit})(6000/\text{year}) = \$45,000/\text{year}$$

The profit at 6,000 units per year is

$$P = TR - TC = \$48,000 - \$45,000 = + \$3,000/\text{year}$$

At 3,000 units per year (to the left of the BE point), the revenue is:

$$(3,000 \text{ per year})(\$8/\text{unit}) = \$24,000/\text{year}$$

The cost at 3,000 units is:

$$TC = FC + VC = \$15,000 + (\$5/\text{unit})(3,000 \text{ units/year}) = \$30,000/\text{year}$$

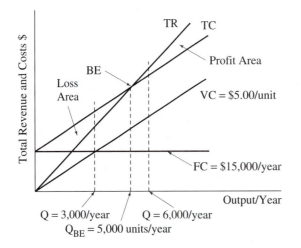

Figure 8–3   Breakeven Graph for Agricultural Products Corp.

The profit at 3,000 units output per year is:

$$P = TR - VC = \$24,000 - \$30,000 = -\$6,000/\text{year (loss)}$$

The location of the BE quantity of 5,000, 3,000, and 6,000 units of output per year are shown on the graph in Figure 8–3. The costs are greater than the revenues (loss) at 3,000 units per year, and the revenue is greater than the costs at 6,000 units per year.

Unit Costs

The previous calculations and graph were based on total costs and revenue. It is also useful to calculate the **unit costs** and unit revenues. To calculate the unit values, the total variable cost, total fixed cost, and total revenues are divided by the quantity produced and sold. For example, the cost per units at 3,000 units per year is:

Total variable cost/3,000 = $15,000/3,000 = $5.00 per unit

The fixed cost per unit at this output is

$15,000/3,000 = $5.00 per unit

The revenue per unit at this output is

$24,000/3,000 units = $8.00 per unit

Note that when the variable cost per unit and the revenue cost per unit are calculated, they are constant over all output values. The fixed cost per unit varies with output, and at 5,000 units output per year is $15,000/5,000 units = $3. The fixed cost per unit at an output of 6,000 units per year is $15,000/6,000 units = $2.50 per unit.

Graphically, the fixed cost per unit and total cost per unit are plotted as curved lines. The plot of the revenue per unit and variable cost per unit are horizontal lines. *The important conclusion from the graph is that as the quantity of output increases, the total cost per unit decreases.*

Calculations can be made for other levels of output to give the resulting graph shown in Figure 8–4.

The graph in Figure 8–4 shows that the total cost decreases as the quantity of output increases. This decreasing cost of output permits the application of different strategies.

1. With the lower cost, the producer-seller is able to reduce the selling price and still maintain overall total profit. The lower price, given the demand curve, permits a larger quantity to be sold. This leads to even further cost decreases. This strategy assumes that the producer has the excess capacity with which to increase output and that there are sufficient additional buyers.

2. Another strategy for the producer is to increase output and gain a lower unit cost, but to maintain the original price and increase advertising and promotion of the product. This strategy keeps the price at the original level while attempting to increase the quantity sold.

Unit Costs

When total fixed costs are constant and independent of output, unit costs decrease as the output increases.

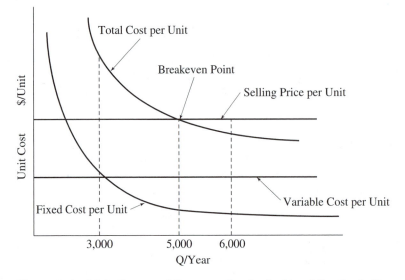

Figure 8–4   Unit Costs and Revenue for Agricultural Products Corp.

Example of Total Cost and Quantity of Output: Relax and Enjoy Co. For Relax and Enjoy Co., a plastics company that produces molded lawn chairs, the total fixed costs for the year, from accounting records, are $4,500,000. The material and labor costs are $6 per unit. The company is currently producing 500,000 units per year on a two-shift-per-day basis. What are the costs of production and what is the current unit cost?

$$\text{Total cost} = FC + VC = \$4,500,000 + (\$6.00/\text{unit})(500,000 \text{ units per year})$$
$$TC = \$4,500,000 + 3,000,000 = \$7,500,000 \text{ per year}$$
$$\text{Total cost per unit} = TC/\text{units per year} = \$7,500,000/500,000 = \$15.00 \text{ per unit}$$

If the company is currently selling the units to wholesalers for $16 per unit, what is the total revenue and total profit from the current two-shift production?

$$\text{Total revenue} = (\text{selling price})(\text{units per year}) = (\$16.00)(500,000 \text{ units per year})$$
$$= \$8,000,000 \text{ per year}$$
$$\text{Total profit} = TR - TC = \$8,000,000 - \$7,500,000 = \$500,000 \text{ per year}$$

Relax and Enjoy Co. is considering adding a third shift as the demand for their product increases. They estimate that the fixed costs would remain constant at $4,500,000. The variable cost per unit would continue to be $6 per unit. What would be the new unit cost if the third shift added 175,000 units per year to the output? What would be the total revenue and profit if the selling price of the new output is the same, $16 per unit?

$$\text{New total cost} = FC + VC = \$4,500,000 + (\$6.00/\text{unit})(500,000 + 175,000)$$
$$TC = \$4,500,000 + \$4,050,000 = \$8,550,000 \text{ per year}$$
$$\text{New unit cost} = TC/\text{units per year} = \$8,550,000/675,000 \text{ units per year}$$
$$UC = \$12.67 \text{ per unit (rounded)}$$
$$\text{New total revenue} = (\text{selling price})(\text{output per year}) = (\$16.00)(675,000 \text{ units/year})$$
$$TR = \$10,800,000 \text{ per year}$$
$$\text{Total profit} = TR - TC = \$10,800,000 - \$8,550,000 = \$2,250,000 \text{ per year}$$

The result from expanding the output from 500,000 to 675,000 units per year, a 35% increase, is an increase in profits from $500,000 to $2,250,000, a 350% increase! This demonstrates the concept of potential profit improvement by increasing output. This is not a completely realistic situation, however.

More realistically, the fixed costs would increase due to increased management costs, increased maintenance, and related fixed costs. For example, the fixed costs would increase from the present $4,500,000 to $5,000,000. Also, it is likely that Relax and Enjoy Co. would have to offer its wholesale buyers a reduced price to sell the additional output (downward-sloping demand curve). Assume that the price would be reduced from the existing $16 per unit to $14.50 per unit, a reduction of $1.50 per unit in selling price. The new revenues and profits are:

$$\text{Total revenue} = (\text{selling price})(\text{units/year}) = (\$14.50)(675{,}000) = \$9{,}787{,}500/\text{year}$$
$$\text{Total cost} = FC + VC = \$5{,}000{,}000 + (\$6.00/\text{unit})(675{,}000) = \$9{,}050{,}000/\text{year}$$
$$\text{Total profit} = TR - TC = \$9{,}787{,}500 - \$9{,}050{,}00 = \$737{,}500$$

The output increased by 35% and the profit increased from $500,000 to $737,500, a 48% increase. This situation is more likely than the previous case in which selling price and fixed costs were constant even when the output increased. This calculation can be shown on a breakeven graph of total revenues and costs versus quantity. (See Figure 8–5.)

On the total revenue and cost graph in Figure 8–5, the *TR* line increases at a slower rate as the price of the chairs is reduced. The *FC* line increases as the output is expanded to 675,000 units per year and this causes the *TC* line to increase also. The resulting profit increases from the original $500,000 to the new value of $737,500 per year.

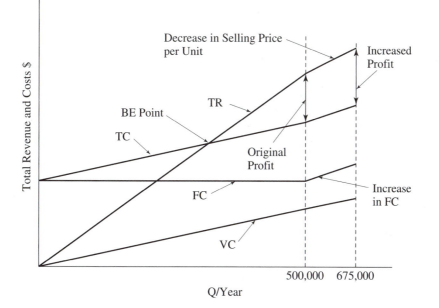

Figure 8–5   Relax and Enjoy Co.

Expanding output permits lower unit costs, which in turn result in a higher volume of sales and possible higher profits.

NONLINEAR BREAKEVEN

The disadvantages and limitations of the linear breakeven example can be improved with a more realistic nonlinear case. In a production situation, there will be a maximum capacity that cannot be exceeded in the short run. This maximum may be stated in terms of the number of shifts that are worked per week, the number of hours available per week, the number of shifts per day, or the maximum number of units per year that can be produced. On the **nonlinear breakeven** graph in Figure 8–6, this maximum capacity is represented by a vertical dashed line.

Fixed Costs in the Nonlinear Case

As this maximum capacity is approached, the fixed costs are no longer independent of the quantity of output, and they will increase (as they do for Relax and Enjoy Co.). An example of this is that maintenance costs do not stay constant as maximum capacity is approached, but *increase*. This is because the equipment is used more hours per week or shift, which causes maintenance costs to rise.

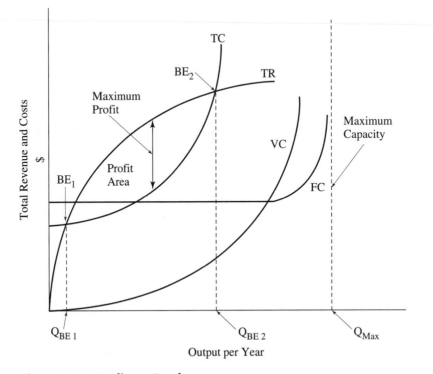

Figure 8–6   Nonlinear Breakeven

Variable Costs in the Nonlinear Case

The variable cost is made up of material and labor costs. As more units are produced, more material is purchased, and the unit cost of the material from suppliers will usually be less due to quantity discounts for buying increased amounts of material. However, unit labor costs will also increase as output goes up because of the additional costs of labor required for the second and third shifts. This shift premium or shift differential pay is extra pay for working the shifts other than the day shift. Also, the labor efficiency sometimes is lower on the second and third shift. Of the decreasing material costs and the increasing labor costs, the labor costs are usually the larger portion, so that the net change to the variable cost curve is an *increase in slope* as it moves away from the origin. When the increasing slope of the variable costs curve is added to the increasing slope of the fixed costs, the *total cost curve has an even greater increasing upward slope*.

Revenue for the Nonlinear Case

The revenue function is also nonlinear. In order to sell more units, either the price per unit must be reduced, due to the demand curve, or additional advertising and promotion costs must be incurred in order to sell the additional units. These additional costs have the effect of decreasing the net selling price. As more units need to be sold, additional advertising and promotion dollars must be spent and the total revenue curve *decreases in slope* due to the decreasing net selling price. The curve starts at the origin and goes up to the right with decreasing slope.

Profit in the Nonlinear Case

The result is two intersections of the total cost and total revenue curves giving two breakeven points. To the left of breakeven point 1, profit is negative because cost is greater than revenue. Between breakeven points 1 and 2, profit is positive because revenue is greater than cost. And to the right of breakeven point 2, profits are again negative. Thus there are three areas of profit and loss. The profit area is between the two breakeven points, and the loss areas are below and above this profit area. Although not perfect, this representation is a better description of actual breakeven situations for a particular product or the total firm than the linear model.

There may be a maximum or optimum profit point. Moving right or left from this point reduces profit. In an actual case, it is difficult to determine this optimum point, but the concept is important to understand. If the firm is operating at an optimum point and increases output further, profits may actually go down even though output is above the breakeven point. In a manufacturing situation, the concept of breakeven is useful to explain why, even as quantity increases, *profits may decrease*. Or sometimes, as quantity of output is reduced in the short run, profit increases. The cost and revenue curves shown here are smooth and continuous. In the actual situation, the lines may be broken or discontinuous over time (as they are for Relax and Enjoy Co.). For example, the slope of the line may change abruptly as the quantity is produced by the second shift, and this change in costs would be shown as an abrupt change in slope of the cost curve.

Semifixed Costs

A third category of costs are *partially fixed and partially variable*. These are usually called **semifixed** or semivariable **costs**. Management salaries are fixed in some situations, and at other times may include overtime pay and bonuses. Advertising is usually fixed, but sometimes in order to increase sales advertising may be increased. Some taxes, such as those on land and buildings, are fixed, but there are taxes based on profits and property taxes, which are dependent on output to some degree. Legal costs are usually thought of as fixed, except that as more product is sold, there may be more warranty claims and legal costs of liability. Power costs may have a fixed portion and also a variable portion. For simplification, the costs can be grouped into either fixed or variable categories.

Breakeven Analysis Conclusions

Design and process engineers, managers, technicians, purchasing buyers, cost estimators, marketing and sales staff should have a basic understanding of breakeven analysis. It is an important tool that is used frequently.

The main conclusion about breakeven is that *profits are dependent on the quantity of output.* There is a minimum output that yields a loss below the breakeven point or no profit at the breakeven point. And there is an output above which profits are maximized by maximizing output. The unit cost of a product or service decreases as the quantity increases due to the smaller portion of fixed cost per unit.

Linear and Nonlinear Breakeven

For linear breakeven, maximum profits occur at maximum output. For nonlinear breakeven, maximum profits occur at an optimum output, which is less than maximum output.

The discussion of breakeven has demonstrated the following:

- Costs can be categorized into fixed and variable costs; the sum of these is total cost. In linear breakeven, fixed costs are constant over all ranges of output. Variable costs and revenue are linear. Variable costs include material and labor. Fixed costs include all other costs.

- Breakeven is defined as total costs = total revenue.

- Profit is positive above the breakeven point and a loss (negative profit) below the breakeven point.

- Profits can be increased by increasing output above the breakeven point. This is because fixed costs are constant over ranges of output.

- In an actual case, fixed costs are only constant over short ranges of output, and variable costs and revenue are not linear. This more general case is called nonlinear breakeven.

- In nonlinear breakeven, there may be multiple breakeven points and a limited range of profitable outputs.

- One of the causes of fixed costs increasing at higher outputs is capacity limitations. As maximum capacity is approached, fixed costs increase.

- One reason why variable costs increase as output increases is the additional labor costs of additional shifts or overtime.

- As output is increased, selling price may have to be decreased in order to sell the increased output. This causes the total revenue curve to decrease in slope.
- Capacity can limit profitable expanding output.

CHANGES IN FIXED AND VARIABLE COSTS AND TOTAL REVENUE

Costs and revenues are dynamic and changing, but if a single cost or revenue is isolated, its effect on breakeven can be seen. The six graphs shown in Figure 8–7 isolate the fixed and variable costs and the selling price of the product or service to show the effect of changes on the breakeven point.

Increase in Fixed Costs

An increase in FC is shown on graph a by the horizontal FC line shifting vertically upward to FC'. An increase in fixed costs would result from the increase of overhead costs of annual equipment, rent, salaries, advertising, and other costs that are independent of output. This causes an increase in the TC with a vertical shift upward to TC'. The result is that BE point moves to the right, $Q_{BE\,1}$ to $Q_{BE\,2}$. A larger quantity must be sold in order to break even.

Decrease in Fixed Costs

A decrease in FC is shown on graph b by FC shifting downward to FC'. This causes a shift downward from TC to TC'. It results from decreases in overhead costs, salaries, rent, advertising or other decreases in costs that are independent of output. The breakeven point decreases and moves left from $Q_{BE\,1}$ to $Q_{BE\,2}$.

Increase in Variable Costs

An increase in VC is shown on graph c by an increase in the slope to VC'. This could be the result of increases in unit material costs or unit labor costs. Material suppliers' price increases, an increase in scrap rates, or other material-related costs could be the cause of the material cost increase. Labor cost increases could be from higher wage rates, decreased productivity, or other labor-related cost increases. The increase in VC slope causes an increase in the slope of TC, which in turn causes the BE point to move to the right from $Q_{BE\,1}$ to $Q_{BE\,2}$. A higher output must be produced and sold to make the same profit.

Decrease in Variable Costs

A decrease in VC is shown on graph d. It results from decreases in unit material labor costs, which could be caused by lower material prices, reduced scrap, reduced wage costs, improved productivity, and other material or labor costs. A decrease in the TC slope results in the BE point moving to the left from $Q_{BE\,1}$ to $Q_{BE\,2}$.

Increase in Total Revenue

When the selling price of the product or services increases, the slope of the TR line (see graph e) increases to TR'. The intersection of the TR and TC lines moves to a new BE point to the left and the BE point moves from $Q_{BE\,1}$ to $Q_{BE\,2}$. Profits would increase.

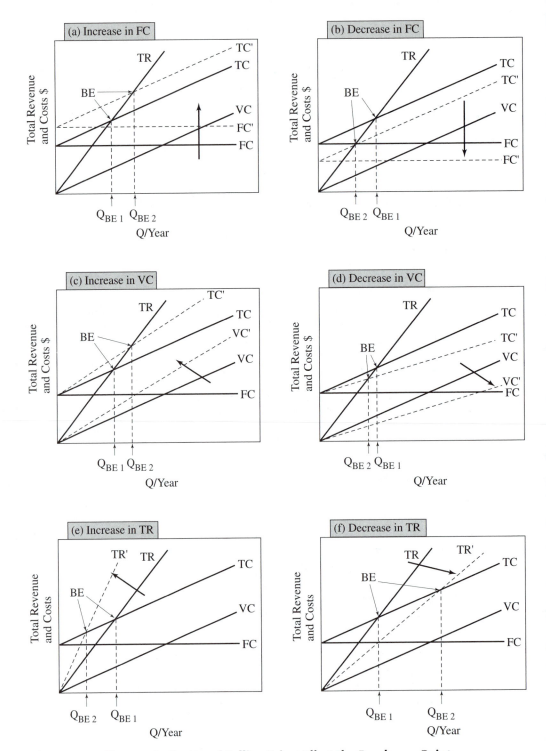

Figure 8–7   Changes in Costs and Selling Price Affect the Breakeven Point

Decrease in Total Revenue

A reduction in the selling price of the service or product results in the slope of the *TR* line decreasing from *TR* to *TR'* (see graph f). The breakeven point moves to the right from $Q_{BE\,1}$ to $Q_{BE\,2}$ and profits decrease.

It is difficult to isolate a single cost or price change. The changes in costs and revenues of a product or service are dynamic and continually changing. Material cost may increase or decrease due to vendors' price changes. Labor costs generally increase over long periods, but are constant for short periods. As new equipment is added and old equipment is retired, the fixed costs increase or decrease over time. Selling prices may increase or decrease as the product meets the competition. By isolating the changes in costs and revenues, the effects on the breakeven point and thus the profit are seen. Also, as discussed previously, the actual breakeven graph for most products and services is nonlinear. The shifts in BE points can be demonstrated in the nonlinear cases as well.

To increase profits, the strategy is to increase selling price, reduce variable and fixed costs, or use a combination of these. In these cases, the breakeven point moves to the left and the profit area to the right of the breakeven point also increases. Some limitations to this strategy are production capacity, market demand, and nonlinear costs and revenues.

Area Graph

In addition to the total cost and revenue and unit cost and revenue graphs, another graph shows breakeven information in a different way. The **area graph** is plotted for a particular output level. It shows the variable costs and fixed costs as areas by plotting the quantity of output and sales on the horizontal axis and the *unit* variable and fixed costs and *unit* selling prices on the vertical axis. Multiplying the unit costs and the output results in areas representing total variable costs, total fixed costs, total revenues and profits. This graph is shown in Figure 8–8.

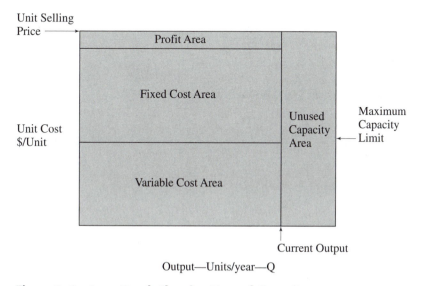

Figure 8–8   Area Graph Showing Unused Capacity

The unit cost is divided into variable and fixed costs, and the difference between the total cost and revenue is profit. The profit per unit times quantity gives the profit area. The potential unused capacity area is the difference between maximum output and current output times the unit selling price. Figure 8–9 shows the diagram if the firm could expand into the unused capacity area.

To expand to the maximum capacity, the variable costs of material and labor would be increased. The additional area of variable cost is shown. If the fixed costs are truly fixed and already accounted for, they do not increase as the output expands into the unused capacity area. The area between the fixed cost area and the maximum capacity limit now becomes additional profit. Potentially, this area is even larger than the original profit area. In an actual case, of course, the fixed costs are neither fixed nor limited. Semifixed costs will expand some amount into this additional profit area, but may not completely fill it.

An area graph gives a different view of the breakeven, capacity, and profit situation. One strategy is to expand output by increasing variable costs, with a small increase in fixed costs, so that overall profit will increase. Another strategy is to sell the additional capacity at a lower cost than the price of the original output. An example of this strategy is when a "name brand" producer of products sells other "no name" or "private label" products at a lower cost than the name brand products.

Private Label Strategy

Sears and others have built their business by going to "name brand" manufacturers and buying the same or similar products at a lower cost. The manufacturer gains because it is using its excess capacity and the seller puts its own name on the product and sells it for less than the name brand product. There may be other differences, such as offering less service and fewer warranties than the name brand. This strategy, with many variations, has been used by many orga-

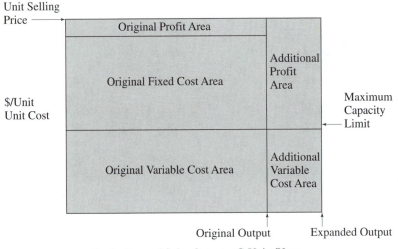

Figure 8–9   Area Graph—Expanding Output to Maximum Capacity

nizations. It may be necessary to reduce the selling price in order to increase sales. Often, even with a price decrease, the total profit received is greater at the lower selling price, since a higher quantity is sold. Each product and its market can be analyzed to determine if this strategy will be profitable. This strategy has worked successfully with appliances, tools, clothing, and other consumer products.

The area graph shows graphically the advantage of increasing output into the unused capacity area, which results in additional profits.

Agricultural Products Corp. Area Graph. Let's return to Agricultural Products. Convert the costs, revenues, and profits for one of their products into an area graph.

Variable costs	$5 per unit
Fixed costs	$15,000 per year
Selling price	$8 per unit
Output	6,000 units per year
Maximum capacity	10,000 units per year

The area graph for this product is shown in Figure 8–10.

The graph shows the unit variable costs of $5 per unit times the output of 6,000 units per year for a total variable cost of $30,000. The fixed costs per year are $15,000. With a selling price of $8 per unit times the output volume of 6,000 units per year, the total revenue is:

$$\$8 \text{ per unit } (6{,}000 \text{ units per year}) = \$48{,}000 \text{ per year}$$

The profit is:

$$P = TR - TC = \$48{,}000 - (\$15{,}000 + \$30{,}000) = \$3{,}000$$

Expanding Output at Agricultural Products Corp. If the output can be expanded into the unused capacity of the facilities and 10,000 units per year are produced and sold, then the

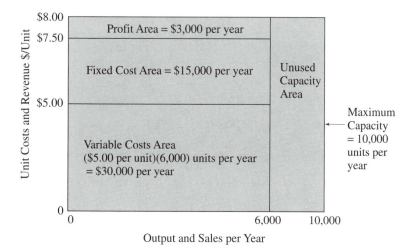

Figure 8–10   Agricultural Products Corp.

profits for the organization can be increased. Producing and selling 10,000 units per year results in the following (and see Figure 8–11):

Total variable cost increases by an additional $20,000 to $50,000, or,

$$(10,000 \text{ units per year})(\$5.00 \text{ unit variable cost}) = \$50,000 \text{ total variable cost at} \atop \text{maximum capacity}$$

Fixed costs remain the same at $15,000 per year. Total revenue becomes

$$(10,000 \text{ units per year})(\$8.00 \text{ selling price}) = \$80,000$$

The new total profit becomes

$$\begin{aligned} P &= TR - TC \\ &= TR - (FC + VC) \\ &= \$80,000 - (\$15,000 + \$50,000) = \$15,000 \end{aligned}$$

The profit is the sum of the original profit, $3,000, and the profit from expanded output, $12,000 = $15,000 per year.

In the actual case, the *FC* area of the expanded output may be larger than the original, and the selling price per unit may be reduced in order to sell more units. This is demonstrated next (see Figure 8–12).

Reducing Prices and Expanding Output at Agricultural Products. The sales manager at Agricultural Products states that in order to increase sales from 6,000 to 10,000 units per year, the selling price of the product will have to be decreased to $7.50 per unit. The manufacturing manager states that the expansion from 6,000 units to 10,000 units per year would likely increase the semifixed costs by $3,000 due to increased maintenance and decreased efficiency. The fixed costs at an output of 10,000 units per year would now be a total of $3,000 + $15,000 = $18,000.

The selling price would be $7.50 per unit. The total variable costs are now $50,000, the total fixed costs are $18,000, and the total revenue is (10,000 units per year)($7.50 selling price)

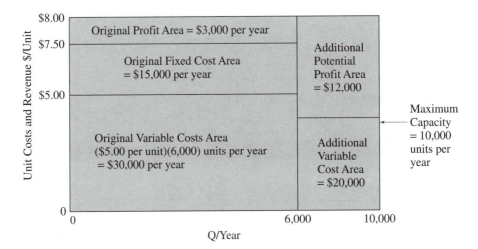

Figure 8–11   Expanded Output at Agricultural Products Corp.

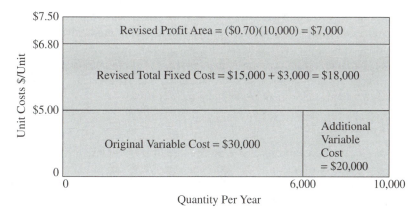

Figure 8–12   Reduced Price and Increased Fixed Costs at Agricultural Products

= $75,000. The profit is the total revenue minus the total costs or $75,000 − ($50,000 + $18,000) = $7,000 per year. This is a $4,000 increase in profits from the original profits of $3,000 per year. This situation of increased fixed costs and lower unit prices is likely to occur for Agricultural Products' expanded output.

CAPACITY LIMITS, EXPANSION, AND PROFITS STRATEGY

Our discussion of breakeven has shown that expanding output is generally a good strategy for increasing profits. There are limitations to this strategy, however.

1. *Market limitations.* If the additional output cannot be sold, the current output is more efficient. Try to expand other products and services.

2. *Profit limitations.* If the expansion of output causes fixed and variable costs to increase too much and/or revenue to decrease so that profit decreases, the current output may be the better alternative. Increasing market share while reducing profits can capture the market, but can be risky.

3. *Capacity limitations.* In the short term, the physical capacity of the facilities and the organization may limit the expansion of output. In the longer term, additional equipment, buildings, and organizational expansion can take place.

The first two limitations were discussed in previous sections. The third limitation to quantity expansion, capacity limitation of output, is considered here. The limit of **capacity** on the overall organization and its output in the short run is not only based on physical facilities. The organization may not be able to financially support increased output. There may be little available internal cash or external funds. Giving additional credit to customers may put excessive strain on the working capital. The existing sales organization may not be able to expand sales without reducing service to existing customers. Management's and technical staff's time and abilities may be limited in the short run as well. There must be a balance of physical output capacity, equipment, and labor with the rest of the organization's capacity. Successful expansion of output, assuming there is demand for the service or product and the additional sales are profitable, must be organizationwide. The increased output, however, places additional strain on the total organization, not just the production area. In addition, productivity may decrease due to poor planning. This is part of the reason for fixed costs increasing as output increases. Increased output that adds profits, increases market share, serves more customers, and provides more jobs is a very positive strategy; but the expansion and growth must be planned and managed efficiently by both senior management and technical staff.

Balancing Capacity

Operating capacity is often expressed as a percentage of the maximum capacity. For example, if the maximum capacity of a firm is 100,000 units per year, and the firm is producing 82,000 units per year, the percentage of utilization is 82%. Breakeven quantity is sometimes expressed as a percentage of maximum capacity.

In the manufacturing area, there are additional considerations of capacity limitations. Frequently, in the production process, there are multiple departments or equipment through which the product or service must flow. The individual components in the process may not be in balance with respect to their capacity and efficiency. For example, consider the flow of materials through equipment A and then through equipment B to make the finished component.

In the first case, with machine A producing more than machine B, excess output must be stored between A and B as in-process inventory. Thus inventory is sitting idle, capital must be invested into inventory, storage space must be provided, and the inventory system must manage additional inventory. We will see in the next chapter that the holding cost for this inventory is typically an additional 20% to 30% of unit cost per year. (See Figure 8–13.)

Frequently, management of an imbalance in the manufacturing area is to work both pieces of equipment the same amount of time (the example could be departments, or people). The justification for this is that idle machines (or departments or people) are costly, and keeping them working is the best decision. The philosophy of "look busy, even if it results in producing excess inventory" becomes the rule. Then a bottleneck is created when one department, worker group, or piece of equipment produces more output than the next, which has a slower productivity rate. This requires storage of the intermediate or in-process inventory. These bottlenecks work at maximum capacity to try to keep up with the other departments. The result is large amounts of inventory waiting to get through the bottleneck.

Imbalance in A and B Output When Both Work Same Amount of Time

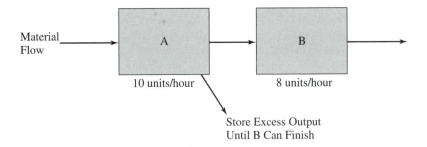

Figure 8–13   Capacity Imbalance

To manage this imbalance in capacity, there are a number of possibilities.

1. One solution is to work machine B longer than machine A. This is a good solution, but it is not always possible. The slow machine B, working an additional shift, has additional costs other than the direct labor. These additional costs would include supervision, opening the facilities just for the one operator, plant security, and additional scheduling effort. This may cause increases in both variable and fixed costs.

2. Another solution would be to purchase an additional machine B. This is a longer term solution. The equipment would be justified on both the need for more output and the reduced costs by elimination of overtime and/or excess inventory costs. Machine B would now have the capability of producing twice as many units or 16 units per hour. Now the bottleneck has switched to machine A.

3. Try to increase the efficiency and productivity of machine B. This seems obvious. Often low output is a result of time-consuming maintenance and changeover. Often changeover time is accepted as a given when in fact it can be reduced dramatically. An industrial engineer, S. Shingo, at Toyota, reduced setup times from many hours to a few minutes on large presses. There will be more discussion of this technique in the next chapters.

4. Work machine A less and pace it with machine B. A valid approach when demand for the output of both machines is low, but usually the problem is to increase output rather than reduce it. This is also a valid solution when machine A can be assigned to other production. This requires flexibility in equipment and the support services.

There may be other solutions to this internal **capacity imbalance.** Imbalances are difficult to avoid, especially when there are not just two machines, but hundreds of imbalances in the organization. Some people, departments, and machines are overworked; others are underworked. As output changes, internal capacity differences may improve in some areas and get worse in others.

The goal is to try to maximize the output and sales of the process or organization while keeping the expenses of the process as low as possible. Sometimes it appears that if all components are working at their individual maximum capacities and when in-process inventory is large, the process is successful. The monthly production reports indicate that all departments are at maximum capacity and there is lots of inventory! But high-capacity utilization of each producing unit and large in-process inventory as a result of the imbalances are a costly solution. In this situation, the overproduction causes excess inventory. The investment in inventory and the additional storage costs increase costs. Recalling the definition of return on investment or return on assets,

$$\text{ROI (or ROA)} = \frac{\text{Revenue} - \text{Costs}}{\text{Investment}}$$

Capacity

The organization can usually maximize profits when working at high capacity. Care must be taken not to have units out of balance with other units. Imbalance can cause increased in-process inventory, which may be very expensive and reduce overall ROI.

To maximize the ROI, revenue and sales should be maximized, costs minimized, and the investment minimized. An imbalance in capacities, if not managed properly, may cause increased costs and higher investment in inventories than necessary. Lower profits or increased inventory investment contribute to reduced ROI. Frequently, to meet increased product demand, the organization will schedule every component at its maximum individual capacity because "everyone should work to contribute to the increased output," while, in fact, this approach contributes to higher costs, higher inventory investment, and lower ROI. There is also the risk that the expanded sales will not be met in a timely manner because the imbalances in capacity cause conflicts in the scheduling system and delivery dates will be missed.

A partial cause of the mistake of working all equipment, departments, and individuals at maximum capacity may be the productivity reporting system. Management may consider the goals of the organization being met when all components are working at or near their capacity. The cost accounting and other reports may emphasize and reward capacity utilization and punish idle capacity, both of which cause overproduction of inventory and imbalance in capacity. If the organization is to improve profits, reduce costs, improve customer satisfaction, and emphasize quality improvement, then overproduction, excess inventory and other assets must be reduced. The basis for flexible manufacturing, just-in-time deliveries, and cycle time reduction is the monitoring and control of bottlenecks and capacity imbalances. We will discuss these topics further in the next chapter on minimum costs.

Breakeven and Time Value of Money

One of the most useful tools for an engineer, technologist, or manager is the combination of time value of money and breakeven. Process engineers use it for process selection, managers use it for capacity analysis, designers use it for material and design selection, and purchasing buyers use it for make or buy decisions.

BREAKEVEN AND TIME VALUE OF MONEY

Breakeven or cost-volume-profit (CVP) relationships, concepts, and calculations can now be combined with time value of money methods. Often, when making breakeven financial decisions, the revenues, costs, and investments do not occur at the same point in time, so time value of money techniques must be applied. For example, an investment in equipment that will expand output has a life greater than one year. The revenues, costs, and profits associated with the process or service may occur annually. In order to analyze the financial data and arrive at a decision, dollars over time and equivalence have to be used.

Matrix Corp. has a product that can be sold for $57 per unit. Material and labor costs are $39 per unit. The cost of the equipment to manufacture the product is $85,600, with a life of 7 years and salvage value of $8,000. Annual operating expenses are $5,000 per year. Determine the breakeven quantity if the desired ROI is 15%.

$$P = \$85,600, A_O = \$5,000/\text{year}, S = \$8,000, n = 7, i = 15\%, VC = \$39/\text{unit},$$
$$\text{selling price} = \$57/\text{unit}, A_{\text{Total}} = \text{unknown}, \text{BE} = \text{unknown}$$

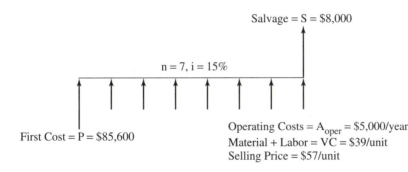

Salvage = S = $8,000

n = 7, i = 15%

First Cost = P = $85,600

Operating Costs = A_{oper} = $5,000/year
Material + Labor = VC = $39/unit
Selling Price = $57/unit

Figure 8–14   Matrix Corp.—TVM and Breakeven

The TVM diagram is shown in Figure 8–14.

There are two steps to this problem. The first is to determine the total fixed costs per year using time value of money calculations. The second is to determine the breakeven quantity using breakeven calculations.

$$A_T = A_O + A_P - A_S$$
$$= \$5,000 + \$85,600(A/P, n = 7, i = 15\%) - \$8,000(A/F, n = 7, i = 15\%)$$
$$= \$5,000 + \$85,600(.2404) - \$8,000(.0904)$$
$$= \$5,000 + \$20,578 - \$723$$
$$A_{Total} = \$24,855 \text{ per year} = \text{fixed cost/year}$$

Solving for the breakeven quantity, at BE,

$$TR = TC$$
$$TR = (\$57 \text{ per unit sold})(Q \text{ units per year}) = TC = VC + FC$$
$$(\$57/\text{unit})(Q/\text{year}) = (\$39/\text{unit})(Q_{BE} \text{ units/year}) + \$24,855/\text{year}$$
$$\$57\ Q = \$39\ Q + \$24,855$$
$$18\ Q = 24,855$$
$$Q = 1,381 \text{ units per year at breakeven}$$

This is shown in the graph in Figure 8–15.

ISOCOSTS AND BREAKEVEN BETWEEN PROCESSES

The previous discussion covered breakeven between revenues and costs. This is the true meaning of breakeven (revenue = costs), but sometimes the concept is applied to the comparison of two or more different process to determine the most economical process. The point at which the cost of one process equals the cost of the other process can be called the **isocost** point or equal cost point. Manufacturing engineers have the responsibility of determining the best of a number of alternative ways a product can be produced. Design engineers have the responsibility of selecting the best of alternative designs that both meet the specifications and are economical. Both of these decisions consider economics, quality, ease of manufacture, customer use, existing equipment, make or buy, costs, and many other factors.

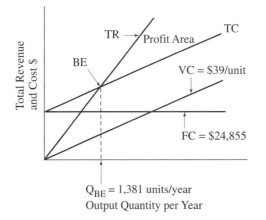

$Q_{BE} = 1,381$ units/year
Output Quantity per Year

Figure 8–15   Matrix Corp.—Solving for Breakeven

No revenue is considered here because the goal is to select the alternative that has the lowest cost rather than to maximize profit between revenue and cost. The design and process engineering personnel do not normally have the responsibility of establishing selling prices. This is usually a marketing function. If the least cost alternative is found by the manufacturing function, and the highest price that the market will pay is established, then profit will be maximized. Often, the quantity of output decision is not within the production area's responsibility, but is given by the customer and/or the marketing function. Using isocosts, managers and process engineers can determine the most economical process for the required output.

Atkinson Co.: Selecting One of Two Processes. A process engineer at Atkinson Co. is deciding between two processes. Both processes will produce quality products, but they have different fixed and variable costs. Process A uses less direct labor, but more equipment. Process B is more labor intensive and has less equipment and more manual operations. For process A, the variable cost is $10 per unit and fixed costs are $25,000 per year. For process B, the variable costs are $15 per unit and the fixed costs are $18,000 per year. Determine the breakeven quantity.

At the isocost (IC) point, the costs are equal.

$$TC_A = TC_B$$
$$VC_A + FC_A = VC_B + FC_B$$

Substituting,

$$(\$10/\text{unit})(Q \text{ units/year}) + \$25,000/\text{year} = (\$15/\text{unit})(Q \text{ units/year}) + \$18,000/\text{year}$$
$$15\,Q - 10\,Q = 25,000 - 18,000$$
$$5\,Q = 7,000$$
$$Q = 1,400 \text{ units per year for equal process costs}$$

This is the isocost point for the two processes. At an output of 1,400 units per year, the cost of the production is the same using either method. If the required output is greater or less than 1,400, the costs are not equal and one process will have a lower cost. This can be seen graphically. Figure 8–16 shows graphs of the two processes, A and B. The costs of the processes are combined into one graph (Figure 8–17). This final graph is not really a "breakeven" graph since breakeven is between revenue and costs.

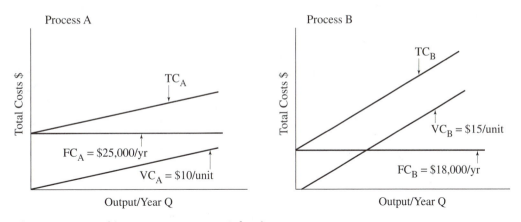

Figure 8–16   Atkinson Co.—Process Selection

From these graphs, we see that the fixed cost of A is greater than the fixed cost of B, and that the slope of the variable cost line for B is greater than the slope of the variable cost line for A. The intersection of TC_A and TC_B is the isocost point. To the right of 1,400 units per year, process A is less costly. To the left of 1,400 units per year, process B is less costly.

If the required quantity is greater than 1,400 units per year, process A has the lower costs. For example, calculating the costs of each process at 2,000 units per year,

$$Q = 2,000 \text{ units/year}$$
$$TC_A = FC_A + VC_A = \$25,000/\text{year} + (\$10/\text{unit})(2,000 \text{ units/year}) = \$45,000/\text{year}$$
$$\text{Cost/unit for A} = \$45,000 \text{ per year}/2,000 \text{ units per year} = \$22.50/\text{unit}$$
$$TC_B = FC_B + VC_B = \$18,000/\text{year} + (\$15/\text{unit})(2,000 \text{ units/year}) = \$48,000/\text{year}$$
$$\text{Cost/unit for B} = \$48,000 \text{ per year}/2,000 \text{ units per year} = \$24.00/\text{unit}$$

A similar calculation could be made for quantities of less than 1,400 units per year, which would show that process B is the least costly.

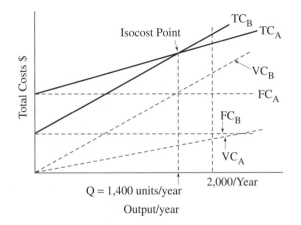

Figure 8–17   Atkinson Co.—Isocost Graph for Processes A and B

Quality Plastics: Isocosts and TVM. The previous Atkinson Co. example showed breakeven between two processes, but it did not consider the time value of money. In this next example, Quality Plastics combines isocosts and TVM for two products.

For process 1 at Quality Plastics, which is labor intensive and uses less equipment, material costs are $4.87 per unit, labor costs are $7.50 per unit, equipment costs are $10,000, salvage is $1,000, equipment life is 10 years, and equipment operating expense is $1,000 per year. For process 2, material costs are $3.60 per unit, labor costs are $5.37 per unit, equipment costs are $22,000, salvage is $2,500, life of the equipment is 10 years, and equipment operating costs are $2,000 per year. Determine the isocost quantity. ROI is 12%.

The solution to this problems has two steps. First, the time value of money portion is solved to determine the fixed cost for each of the processes. Second, the isocost or breakeven portion of the problem is solved. Figure 8–18 shows the TVM diagrams for the processes. The Total Annual Cost is equal to the Fixed Cost per Year.

First, find the fixed costs using the TVM functions and the annual cost calculations. For process 1,

$$A_T = A_O + A_P - A_L$$
$$= \$1,000 + \$10,000(A/P, n = 10, \ i = 12\%) - \$1,000(A/F, n = 10, i = 12\%)$$
$$= \$1,000 + \$10,000(.1770) - \$1,000(.0570)$$
$$= \$1,000 + \$1,770 - \$57$$
$$A_T = \$2,713 \text{ per year} = \text{fixed cost for process 1}$$

For process 2,

$$A_T = A_O + A_P - A_L$$
$$A_T = \$2,000 + \$22,000(A/P, n = 10, i = 12\%) - \$2,500(A/F, n = 10, i = 12\%)$$
$$= \$2,000 + \$22,000(.1770) - \$2,500(.0570)$$
$$= \$2,000 + \$3,894 - \$143$$
$$A_T = \$5,751 \text{ per year} = \text{fixed cost for process 2}$$

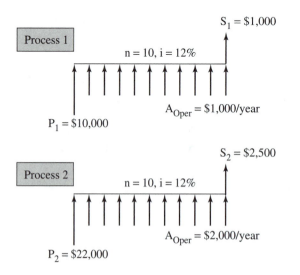

Figure 8–18   Quality Plastics—TVM Diagrams

Solving for the BE/isocost point,

$$TC_1 = TC_2$$
$$VC_1 + FC_1 = VC_2 + FC_2$$
$$(\$12.37/\text{unit})(Q \text{ units/year}) + \$2{,}713 = (\$8.97/\text{unit})(Q \text{ units/year}) + \$5{,}751$$
$$3.40\,Q = \$3{,}038$$
$$Q = 894 \text{ units per year for equal cost between 1 and 2}$$

The graph of this calculation is shown in Figure 8–19.

Below an output of 894 units per year, process 1 is the lowest cost process, and above 894 units per year, process 2 is the lowest cost alternative. This type of analysis would assist the process engineer in selecting the lower cost of two processes depending on the required production output, which would be given by the customer or the seller's sales department.

Make or Buy Decisions

Breakeven analysis combined with time value of money can be useful to analyze and select materials, processes, designs, and suppliers. There can be many isocost points between many alternatives. There are also many noneconomic factors to consider in the make or buy decision. Supplier quality and response time, outsourcing policy, and internal capacity utilization are important factors.

Make or Buy?

When a product or service is purchased from an outside vendor, its cost is compared to the cost from other vendors and to the cost of producing the product or service in-house. The decision to purchase from an outside seller compared to producing a product internally is called the **make or buy** analysis and selection decision.

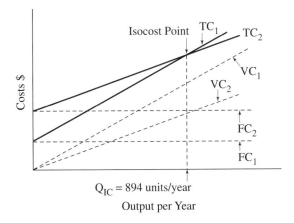

Figure 8–19   Isocosts at Quality Plastics

Make or buy is a special case of isocost analysis. In the make alternative, the variable and fixed costs are included. However, for the buy alternative, there is *no fixed cost*; only the variable costs are calculated by the buyer. To determine the isocost point for the make and the buy alternative, the variable and fixed costs of the make alternative are set equal to *only the variable cost* of the buy alternative. The graph of the problem is shown in Figure 8–20.

The graph shows the normal lines to represent *VC*, *FC*, and *TC* for the make alternative, but the buy alternative has no fixed cost. Suppliers usually quote prices in cost per unit and have no fixed costs in the price. The *VC* of the buy alternative equals the *TC* of the buy alternative since the *FC* of the buy is zero.

There is no revenue function in this case. The selling price is constant whether the item is purchased or made. The isocost point is between the two alternatives. The quantity to the right of the BE point would be the least cost if the item is made, and the quantity to the left of the BE point would be the least cost if the item is purchased.

Make or Buy an Electronic Circuit Board: Precision Electronics. To make an electronic component part, material can be purchased for $4.98 per unit, labor costs are $5.60 per unit, annual operating costs are $2,000 per year, and equipment costs are $35,000 with salvage revenue of $4,000 and an estimated life of 8 years. There are 5,000 units required per year of the component part. Desired ROI is 20%. The part can be purchased for $14.50 per unit. Determine if the part should be purchased or made in-house.

The first step is to solve for the fixed cost of the make alternative. The TVM diagram is shown in Figure 8–21.

Solving for the annual cost, which equals the fixed cost,

$$
\begin{aligned}
A_T &= A_O + A_P - A_L \\
&= \$2,000 + \$35,000(A/P, n = 8, i = 20\%) - \$4,000(A/F, n = 8, i = 20\%) \\
&= \$2,000 + \$35,000(.2606) - \$4,000(.0606) \\
&= \$2,000 + \$9,121 - \$242 \\
&= \$10,879 \text{ per year} = \text{fixed cost/year to make}
\end{aligned}
$$

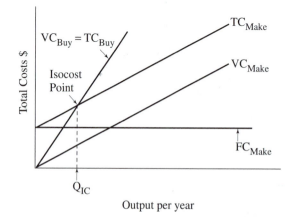

Figure 8–20   Isocost Make or Buy Decision

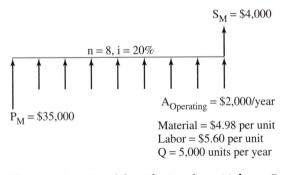

Figure 8–21   Precision Electronics—Make or Buy

Cost to make 5,000 units per year,

$$TC_{Make} = VC + FC = (5{,}000 \text{ units/year})(\$4.98/\text{unit} + \$5.60/\text{unit}) + \$10{,}879/\text{year}$$
$$= 5{,}000(\$10.58) + \$10{,}879 = \$52{,}900 + \$10{,}879$$
$$= \$63{,}779 \text{ to make } 5{,}000 \text{ units per year}$$

Cost per unit = $63,779 per year/5,000 units per year = $12.76 per unit to make the product

Since the cost of the purchased item is $14.50 per unit, it is *less costly* to make the component.

How many items would be required in order to make the purchased alternative the least expensive? Set the costs of the two alternatives equal to each other to determine the breakeven quantity.

$$TC_{Buy} = TC_{Make}$$
$$VC_{Buy} + FC_{Buy} = VC_{Make} + FC_{Make}$$

(Note that FC_{Buy} is zero.)

$$(Q \text{ units/year})(\$14.50/\text{unit}) + 0 = (Q \text{ units/year})(\$10.58/\text{unit}) + \$10{,}879 \text{ per year}$$
$$(\$3.92 \text{ per unit})(Q \text{ units per year}) = \$10{,}879$$
$$Q = 2{,}775 \text{ units per year for isocosts of the make and buy alternatives}$$

If less than 2,775 units per year were required, the buy alternative would be the least costly. If more than 2,775 are required (as in this case), the make alternative is the least costly. The isocost diagram is shown in Figure 8–22.

There are other considerations in the make or buy decision besides cost: quality, delivery, available labor, outsourcing policy of the organization, capacity utilization, and other factors. Cost is an important consideration, but not the only one.

Calculating Three Isocost Points: Tin Cup Manufacturing Inc. A manufacturing engineer has completed the time value of money analysis on a process that can be used to produce a subassembly. The automated process has fixed costs of $90,000 per year and variable costs of $4.90 per unit. The existing process for the subassembly that is currently being used has fixed costs of $40,000 per year with variable costs of $8.50 per unit. The existing process is worn out and ready to be replaced. The alternatives are:

1. Replace the existing process with identical equipment.

2. Replace with an automated process having costs as detailed above.

3. Buy the subassembly from a supplier at $14.50 per unit for quantities ordered.

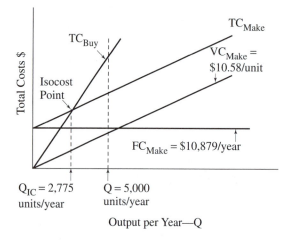

Figure 8–22   Precision Electronics—Isocost Diagram for Make or Buy Decision

The analysis of this situation requires the calculation of three isocost points. The first isocost point is between the existing process and the automated process. At the IC point,

$$TC_{Existing} = TC_{Automated}$$
$$VC_{Exist} + FC_{Exist} = VC_{Auto} + FC_{Auto}$$

Substituting,

($8.50/unit)($Q_{E-A}$ units/year) + $40,000/year = ($4.90/unit)(Q_{E-A} units/year) + $90,000

Solving for $Q_{Ex-Auto}$,

Q = $50,000/year/$3.60/unit = 13,889 units per year for the existing/automated alternative

The second isocost point is between the existing and the buy alternatives. At the IC point,

$$TC_{Existing} = TC_{Buy}$$
$$VC_{Exist} + FC_{Exist} = VC_{Buy} + FC_{Buy}$$

Substituting,

($8.50/unit)($Q_{E-B}$ units/year) + $40,000/year = ($14.50/unit)(Q_{E-B}/year) + $0

Solving for Q,

Q = $40,000 per year/$6.00 per unit = 6,667 units/year for the existing/buy alternative

The third isocost point is between the automatic and buy alternatives. At the IC point,

$$TC_{Auto} = TC_{Buy}$$
$$VC_{Auto} + FC_{Auto} = VC_{Buy} + FC_{Buy}$$

($4.90/unit)($Q_{A-B}$ units/year) + $90,000/year = ($14.50/unit)(Q_{A-B} units/year) + $0

Solving for Q_{A-B},

Q_{A-B} = $90,000/year/$9.60 per unit = 9,375 units per year for the automated/buy alternative

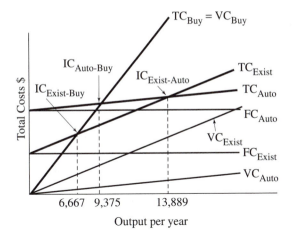

Figure 8–23   Tin Cup Manufacturing Inc.

A graph of the above calculations and IC points is shown in Figure 8–23.

On the graph, the total cost lines intersect at three isocost (IC) points. The solution for selecting the best process depends on the output of the subassembly that is required. The graph shows that from zero to 6,667 units per year, the buy alternative is the lowest cost; from 6,667 to 13,889 units per year, the existing process is the lowest cost; and for quantities greater than 13,889 units per year, the automatic process is the lowest cost.

In many actual situations there may be even four or more alternatives. The calculations are made in a similar manner for these multiple alternative decisions.

Quantity Discounts for Purchased Products and Services

One factor that is an important part of the make or buy decision is *quantity discounts* offered by vendors. Up to a certain quantity, the unit price is one amount. If an order is for an amount greater than the specified quantity, the unit price of all the units decreases. For example, purchasing up to 499 units results in a price of $10 per unit. If the order is for 500 up to 999 units, the unit price of all the units is $9.50. For orders above 1,000, the units cost $9 per unit.

The reason why the supplier gives a quantity discount can often be traced to its internal costs and the desire to sell larger quantities above its breakeven point. As the quantity of output and sales increases, the profit increases, as we have seen, and the supplier passes along some of the reduced cost to the customer in the form of lower unit prices. This reduces the material costs for the customer, which in turn reduces its total unit costs and prices to its customers, and the cycle continues.

Example: Jane's Output Increase Decision. An existing process has fixed costs of $25,000 per year and variable costs of $13.80 per unit. The product is currently being sold for $21.50 per unit. The output and sales are 8,000 units per year, which is the maximum output for the existing process. Jane, the process engineer, is considering improving the process by adding automatic loading and unloading equipment that would increase output to 9,000 units per year. The marketing department believes that this new quantity can be sold for a

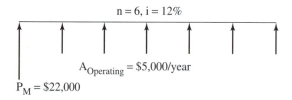

$$n = 6, i = 12\%$$

$$A_{\text{Operating}} = \$5,000/\text{year}$$

$$P_M = \$22,000$$

Figure 8–24   Jane's Output Improvement Project

maximum of $20 per unit. The handling equipment costs $22,000 initially, has operating costs of $5,000 per year, a life of 6 years, and no salvage value. The desired ROI for the firm is 12%. Is the addition of the automatic handling equipment warranted? See the TVM diagram in Figure 8–24.

The annual profit of the current process is:

$$P = TR - TC = TR - FC - VC$$
$$= (8,000 \text{ units/year})(\$21.50) - \$25,000/\text{year} - (8,000 \text{ units/year})(\$13.80/\text{unit})$$
$$= \$172,000 - \$25,000 - \$110,400 = \$36,600$$

The annual cost of the automatic handling equipment is calculated as:

$$A_{\text{Total}} = A_P + A_{\text{Operating}} = \$22,000 \ (A/P, n = 6, i = 12\%) + \$5,000$$
$$= \$22,000(.2432) + \$5,000 = \$5,350 + \$5,000$$
$$= \$10,350/\text{year}$$

The revised profit using the higher output of 9,000 units and the reduced selling price is:

$$TR - TC = TR - FC - VC = P$$
$$= (9,000 \text{ units/year})(\$20.00) - (\$25,000 + \$10,350) - (9,000 \text{ units/year})(\$13.80)$$
$$= \$180,000 - \$35,350 - \$124,200 = \$20,450/\text{year}$$

Jane, the process engineer, should reconsider the purchase of the automatic handling equipment. Profits would decrease from $36,600 down to $20,450 as a result of the additional costs, lower selling price, and increase in output.

Ammerman Products Corp.: Make or Buy? Ammerman Products Corp. currently purchases a product from a vendor for $40.80 per unit. Ammerman is considering making the product. The equipment to make the product would cost $125,000, with operating costs of $10,000 per year and material and labor costs of $33 per unit. The firm desires to earn 15% ROI. How long must the equipment be used in order for the make alternative to be economical if 5,000 units per year are required?

To calculate the number of years, first solve the isocost portion of the problem.

$$TC_{\text{Buy}} = TC_{\text{Make}}$$
$$(5,000 \text{ units/year})(\$40.80) = FC_{\text{Make}} + VC_{\text{Make}}$$
$$\$204,000 = FC_{\text{Make}} + (5,000 \text{ units/year})(\$33.00/\text{unit})$$

or

$$FC_{\text{Make}} = \$204,000 - \$165,000 = \$39,000 \text{ maximum for an isocost at 5,000 units/year}$$

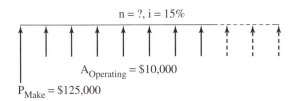

Figure 8–25   Ammerman Products Corp.—Make or Buy?

Figure 8–25 shows the diagram. Solving for the unknown equipment life,

$$FC_{Make} = A_{Total} = A_P + A_O = \$39,000/year = A_P + \$10,000/year$$
$$A_P = \$39,000 - \$10,000 = \$29,000/year$$
$$\$29,000 = \$125,000(A/P, n = ?, i = 15\%)$$
$$(A/P) = 29,000/125,000 = .2320$$

Using the 15% tables and the A/P factor, n must be 8 years or greater (the exact answer using a calculator is 7.4 years). If the life of the equipment for the process is less than 8 years, the annual cost, which equals the FC per year, would be greater than $29,000 and the buy alternative would be the least expensive. This is an example of an isocost calculation based on the breakeven of the number of years.

LEGISLATION: BREAKEVEN AND ASSET- OR LABOR-INTENSIVE ORGANIZATIONS

Our discussion of breakeven and of fixed and variable costs can be extended to the topic of asset- or labor-intensive organizations and industries that have affected public policy with respect to monopoly and competition. Over the last century in the United States, we have seen the breakup of monopolies in the late 1800s, the encouragement of federal- and state-created monopolies in the early 1900s, and finally a return to the policy of discouraging and breaking up monopoly industries beginning in the 1980s.

In the late 1800s, the Sherman Act legislation targeted the petroleum and railroad industries as having too much financial concentration and noncompetitive determination of prices. Monopolies were broken up to encourage competition. The opposite occurred in the early 1900s. As the power, water, gas, telephone, airline, taxicab, trucking, and mail service industries grew, they were generally given monopoly authority by federal, state, and local government legislation. Recently, the telephone, airline, trucking, cable television, and power industries have been required to reduce their monopoly authority.

Although much of the legislation process is politically determined, there is a economic basis for both permitting monopolies and breaking up some monopoly-structured industries. The roots of the legislation are in the concept of variable costs, fixed costs, and breakeven.

With a monopoly, either there are no competitors or the competitors are small and have little impact on the monopoly. The single seller of products and services can set prices as it chooses. In a controlled or legislative-created monopoly such as a power company, the prices are monitored by a commission or agency. There is little advantage for the monopoly to be efficient and reduce its costs; it is easier to ask for price increases from the regulating agency or commission. If costs are high, profits and ROI are low for the organization, and it is likely that based on the low ROI, the commission will grant a price or rate increase.

In a competitive industry, where no one supplier has control of prices, there is motivation to keep prices low and ROI high. Efficiency is rewarded with more customers and increased ROI. Inefficiency is punished by low ROI and even the potential of going out of business. What is the justification for policy that creates monopolies and for legislation that breaks up monopolies? The answer is partially found in the concept of breakeven.

Asset-Intensive Organizations

There are two general types of organizations that can be described using the concept of breakeven. The first is the **asset-intensive firm** with high fixed costs and relatively low variable costs. Examples of this type of firm would be a utility company, an airline, an automated manufacturing plant, a steel manufacturer, a chemical firm, a railroad, or an oil refining company. In this type of firm, the high fixed costs and relatively low variable costs cause a total cost curve with a low or flat slope. When this total cost line intersects with the total revenue line to form the breakeven point, a relatively large angle, alpha, is formed between the *TR* line and the *TC* line.

This large angle, alpha, causes the asset-intensive firm to be very "profit sensitive." That is, *a small change in the breakeven quantity, either up or down from the BE point, causes a relatively large change in profits.* This is due to the rapid divergence between the *TR* and the *TC* functions when moving away from the BE point.

An example of this situation is the commercial airline flight that has exactly the number of passengers to break even, neither making a profit nor suffering a loss. Then, at the last moment before takeoff, one more passenger buys a ticket. The additional costs of adding this one passenger are very small: luggage handling, some additional fuel, a meal. These are all variable costs, not additional fixed costs. Since these additional costs are small and the passenger pays the same ticket price as everyone else, most of the ticket revenue is profit. Thus, on the graph, one additional passenger increase in *Q* causes a relatively large increase in profit. The opposite would occur if the flight was at breakeven and a passenger canceled at the last moment. A relatively large loss would occur. Another example is if a power company added one more electrical appliance on the electrical system. The additional cost is very small while the additional revenue is the same as other electrical rates. For asset-intensive organizations with high fixed costs, *a small change in Q causes a large change in profit.* (See Figure 8–26.)

Labor-Intensive Organizations

A **labor-intensive firm** is one that has relatively high amounts of labor and low amounts of fixed costs. Examples of this type of firm would be all types of service firms, such as engineering consulting, legal offices, medical offices, and manual assembly operations.

Graphically, for this type of organization, the slope of the variable costs is high due to a high proportion of labor costs. The amount of fixed costs is typically low, so the resulting total cost line has a high or steep slope, and the intersection with the revenue curve to form the BE point causes a small angle, alpha, between the total revenue and total cost lines.

This small angle, alpha, causes a change in the quantity to result in a relatively small change in profit. These types of firms are not as profit sensitive as asset-intensive, high fixed cost organizations. For example, a dentist who is operating at breakeven and takes on an addi-

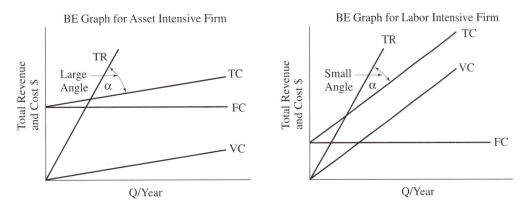

Figure 8–26   Breakeven in Asset- and Labor-Intensive Organizations

tional patient increases the quantity of service offered. But, because the service is labor intensive, the additional profit is small.

In a manufacturing organization, the component manufacturing departments can be asset intensive with large investments in automated equipment and proportionally small labor costs. The firm's assembly department may be very labor intensive with small investments in equipment and lots of hand operations. In the same firm one area is very sensitive to quantity changes and another is not. The decisions made by management of one department would likely be different from those of the other department.

Monopoly Justification

Monopoly production of a good or service occurs when there is a single producer. Generally, two types of monopolies exist. One type of monopoly or near monopoly can exist because the producer is efficient, high quality, has patents, and/or has the marketing strength to beat the potential competition and win high market share. This firm maintains its single seller position as long as it can maintain its market advantage. It could be called a natural monopoly created by competition.

The second type of monopoly is created by local, state, or federal legislation. Examples of this second type would be utilities, first-class mail service, railroads, and, in the past, airline and trucking firms. This is the legislated monopoly. The firm has the obvious advantage of being able to set the price of the service or product without respect to any competition. For this reason most government-created monopolies usually have a regulatory agency that has the responsibility of monitoring the prices of the monopoly. A state Utility Regulation Commission or the Interstate Commerce Commission are examples of this type of agency.

When airlines were being formed in the early part of the 1900s, they had a small customer base. So the airlines argued that they each would like some type of "monopoly protection" of routes between cities. They said that if many airlines were permitted to fly between the same cities, given the small customer base, no one airplane would be full enough to be above their breakeven point. The federal government accepted this argument and established the legislation giving a federal agency the responsibility of approving fares and granting routes to the air carriers. Many carriers were also required by the agency to establish operations in small cities. These smaller locations generally were not profitable because of few customers, but they were required in order to obtain the more profitable routes. Thus, an airline could be sure that it was the only carrier between two points and this would help ensure that it would be above the breakeven point. Recall that an asset-intensive firm's profits are very sensitive to the breakeven point because of the wide divergence angle between the total cost and the total revenue lines. This same argument can be applied to the trucking, power, telephone, natural gas, and other asset-intensive industries.

This situation with the airlines continued into the 1970s. The customer base had grown, the fare regulation was tight, and requested pricing changes were slow, often taking months to be granted. The industry was limited to a few large airlines with guaranteed profits. There was little entry of new airlines into the industry. Likewise, there were little, if any, instances of firms moving out of the industry due to their low profitability or inefficient management. The monopoly authority rewarded the organizations equally whether they were efficient or inefficient. And, as with many monopoly industries with secure profits, there was little incentive for firms to be efficient and reduce costs in order to increase profits or to do any innovative marketing to attract customers and thus additional profits. If profits were "low" by regulation standards, the rates could be increased by the regulatory agency. If profits were "too high," the firm's fares were reduced by regulation. A static, noninnovative industry, with no rewards to the cost-efficient firm and no penalties to the high-cost, inefficient firm was the result. Similar situations have existed in utilities, trucking, and other regulated monopoly industries.

Deregulation

In the late '70s and early '80s there was discussion among the major airlines concerning deregulation. This discussion was primarily started by the larger, more efficient airlines. Their reasoning was that the efficient firms were being penalized because they were not permitted to grow and become more profitable due to regulation restraints. The inefficient firms were protected from their own mistakes and poor management by regulation. The efficient firms argued to Congress that the public interest would be best served by deregulation. Congress listened, had many hearings, and agreed. Deregulation began. Small communities that were served by the large airlines complained that with deregulation, the large airlines would not be required to serve their towns, and they would be without service.

After Deregulation

The large, efficient airlines expanded into new markets after deregulation. The inefficient firms failed, declared bankruptcy, and were reorganized. The large airlines did choose not to serve the small towns because of the low revenue and high cost. But deregulation also meant that many new, small "commuter airlines" were formed that were not permitted prior to deregulation. These feeder airlines served the small towns with more flights and lower fares to more destinations than before deregulation. The air service to the small towns was improved.

The inefficient firms went out of business or merged with efficient firms. The regulating agency was gone, saving taxpayers' expenses. Fares went down, not up. More flights were added. Special fares were created that were not permitted by the old regulations. Both the passengers and the airlines were better served. Fare changes could be made quickly by the airlines without regulation interference. Flights below the BE point could be filled quickly with price reduction, special fares, and promotions.

The basis for the improved conditions was the *profit sensitivity of the asset-intensive firms operating above their breakeven point.* An understanding of breakeven was the key to the deregulation. After the success in the airline industry, deregulation has continued in the trucking industry, which was in a confusing and expensive regulation environment. And then there was the deregulation of the telephone industry. It is difficult to deregulate an industry after years of secure monopoly profits. There are dislocations to employees, suppliers, and communities, but the net result is better service and lower prices. Breakeven analysis helped the situation to be seen more clearly.

An argument similar to that for the airline industry can be made for the deregulation of the telephone industry, the trucking industry, the power industry, cable television, and other state-created monopolies.

SUMMARY

This chapter has introduced breakeven and isocost analysis, an important economic analysis and selection tool. When alternatives are analyzed and the selection of one process, supplier, material, method, equipment, or other alternative is required, it is common to apply breakeven and isocost analysis. When combined with other techniques such as time value of money, breakeven/isocost analysis is very useful in the analysis and selection of alternatives. Breakeven is an important analytical tool for engineers, managers, technicians, and purchasing department buyers.

The topics introduced in this chapter include the following:

- Linear breakeven concepts including breakeven calculations and graphs.
- Unit cost calculations and linear graphs.
- Area breakeven graphs for profit-increasing strategies.
- Nonlinear breakeven concepts.
- Variable, fixed, and semifixed costs.
- Breakeven point changes caused by increases and decreases in variable and fixed costs, and increases and decreases in revenues.
- Capacity limits, maximization, and optimization.
- Breakeven and time value of money analysis.
- Isocosts and time value of money of two processes.
- Isocosts and time value of money calculations.
- Make or buy analysis.
- More than two alternative isocosts analysis.
- Quantity discounts for buy alternatives.
- Asset- and labor-intensive organizations and industries.
- Breakeven and monopoly justification.

QUESTIONS

1. State the definition of breakeven.
2. What are the typical components of variable cost?
3. What are the limitations of the linear breakeven case?
4. What are the profits for the linear case when operating above and below the BE point?
5. What are the typical components of fixed costs?
6. Sketch a typical nonlinear breakeven case.
7. Why does the FC increase as maximum capacity is approached in the nonlinear case?
8. Why does the slope of the variable costs increase as a function of Q in the nonlinear case?
9. What are the causes of the revenue function decreasing in slope as a function of Q in the nonlinear case?
10. What are the profit areas on the nonlinear BE graph?
11. How can an increase in output cause profits to decline?
12. How could there be more than two breakeven points?
13. What is the condition that causes the make or buy situation?
14. Sketch the general breakeven or isocost graph for two processes.
15. How can manufacturing and design departments make process decisions independently of the sales department's pricing the product and still have profits maximized?
16. With respect to the $TC = TR$ linear breakeven problem, what happens to the BE point when the following occur?
    a. i increases
    b. n decreases
    c. material costs decrease
    d. labor costs increase
    e. salvage value decreases
    f. selling price decreases
    g. fixed costs increase
    h. selling price increases and fixed costs decrease
17. How are unit costs calculated?
18. On the unit cost graph, why are variable costs per unit horizontal and fixed and total costs per unit downward sloping?
19. What is meant by the term "semifixed" costs?
20. What noneconomic factors must be considered in making breakeven decisions?

21. State the general conclusions about breakeven analysis.
22. Give examples of labor-intensive firms and of asset-intensive firms.
23. What are the arguments for state and federal regulated monopolies? For competition?
24. Why was the airline industry deregulated?
25. What is meant by the statement, "State or federal regulation is not required of low fixed cost organizations"?
26. How does an area graph show costs, revenue, and profit differently from the total cost and revenue graphs?
27. What is the advantage of using an area graph? How does it relate to capacity utilization?
28. Usually in mechanization/automation situations the fixed costs increase significantly. Discuss this in terms of breakeven.
29. Why do unit costs decrease as output increases?
30. What are the causes of overall organization capacity limitations?
31. Why is the technique of working all elements in a process at their individual maximum capacities not always successful?
32. What are some possible ways to manage internal capacity limits and imbalances (bottlenecks)?
33. How can capacity imbalance cause low ROI?
34. How is the fixed cost calculation made when combined with TVM?
35. How is TVM related to make or buy decisions?
36. What is the effect on breakeven and profits from a price increase by a power company compared to a fee increase by an engineering consulting firm?
37. What is the justification for the government control of a utility that is a monopoly?
38. If, in a competitive industry, some firms reduce costs and prices, what will other firms do? What occurs in a monopoly industry when costs are reduced?
39. Is it possible to deregulate other utilities than the telephone industry, such as power, water, cable TV?
40. How can a company's financial statements be used to determine if the firm is a "profit sensitive" firm?

PROBLEMS

1. For a linear case, the fixed costs for a process are $100,000 per year, material and labor costs are $15 per unit, and the selling price is $22.50 per unit.

    a. What is the profit amount if 10,000 units are sold per year?
    b. What if 16,800 units are sold per year?
    c. What is the BE point?

2. Material costs are $25.50, labor is 3.5 hours per unit, wages are $7.50 per hour, fixed costs are $78,500 per year, and the selling price is $61.50 per unit.
   a. Determine the breakeven point.
   b. What is the unit cost at breakeven?
   c. Using a spreadsheet or calculator, calculate the total costs and revenue and plot them.
   d. Using a spreadsheet or calculator, calculate the unit costs and revenue and plot them.
   e. What occurs to the BE point if material costs decrease to $22.50 per unit and fixed costs increase to $80,000/year?

3. The Flowers and Plants Shop has 7 delivery vans. They are considering converting the vans' existing gasoline engines to the use of compressed natural gas (CNG). The conversion costs $2,200 per van. The desired ROI is 12%. The cost of the CNG equivalent in BTU content of a gallon of gasoline is $.70. Gasoline is currently $1.05 per gallon. The fuel mileage of 12 miles per gallon is the same for natural gas. The average mileage of each van is 20,000 miles per year.
   a. How many years do the vans have to be kept in order to justify the conversion?
   b. What are some noneconomic advantages of the conversion?
   c. What are some disadvantages of the conversion?

4. Richard can have his lawn cut by a neighbor's son for $20, or he can purchase a mower for $350 that will last 4 years and cut it himself. The operating costs of the mower are expected to be $30 per year, and he desires an ROI of 10%.
   a. How many times does the lawn need to be cut to make the purchase of the mower justified, not considering the value of Richard's time?
   b. What if Richard takes 2 hours to cut the lawn and his time is valued at $10 per hour?

5. Peggy just purchased a dog and needs to fence in her backyard. The perimeter of the yard is 340 feet. The fencing company has quoted a price of $10.50 per foot plus $35 each for two gates. Peggy can purchase the materials for $1,400 delivered and hire a helper for $6 per hour. She expects that it would take her and the helper 30 hours to install the fence. A power posthole digger can be rented for $30 per day. She is also considering whether to install an electronic underground fence. The quotation for it is $2,780, complete with controls and sensors installed. What

should she do? Plot the graph of the costs of the alternatives.

6. Duggens Manufacturing has the capacity of 100,000 units per year production. The current variable costs are $13 per unit, fixed costs are $150,000 per year, and the selling price is $15 per unit. They are currently producing and selling 82,000 units per year.
   a. Determine the current profit per year.
   b. Draw an area diagram.
   c. 10,000 additional units can be produced and sold if the price is reduced to $14.40 for all units and material costs decrease by $.50 per unit. Determine the new profit.
   d. Draw the new area diagram.
   e. Draw the original and final unit cost graphs.

7. A restaurant sells 60,000 Special Burgers per year. The secret special sauce is the reason for the burger's popularity, the owner feels. The current fixed costs of the restaurant are $70,000 per year, and the variable costs for the Special Burger are $2.50 with the selling price of $4.50 per sandwich.
   a. What is the current profit per year and the breakeven point?
   b. Currently the restaurant is open 6 days per week. The owner is considering opening one more day per week. If additional fixed costs would be $5,000 per year, how many more Special Burgers would she have to sell to justify the extended open time?
   c. Draw a total cost and revenue graph of the calculations.

8. For process A, the fixed costs are $10,000 per year and variable costs are $10 per unit. For process B, which will produce the same quality parts as process A, the fixed costs are $25,000 per year and the variable costs are $5 per unit. Calculate the isocost point, and plot a graph to scale manually or using spreadsheet software, if available.

9. An organization is trying to decide if it should purchase cars or pay mileage to its salespeople. They could pay $0.23 per mile for the use of personal cars. They could purchase cars for $12,000, having a life of 3 years; salvage of $5,000; annual cost of insurance, maintenance, and repairs of $2,000 per year; and operating expenses of $0.11 per mile. Determine the number of miles driven per year in order for the purchase to be the least cost alternative. What noncost factors should be

considered? ROI is 12%. Sketch a graph of the costs.

10. An engineering consulting office is considering whether to purchase or lease a computer printing system. The small system can be purchased for $40,000 with a life of 4 years and salvage value of $2,000. Operating expenses are expected to be $60 per day and annual maintenance to be $500 per year. An identical system can be leased for $125 per day, including maintenance and operating expenses. Determine the number of days per year that the system would have to be used to make the purchase economical. The desired ROI of the firm is 15%.

11. A rancher is considering an irrigation system for his fields. Two systems are under consideration.

	Batx System	Abex System
Power cost/hr	$.12	$.18
Motor and pump	$10,000	$6,000
Pipe cost	$20,000	$13,000
Salvage value	$300	$100
Life—years	12	12

The desired ROI is 15%.

a. Determine the usage required for each system to be economical.

b. Draw an isocost/BE diagram.

12. A chemical firm is considering purchasing pumps for its processes. It desires an ROI of 20%. The steel units are less expensive but require more frequent maintenance due to the corrosive liquids being pumped. The life of all systems is 5 years. The following costs apply.

	Carbon Steel	Stainless	Alloy 37
Initial cost	$150,000	$275,000	$400,000
Oper. cost/yr	$75,000	$40,000	$35,000
Annual maint.	$100,00	$50,000	$20,000
Salvage value	$500	$1,000	$10,000
Variable cost/hr	$1.10	$.95	$.72

The pumps run continuously except when they are down for repairs.

a. Determine the isocost points for all the alternatives.

b. Plot the costs on a graph using manual or spreadsheet methods.

c. Indicate the ranges of economical use for the pumps.

13. A simple circuit board can be produced either mechanically or by hand. The hand method requires $12 of material and takes 2.8 hours per unit with a labor cost of $9 per hour. The equipment has a fixed cost rate of $120,000 per year and variable costs of $24.50 per unit. The customer requires 12,000 units. If the equipment can be used for other products when not producing these units, which process is the least costly? What if the equipment could *only* be used for production of this unit?

14. A firm has the capacity to make 100,000 units per year. It is operating at 80% of maximum capacity. The product can be sold for $1.10 per unit. Fixed costs are $25,000 per year, and variable cost is $0.70 per unit. Determine the current profit per year on the product and find the breakeven capacity.

15. Jim is considering a solar-powered water heater for his home, which is currently under construction. A conventional gas water heater is $550 installed. It has an estimated life of 15 years. The solar system, including the panels, storage tank, pump, and controls, costs $2,000 to install. The estimated cost of gas for the traditional system is $300 per year, and the cost to run the pump on the solar system would be $100 per year. The solar system is expected to last 25 years. Which system is the most economical? What other factors should be considered? ROI = 15%.

16. A carbon steel pressure vessel costs $45,000 initially and would last 6 years. A stainless steel vessel would cost $90,000. If the firm is doing some corrosion testing, how long does the stainless steel vessel need to last in order to make its purchase economical? The desired ROI of the firm is 15%.

17. A manufacturing engineer is considering a make or buy decision. The first cost of the equipment required to make the product is $120,000, with a life of 9 years and salvage value at that time of $10,000. The material and labor costs for the make alternative would be $23.60 per unit. The firm requires 50,000 units per year. If the desired ROI is 12%, what is the maximum to pay a vendor for the product?

DISCUSSION CASES

Corbin Robot Co.

Corbin's sales of robots are expanding rapidly. It has two basic areas in the manufacturing department. One area builds the components of the robots and the other is the assembly area. Both areas are working two shifts per day and some overtime on the weekend as the schedule requires. The third shift is currently used for maintenance of the facilities, but sometimes overtime work takes place during the third shift.

Inventory of finished goods is very small because of expanding sales. Raw material inventory is difficult to manage because of the changing production schedules. In-process inventory is very high. There are large quantities of component parts waiting for subassembly and final assembly. Much of the scheduling problem is because of the large in-process inventory.

Corbin is considering adding facilities for both manufacturing and warehouse space to support the expanding sales. The existing plant is located on 20 acres and there is room for expansion. The company is located in a small midwestern town, and Corbin is having difficulties in getting additional skilled employees.

John Corbin, the son of the founder, is trying to manage the expansion. He says, "Increasing sales are a good problem to have, but I'm not sure we are doing the best we can. I want to expand our output, but there is the possibility that we could overexpand and then sales might fall off and we would have lots of excess capacity. We are working at our maximum, I think. There is not much room for more output under the present conditions. All the employees are working flat out and I appreciate their efforts. The schedule and inventory situation is very confused. We are late on some deliveries because we can't find the parts to assemble even though we know we have produced them. I'm not sure what to do. I know that if we can expand into the third shift our fixed costs will increase some, but I also know that if we can use our extra capacity, we can increase our profits and increase the profit sharing for our employees."

1. What are the main problems at Corbin?
2. How does the concept of breakeven and capacity utilization apply to Corbin?
3. Can return on investment be increased without expanding output?
4. Do the individual capacities of the departments and equipment seem to be in balance?
5. How could John Corbin identify the internal bottlenecks?
6. What other factors should be considered?
7. What should John do?

Hardwick Manufacturing

An electronic product that Hardwick produces currently has fixed costs per unit of $19 and variable costs of $25, with a wholesale price of $61 per unit. The product is sold in hundreds of retail stores throughout the world. Hardwick has approximately 60%, or 300,000 units per year, of the world retail market.

Joe Hardwick formed the company 8 years ago and is the president. An electronics marketing firm has suggested to Joe that the organization could increase its market share to

70%–75% if it could reduce its wholesale price by $10. Joe estimates that the organization could invest an additional $1,500,000 into processes that would have a 5-year life with no salvage value, and reduce variable costs by $5.50 per unit. He is considering selling additional stock to make a number of investments into other products. Should he consider financing the additional costs to increase output of the product?

1. Is the expansion of market share economical?
2. What if the market share only increases to 60%–65%? To 90%?
3. What if the life of the product is only 3 more years?
4. What are some noneconomic factors for Joe to consider?
5. What if Joe desires 10% ROI? 20% ROI?

Games Sporting Goods Co.

Alex Roberson, owner of Games Sporting Goods, has been in business for 12 years and has a profitable organization. He is always interested in increasing sales and expanding output. Currently the store is open 6 days per week. Alex is considering opening for 7 days per week, since the competition's stores are on a 6-day week. Another alternative that he is considering is to open 1 hour earlier than the current 10 A.M. and to stay open 1 more hour at the end of the day until 8 P.M. The store could be open on Sunday from noon to 6 P.M.

Alex has discussed his ideas with the employees. They are reluctant to work on Sunday, although they realize that they would get time off on other days. The scheduling at the company has always been flexible and the company tries to keep the schedule flexible to accommodate working mothers, part-time students, and other salespeople. Alex has not surveyed the customers about these ideas.

1. With respect to breakeven and capacity, what are the advantages of Games Sporting Goods being open more hours during the week?
2. What are the advantages and disadvantages of extending the opening hours from the employee viewpoint and from the management viewpoint?
3. Will additional hours per week increase total sales, or will current customers decide to buy during the new hours? (This would keep total sales constant since the same customers would shop at other times.)
4. With reference to an area diagram, what is Alex trying to do?

U.S. Postal Service

The postal service has been trying to automate facilities for collecting, sorting, and delivering the mail. They are trying to upgrade special services such as overnight delivery and to compete with private services such as Federal Express, United Parcel, and others. Some of these competing services would also like to be able to deliver first-class mail, which the postal services now handles exclusively. The Post Office says that because of the high equipment investment, if others were permitted to handle the first-class items, the Post Office would operate below its breakeven point.

Postage rates for first class have been going up. Another problem is that the Post Office feels that the delivery of first class mail is a federal government responsibility. Letters should not be turned over to private carriers that might not protect the privacy of the information.

1. Is the argument of breakeven and capacity a correct argument for this application?
2. Compare the argument for breakup of the airline monopoly to the Post Office.
3. Does it appear that the unit costs of first class mail are decreasing as the output expands? Why?
4. Is the Postal Service a unique situation? Are the concepts of breakeven and capacity utilization applicable?
5. What other considerations apply to this situation?

Abby Pet Products Inc.—Variable Costs

Jim and Ann Stuart have an opportunity to sell some of their products to a large mail order company. Abby is profitable. Their products are successful and they have a good reputation with their customers. The additional output would use some additional unused equipment time. The mail order company wants Abby to put its name on the products, instead of "Abby." The mail order company wants to purchase the products at a lower cost than the products are sold to Abby's own distributors.

Ann feels that the additional output would increase fixed costs as well as variable costs. She also believes that the mail order company's sales would reduce Abby's sales. Jim feels differently. He wants to increase output and to price the output to the mail order company at a lower level than Abby sells to their own distributors. He says, "We can price the orders to the mail order firm so that all the variable costs are met, and then have some additional dollars that can contribute to fixed costs. Let's just cover our variable costs and then any additional income from the order can apply to our fixed costs—let's go for it!"

1. Is Jim's position correct?
2. Are Ann's concerns important?
3. What are the advantages and disadvantages of the order from the mail order company?
4. Should they accept the order? What should be the terms?
5. How do breakeven, capacity, and profits apply to this decision?

Sticks and Bricks Construction Inc.—A Second Floor?

Ben Furguson, civil engineer and builder, is preparing a bid for a small single-story commercial building of 10,000 feet. His initial estimate is $750,000 for the construction. The user of the building is a growing retail business. Although not requested to do so, Ben is considering suggesting an additional alternative to the buyer. The single-story building could be made into a two-story building. The second story could be used for future expansion, office space, or leased out to other users. The advantage, Ben feels, is that by increasing the size and strength of the footings and columns of the single-story building, the two-story building would cost less than twice the single-story structure. The additional footings and columns would add $50,000 to the cost of the first floor. The additional cost of the second floor would be $500,000 for 10,000 square feet.

The alternatives could be as follows:

1. Single-story building as requested, $750,000.
2. Single-story building with upgraded footings and columns, $800,000. The second story could be added later.
3. Two-story structure for $750,000 + $50,000 + $500,000 = $1,300,000.

   The buyer is considering the alternatives.

1. Should the buyer consider the additional alternatives Ben has suggested?
2. What would the retailer's lender bank say to these alternatives?
3. How do the alternatives relate to breakeven and capacity analysis?
4. Draw an area diagram of the alternatives.
5. What other factors should the buyer consider?

Chapter 9　Minimum Cost Analysis

KEY TERMS

Acquisition Costs The cost of purchasing material from an outside vendor, including negotiating, expediting, transportation, receiving, inspection. It can also be the cost of setup and changeover of equipment when the components are manufactured in-house.

Bottleneck An equipment or step in a process that takes longer than the preceding and following activities so that inventory is accumulated as it waits to be completed. This results in high in-process inventories, low ROI, scheduling delays, and slower customer response times.

Economic Lot Size (ELS) Same as EOQ.

Economic Order Quantity (EOQ) The quantity of materials per order that results in the minimum total cost when the acquisition (purchase or setup) and storage costs are added together.

Inventory Improvement Cycle The steps of reducing purchasing and setup costs, resulting in lower costs, higher ROI, reduced inventory, and increased customer satisfaction.

Just-in-Time (JIT) A term applied to an acquisition system with outside suppliers or internal producers delivering the required material upon request by having production systems that can respond quickly to the request without relying on inventory. It is the opposite of long delivery lead times caused by inflexible manufacturing systems or high inventory stocks.

Minimum Cost The sum of two costs, one that increases and another that decreases with respect to a key variable. The sum of the two costs declines to a minimum point and then increases with respect to the key variable.

Pull Inventory System A flexible and responsive manufacturing system that can respond to customers' requirements using JIT purchasing and quick changeover with low inventory storage quantities.

Purchase Quantity Discount The price reduction given by a supplier for the customer buying a large quantity of material.

Push Inventory System An inflexible system of forecasting sales, producing to that forecast, and storing the inventory until it is sold.

Setup Reduction The improvement of changeover time on equipment that permits the process to be run more frequently with smaller orders per run, resulting in low inventory levels and fast customer response times.

Storage or Holding Costs The costs of holding inventory, including the time value of money (TVM) on the investment, space costs, insurance, taxes, materials handling, and related costs.

LEARNING CONCEPTS

- Minimum cost concepts.
- EOQ to determine the quantities of manufactured lot sizes for purchased material and supplies.
- Implementation of just-in-time (JIT) techniques.
- Setup time reduction.
- Push versus pull systems, bottlenecks, inventory turnover improvement, capacity utilization, and increasing ROI.
- Minimum cost techniques applied to improving quality and maintenance costs.
- Minimum cost analysis of various technical applications, including electrical conductor sizing, piping sizing, structural design, and heat transfer.
- Inventory improvement cycle.

INTRODUCTION

An important economic tool for design engineers, process engineers, managers, purchasing buyers, and other employees in the organization is *minimum cost analysis.* By recognizing cost areas that have opposing costs, one increasing and one decreasing, and determining the minimum of the sum of the two costs, the costs of products and processes can be reduced.

Minimum cost analysis is the basis for understanding and applying such process engineering and management techniques as economical purchase of quantities of materials from suppliers, economical quantities of units to produce per production run, just-in-time purchasing and scheduling, inventory reduction and ROI improvement, bottleneck improvement and capacity maximization, reducing equipment changeover times, improving product delivery times, and minimizing maintenance costs and cost of quality. Minimum cost analysis can also be used in design applications.

Creative application of minimum cost analysis can improve the organization's profits and lead to asset reduction, improved ROI, faster deliveries and customer response times, and capacity utilization. It may not be a "cure-all," but it does have potential applications that can cause significant financial and technical improvements.

MINIMUM COST ANALYSIS TECHNIQUES

Often, design engineers, process engineers, and manufacturing managers encounter problems that don't seem to fit the standard models of economic analysis. Some cost analysis problems have two opposing costs with no single answer. One cost *increases* as a key variable changes, and another cost *decreases* as the same variable changes. Although this **minimum cost** technique is not encountered as often as other analysis techniques, such as TVM and breakeven, it is an important decision making tool. It has applications in a number of design areas. Some that we will discuss in this chapter are electrical power distribution, pipe sizing for fluid flow, heat transfer, structural analysis, inventory control and management, and optimum quality costs.

Typical Minimum Cost Applications

Process Engineer	Design Engineer
EOQ	Electrical
JIT	Structural
Quick Change	Fluid Flow
Quality Costs	Heat Transfer
Maintenance	
Purchasing	
Inventory	

For example, in order to reduce the heat loss and energy costs of a building, insulation is added to the walls and ceilings. This reduces future expenditures on energy costs, but the purchase and installation of the insulation materials increase investment costs in the building. There is an optimum balance of costs, a minimum cost point, between insulating and heat loss. For another example, for a given power load on an electrical line, the resistance to the electrical energy flow decreases as the diameter of the conductor increases. In order to increase efficiency and reduce the cost of power distribution losses, the diameter is increased in size. But if the diameter of the conductor increases, the material and installation costs increase. There is an optimum cost, or minimum cost point, between the conductor and the power loss costs.

In process engineering, there are optimum quantities of material to be manufactured per setup. There is also an optimum number of times per year to purchase from vendors. These are determined by using minimum cost techniques.

Difference between Minimum Cost and Breakeven Analysis

In the last chapter we studied breakeven analysis applications. In the practice of economic analysis and selection, it is sometimes difficult to differentiate between breakeven and minimum cost analysis. There is an easy method to differentiate between the two analysis tools.

With breakeven, isocost, and make or buy analysis, costs *increase* with respect to the key variable such as output quantity per year. These are variable costs. Other costs, fixed costs, are independent or constant with respect to the cost of the key variable.

With minimum cost analysis, there are two costs, one *increasing* and one *decreasing* with respect to the key variable, such as order quantity, percent scrap, conductor or pipe diameter, and others. Having one cost increasing and one cost decreasing identifies minimum cost situations.

> **Minimum Cost**
>
> One cost increases and a second cost decreases with respect to the key variable.
>
> **Breakeven**
>
> One cost is constant and a second cost increases with respect to the key variable.

Product Design and Minimum Costs

> Technical designers must create a sound technical design that is reliable, that can be successfully manufactured at a minimum cost, and that will satisfy the customer's expectations.

Technical designers have the responsibility of creating a product that will perform the customer's required functions for the specified period of time. In addition, the product must be designed for efficient manufacturing at the necessary quality level. It is this latter responsibility that involves minimum costs. It is not enough to design a technically good product; it must be economical as well. The concept of minimum cost is frequently encountered when designing and manufacturing a product. Electrical transmission systems, electronics, structural designs, hydraulic and piping designs, heating and heat exchange systems, quality systems, and many other designs have cost functions that are directly related to minimum costs. We will look at these examples later in this chapter.

Process and Manufacturing Engineering and Minimum Cost

After the product is designed, process engineering begins. From the product's design, materials required for the product are determined and components are ordered. The steps in the manufacturing process are determined and the equipment and departments that will manufacture the product are assigned and scheduled. Special tooling and equipment is designed, ordered, and installed. Methods of manufacturing and inventory levels of raw, in-process, and finished inventory are determined. Schedules are developed and production run sizes are established. Quality control, sampling, gaging, and inspection procedures are established. Manufacturing begins.

Process and manufacturing engineers have the technical responsibility of planning and controlling the manufacturing process as well as the responsibility for cost analysis, cost control, and planning. Establishing the amount and type of quality control often involves minimum cost analysis. The determination of equipment setup and number of parts per run is also determined by minimum cost. The ordering of raw inventory and components from suppliers for the product is based on **economic order quantity** (EOQ) calculations, which is a minimum cost case. Scheduling delivery frequency and quantities of products from vendors and scheduling internal production setups are minimum cost decisions also. As will be seen, response and delivery times to customers are based on minimum costs.

Concurrent Engineering and Minimum Cost

The traditional technique for engineering a new product or revising an existing product is to design the product and then determine the departments, equipment, processes, and personnel that will manufacture the product. It is a linear sequence of steps from design to process engineering. The time to complete this sequence of events may be a few days or a number of years. Contemporary engineering management combines the two functions and they occur simultaneously. This is *concurrent engineering*.

The larger and more complex the product, the longer the time required to complete the design and process engineering functions following the traditional linear sequence. Concurrent design and process engineering can reduce this time, often to less than half.

When Boeing, the aircraft firm, applied the concept of concurrent engineering to the design and production of the 777 aircraft model, the time required for design and process engineering was reduced by several years. In this approach, the process engineers worked with the designers from the beginning. The process engineers could alert the designers to any potential manufacturing problems so the design could be revised immediately rather than waiting until it was completed, when the changes would be more costly and time consuming to implement.

> The best place to improve the product and reduce costs is in the design and process planning stage of a product's life, not during production.

Boeing even included potential customers in the early stages of product and process design. This allowed maintenance personnel, for example, to identify any potential maintenance problems that might occur with the aircraft. Those problems could be identified and corrected while the product was still in its early planning stages rather than waiting until the product was built. Concurrent engineering reduces design and process planning time and cost. The techniques of concurrent engineering require a number of applications of minimum cost analysis. Additionally, it gives the customer a better product at a lower price and in a shorter delivery period.

Inventory and Minimum Costs

> For many organizations, inventory is the second largest asset investment category, after fixed assets, appearing on the Balance Sheet.

Inventory EOQ applies to two areas. The first is making products within the organization for internal use, such as for setup time reduction, or to sell to customers. The second is purchasing components from suppliers outside the organization, such as JIT purchasing. Minimum cost analysis can be used to determine order quantity and frequency for internal and external materials.

Just-in-time (JIT) is an EOQ concept of having inventory and supplies available just before they are required rather than holding them in storage for long periods waiting to be used. Often thought to have originated in Japanese manufacturing facilities, JIT concepts of purchasing and manufacturing have roots in the United States. JIT is another application of EOQ analysis. We will see that it is a special case of the more general minimum cost analysis techniques.

Some manufacturing equipment must be changed over from producing one product to another product in the production schedule. This changeover may be very costly if it leads to time delays. A number of manufacturing firms have developed techniques to reduce the changeover time and decrease the costs of equipment downtime. The benefits of the changeover reduction are decreased product costs, increased production capacity, and reduced inventories. This changeover analysis, sometimes called setup reduction, also has its roots in minimum cost analysis and TVM.

Inventory Economic Order Quantity

Whether to purchase inventory from suppliers or to manufacture components for internal use (or to do both) is a major decision in the production system. Inventory purchase, manufacturing, storage, movement, monitoring, shipping, and other such costs are significant. It is important to analyze them carefully.

From the viewpoint of the sales and marketing departments, the larger the inventory, the better. A salesperson is reluctant to say to a customer, "We don't have that in stock. It will be 4 to 6 weeks before we can ship." A sale could be lost to a competitor. For the finance/accounting department, inventory is a large and costly investment. The best policy, in their view, is a small inventory investment. Scheduling and inventory control personnel see inventories as something to order, monitor, and schedule. They, too, want ample inventory so that there will be fewer constraints and stock-out problems. The purchasing department views inventory as material that must be ordered, stored, and paid for. Customers want the product to always be available when they want it. Inventory, then, is vital to many parts of the organization. Keeping the correct levels of inventory and managing it economically can be a major determining factor of an organization's financial success. Because some costs increase as the amount of inventory ordered increases, and other costs decrease, the analysis must find the optimum or best combination of these opposing inventory costs. The result of balancing the costs is a minimum cost solution called economic order quantity.

Some organizations periodically attempt to improve the Balance Sheet and Income Statement by arbitrarily decreasing inventories to lower their investment in material and free some cash for other uses. Because inventory represents a large financial investment, it is attractive to management as a source of funds.

Often, when the organization reduces inventory levels, the amount of orders to vendors, in-process inventory, and finished goods are all decreased. Generally, the reduction of investment in these assets is healthy because it increases the ROI of the organization. But the difficulty with arbitrary inventory reduction is that although it may make cash available, it may also cause scheduling and delivery problems, stock-outs, and related problems for both internal and external customers. Prevention and management of inventory problems, therefore, is preferred over arbitrary reduction of inventory. Minimum cost techniques can assist in this process.

Example of EOQ Application: Harlow Corp. Jenny Cox, a project engineer at Harlow, is responsible for the design, manufacturing, and customer scheduling of an electronics component, RL 12-40, that Harlow has been producing for a number of years. Harlow has a reputation for high quality, reasonable prices, and the ability to deliver orders on time. The firm manufactures and sells many electronic components, but it is recognized as a leader in RL 12-40 sales. The product is designed in different sizes and different configurations depending on the customer's requirements. Orders range from prototype orders of two or three dozen units up to a thousand units or more. Some customers have standing orders with Harlow for RL 12-40 units with periodic shipments.

Jenny is trying to improve the ordering of raw materials, equipment downtimes for changeover, and scheduling and delivery of orders. One order that she is currently working on is for a regular customer. The customer needs 12,000 units of RL 12-40 over the next year. Jenny is considering the application of EOQ analysis to determine the best schedule to apply to this order.

Setup costs—One solution for this order is to set up and run all 12,000 units at one time. This would be approximately a five-week production run. The items would then be stored in finished goods inventory and shipped to the customer as requested during the year. The advantage is that there would be one setup and one changeover of the manufacturing equipment for the year. The cost of setup and changeover is approximately $350 per setup and it requires two workers spending one and a half days (2×12 hours/worker = 24 hrs) to convert the equipment configuration to the new order from the previous order.

Holding or storage costs—When the product is produced, it is stored in the finished goods warehouse. The **holding costs** include the value of the warehouse space and the cost of the investment in inventory based on Harlow's target ROI for investments. The ROI is 10%. Other costs include insurance and taxes on inventory, spoilage and damage, material handling costs, and related storage space costs. The disadvantages of low manufacturing frequency are high holding costs and large inventory storage.

The RL 12-40 product has a value of $50 per unit, and the organization has a desired ROI of 10%. The more items that are in storage for a longer period, the higher the investment cost. For example, if 12,000 units are manufactured and stored, and if these units are dispersed over the year, the *average* quantity in storage during the year is 12,000 units/2 = 6,000 average units in storage during the year. If the value of the units is $50 per unit, the average material value of the units in storage is (6,000)($50) = $300,000. If the target ROI is 10%, the average cost of capital to hold these items is ($300,000)(.10) = $30,000 per year holding cost.

$$\text{Average Quantity Stored per Year} = \frac{\text{Order Quantity}}{2} = \frac{Q}{2}$$

The return on investment is only part of the cost of holding the inventory. Typically, the costs of capital plus other costs are in the range of 20% to 30% per year per part value or more. A graph of usage rates is shown in Figure 9–1.

The graph shows an initial production of 12,000 units, storage, a constant usage rate over 12 months with no other setups and runs, and depletion to zero of the RL-12-40 units at the end of the twelfth month. It also shows the initial production of 6,000 units, storage, usage over six months, depletion to zero at the end of the sixth month, and repeating the cycle for the next six months. Finally, a third alternative is shown of a production order quantity of 3,000 units, repeated four times. All three alternatives produce the required 12,000 units for the year, but there are different numbers of setups and storage quantities for each alternative.

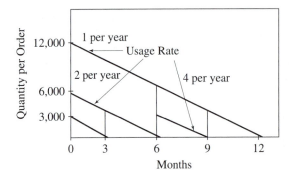

Figure 9–1   RL 12-40 Order Quantity Alternatives

Determining economic order quantity—First we will look at storage costs:

Q or EOQ = Order or run size—units per order

$Q/2$ = Average order quantity stored—units/year based on constant or linear usage rate

Holding cost % per year = H = ROI + other costs

PV = Part value—$/unit

Holding cost per year is usually stated as a percentage of the part value per year. It is not uncommon for this percentage to be 20% to 30% per year *or more*, depending on the product, facilities, and target ROI.

$$\text{Annual storage or holding costs} = (Q/2)(H)(PV)$$

Next we'll look at setup costs:

AU = Annual usage or requirement—units per year assuming constant rate of use

SU = Setup cost for each equipment changeover—$/setup

f = Order frequency—number of setups per year

$$AU = (f)(Q) \text{ or } f = AU/Q$$
$$\text{Annual setup costs} = (f)(SU) = (AU/Q)(SU)$$

It can be shown that when annual holding or storage costs are equal to the annual setup costs, the lowest or minimum cost is achieved.

$$\text{Annual Storage Cost} = \text{Annual Setup Cost}$$

$$\left(\frac{Q}{2}\right)(H)(PV) = \left(\frac{AU}{Q}\right)(SU)$$

This equation can be rearranged to give the following:

$$Q = \sqrt{\frac{(AU)(SU)(2)}{(H)(PV)}}$$

For the RL 12-40 component, a solution can be found using the above formula:

$$AU = 12,000, \ SU = \$350, \ H = 20\%, \ PV = \$50, \ Q = ?$$

$$Q = \sqrt{\frac{(12,000)(\$350)(2)}{(.20)(\$50)}} = 916 \text{ units per order}$$

The number of orders per year can be calculated as:

$$f = \frac{AU}{Q} = \frac{12,000}{916} = 13.1 \text{ orders per year}$$

Although this is the exact mathematical solution, Jenny can modify her quantity decision to reflect the customer's and Harlow's schedule requirements, such as using 12 times per year if that is more acceptable to the customer. The difference in total cost between 13.1 and 12 times per year is small, so if 12 times/year or monthly is easier to schedule for the customer and Harlow, it is still near the minimum cost point on the total cost curve. Harlow's and the customer's schedules can become synchronized. If the customer requires a different schedule, some small inventory could be maintained by Harlow to meet the customer's requirements and still use EOQ techniques. The customer may also be willing to modify its schedule if it knows that the supplier is trying to minimize its costs. Perhaps some of Harlow's cost savings can be passed along to the customer as lower prices.

Table 9–1 shows that as the two costs, storage and setup, are added, the Total Costs decline and then increase, giving a minimum cost point. This point occurs at the order quantity of 916 units per order for an order frequency of approximately 13 times per year. For the given costs, this is Harlow's best or optimum point for producing the RL 12-40 component. This is also shown in the accompanying graph, which shows the plot of the annual storage cost, annual setup cost, and total cost. It can be seen from the graph that the total cost curve is relatively flat in the range of 600 to 1,200 units per order, indicating that Jenny has some flexibility in setting the order quantity for the RL 12-40 component.

What-if analysis—What should be done to the optimum order quantity if the customer increases their annual order of RL 12-40 to 20,000 units per year?

In this case, the revised EOQ can be calculated as shown below for $AU = 20,000$:

$$Q = \sqrt{\frac{(20,000)(\$350)(2)}{(.20)(\$50)}} = 1,183 \text{ units per order}$$

This would give the number of setups of

$$f = \frac{AU}{Q} = \frac{20,000}{1,183} = 17 \text{ per year}$$

If the number of setups is increased from approximately one per month to one every three weeks, the new production volume would be optimized.

The RL 12-40 example is typical for the determination of lot sizes and the setup frequency of internally manufactured products. We will revisit this later in the chapter when Jenny tries to reduce the setup costs using some setup reduction techniques. Until then, we will look at other examples and the calculations for setup, order size, and frequency, and for products that are purchased from external suppliers.

Table 9–1   Annual Setup, Storage, and Total Costs for Component RL 12-40

Orders per Year *f*	Quantity per Order *Q*	Annual Setup Cost *(AU/Q)(SU)*	Annual Storage Cost *(Q/2)(H)(PV)*	Total Cost *(AU/Q)(SU) + (Q/2)(H)(PV)*
1	12,000	$350	$60,000	$60,350
2	6,000	$700	$30,000	$30,700
4	3,000	$1,400	$15,000	$16,400
6	2,000	$2,100	$10,000	$12,100
8	1,500	$2,800	$7,500	$10,300
10	1,200	$3,500	$6,000	$9,500
12	1,000	$4,200	$5,000	$9,200
14	857	$4,900	$4,286	$9,186
16	750	$5,600	$3,750	$9,350
18	667	$6,300	$3,333	$9,633
20	600	$7,000	$3,000	$10,000
22	545	$7,700	$2,727	$10,427
24	500	$8,400	$2,500	$10,900
26	462	$9,100	$2,308	$11,408
28	429	$9,800	$2,143	$11,943
30	400	$10,500	$2,000	$12,500
32	375	$11,200	$1,875	$13,075
34	353	$11,900	$1,765	$13,665
36	333	$12,600	$1,667	$14,267
38	316	$13,300	$1,579	$14,879

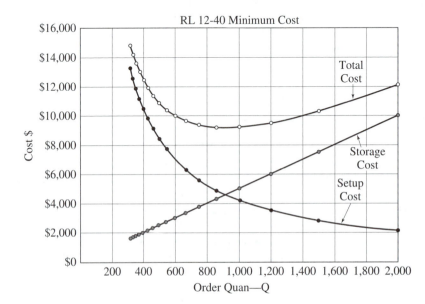

Component S-350 at Harlow: Determine Number of Setups. Jenny is also responsible for another Harlow product, S-350, which contains a large amount of precious metals. The storage costs for this item are 40% of the part value because security storage in a vault is required. If 10,000 units are needed per year, setup costs are $200 per setup, and the part value is $230, what is the optimum number of setups?

$$AU = 10,000, SU = 200, H = 40\%, PV = 230, Q = ? f = ?$$

$$Q = \sqrt{\frac{10,000)(\$200)(2)}{(.40)(\$230)}} = 208 \text{ units per order}$$

$$f = \frac{AU}{Q} = \frac{10,000}{208} = 48 \text{ per year}$$

This requires a schedule of approximately one setup per week. The high frequency and low number of units per order are due primarily to the high storage costs. This is an example of how high storage costs can cause lower order quantities and larger frequency of production runs. The opposite situation would occur if a product requires minimal storage, such as products stored outdoors. In this case, the order quantity would be large and the order frequency small, resulting in fewer setups per year.

EOQ and Customer Requirements

The application of EOQ concepts must always be tempered with the needs of internal and external customers. Jenny at Harlow Corp. can examine the current delivery schedule to see if it agrees with the previous calculations. If Harlow's revised schedule of production does not agree with the customer's needs, the customer's requirements should have priority, if possible. Harlow may have to store inventory so that it is available when the customer requires it. Often when customers learn of the supplier's intent to optimize its costs and scheduling using cost improvement techniques (discussed in later chapters), they will happily cooperate with the supplier, especially if part of the savings is passed on to them as a price reduction.

The next step for the supplier is to further reduce setup costs. Reducing costs using EOQ is one of the repeated steps of continuous improvement until the optimum relationship between the supplier's and the customer's schedule and costs is achieved.

> The customer's needs should come before the supplier's with respect to order quantity and frequency. But often both the supplier's and the customer's requirements can be met simultaneously.

This concept applies to internal as well as external customers. Optimizing the schedule in one department, thus causing problems and high costs in another area, is not beneficial to the total organization. Continuous improvements in setup costs can assist in getting the two schedules synchronized.

EOQ and Purchased Components

Purchasing materials from suppliers, or *outsourcing,* can be analyzed like the EOQ analysis for internally produced parts. After project engineers such as Jenny Cox determine that the component should be purchased outside rather than made inside the organization (using make or buy analysis), purchasing buyers can begin the process of obtaining quotes and awarding the contract to a supplier.

The purchasing department at Harlow Corp. is also applying the concept of EOQ to buy components from vendors. They estimate that the cost of executing a purchase order, including the cost of meeting with the salesperson, phone and fax costs, receiving, and related costs, is typically $125 per order. One raw material item has a value of $5.50 per unit, an annual requirement of 5,000, and a storage cost of 25% of the part value per year. What would be the optimum order quantity if Harlow issues different purchase orders for each shipment? The cost of issuing a purchase order can be substituted for the setup cost in the previous formula so that instead of *SU* in the equation, *PO* is used, as shown below.

$$Q = \sqrt{\frac{(AU)(PO)(2)}{(H)(PV)}}$$

Substituting in the equation and solving,

$$Q = \sqrt{\frac{(5,000)(\$125)(2)}{(.25)(\$5.50)}} = 953 \text{ units per order}$$

$$f = \frac{AU}{Q} = \frac{5,000}{953} = 5 \text{ per year}$$

The order frequency of five times per year, approximately every 10 weeks, and the order size of 953 units per order can be modified to account for standard order sizes, shipping quantities, and other considerations. For example, perhaps the standard carton, pallet load, or truckload contains 1,000 units or 900 units. If so, the order quantity can be modified to meet shipping lot size requirements. Look at the total cost curve in the area of the EOQ. If the curve is relatively flat, changing the actual order size from the exact mathematical EOQ will have a small cost effect.

Purchase order costs may be reduced by working closely with the supplier and giving a long-term contract for the component. By using an annual or longer contract period, negotiating, quoting, and contract-awarding costs can be reduced, which would reduce order size and increase order frequency. This leads to the concept of just-in-time purchasing.

JUST-IN-TIME PURCHASING

If the cost of executing the purchase order can be reduced, the order frequency increases and the quantity per order decreases. This is the basis of the just-in-time (JIT) purchasing system. This reduces stored inventory and shortens response time to the customer.

From the previous example, the next question could be, "What if a single purchase order is issued to a supplier for the entire year?" There would still be some costs of issuing the PO, expediting, and receiving the order, but this would be less than the $125 per order. If the average cost of ordering and receiving the order decreased to $50 per order, what would occur?

$$EOQ = \sqrt{\frac{(AU)(PO)(2)}{(H)(PV)}} = \sqrt{\frac{(5,000)(\$50)(2)}{(.25)(\$5.50)}} = 603 \text{ units per order}$$

Using the EOQ formula, $Q = 603$ units per order and the frequency would become 5,000 units per year/603 units per order = 8 times per year (rounded). As the PO costs are reduced further, the order quantity decreases and the number of orders per year increases. This further reduces the customer's need for storage. Over the long term, if this continuous improvement of smaller, more frequent orders continues for all purchased materials and supplies, the warehouse and in-process storage space can be significantly reduced. Existing storage space can be converted to production space as output expands.

Successful just-in-time ordering is based on EOQ analysis.

Working with a vendor to establish long-range ordering plans under a single purchase order creates more frequent orders and smaller quantities per order. The opposite is also true. If a great deal of time and money is spent sending inquiries to multiple vendors, analyzing the quotes, and negotiating the price, delivery, and quality each time the order is placed, the order frequency will be low and the number of units per order high due to the increased cost of ordering. This is the opposite of JIT. This approach is being replaced by methods using single suppliers who have long-term commitments to their customers. A supplier who has a long-term relationship with a customer is willing to invest more into new equipment and training employees. This generally results in higher quality products and services and lower long-term costs. If these lower costs are passed to the customer as lower prices, both the supplier and the customer benefit. This system requires transferring more responsibility for quality, quantity, and scheduling to the supplier. In order for the supplier to accept this additional responsibility, a long-term commitment must be made. This is the basis for JIT purchasing.

The quality expert Edwards Deming stated that multiple suppliers are costly and lower the quality of the incoming material. He suggested that buying material from more than one supplier caused higher variation because of the multiple lots of incoming raw material rather than a single batch of material from a single supplier's process. Typically, the greater the variation in raw material, the more difficult it is to produce a high-quality product that meets specifications. A single supplier can also provide lower purchase order or acquisition costs for the customer. These factors, along with EOQ concepts, are responsible for many changes in the contemporary purchasing relationships between vendors and customers. More responsibility for quality is transferred to the supplier, and longer term relationships are established. As one purchasing manager commented, "It becomes a marriage rather than a brief affair."

As suppliers and customers become "partners," costs decrease, quality increases, and schedules become more synchronized. The results of this effort are better profits for both supplier and customer as well as higher customer satisfaction.

The increased emphasis by customers on more frequent shipments, lower quantities per order, and longer term vendor-customer relationships encourages vendors to make changes in their scheduling practices and manufacturing processes. The vendor must be more responsive and have the ability to produce smaller orders efficiently or have large inventories to meet the customer's requirements. If the vendor continues to produce the same large quantities, store them, and then meet the customer's request for smaller, more frequent orders from inventory, the vendor is going to have high storage costs. These costs are passed on to the customer with higher prices. The most efficient way for the supplier to respond to the customer that wants smaller, more frequent orders and to avoid the high inventory storage costs is to adjust its own production system so it is more flexible and responsive. This means reducing setup costs and changeover time. The customer's desire to have smaller inventories and more flexible manufacturing causes the supplier to apply minimum cost analysis to setup times and their own supplier order system. There is a ripple effect as each customer requests more frequent deliveries and smaller order quantities from their suppliers. This can lead to lower optimum costs for both the customer and the supplier.

> If we want lower costs and higher quality, customer-supplier relationships should be based on "we," not "us vs. them."

Personal Applications of EOQ

There are personal applications of these inventory/purchasing concepts. For example, if a family consumes one pound of coffee per week, what is the best purchasing policy for this grocery item? One option is to purchase 52 pounds of coffee once a year. This has the advantage of requiring only one trip to the supermarket and thus low costs of purchasing because of reduced time visiting the market, lower transportation costs for car and gasoline, etc. Perhaps the seller is even willing to give a lower price per pound because of the increased order quantity. The disadvantages of this single purchase are the requirement of more cash to purchase the 52 pounds at one time, storage of the product for the year, the potential for damage and spoilage, and change of taste during the year. The storage space could likely be found in the basement, garage, or attic for the coffee, but what if this once-a-year purchase policy is followed for *all* purchases by the family? Then it is likely that storage space would become scarce in the home and additional space would have to be added at additional cost. The initial cash requirement is probably the major disadvantage. If the purchase were made using a credit card, interest would be charged on the loan. The same applies to a bank loan. What about more frequent purchases of smaller quantities, such as daily purchases of a few ounces for that day's coffee requirement? This has the obvious advantage of almost no storage space requirement, but making daily trips to the market would be costly and inconvenient. Perhaps if the market delivered on a daily basis, this alternative could be considered. But the unit cost of the coffee on a daily basis would likely be higher than for weekly or yearly purchases. An optimum order frequency is required.

When we purchase gasoline for the car, buy groceries, use "1-hour" dry cleaning services or while-you-wait shoe repair, 10-minute oil change or "1-hour" film processing, we are using JIT ordering. Most individuals do not make actual EOQ calculations, but they do think about the same factors as in EOQ analysis. Often, by trial and error, we arrive at the best order quantity and frequency for ourselves.

> Large organizations are finally catching up with "while-you-wait" shoe repair shops, dry cleaners, and 10-minute oil changes with their "new" JIT systems.

QUANTITY DISCOUNTS AND EOQ

Frequently, a supplier will offer a reduced price for a product if the customer orders a minimum quantity or minimum dollar value. The buyer then must base the purchasing decision on the EOQ or on the quantity that will provide a discount. The following example demonstrates the solution. The annual usage of a product is 30,000 units per year, storage costs are 25% of the part value, the part value is $12 per unit, the cost for the acquisition of the order is $50 per order, and the supplier offers a quantity discount of 2% on orders greater than 1,500 units per order. The EOQ calculation is:

$$Q = \sqrt{\frac{(AU)(PO)(2)}{(H)(PV)}} = \sqrt{\frac{(30,000)(\$50)(2)}{(.25)(\$12)}} = 1,000 \text{ units per order}$$

The order frequency using 1,000 units per order is $30,000/1,000 = 30$ times per year.

The order frequency using the discount quantity of $Q = 1,500$ units per order would require ordering $30,000/1,500 = 20$ times per year. This is a higher quantity per order and a lower order frequency.

The part value based on the revised discount price is now $(.98)(\$12) = \$11.76$ per unit.

The EOQ is 1,000 units per order, but the 2% discount is based on 1,500 or more units per order.

The alternatives can be analyzed as shown in Table 9–2.

The annual savings from the quantity discount price of $Q = 1,500$, in this case, is less than the annual cost for $Q = 1,000$ units. Therefore, it would be more economical to order the higher quantity per order so the discount can be obtained. If the discount quantity had been higher and/or the discount percentage had been lower, it is possible that the EOQ order quantity would

Table 9–2 Comparison of Costs with and without Quantity Discount

	1,000 Units per Order $/Year	1,500 Units per Order $/Year
Storage Cost = $(.25)(Q/2)(PV)$	$1,500	$2,205
Acquisition Cost = $(f)(PO)$	$1,500	$1,000
Material Cost	$360,000	$352,800
Total Annual Cost	$363,000	$356,005

have been the less expensive. Each case must be analyzed by itself. No generalization can be given, such as, "Always take quantity discounts or never take quantity discounts."

CAUTIONS CONCERNING EOQ

As with most mathematical solutions, cautions must be applied. The following are other factors that should be considered when making EOQ calculations.

- **"Lumpy" demand.** The usage and demand for the product or service might not be constant or linear, as in the assumption of the EOQ calculations. Often, the usage rate is close enough to be approximated by assuming constant usage, however.

- **Minimum order quantity.** Sometimes a supplier will require a minimum quantity per order. This quantity might be above or below the EOQ quantity, and should be considered in the cost analysis.

- **Transportation costs.** The shipping and freight charges might or might not be included in the quoted price. These costs are often as high for small orders as for large orders due to the capacity of the shipping container.

- **Unit loads.** There may be a standard quantity such as a pallet load containing a number of units or a carton that has a fixed quantity, and this quantity may be different from the EOQ.

These and other factors can be used to "fine tune" the purchasing decision.

RETURN ON INVESTMENT AND JUST-IN-TIME

The concept of JIT ordering from external sources and internal scheduling is often thought to be a recent development. Sometimes, credit is given to the Japanese for discovering the concept. Actually, there were examples of JIT being used in the U.S. well in advance of Japanese applications.

> Just-in-time is not a recent development.

In the late 1960s, a major U.S. steel firm ran a number of full-page advertisements in a national financial publication with the headline, "Let Us Be Your Steel Warehouse." The ad explained that if a customer committed to a minimum tonnage of steel purchases each year, the steel firm would build a warehouse near the customer (across the street if possible) so that the customer could call the supplier in the morning and have the required steel items delivered in the afternoon. This would eliminate the need for the customer to stock any of its own steel. A regional sales manager for the steel company commented that *no customers* took advantage of the steel firm's offer. He said that the customers believed that the supplier would increase the prices of the steel to cover the new warehouse costs and that they did not want to be committed to a single supplier for long-term steel supplies. The steel firm demonstrated that the additional cost of the steel to the customer would be less than the cost of having their own warehouse, but still no customers wanted to adopt the plan. Today, there are many successful arrangements of suppliers providing warehouse services to customers. Letting suppliers provide material on a JIT basis and reducing the number of suppliers is not a recent practice.

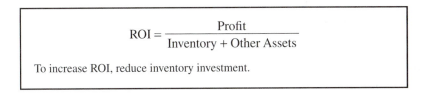

$$ROI = \frac{Profit}{Inventory + Other\ Assets}$$

To increase ROI, reduce inventory investment.

Dairy companies and bakeries have provided daily milk and bread deliveries to homes as early as the late 1800s. Even in the 1950s, a major midwestern grocery chain was applying EOQ and JIT concepts, although they didn't use those terms. They received deliveries of groceries and nonmeat or produce items on a two-week basis. Each stock clerk was responsible for an aisle of groceries in the store, such as canned goods or cereal. The clerk would place an order one week with the central warehouse and then receive the order the following week. As a result, overstocking occurred and extra inventory was held in the back room of the store. Inventory would be replenished from the storage area and placed on the sales floor as customers purchased the items. This stocking process occurred an average of 26 times or every two weeks during the year. The result was high stock levels of static inventory at hundreds of store locations. By increasing the delivery schedule to weekly or 52 times per year and giving small monthly bonuses to the stock clerks for keeping their "back room inventories" to a minimum, the total inventory on hand was reduced and the inventory turnover was increased by significant amounts. The total investment in inventory by the grocery chain was reduced by 35%—a significant savings, and customer requirements were met. By spending more on training and rewarding the stock clerks with a small amount of money, the inventory investment was reduced significantly. ROI was increased.

Suppliers are generally willing to supply the product in any form and quantity that the customer wants if there is potential to make a profit. If the supplier has to store the inventory for the customer, it is likely that the vendor will charge more than if a larger quantity was ordered and the vendor did not have to store the inventory. Quick response time by suppliers gives customers the required service, and suppliers keep prices down by having reduced changeover times and flexible manufacturing systems.

If suppliers do not want to store customers' inventory, they must have a flexible manufacturing system that will permit quick changeover from one process to the next. This is the true quick change or just-in-time manufacturing system.

By reducing inventory, total assets are reduced and ROI is increased; however, inventory should not be reduced so much that the organization's competitive position is affected.

Japanese Applications of JIT

In the late 1970s and early 1980s Japanese auto manufacturers were gaining significant increases in market share of auto sales in the U.S. American visitors to Japanese auto plants looked for the secret of higher quality production at lower cost. One of the visitors at a major auto producer's plant noted that warehouse space was very limited. How could they successfully schedule production without more storage space? The observers were shown large doors that opened to the outside of the plant. They were told that the material suppliers had keys to those doors. During the early morning hours when the plant was not working, the suppliers would drive their trucks into the plant and unload the next day's inventory requirement next to the equipment that would be using the material. This was an example of JIT with an inventory turnover of more than 150 times a year! Not all goods were delivered in this manner, but significant quantities were. The visitors returned to their U.S. plants with hopes that a similar plan could be implemented at their facilities.

One of the reasons for the Japanese suppliers' JIT ability was that they had developed flexible manufacturing systems that could change from one product to another quickly and at a low cost. Their internal inventory and JIT systems were very responsive and flexible. Without this ability, the suppliers could not have met their customers' requirement for frequent deliveries without accumulating large quantities of stored inventory.

What was the driving force of the Japanese applications of JIT? One reason was the high cost of land, which resulted in high inventory storage costs. In the U.S., a warehouse can often be built by purchasing adjacent land, perhaps a field, at agricultural land prices. In Japan, land prices can be hundreds or even thousands of times the cost of U.S. land prices. High storage costs usually result in higher frequency of ordering and smaller quantities per order.

In Japan, distances between suppliers and customers are often shorter than those in the U.S. A customer might be only a few hours away from a supplier. In the U.S., a customer in California and a supplier in New Jersey are usually a number of days apart when goods are delivered by motor vehicles. An out-of-stock situation isn't usually as critical when suppliers are in close proximity.

In Japan there is a sense of partnership between customers and suppliers. Historically, in the U.S. there has been an emphasis on price rather than on delivery, quality, and service. For a supplier to charge a higher price in order to provide better delivery, service, and other benefits to the customer sounds logical, but too often the customer is under a great deal of pressure internally to buy the product at the lowest price. Many American customers have changed suppliers just to obtain a very small reduction in the purchase price. In the process they often sacrifice quality and delivery advantages.

In the U.S. relations between management and unions are also different than in Japan. In the past, Japan has had union problems, but today these are rare occurrences. A large U.S. automobile firm had established a just-in-time relationship with its supplier plants. One of the electronic plants that supplied eight other assembly plants went on strike and within seven days the assembly plants were forced to stop production because they were out of electronic components. The firm has since modified its just-in-time relationships between the component plants and the assembly plants.

SETUP TIME REDUCTION

One of the key elements in a flexible manufacturing and JIT system is minimum **setup** or changeover times of equipment from one order to another. To reduce the EOQ for an in-house manufactured component, the annual usage, storage percentage, or setup time must be reduced. For the EOQ calculations that have been used in this chapter, if the setup costs can be reduced, the EOQ can be reduced. In the past, setup costs were taken as a fixed cost that could not be reduced. This is not true in most cases. Rarely were the setup or changeover times analyzed to determine effective reduction methods. This situation has changed in many organizations.

Shigeo Shingo, an industrial engineer who developed the setup reduction and quick changeover system for Toyota, has received considerable attention. The Toyota/Shingo approach has achieved reduction in setup times, smaller lots, and faster response times. In one case, this approach reduced a large press setup from more than 4 hours to 10 minutes. Other applications have reduced setup times from over 24 hours to 3 hours and less. This significantly reduces the EOQ and increases order frequency. The result is lower inventory stored and quicker response time to the customer or next department in the process.

At a Toyota automotive facility, Shingo was able to decrease the time required to change dies and other tooling between various operations. Shingo observed that the setup workers did not always have the tools they needed when they began their work. They had to find the tools and this increased the downtime of the equipment. By placing all the required tools near the work area *before* the changeover began, the downtime was reduced. Shingo designed simple bolts, connectors, and other fasteners to decrease the time required for unbolting the old equipment and tightening the fasteners on the new setup. He created a second set of dies for large presses that could be precisely adjusted off the press while the equipment was still running. By shutting down the old run, removing the old dies, and quickly inserting the preset new dies, the changeover was made in minutes rather than hours. This required a significant investment in the off-press equipment. However, it still proved economical when considering the costs of extended changeover periods. Shingo simplified the on-press methods and techniques, and he simplified the operation by doing as much of the work as possible while the equipment was running. As a result, overall setup time was reduced from hours to minutes. This permitted frequent changes of equipment and improved responsiveness to the assembly departments (the internal customers) that required components from the presses while maintaining limited quantities for storage. Inventories were lowered, and there was an increase in the capacity and availability of the equipment. It must be remembered, however, that this is a *long-term* improvement activity, not a quick fix. It often takes years, not weeks, to reach these reduced setup times, after making many small, continuous improvements.

When first considering the concept of just-in-time, many organizations feel, "It might work for *other* firms, but it won't work for us." However, after they realize setup time reduction, smaller lots, more frequent supplier deliveries, and other applications of JIT, they achieve large financial benefits. They can enjoy lower inventories, higher ROI, faster response time to customers, smaller manufacturing facilities, and better quality control through smaller lots, resulting in better service and price for customers and lower manufacturing costs for producers.

Reducing Setup Costs

Setup Time Reduction at Harlow Corp. Jenny Cox, the project engineer for the RL 12-40 component at Harlow Corp., is trying to reduce the setup costs of the component so that the EOQ can be reduced and the order frequency can be increased.

Currently, the annual usage for the RL 12-40 is 12,000 units per year. It has a part value of $50 per unit, a storage cost of 20% of part value per year, and setup costs of $350 per changeover. In the previous Harlow Corp. example, the optimum order quantity for the RL 12-40 component was calculated at 916 units per run. The number of setups per year is $f =$ 13 times per year, as shown in Table 9–1.

The setup reduction team, composed of Jenny, the equipment operator, setup personnel, and the supervisor, studied the process and identified areas for potential savings. They found that locating the proper tools and equipment to make a change requires 30 minutes. By performing some of the "get ready" activities *before* the process is shut down, changeover time can be saved. Redesigning some of the fixtures for making changeovers and final adjustments further reduces setup time. By studying methods used in the existing changeover, more improvements can be made. Overall, the costs of making a changeover were reduced from $350 to $150. Jenny determined that this had an effect on EOQ, order frequency, and machine capacity utilization. The revised calculations are shown below.

$$Q = \sqrt{\frac{(12,000)(\$150)(2)}{(.20)(\$50)}} = 600 \text{ units per run}$$

The revised order frequency becomes $AU/Q = 12,000/600 = 20$ times per year. Reduced setup costs and their effect on reducing order quantities are shown on the graph in Figure 9–2.

On the graph, the quantity per order is reduced from 916 to 600 units and the number of orders per year increases from 13 to 20. An additional benefit is that the overall utilization and capacity of the equipment is increased, because the reduced setup time can provide additional actual production time.

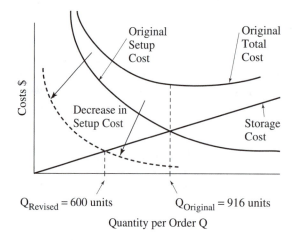

Figure 9–2   Revised RL 12-40 Setup Costs

PUSH AND PULL INVENTORY SCHEDULING AND EOQ

In a **"pull" inventory system,** a supplier does not manufacture according to a forecasted amount. Rather, the supplier waits for the customer's order rather than producing in anticipation of the order and storing the output until the customer requires it. By making an order, the customer "pulls" the product from the supplier. This is possible when changeover times are low and response time is fast and flexible.

The opposite is a **"push" inventory system,** which requires a supplier to forecast the customer's needs. The required units are produced in advance of the customer's request, "pushed" from manufacturing into inventory, and stored until a delivery is made to the customer. This system is used when the changeover times are high. Large amounts of inventory accumulate, and there are long changeover times. Minimum cost analysis is a tool that can assist with the creation of a pull system.

Push and Pull

A "push" inventory system forecasts customer requirements, manufactures that quantity, and pushes the inventory into the warehouse for storage until an order is received, resulting in high inventory quantities.

A "pull" inventory system waits for the customer to request the material and then rapidly produces the products, which results in low inventory storage quantities.

To move from a push to a pull system, changeover times must be reduced so that response times are short and inventory is reduced significantly.

CAPACITY, BOTTLENECKS, AND MINIMUM COSTS

Equipment with long and expensive changeover times, used to produce large quantities of inventory, can cause **bottlenecks.** When the setup costs are high for internally produced products and when purchase and acquisition costs are high for externally acquired materials, the EOQ formula indicates that larger quantities per order are economical. A large inventory is produced and stored to avoid costly and time-consuming changeovers. Infrequent changeovers cause delays and bottlenecks, which delays other orders. Orders must wait and inventory accumulates— a classic bottleneck situation.

Bottlenecks

If all equipment and workers in the process are used at the same time, the slowest part of the process, the bottleneck, causes other process components to produce excess inventory that must wait for the bottleneck. The pace of the process flow is determined by the slowest component.

This bottleneck then sets the pace for the whole department, process, or product. More efficient equipment stands idle while waiting for the bottleneck, since that is often more economical than producing inventory that will accumulate and increase storage costs. This bottleneck cycle can be broken by reducing setup time and costs, and by improving the efficiency of the slow equipment. Sometimes the equipment setup or efficiency can't be improved. In those cases, the equipment can work longer using additional shifts or overtime, or an outside supplier can be used to produce the required output. Of course, one solution is to purchase additional bottleneck equipment to increase capacity. This solution would only be acceptable if all other methods have been attempted and if it is economical.

If excess inventory is produced and must wait for bottleneck equipment, there is a reduction in inventory turnover, storage costs are increased without the benefit of higher output, and ROI declines. Costs are increased without increasing output or customer service. Increased capacity utilization can only be obtained by producing higher quantities of inventory. Having all departments and equipment working gives the appearance of improved capacity utilization. In the past, high stocks of inventory were an indicator of a wealthy organization. But contemporary process engineering and manufacturing management is rewarded for low inventory, high turnover, high ROI, quick customer response time, ability to reduce bottlenecks, and ability to produce economic quantities per order.

CONCLUSIONS ON JIT, REDUCED SETUP TIME, AND EOQ ANALYSIS

When we summarize the benefits of these inventory techniques, we realize their power to affect the total organization. By implementing the concepts of JIT purchasing, EOQ, and quick changeover, manufacturing managers, purchasing buyers, and process engineering staff can:

- reduce inventory levels
- increase productivity
- increase inventory turnover
- reduce costs
- increase ROI
- reduce customer delivery times
- increase the organization's overall financial results

Figure 9–3 shows the benefits of implementing these inventory techniques.

QUALITY AND MINIMUM COSTS

In the past few years, the application of quality principles has had a significant effect on the ways many organizations run their business. Not since the Industrial Revolution and the rise of the factory system has a single concept changed organizations the way quality management practices have. A significant part of the quality system is the analysis and improvement of the costs of quality output.

Designers, engineers, and management are concerned about these quality costs. Most causes of the high costs can be traced to the design, management, process, or material suppli-

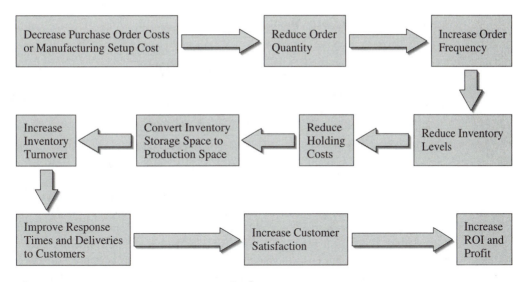

Figure 9–3   Inventory Improvement Cycle

ers. These costs can be a large portion of an organization's expenses. For the organization that has no quality improvement program, the sum of the quality costs can be as large as 20 to 25% of the organization's sales. Achieving quality costs under 5% of sales requires a concentrated effort and is a long-term effort. This minimum cost application and improvement is discussed in future chapters.

Minimum Costs and Maintenance Costs

Maintenance Costs and EOQ at Harlow Corp. Plant engineer Harry Coombs at Harlow Corp. is responsible for maintenance, utilities, repairs, equipment installation, and related tasks. He is attempting to apply minimum cost practices to equipment maintenance. On one piece of equipment, the large Masco press, the bearings, hydraulic system, pumps, and computer interface equipment are inspected, repaired, and adjusted every six months. The process takes three days, or 72 hours of work. The total cost is $3,500, which includes materials. Not included is the additional cost of lost production time, which is valued at $800 per day. This makes the total cost equal to $5,900 per repair every six months ($3,500 + 3 × $800). This is $11,800 per year.

Harry has just returned from a series of seminars on preventive maintenance. One of the presenters from Masco, the equipment manufacturer, suggested that Harlow should follow the new recommended maintenance protocol of performing weekly inspections on the eight major systems in the unit and using bi-monthly maintenance routines. The maintenance could be done in four hours by two workers, and would yield fewer major repair costs and less down time. The total cost of this maintenance including materials is $4,000 every six months. All activities can be performed during nonproduction times. A summary of the annual costs are shown in Table 9–3.

Table 9–3   Harlow Corp. Maintenance Costs

	Present Schedule Costs	*Proposed Schedule Costs*
Labor and material costs for 6 months	$3,500/6 mo	$4,000/6 mo
Downtime costs	$2,400/6 mo	none
Total costs for 6 months	$5,900/6 mo	$4,000/year
Total costs for 1 year	$11,800/year	$8,000/year

Because the new schedule requires additional labor hours during nonproduction periods, Harry will have to change the schedule of the repair and maintenance personnel. The equipment still requires a major overhaul every three years during the plant's annual two-week shutdown for vacation. These costs remain the same for both the old and new schedules. By applying this approach, the total labor costs increase but the materials and costs of downtime decrease. In addition to the lower costs, this approach will let workers spot trouble sooner and make repairs at a lower cost. Harry plans to implement this approach and monitor the costs carefully over the next year to see that the equipment manufacturer's recommendations really work. He wants to apply this same concept to other equipment to see if total maintenance costs and repairs can be reduced.

A plot of the costs for the equipment is shown in Figure 9–4. Harry knows the graph of these costs does not match the typical minimum cost graphs because only two points of the maintenance costs are known rather than many points. He thinks that the labor cost will increase with respect to the key variable, repair costs; that the other costs, including downtime, will decrease; and that the total cost will decrease. Although it's uncertain whether the total costs would decline to a minimum level and then increase, as in the standard minimum cost model, it is likely.

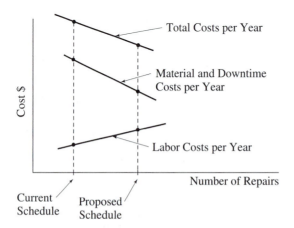

Figure 9–4   Masco Press Maintenance Costs at Harlow

PRODUCT DESIGN AND MINIMUM COST APPLICATIONS

The product design department can also use the economic tool of minimum cost analysis. The design engineer may encounter minimum cost applications, for example, in sizing electrical conductors and piping, selecting insulation materials, or sizing structural members. We will only look at a few examples here, but there are other minimum cost applications.

Electrical Conductor Sizing

As the diameter of an electrical conductor increases, the resistance decreases. The greater amount of material in the conductor also causes the cost of material to increase. Thus, with respect to the increase in conductor diameter, the material cost increases and the power loss, I^2R, decreases, as shown in Figure 9–5a. The sum of the two cost curves gives a total cost curve that has a minimum point indicating the optimum conductor diameter.

Fluid Flow and Pipe Sizing

Another application of minimum costs is determining the size of a pipe based on fluid flow. As the diameter of the pipe increases, the resistance to the flow decreases because of a decrease in the friction factor. And, as the diameter of the pipe increases, the cost of material, labor, and installation increases. The sum of these two costs gives the total cost curve, which has a minimum point at the optimum pipe diameter size, as shown in Figure 9–5b.

Heat Transfer and Insulation Thickness

Minimum costs apply to the situation of determining the amount of insulation to add to a home, building, or industrial heating/air conditioning system. As the thickness of the insulation increases, the heat loss and heat cost across the insulation boundary decreases, and as the insulation thickness increases, the cost of material, labor, and installation increases. At some point the two costs are equal and the optimum thickness is realized. A greater amount of

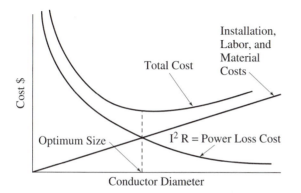

Figure 9–5a   Electrical Conductor Sizing

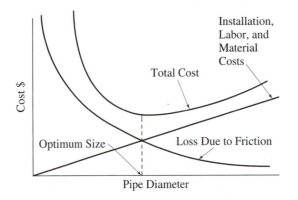

Figure 9–5b   Pipe Sizing

insulation thickness would result in decreased heat loss, but would also cost more. A reduced insulation thickness would be less expensive, but would result in a higher heat loss. This relationship is shown in Figure 9–5c.

Structural Design

There is a minimum cost relationship between column spacing and the size of the horizontal beams supporting floors and roofs. As the spacing of the columns becomes wider, the beam size required to span the distance between columns increases because the required beam length increases. To support the same load, the beam's cross-sectional area must be increased. As the column spacing increases and the cost of the columns decreases, the beam cost increases because of the increased strength requirement. This is shown in Figure 9–5d.

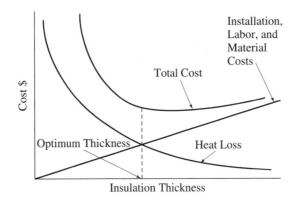

Figure 9–5c   Insulation Thickness

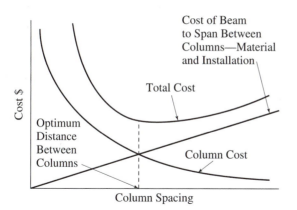

Figure 9–5d   Column Spacing

SUMMARY

Minimum cost analysis is an important tool for managers, engineers, and technologists. There are applications of minimum cost in many parts of the contemporary organization, even though its occurrence may not be as obvious as other economic tools such as breakeven. The financial rewards are high for those who successfully use this tool in the pursuit of higher ROI, efficient and cost-effective purchasing, product design, capacity utilization, setup time reduction, cycle time improvements, quality cost analysis, and for many related applications.

The basis for understanding and applying the concepts of JIT, EOQ, quick changeover, quality cost analysis, balanced manufacturing, capacity and bottleneck applications is found in minimum cost analysis. Like the other financial and economic tools, minimum cost analysis is part conceptual and part mathematical. No financial problem can be completely reduced to dollars and cents. Managers and technical staff must also use their experience, judgment, and non-financial abilities combined with the mathematics.

Topics in this chapter have included:

- Economic order quantity

- EOQ mathematical and graphical analysis

- EOQ application to purchased material

- Inventory planning and management using minimum costs

- Just-in-time purchasing

- Personal applications of EOQ

- Differences between breakeven and minimum cost as cost analysis tools

- Process and design engineering applications of minimum costs.

- EOQ and quantity discounts on purchased material

- Cautions and limitations in the application of EOQ

- Just-in-time applications in the U.S. and Japan

- Setup time reduction and quick changeover
- Push and pull inventory systems and EOQ
- EOQ and the costs of product and service quality
- Bottlenecks, capacity, and EOQ
- Maintenance costs and EOQ
- Examples of minimum cost analysis in design

This chapter has emphasized the relationship between optimum costs and the design of products and processes. The engineer, technician, manager, purchasing buyer, maintenance supervisor, quality manager or engineer, and salesperson has the responsibility for managing, creating, and financially improving products and processes. Understanding minimum costs is vital for the organization that wants to deliver the highest quality of products and services to customers at the lowest cost, and to receive profits from its efforts.

QUESTIONS

1. Define the following with respect to economic purchase quantities:
   a. Annual ordering cost.
   b. Annual storage or carrying cost.
   c. Annual total cost.
2. What are the assumptions used for EOQ calculations?
3. What is the EOQ equation?
4. How could the concept of EOQ be applied to a family that makes twice weekly trips to the supermarket? Identify the storage costs and the purchase costs. Why isn't it usually economical to purchase a year's supply of groceries at one time?
5. What is the minimum cost application for someone who lives in the bush country of Alaska and has food and supplies delivered by aircraft?
6. What are the EOQ applications of a computer malfunction aboard an earth orbital space station?
7. How can it be determined whether a design or manufacturing problem is a minimum cost or breakeven problem?
8. What is the major component of inventory storage costs?
9. What are the cost components of inventory purchasing costs?
10. For EOQ, determine what happens as:
    a. Interest rate increases.
    b. Annual usage decreases.
    c. Part value increases.
    d. Setup decreases.
    e. Purchase order costs increase.

11. What are the differences in the application of EOQ to the purchase of scrap steel vs. gold for electronic components?
12. What is JIT purchasing?
13. Why is it important to consider inventory costs in the manufacturing process?
14. Identify the following:
    a. Inventory turnover
    b. Storage cost
    c. Acquisition cost
    d. Setup cost
    e. Minimum cost
    f. Annual usage
    g. Order frequency
15. How does the EOQ change when the cost of the material is included if no quantity discounts are considered? If the quantity discounts are considered?
16. Sketch a typical graph that relates storage costs to acquisition costs of purchased material.
17. What are some cautions or possible limitations to EOQ applications?
18. What are the effects of the following on EOQ?
    a. Lumpy demand.
    b. Minimum order quantities.
    c. Transportation costs.
    d. Quantity discounts.
19. What is the financial basis for JIT of internally produced items?
20. What are the cost benefits of JIT to a producer? The benefits to customers?

21. Was JIT "invented" in Japanese manufacturing firms? Comment.
22. What are some differences between the application of JIT in the U.S. and Japan?
23. What is quick changeover and how does it relate to EOQ?
24. How is setup reduction able to increase customer responsiveness?
25. Sketch the minimum cost graph for reduced order quantity that occurs when setup times are reduced.
26. What is a bottleneck in the production system? How does it relate to minimum costs?
27. Why is it important not to produce excess inventory prior to or after a bottleneck in the process?
28. What are some techniques that can be used to reduce bottlenecks?
29. What are some examples of personal applications of minimum costs that you use? (Such as 1-hour photo processing.)
30. Sketch and explain the inventory improvement cycle.
31. How do minimum costs apply to the maintenance of a personal car? Do more frequent oil changes lower or increase total maintenance costs?

32. Can early or increased maintenance prevent major repairs? With respect to maintenance/repair costs and minimum costs, comment on the statement, "If it ain't broke, don't fix it."
33. Designers have the responsibility of designing a product that will provide the service required by the customer, but why are costs also a design responsibility?
34. What are some typical product design applications that require the use of minimum cost analysis?
35. Show how minimum cost analysis can affect the designs of the following systems:
    a. Warehouse insulation.
    b. Power distribution system in a manufacturing facility.
    c. Piping design in a petroleum refinery.
    d. Column spacing and beam sizing of an addition to an existing manufacturing facility.
    e. Air cooling and heating system insulation in a home.
    f. Inventory turnover for a retail store.
    g. Maintenance of assembly line equipment.
    h. Quality of an existing product or service.

PROBLEMS

1. If the annual usage is 25,000 units, purchase order costs are $150 per order, part value is $5 per unit, and holding costs are 30% of the part value, determine the best order quantity and order frequency. Graph the results using software or manual methods.
2. A small retail store sells approximately 5,000 pieces of clothing per year. The shop currently places four orders per year to coincide with the four seasons. The cost of placing the orders, which requires taking a physical inventory to determine which sizes and colors are already in stock, is approximately $175, including freight and delivery costs. The store estimates that storage costs, which include the sales floor as well as the back room storage, are 20% of inventory value. The average value is $21 per piece of clothing. Is the current order system the best? State your recommendations.
3. To set up for a particular part costs $180 per run. The part value is $25 each. Fifty thousand units per year are required. The holding costs are 25% of part value.

    a. Solve using any method for the EOQ and the order frequency.
    b. Set up a table and graph solution from part a.
    c. Show what would happen if the setup costs go to $300 per run.
    d. What if the usage increases to 75,000 per year? Decreases to 25,000 per year?
4. A raw material has a cost of $20 per unit and there is an annual requirement of 10,400 units per year. Storage/carrying costs are 25% of part value. The cost of placing and receiving an order is $60. Determine:
    a. The number of orders to place economically per year.
    b. The EOQ.
    c. Annual ordering cost.
    d. Annual holding cost.
    e. Total annual cost.
    f. Maximum and average quantity in storage.
5. For problem 3, what would occur to the EOQ and order frequency if the unit value increased to $50 per unit?

6. For problem 3, what would occur if the unit costs remain at $25 per unit but the setup costs decrease to $45 per order?

7. An electronics firm uses 5,000 ounces of silver each year, valued at $6.35 per oz. Because of security, the cost of storage is $4.00 per oz. per year. Purchase costs are $270 per order. Determine the order frequency.

8. A firm has been ordering on the basis of EOQ for a product that has an annual requirement of 22,000 units per year. The unit cost is $40 per unit. Ordering and receiving costs are $75 per order, and the annual carrying costs are 30% of the part value in storage. The supplier is offering a 5% price discount on orders greater than 800 units per order.
   a. Determine the EOQ without considering the quantity discount.
   b. What is the total annual cost including the value of the product without a discount?
   c. What is the total annual cost including the value of the product if the order quantity goes to 800 and the discount is taken?
   d. Should the customer order the discount quantity?

9. A part has a unit cost of $10 per unit and an annual requirement of 5,500 per year. The supplier is now offering a quantity discount on the part. Purchase order and receiving costs are $25 per order and the storage costs are 20% of the part value. The discount would be 3% off the normal price for more than 2,000 units per order. Calculate:
   a. The EOQ and order frequency without considering the discount.
   b. The total annual costs including part costs using the EOQ.
   c. The order frequency if the discount quantity is used.
   d. The total annual costs including the part value if the discount quantity is ordered.

10. A firm is currently using a 25-ton press that requires an average of 4 hours to change from one set of dies to another. It is estimated to cost $500 to change over, including direct labor costs, overhead, and loss of production time. Typical order quantities from customers are 5,000 units and storage costs are approximately 30% of the part value, which is $75 per unit. The industrial engineering department believes that the setup time can be reduced to 2 hours using the principles of setup reduction devel-

oped by Shingo at the Toyota plants. What will be the effect on EOQ, costs, capacity utilization, and deliveries to customers if the setup time can be reduced? Show graphs of the solution using software or manual methods.

11. Jim Harrison, a process engineer for a welding department, wants to apply economic setup analysis to his process. The process currently produces 23 different weldments for the assembly area. He would like to reduce the in-process inventory of the weldments waiting for assembly. One item, part ET-16, has an annual requirement of 20,000 per year, part value of $6.70 per unit, setup/changeover times of 4.5 hours at $50 per hour, and storage costs of 25% of averaged stored part value.
    a. Determine the best order quantity and order frequency for this item.
    b. If Jim could find ways to reduce the setup time to 2.5 hours, what would the revised order quantity and frequency become?
    c. If the revised setup times could be implemented, what would this do to the average inventory being stored? To the total annual costs?
    d. Sketch the results of this analysis on a graph showing all costs.

12. An organization is attempting to apply concepts of minimum costs in its preventive maintenance program. Currently, equipment requires major repairs every year at a cost of $10,000 for materials and $4,000 for labor. In addition, the lost production time of 4 days is valued at $1,000 per day. The firm is considering using a monthly maintenance program that has labor costs of $150 per month and material costs of $500 per month. The downtime cost on the monthly schedule would be reduced to 1 hour at $50 per hour. The maintenance manager estimates that by using this monthly maintenance program, the annual overhauls can be eliminated and replaced with overhauls every three years that cost the same as the current annual overhaul. Should the organization use the revised maintenance system? What other considerations are there?

13. Julia, a maintenance manager, has responsibility for 10 delivery vans. The average maintenance and emergency repair costs of the vans have been $600 per van per year. Julia has studied the maintenance costs of two vans that were randomly selected this past year that received inspection, adjustments, oil changes,

and repairs as needed on a monthly basis. The costs were an average of $35 per van per month including emergency repairs. Based on this information, should Julia apply the monthly program to the other delivery vans? What other items should be considered?

14. Select a design problem within your field of study or on the job (such as structural, heat transfer, electrical distribution, fluid flow, or another application) that has minimum cost characteristics and apply minimum cost analysis to the problem.

DISCUSSION CASES

Perry Co.

Ann Higgens is in the purchasing department of Perry Co. She is considering implementing an EOQ system for some items. This is a long-term project, and she realizes that the benefits will not be immediate. She has identified six suppliers that have been with Perry for some time, and that would be willing to work with Perry on such a system. The six suppliers represent approximately 20% of the firm's annual purchased materials. Total annual purchases for Perry are approximately $10,500,000 per year.

The order details for the six firms during the past year are shown below.

Last Year's Orders, Frequency, and Dollar Value for Six Suppliers

Supplier	Component	Annual Usage (Units)	Part Value $/Unit	Annual Purchase Cost $/Yr	Number of Orders Last Year
A	R-1369	10,000	$40	$400,000	10
B	TM-45	30,000	$10	$300,000	3
C	WC-60	4,000	$56	$224,000	20
D	Q-12	2,000	$230	$460,000	5
E	3457	3,000	$70	$210,000	12
F	12-T	18,000	$26	$468,000	15
			Total	$2,062,000	

Ann has calculated that placing and receiving an order costs approximately $55 per order. This cost includes the purchasing department's salaries, fringe benefits, overhead costs, supplies and materials, computer costs, receiving, expediting orders, phone and fax costs, receiving inspection, and related costs.

The cost accounting department did a study for Ann to determine the cost of storage for Perry. Currently the costs are approximately 20% of the average value of the items in storage per year. This means that to store a $100 item for one year costs $20. These costs include the time value of money on the stored inventory, materials handling costs, insurance, taxes, space charges for the warehouse, labor and management salaries, and related costs.

Ann hopes to reduce the costs of ordering and the costs of storage for these six items, and to set up a system of ordering that will serve as a model for dealing with other suppliers. Once the system is under way, she would like to work with suppliers so that the order cost per purchase order can be reduced. This would require reducing incoming inspection costs based on the supplier's improved quality, and reducing the purchasing department's expenses by streamlining the process so that fewer employees are required. Two buyers will retire from the purchasing department in the next two years. Also, purchase order costs could be reduced if the number of suppliers could be decreased over the next few years. She has heard about "single source purchasing," but she isn't sure if it would work at Perry.

Ann feels that if she could work with the inventory control people, she could help them to reduce their required minimum inventory quantities if she could convince them that the suppliers could become more responsive. This would, in turn, reduce the inventory storage costs by reducing the total warehouse space for raw storage.

1. Are the potential savings high enough to make the plan workable?

2. What requirements should Ann place on the suppliers for this application?

3. What coordination will be required with the inventory control department?

4. Are there any unknown factors that Ann needs to learn about as the plan is tried?

5. What are the EOQ and frequency of ordering for the six components?

6. Is it feasible to reduce the number of suppliers?

7. What other factors should be considered?

Hometown Products Co.

Tim Samules is a manufacturing manager for Hometown Products Co. Hometown makes and sells traditional furniture made from wood grown in the area. The furniture is sometimes called twig or log furniture because it is not finished, but is left with bark and imperfections on the finished piece. The firm sells from its showroom attached to the small manufacturing area, and it also has 20 dealers in surrounding states that sell a few pieces of the furniture each month. Sales are growing and the firm is profitable.

Tim is the manager of the manufacturing area that employs between 10 and 25 employees depending on the demand and need for products. He purchases the raw logs, schedules the orders in the shop, and does other manufacturing-related activities. Once he decides on what products to make, he gives the orders to one of the two "lead men" who assign the job to the workers. The lead men work with the other employees assembling the products. Each piece of furniture is unique and represents the individual craftsmanship of the person that made it. A new employee is hired based on the recommendation of an older employee. New workers help older employees for 12 to 18 months until they are ready to build furniture on their own. Techniques, shortcuts, designs, and tools are passed on from the older workers to the younger. There are father-son combinations in the plant, as well as other family relationships. Tim's grandfather started the company and Tim has worked there from an early age and during the summers when he was getting his industrial engineering technology degree. He wants to apply contemporary manufacturing techniques to the company because he knows that they need to become more efficient to meet the growing sales.

It usually takes four to five weeks from the time an order is taken in the showroom or from one of the dealers until it is shipped. There is a small inventory of the popular items such as chairs, tables, and loveseats, but approximately 90% of the orders are made to order. Sometimes customers come into the work area and pick out the wood and logs to be used in their furniture. Tim wants to reduce the cycle time of the orders. He knows that he could hire more workers, but because of the seasonal nature of the business he would have to lay off a large number of workers during the slow season. Also, there are not many more skilled potential employees in the area. He knows that some of the applications of EOQ, JIT, and setup time reduction cannot be applied exactly to the problems at Hometown, but he wonders if some of the concepts can be incorporated into the production planning. Storage space of finished product is very limited, and Tim is reluctant to build a finished goods storage area due to the cost. Currently, the raw logs are purchased from three different local suppliers that contract with local landowners and then harvest the logs when the weather is good. These are delivered to Hometown and stacked in the area behind the showroom and plant. In peak times the logs cover a two- to three-acre area piled according to size and wood type in stacks approximately four to five feet high. Workers select their own logs from the field depending on age, size, moisture content, wood type, and related features.

1. Is there anything that Tim can do to improve the order time, the inventory situation, and the production system related to JIT, EOQ, or cycle time reduction? Or is this manufacturing system not modern enough to apply these concepts?

2. What are the magnitudes of the storage costs for the raw material? For the finished goods? What are the magnitudes of the setup and "get ready" costs for the orders?

3. Is there any solution to the 4–5 week delivery times? What do you recommend?

4. Hometown is a job shop rather than a continuous manufacturing plant. Do the techniques discussed in this chapter apply to job shops?

5. What potential problems are going to be created as sales grow? How can Tim anticipate the problems and minimize their impact?

Abacus Products

At a recent sales meeting, many of the salespeople complained that customers were requiring faster and faster deliveries. Some orders were lost to the competition because delivery and response time was too slow. Typically, it takes six weeks from the time the orders are received in the order department until they are shipped. Rush orders can be done in one week. They go ahead of all other orders and cost the customer approximately 50% more than regular delivery.

Harry, the sales manager, has a meeting with Linda, the manufacturing manager. Harry explains that "something must be done" to improve deliveries. Linda knows there is a problem, but doesn't know how to approach the problem. Setup times are between three and six hours for each order. Some changeovers require two operators for half a day's work. The run times typically take between five and ten days per order. Most of the delays in the order process come from waiting for some of the equipment to become available. Typically, an order goes through seven to ten different pieces of equipment and processes before final inspection and shipping.

Linda has tried storing some of the more popular items in the product line so that they could ship from inventory, but because there are so many variations in the product features, a product can rarely be shipped from stock without going back through three or four of the steps in the processes. Also, many items had to be scrapped after no orders were received for three years, and the model was discontinued. She feels that the inventory holding costs are too high.

Some component parts that are used in the product are difficult to obtain from suppliers sooner than two to three weeks. Linda has worked closely with the purchasing department to find faster suppliers, but most are small shops that have low cash flow to finance their inventories, and most produce to order.

Linda feels squeezed between the sales department's requirements for shorter and faster deliveries, and the long production times and slow suppliers. She knows about JIT and EOQ generally but doesn't know if they apply to her situation. Competition is becoming tougher. Customers are more demanding.

1. How should Linda proceed to study the delivery problem?
2. Do the techniques of EOQ and JIT apply to this situation?
3. Does reduced setup time analysis have any application here?
4. Does the inventory improvement cycle apply to this situation?
5. What other things should Linda consider?

Harrison Co.

Fred Fischer is a manufacturing engineer for Harrison Co., a designer, builder, and fabricator of pressure vessels for chemical, fertilizer, petroleum, cryogenic, brewing, food processing, municipal water, and military applications. The company makes heat exchangers, processing tanks, columns, vats, and related equipment from carbon and stainless steel, other ferrous and nonferrous materials including titanium, and many of the so-called exotic alloys for highly corrosive chemicals. Customers range from small food processors and retail gasoline outlets to international Fortune 500 firms and aerospace firms. Sales are international, and it is believed that at least one of Harrison's pressure vessels is in every country of the world.

Large presses are used to form the cylindrical shells and to "dish" the curved heads or ends of the vessels. These are rolled and pressed into the correct configurations, assembled and welded, and nozzles added for piping inlets and outlets for the customer. Many have complex internal baffles, tubes, platforms, and heat exchanger assemblies. Deliveries of raw material range from two to three weeks for common steel items to four to six months for specialized alloys. Manufacturing time ranges from one to two weeks for simple items to eight to ten months or longer for complex designs. Harrison will assemble, weld, and install their products anywhere in the world at the customer's request. Sizes range from a few feet in length and diameter to on-site constructed items that are up to 150 feet in diameter and 200 feet long.

There is a great deal of automation of the processes at Harrison. They have over 25 welding robots and assembly robots. Work is scheduled using the queue or "backlog method." As soon as material is delivered from the supplier, it is put into the line or backlog waiting for press work. When its turn arrives, the components are pressed, and then it is put into the "assemble backlog" and waits in a queue for its turn. Backlog queues are from one week to three to four months. The presses are generally behind schedule. The final activity for many of the products is the heat treating furnace. These are located outside the building and require extensive setup to accept the pressure vessels. Often four to six employees working several days to prepare and convert the furnaces for large jobs. There is a great deal of in-process inventory waiting for processing by the presses. It is difficult to walk through the crowded manufacturing area.

The treasurer has been complaining lately that because of increased sales, the levels of inventories are the largest in the company's history. Cash flow is very tight, even with outside financing, which, he says, reduces the profit margin because of the interest paid.

The sales department has believed for some time that the business was becoming much more competitive. In the past, most competition came from domestic producers. Now, Pacific Rim, post-Soviet, and European competitors are bidding on domestic jobs. Harrison has a reputation for quality, but Harrison's prices tend to be higher and deliveries are longer than most other producers.

Purchasing buyers spend most of their day locating and expediting material from suppliers by fax, phone, and Internet. They pride themselves on locating scarce material and getting it delivered under difficult conditions. Although it requires a lot of work, they usually find the lowest prices as well as difficult-to-locate materials. Harrison is known in the industry as fair, but tough to deal with as far as prices are concerned.

Cost accounting issues cost data to all department heads, manufacturing managers, process engineers, purchasing buyers, and other key staff weekly and monthly. The information is about each project's unit costs, department equipment capacity utilization, material cost labor productivity, and related financial information. These are important criteria used to measure the performance of the project, equipment, engineering, management, and personnel.

There have been more problems with missing the promised delivery dates recently. The scheduling department gives the sales department the estimated delivery times for materials and the manufacturing time when the salesperson quotes the job to the customer. If the contract is received, the customer is updated twice each month by the salesman about the progress of the project including the current delivery schedule status. The deliveries have been slipping badly for the past six to eight months and a high percentage have been late. Current deliveries for new projects are the longest that they have been for many years.

Fred has been asked by the president, Alan Harrison, to look into the scheduling and delivery problems and make a recommendation to him.

1. What should Fred consider?
2. What recommendations should Fred make to Alan Harrison?

Chapter 10 Replacement Analysis

> . . . every man is the architect of his own fortune.
>
> —Sallust's speech to Caesar

KEY TERMS

Book Value The current theoretical value of the asset as determined by calculating the first cost of the asset less the accumulated depreciation since the asset was purchased. Book value is the current value in accounting records and on the Balance Sheet.

Challenger The new equipment that is "challenging" the existing equipment in terms of costs.

Defender The existing or old building and equipment that are currently used.

Economic Life The life based on the calculation of the asset's annual cost. Economic life is the duration into the future when the annual cost of the asset is less than other alternatives. It may be less than or equal to the physical life. Economic life might be equal to one year in some cases.

Equipment and Building Life Life may be based on physical, accounting, owned, useful, or economic life estimates and calculations.

Market Value The current value of existing or new equipment or buildings as determined by the sale of an asset, based on comparable sales, an appraiser's estimate, cost of reproduction of the asset, or as established by the present worth of future income streams of the asset.

Nonowner Viewpoint The analysis of existing alternatives based on the current market value and the assumption that the asset is not owned. This approach neglects sunk costs and assumes the asset is being purchased today.

Salvage Value The future market value of an asset.

Sunk Cost Past costs of purchasing, operating equipment, buildings, or other assets. Sunk costs are ignored in the decision-making process concerning replacement of existing equipment.

What-If or Sensitivity Analysis This form of analysis selects different values for life, market value, salvage value, operating costs, and other factors in determining the effect on the outcome found by the original replacement analysis.

LEARNING CONCEPTS

- Understanding the reasons for replacing fixed assets such as equipment and buildings.

- Applying the concepts of nonowner viewpoint and neglect of sunk costs in the analysis of the costs of existing buildings, equipment, and fixed assets.

- Understanding various methods for determining the current market value of fixed assets.

- Calculating and applying the concepts of economic life to replacement analysis situations.

- Applying sensitivity and what-if calculations to replacement analysis, and monitoring investments, projects, equipment, buildings, and other investments to determine if replacements should be made.

- Understanding the application of replacement analysis as it relates to productivity and maintenance improvements, reduction of scrap and operating costs, and personal financial analysis.

> If it ain't broke, don't fix it (*unless* costs can be reduced and/or quality can be improved).
> —Revised American saying

INTRODUCTION

Our discussions up to now have been based on the assumption that each alternative is new and existing equipment is not in service. In this chapter, we will consider the case of fixed assets, such as equipment or buildings, that are already providing a service or product, and whether they should be replaced. This chapter considers the analysis of existing equipment, buildings, and processes and compares current costs with the costs of assets that may replace the existing assets and investments.

Old or existing equipment is replaced for many reasons, including deterioration, reduced performance, new requirements, costs, equipment failure, obsolescence, or cost-effective leasing alternatives. Replacement analysis uses a TVM diagram. However, it is different from the TVM diagrams used previously. The nonowner viewpoint is introduced. Past or sunk costs of the existing equipment are neglected, current market value of the old asset is estimated, and the nonowner viewpoint is assumed and applied.

Fixed assets have different lives—physical, accounting, owned, useful, and economic life. Economic life is most important for replacement analysis. Economic life is the period when the asset or investment has the lowest cost compared to itself and other alternatives. As equipment ages and approaches the end of its physical life, the maintenance and operating costs increase. Eventually, the costs of keeping and maintaining the asset next year will be more than for this year. When this occurs, the end of the economic life has been reached and the equipment, building, or asset should be replaced.

A number of examples in this chapter demonstrate replacement analysis based on economic life calculations, productivity improvements, scrap cost reduction, maintenance cost improvement by replacement, and personal finance examples. In the examples, the application of what-if or sensitivity calculations helps determine the outcome of the replacement decision. Finally, the concept of continuous monitoring of costs to determine if an asset should be replaced is introduced.

REASONS FOR REPLACEMENT

Fixed asset investments, including buildings, storage facilities, production equipment, tooling, computer hardware, material handling equipment, delivery trucks, and other equipment are continually being replaced. Individuals replace their cars, renovate their homes, buy personal computers, move into new homes, and buy new audio and video equipment. For both corporations and individuals, comparing existing equipment with new assets is ongoing.

Reasons for replacing equipment, buildings, and other assets include:

1. **Deterioration**—The equipment is worn out. Maintenance and operating costs are higher than the cost of replacement. The variation in output is increasing, and the equipment is difficult to adjust or align.

2. **Reduced performance**—Reliability of the equipment has decreased, quality of products is declining, or productivity is decreasing.

3. **New requirements**—Designs and process requirements have changed. The old equipment cannot meet the tolerances and specifications of currently designed products; products have higher quality requirements, including reduced variation of output; and old equipment cannot meet new productivity or capacity requirements. As a result, unit costs are higher.

4. **Costs**—Compared to the new equipment available, existing equipment has high costs because of increased operating costs, inefficiency, repairs, power, labor, material, and related costs.

5. **Technological obsolescence**—Newer equipment is superior. The old equipment is slower and has fewer desired features. For example, new PCs have more and faster memory, CD-ROM, sound, and other capabilities.

6. **Abrupt failure**—This includes a major breakdown; destruction from fire, flood, earthquake, or other natural cause; and failure due to human causes such as explosion, improper use, or major accidents.

7. **Leasing**—The decision to lease may be advantageous because of the organization's available financial resources and cash flow. The money that would have been invested in the purchase of equipment can be spent on other projects. Limited availability of funds may be more important than the cost of the service provided by the equipment. A higher lease

cost may be acceptable if the organization's funds are low or other cash sources are un-available.

8. **Purchasing**—Equipment is not required when the organization decides to obtain the product or service from an outside supplier.

9. **Services no longer required**—Products or services are discontinued.

Replacing Existing Fixed Assets—Equipment and Buildings

All of the examples of alternative comparison and selection up to now have been based on new alternatives. However, this is not typical. Usually, a product or service is being produced and the question is, "Should the old equipment, methods, and processes be replaced by new methods and equipment?" Often the newer equipment is more efficient, has higher productivity and lower costs, or produces a higher quality service or product than the older equipment. Cost comparisons of the old versus the new assets must be made.

> It is normal to have processes, equipment, and buildings in service that must be economically compared to new alternatives. This replacement analysis requires some special assumptions and calculations.

Replacement analysis requires a different understanding and application of TVM. Previously, the analysis of two or more alternatives has considered time value of money, operating expenses, and comparison of present value, annual amount, future value, or rate of return. In these examples, equipment and buildings were not yet owned. In replacement analysis, *there is already some use of the equipment that is delivering the service or product.*

The analysis is further complicated by the fact that there have been historical costs and investments into the existing process, the existing equipment has some current value, the asset is being depreciated in the accounting system, and the existing asset's physical life might be different from its economic or accounting life. Further, the analysis is made more difficult by the calculations not agreeing with what seems to be the intuitive or commonsense approach to analyzing replacements. For some, this replacement concept is very difficult to accept.

Sometimes the replaced asset is kept as a spare or backup for the new asset. Although the old asset is being replaced, it actually becomes a secondary piece of equipment rather than the primary unit. The discussion begins with a summary of replacement analysis and examples. This discussion relies on understanding time value of money, accounting, and depreciation concepts, and some new ideas such as nonowner viewpoint and economic life.

Replacement Analysis Steps

1. Draw the TVM diagram for the replacement situation.

2. Identify the sunk costs of the existing equipment and its book value.

3. Apply the nonowner viewpoint.

4. Determine the relevant market value of the existing equipment.

5. Estimate the future residual salvage value and other costs of the existing equipment.

6. Estimate the life of the existing equipment.

7. Determine the economic life and annual costs of the equipment.

8. Calculate the annual costs comparison of the old asset and the proposed asset.

9. Consider noneconomic factors.

10. Make the decision to obtain the new asset or retain the old asset.

Example: Equipment Replacement Analysis at Sandor Corp. Don Malard, project engineer at Sandor, is considering replacing an existing computer-controlled machine. The existing equipment was purchased for $50,000 three years ago. Annual operating expenses are $1,000 per year. An estimated five more years of equipment life remain. Salvage of $5,000 will be realized at the end of five years. Book value according to the accounting records is presently $30,000, current market value is estimated to be $25,000, and desired return on investment is 15%. A supplier of new equipment has offered Don a trade-in price of $40,000. Because the equipment has been in service for three years, Don is not sure how to conduct an analysis using time value of money techniques. The diagram for this situation is shown in Figure 10–1.

TVM diagram—The TVM diagram in Figure 10–1 is different from previous diagrams. The past service life from when the equipment was purchased to the present is shown as a *dashed line* indicating the past costs. The future service that is available is shown as a *solid line*. The initial cost of the equipment, $50,000, is also shown as a dashed line. The present time is indicated as t_0. The past service life is shown as $n = 3$ and the future service as $n = 5$ for a total of eight years of service. The annual operating costs and residual salvage revenue are shown in the usual manner as solid lines. The market, book, and trade-in values of the equipment at the present time are indicated on the TVM diagram below t_0.

The standard approach that has been used to calculate the total annual cost is to consider the first cost of $50,000 taken to an annual amount, add the annual cost of the maintenance/operating costs, and reduce the amount by the annual value of the future salvage value. This is the standard method of calculating annual cost when considering only new, not replacement,

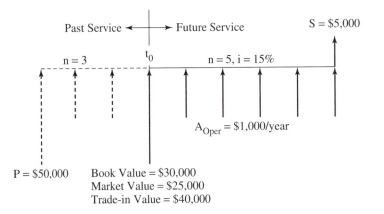

Figure 10–1   Equipment Replacement Analysis at Sandor Corp.

analysis. With replacement analysis, historical costs, book values, market values, and other costs must be considered.

Sunk costs—How does the fact that the equipment was purchased for $50,000 three years ago affect the analysis? The money spent three years ago is considered a historical or sunk cost. **Sunk cost** is money spent in the past that has no effect on the current analysis. When the equipment was purchased, the accounting system recorded that the cash account, an asset, decreased by $50,000 and the equipment account, also an asset, increased by $50,000 to balance the Accounting Equation. This entry transferred dollars from one asset account to another. That entry occurred three years ago. Since then, the accounting system has recognized the decline in the asset's value and applied depreciation expense to the asset. In the organization's accounting system, depreciation reduces the value of the equipment shown on the Balance Sheet, and simultaneously increases costs as depreciation expense on the Income Statement. The net effect is the reduction of the asset over time and the increased costs on the Income Statement, which in turn reduces profits and may reduce income taxes for the period.

> Sunk costs are historical costs that have no relevance to the present or future analysis.

For replacement analysis, sunk costs are neglected, and *only current and future costs* are considered. Past costs are considered in the accounting system, but not for the analysis of present or future costs. This may be a difficult concept to accept, especially when personal expenditures and investments are concerned. Someone may say, "I paid $18,000 for my car three years ago, and it is still useful; I can't afford to sell it for $13,000 today!" Or, "The current price of the security is $13 per share, but I paid $18 per share three years ago; I can't afford to sell it today for $13!" An organization might have the same reaction when selling equipment that cost a large amount in the past and would be sold for a smaller amount today. Often, a statement such as, "We can't afford to sell it at that price because we paid a large amount for it a few years ago!" reflects a viewpoint that does not realistically consider sunk costs.

Book Value

The book value of equipment, buildings, and other fixed assets is the value of the asset as determined by the depreciation entries into the Accounting Equation. The remaining net value is book value. This may or may not represent the market value of the asset. Depreciation methods may be selected for their tax advantages rather than to represent the true market value. For this reason, book value is generally not used as a measure of the asset's value in replacement analysis.

Market Value

How is market value determined? The way to determine the true value is to *sell* the asset. As a realistic substitute for selling, the value may be estimated by engineers, technicians and others using current sales of similar assets. Another method is to rely on an *expert appraiser's* determination of the value. An expert appraiser is someone in the field who is knowledgeable about the market and current values. For plants, structures, warehouses, and office buildings, the esti-mated costs of *duplicating* the asset in its present condition are estimated using standard esti-mating techniques to determine the current value. A final method is determining the *income* and profits that the asset could produce in the future. From this income flow, the current value of the asset can be estimated using the time value of money concept of perpetuity. This is the rental/lease viewpoint that looks at the potential rents or leases of the asset in the future. Then, using TVM techniques, the future income is converted to a *present value* to determine the cur-rent value of the structure or equipment. Typically, a combination of methods is used to deter-mine the market value of assets.

Methods of Determining Market Value

1. Sell the equipment or structure.
2. Compare sales of similar buildings or equipment.
3. Obtain an appraiser's estimate.
4. Obtain an estimate of reproduction costs.
5. Determine the present worth of future income.

To demonstrate the concepts of market value determination, consider a warehouse owned by Sandor Corp. Assume it is an existing warehouse that has 25,000 square feet. The organiza-tion wants to determine its value.

Sell the Equipment. Selling the building is the best way to determine its value. Of course, this is not applicable if Sandor desires to continue using the warehouse.

Comparison Method. Comparing the building to related sales requires information re-garding other sales. Suppose that a similar warehouse of 30,000 square feet, located in the same geographical area and of the same age, recently sold for $1.2 million, and that a struc-ture of of 15,000 square feet, in similar condition and location, sold two years ago for $500,000.

The value of the first current sale in dollars per square foot is:

$$\text{Value} = \text{Dollars per Square Foot} = \frac{\text{Selling Price}}{\text{Building Area}} = \frac{\$1,200,000}{30,000 \text{ ft}^2} = \$40/\text{ft}^2$$

If the industrial buildings have been increasing in value at a rate of 5% per year, the two-year-old sale can be brought forward to the present time using time value of money:

$$F_{\text{Current Value}} = P(1 + i)^n = \$500,000(1.05)^2 = \$551,250$$

The current value of a square foot of warehouse space is:

$$\text{Value/Square Footage} = \$551,250/15,000 \text{ ft}^2 = \$36.75/\text{ft}^2$$

These two values of warehouse cost/square foot are $40.00 and $36.75 based on today's dol-lars. The average of these two values is $38.38 per square foot. Applying this average cost to Sandor's building:

$$(25,000 \text{ sq. ft.})(\$38.38) = \$959,500, \text{ an estimate of the existing 25,000 sq. ft. warehouse}$$

This is an estimate of the current market value of the warehouse. Is this an *exact* answer? No, it is a starting point for determining the warehouse value. This estimated value can be increased or reduced based on the construction, age, location, and other features of the building. Care must be taken when using this method to consider other assets that are similar to the one in question. For example, it would not be appropriate to compare Sandor's warehouse with warehouses in other locations, or new warehouses with many different features. When similar assets are compared, this is a useful technique.

Appraiser's Estimate. An expert's or appraiser's estimate of the building value is made by someone who has the experience and information to make a reliable estimate of value. An external appraiser can often give an unbiased opinion.

Often, technical personnel closely associated with the equipment or building are also able to estimate its value very accurately. They know about other equipment retired and sold, and they are aware of similar buildings. Appraisers outside the organization and internal technical staff typical use a combination of reproduction cost, income, and comparable methods to estimate market values.

Duplication or Reproduction Cost Technique. This method determines the cost of duplicating the existing building in its present condition by using a combination of cost-estimating techniques. Material, labor, overhead, and other costs are estimated to reproduce the existing equipment. The asset's cost must be estimated using current materials and labor rates. It is difficult to adjust for the age of the existing asset when estimating the costs of duplication at today's prices. For relatively new equipment and buildings, the reproduction method is an effective technique.

Income Method. The final method of determining the market value of Sandor's warehouse is based on the income that could be obtained from the asset if the space is rented or leased to a tenant. In this case, current space could be rented for $55,000 per year. This can be determined by either actual lease agreements on the property or by looking at similar lease costs by other owners of similar facilities. Further, assume that a return rate of 5% is used, based on the current rate of return of Treasury bonds or similar low-risk investments.

Assuming infinite income (perpetually) in the future, the current value is:

$$P = A/i = \$55,000/.05 = \$1,100,000 \text{ for the existing warehouse}$$

This estimate is more than the estimate of value using the comparable sales method for the building. By combining the different methods and applying judgment and experience, a reasonable estimate of building values can be determined. It should be remembered, however, that all of the methods, except an actual sale, are *estimates*.

Salvage or Future Market Value

The future residual salvage value is derived from the market value. If there are tear-down, removal, disposal, or related costs, the future selling price is reduced by these costs. Net market value describes the market value less the removal/tear-down costs. The future value of a fixed asset might be greater than, but is usually less than, the current market value. It may even be zero or less than zero, if the removal/tear-down costs are greater than the salvage/scrap value of the material. In TVM calculations, the future amount, when brought to an annual amount, is multiplied by the *A/F* factor. The *A/F* factor is typically small and becomes smaller as the life, n, is longer. Therefore, in some applications, the future salvage value times the A/F factor results in a small amount that can be neglected.

To do what-if and economic life calculations, it is necessary to estimate the salvage values between now and the final salvage value at the end of the physical life. These values can often be estimated using a linear estimate, similar to calculating straight-line depreciation. This method is usually adequate when no other information is known about the future value of the equipment or structure. If certain events affecting the future value are known, these can be considered when making salvage estimates; otherwise a simple linear extrapolation may be acceptable.

Future technological improvements may be estimated with a high degree of certainty for some replacement situations. When this is possible, the future value of new and existing equipment may be affected. The current value of the existing equipment may drop suddenly and the cost of the new, improved equipment may change abruptly. Consider, for example, the price history of electronic equipment such as VCRs, PCs, hand-held calculators, CD-ROMs, and similar equipment.

THE NONOWNER VIEWPOINT

If the past sunk costs are neglected, what costs should be considered? The **nonowner viewpoint** can be used. It is assumed that the present equipment is not owned. If so, the alternatives would be buying the new equipment or buying used equipment. (This old, used equipment is actually the existing equipment that is owned, but we assume that it is not owned for analysis purposes.) What would its purchase price be if it were purchased as used equipment? The price would be the prevailing market value, not the calculated book value or trade-in value. Assume that it could be purchased for its present market value. Then the same TVM techniques would apply as if the investment were being made today at the market value of the existing equipment.

> The nonowner viewpoint neglects past or sunk costs. It assumes that the equipment is not owned and is purchased as used equipment.

To calculate the annual cost using the nonowner viewpoint for the existing equipment, the current market value is considered rather than the original first cost. In Figure 10–1, only the market value is used. First cost, trade-in, and book value are neglected. This diagram uses the present market value as the current cost of "purchasing" the equipment in the used equipment market. The sunk costs, trade-in value, and book value are neglected because they are not relevant to the nonowner viewpoint. The value of the "used" equipment is the current market value, $25,000. The remaining life of five years, the ROI of 15%, operating costs of $1,000 each year, and the residual salvage of $5,000 are the same as originally estimated. The current salvage value is assumed to be a net value after disposal and removal costs, if any.

To demonstrate how an *incorrect solution* can be obtained when the past sunk costs are used, the calculation of the total annual cost using the standard method and the initial cost of $50,000 is shown:

$$A_T = A_P + A_O - A_S$$
$$= \$50,000(A/P, 8, 15) + \$1,000 - \$5,000(A/F, 8, 15)$$
$$= \$50,000(.2229) + 1,000 - 5,000(.0729)$$
$$= \$11,145 + \$1,000 - \$364$$
$$A_T = \$11,781 \text{ per year (based on incorrectly using the sunk costs)}$$

This is the calculation that would have been made when the equipment was originally purchased. Now, since the investment is three years old, the correct analysis is to *neglect* the sunk costs and to base the analysis on the current market value as an investment amount—the nonowner viewpoint.

The revised annual cost calculation using the nonowner viewpoint, which neglects past sunk costs and uses current market value as the purchase price of the used equipment, is shown below:

$$A_T = A_{MV} + A_O - A_L$$
$$= \$25,000(A/P, 5, 15) + \$1,000 - \$5,000(A/F, 5, 15)$$
$$= 25,000(.2983) + 1,000 - 5,000(.1483)$$
$$= \$7,458 + \$1,000 - \$742$$
$$A_T = \$7,716 \text{ per year}$$

This calculation assumes that the equipment is kept for the remaining five years. The original calculation made by Don Malard at Sandor gave an annual cost of $11,781 per year. Now, using the nonowner viewpoint, we have a cost of $7,716 per year for each of the next five years.

The nonowner viewpoint may be difficult to accept. When large amounts of money have been spent in the past it is difficult to neglect those expenditures and focus on the current and future market value of the existing assets. Remember, it is the *current* investment that is being analyzed, not the *historical* investment.

Trade-in Value

The given trade-in value by the supplier in this example is $40,000. This may or may not be an accurate estimate of the true value. Often, as a sales incentive, the price of the product is inflated and the trade-in value is also increased, so the buyer believes that he or she is getting a "better" price. This is confusing and unrealistic. The trade-in value, like the book value, is usually not a good indication of the equipment's value. Market value and/or current salvage value are better estimates of the asset's current value.

EQUIPMENT LIFE

The examples and discussion so far in the text have assumed that the life of the fixed assets, both buildings and equipment, is estimated at the beginning of the asset's use and the lives are fixed. This is the asset's *physical life*. In expanding the application of economic analysis to include replacement analysis, additional asset lives must be considered. Types of asset lives include:

1. The **physical life** is the period until the unit is salvaged, scrapped, or torn down. It is usually the longest of the various lives, and can cover ownership by a number of different owners.

2. The **accounting life** or tax life is the time over which the asset is depreciated. It may or may not reflect the physical life.

3. **Owned life** is the period during which the organization owns the equipment. The equipment could be purchased as used equipment or sold before its physical life is over.

4. **Useful life** is the life over which the equipment or structure will provide useful service. Usually, the useful life is equal to the owned life.

5. **Economic life** of the existing equipment is the time remaining until the asset's cost becomes greater than the alternative's cost. Economic life of the new equipment is the minimum annual cost for the equipment, and is the time at which the life of the existing equipment should be terminated and replaced with the new equipment, which has a lower annual cost. Sometimes the economic life can be estimated with a high degree of certainty; other times, the point at which the old equipment should be replaced is not clear. An attempt, in either case, can be made to determine the economic life.

Economic Life and Annual Costs

The nonowner viewpoint calculation for Sandor Corp. assumed that the existing equipment is kept for the remaining five years of its physical life. Additional calculations can be made to evaluate the annual cost of the equipment using what-if calculations when the asset is held for less or more than five years.

Using What-If Calculations. The salvage values for the end of the years can be estimated as shown below:

End of Year	Salvage Value
1	$12,000
2	$10,000
3	$8,000
4	$6,000
5	$5,000

For example, if the equipment is held for only 4 more years, the calculation would be:

$$
\begin{aligned}
A_{\text{Total}} &= A_{\text{MV}} + A_{\text{Oper}} - A_{\text{Salv}} \\
&= \$25,000(A/P, n = 4, i = 15\%) + \$1,000 \\
&\quad - \$6,000(A/F, n = 4, i = 15\%) \\
&= \$8,757 + \$1,000 - \$1,202 = \$8,555 \text{ per year}
\end{aligned}
$$

Continuing the what-if calculations for $n = 1$, 2, and 3 years, for 1 year of remaining life:

$$
\begin{aligned}
A_{\text{Total}} &= \$25,000(A/P, n = 1, i = 15\%) + \$1,000 - \$12,000(A/F, n = 1, i = 15\%) \\
A_{\text{Total}} &= \$28,750 + \$1,000 - \$12,000 \\
&= \$17,750 \text{ per year}
\end{aligned}
$$

(Note that A_P is greater than P if $n = 1$ due to the end of the year assumptions.)

For 2 years:

$$
A_{\text{Total}} = \$15,378 + \$1,000 - \$4,651 = \$11,727 \text{ per year}
$$

For 3 years:

$$
A_{\text{Total}} = \$10,949 + \$1,000 - \$2,304 = \$9,645 \text{ per year}
$$

Table 10–1   What-If Calculations

Life in Years	A_{MV}	A_{Oper}	A_{Sal}	A_{Total}
1	$28,750	$1,000	($12,000)	$17,750
2	$15,378	$1,000	($4,651)	$11,727
3	$10,949	$1,000	($2,304)	$9,645
4	$8,757	$1,000	($1,202)	$8,555
5	$7,458	$1,000	($742)	$7,716

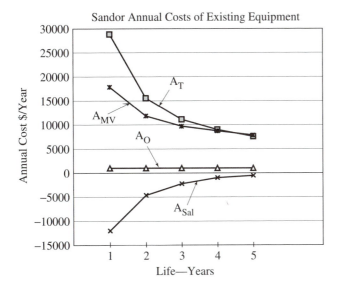

The calculations are summarized in Table 10–1 and the accompanying graph.

Physical Life and Economic Life

The Sandor calculations are based on the what-if analysis of the equipment lasting one, two, three, four, and five years and using the nonowner viewpoint. The physical life has been estimated at five years. From Table 10–1, the lowest cost alternative on an annual cost basis is to keep the equipment for five years. This demonstrates that the economic life and the physical life are the same. The annual cost of the equipment is lowest when kept for five years in this case.

The graph and calculations show that the annual cost of A_{MV} is high when the life is short, and becomes less per year as the life increases. This is due to the decreasing A/P value as n increases. Likewise, the annual value of the salvage decreases as the life increases.

The calculation of the economic life determines if the equipment can be used for the total physical life or if it should be replaced before its physical life is completed.

The following examples demonstrate two other cases of economic life.

Economic Life Less Than Physical Life

Another case of economic life is when the operating and maintenance costs are increasing rapidly as the equipment ages. This causes A_{Total} to increase into the future. The conclusion from this type of analysis is that the equipment should be replaced with less costly equipment as soon as possible, usually within the next year.

Increasing Operating Costs and Economic Life

Using the Sandor example, assume the annual operating cost of the equipment increases over the equipment's life. Rather than a constant $1,000 per year, assume that the first year's operating costs are $1,000 and that they increase by $2,000 each year for the life of the equipment. This is a typical situation; older equipment and buildings usually have increasing maintenance costs as they age.

The salvage values for the equipment at the end of each of the five years are also revised as shown below:

End of Year	Salvage Value
1	$22,000
2	$18,000
3	$12,000
4	$9,000
5	$5,000

The salvage values are higher here than in the original case. Revised annual cost calculations are, for 1 year of life,

$$A_{Operating} = \$1,000 \text{ per year}$$
$$A_{Salvage} = \$22,000(A/F, n = 1, i = 15\%) = \$22,000 \text{ per year}$$
$$A_{Total} = A_{MV} + A_{Oper} - A_{Sal}$$
$$= \$28,750 + \$1,000 - \$22,000 = \$7,750 \text{ per year}$$

For 2 years of life,

$$A_{Oper} = \$1,000 + \$2,000 \, (A/G, n = 2, i = 15\%)$$
$$= \$1,000 + \$2,000(.4651) = \$1,000 + \$930 = \$1,930 \text{ per year}$$
$$A_{Sal} = \$18,000(A/F, n = 2, i = 15\%) = \$8,372 \text{ per year}$$

The total annual costs for the 2-year life alternative are:

$$A_{Total} = A_{MV} + A_{Oper} - A_{Sal} = \$15,378 + \$1,930 - \$8,372$$
$$= \$8,936 \text{ per year}$$

For 3 years,

$$A_{Oper} = \$1,000 + \$2,000(A/G, n = 3, i = 15\%$$
$$= \$1,000 + \$2,000(.9071) = \$1,000 + \$1,814 = \$2,814 \text{ per year}$$
$$A_{Sal} = \$12,000(A/F, n = 3, i = 15\%) = \$3,455 \text{ per year}$$

The total annual costs for the 3-year life alternative are:

$$A_{Total} = A_{MV} + A_{Oper} - A_{Sal} = \$10{,}949 + \$2{,}814 - \$3{,}455$$
$$= \$10{,}308 \text{ per year}$$

For 4 years,

$$A_{Oper} = \$1{,}000 + \$2{,}000(A/G, n = 4, i = 15\%)$$
$$= \$1{,}000 + \$2{,}000(1.3263) = \$1{,}000 + \$2{,}653 = \$3{,}653 \text{ per year}$$
$$A_{Sal} = \$9{,}000(A/F, n = 4, i = 15\%) = \$1{,}802 \text{ per year}$$

The total annual costs for the 4-year life alternative are:

$$A_{Total} = A_{MV} + A_{Oper} - A_{Sal} = \$8{,}757 + \$3{,}653 - \$1{,}802$$
$$= \$10{,}608 \text{ per year}$$

For 5 years,

$$A_{Oper} = \$1{,}000 + \$2{,}000(A/G, n = 5, i = 15\%)$$
$$= \$1{,}000 + \$2{,}000(1.7228) = \$1{,}000 + \$3{,}446 = \$4{,}446 \text{ per year}$$
$$A_{Sal} = \$5{,}000(A/F, n = 5, i = 15\%) = \$742 \text{ per year}$$

The total annual costs for the 5-year life alternative are:

$$A_{Total} = A_{MV} + A_{Oper} - A_{Sal} = \$7{,}458 + \$4{,}446 - \$742$$
$$= \$11{,}162 \text{ per year}$$

The above example is shown in Table 10–2 and the accompanying graph.

From the graph and table, we see that the annual total cost increases as the life of the equipment increases. The lowest A_{Total} is for the first year. This is primarily due to the increasing annual operating costs. It is not unusual for the operating costs, including maintenance and repair costs, to increase as the equipment or buildings become older.

The least costly holding period is one year, not five years (economic life is less than physical life). The longer the equipment is kept, the higher the annual costs become. Thus, Don Malard should attempt to find less expensive equipment to replace the existing equipment sometime this year. The unit cost of the products also increases over time. From a cost improvement viewpoint, he should attempt to find other equipment that is less expensive, or make improvements to this process that will reduce the unit costs. Don should place this process on his list of improvement projects.

Identifying and improving the costs of a process or equipment are part of the normal continuous cost improvement activities of an organization. Economic analysis such as this is not a one-time event, but requires continuous attention.

It must also be remembered that the above analysis is based on estimates of salvage, life, ROI, market and operating costs. What-if calculations or sensitivity analysis should be made to determine the effect of changes in the estimated values on the final decision. There are also nonfinancial effects of these decisions that are difficult to quantify. Experience and judgment must be applied to the financial calculations to determine the optimum replacement decision.

The next example demonstrates the situation when the costs are decreasing for part of the asset's physical life and then begin to increase. It is a minimum cost application.

Table 10–2   Increasing Operating Costs

Life in Years	A_{MV}	A_{Oper}	A_{Sal}	A_{Total}
1	$28,750	$1,000	($22,000)	$7,750
2	$15,378	$1,930	($8,372)	$8,936
3	$10,949	$2,814	($3,455)	$10,308
4	$8,757	$3,653	($1,802)	$10,608
5	$7,458	$4,446	($742)	$11,162

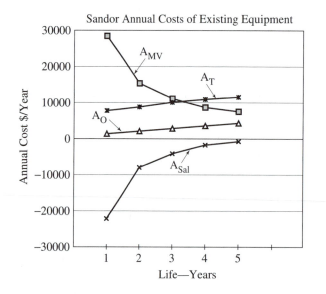

Economic Life as a Minimum Cost: Waste Disposal System at Sandor Corp.

Don Malard has been asked to do some economic analysis of a proposed waste disposal system that has been designed to handle contaminants created by one of the Sandor processes.

The initial cost of the proposed equipment is $100,000, with a life of 12 years, and a targeted ROI of 8%. Because of the nature of the equipment, it will have maintenance costs of $5,000 the first year, and they will increase by $4,000 each year over the equipment's life. Because of the special purpose of the equipment, it has negligible salvage or market value. Don thinks that the estimate of 12 years is reasonable for the physical life of the equipment, but he wonders if the equipment will be economical over that period. Since there will be high maintenance and repair costs to keep the equipment functioning properly, Don wonders if, at some point in the future, the equipment can become too expensive even though it is still functioning.

Figure 10–2 shows the TVM diagram for this problem. The basic calculations for this analysis are similar as for all previous situations, except a what-if analysis is made for years 1

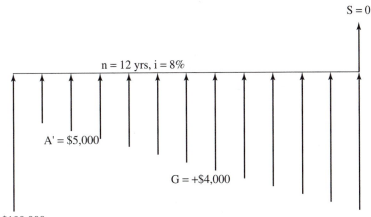

Figure 10–2   Waste Disposal System at Sandor Corp.

to 12. Normal TVM calculations are made and then the question "What are the annual costs if the equipment is kept less than 12 years?" can be asked.

$$A_{Total} = A_P + A_{Oper} - A_{Sal}$$
$$= \$100,000(A/P, n = 12, i = 8\%) + [\$5,000 + \$4,000(A/G, n = 12, i = 8\%)] - 0$$
$$A_{Total} = \$100,000(.1327) + [\$5,000 + \$4,000(4.5958)] - 0$$
$$= \$13,270 + (\$5,000 + \$18,383)$$
$$= \$13,270 + \$23,383 = \$36,653 \text{ per year}$$

Now Don can ask the question, "If the equipment is used for less than 12 years, what are the annual costs?"

For the one-time calculation shown above, hand-held calculators are applicable, but for what-if questions, when many similar calculations are made, spreadsheets are very useful.

> The physical life of equipment is not always the same as its economic life.

Don has estimated the physical life of the waste system as 12 years. From the calculations shown in Table 10–3 and the accompanying graph, the system has the lowest cost per year if the equipment is used for eight years. This is the **economic life** of the equipment, and in this case it is an example of minimum cost. Ideally, if the estimated costs actually occur, Don should replace the equipment in the eighth year with an identical system, or another system that costs less on an annual basis than the current one.

Note that Table 10–3 and the graph give an example of minimum costs. One cost, $A_{Operating}$, is increasing, and one cost, A_P, is decreasing. The total cost, A_{Total}, decreases to a minimum cost at an n of 8 years, and then increases.

Table 10–3   Economic Life of Waste Disposal System

Life Remaining	A_P	A_{Oper}	A_{Total}
1	108,000	5,000	113,000
2	56,077	6,923	63,000
3	38,803	8,795	47,598
4	30,192	10,616	40,808
5	25,046	12,386	37,432
6	21,632	14,106	35,738
7	19,207	15,775	34,982
8	17,401	17,394	34,795***
9	16,008	18,964	34,972
10	14,903	20,485	35,388
11	14,008	21,958	35,966
12	13,270	23,383	36,653

*** minimum cost at 8 years

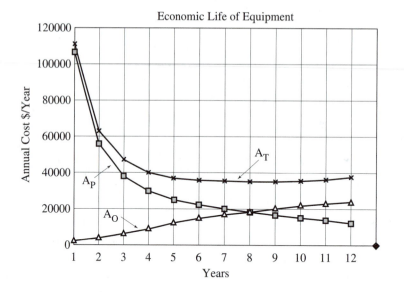

Note that for this example, the annual cost differences between years 7, 8, and 9 are small. Although year 8 is the economic life of the equipment, the additional cost of replacing the equipment in year 7 or 9 is small. As with some other minimum cost cases, there is often a range of values that are close to the optimum. The calculations and graph also show the cost estimates as smooth, continuous functions. Rarely is this the case. Costs often change abruptly.

Three Types of Economic Life Costs

Type 1

The total annual costs of the equipment or structure decrease over its physical life so that the optimum holding period is the physical life of the asset.

Type 2

The total annual cost of the equipment or structure increase from the present into the future so that the optimum solution is to replace the equipment within the next year. This is sometimes called the "one-year case."

Type 3

The total annual costs of the equipment or structure decrease over time and then increase, giving a minimum cost at which the optimum solution is to replace the equipment.

Continuous Monitoring

The costs, revenues, profits, and life for projects and investments change over time. The analysis of the project should occur regularly to monitor the changes, and not assume that the initial estimates are valid throughout the investment's life.

CONTINUOUSLY MONITORING REPLACEMENT DECISIONS

The Sandor examples demonstrate the application of the nonowner viewpoint. Part of the equipment's life is already consumed and part of the life remains.

The Sandor examples are based on the current estimates of future costs. Perhaps a year from now, the existing equipment's costs will be higher or lower. Its life will likely be shorter. The analysis of the equipment, buildings, and other fixed assets is a continuous process, not a one-time event. There should be initial analysis when the new structure or equipment is acquired, and then periodic analysis, at least every year or when costs change significantly. At some point in the future, the existing equipment costs will be greater than that of other available equipment. At that point, the old equipment can be replaced.

There are variations on the above three cases but these three types are typical of actual replacement problems encountered in practice. It is important to remember when performing economic life analysis that the calculations are based on *estimates* of physical life market values, residual salvage amounts, operating costs, desired ROI, and other associated costs and revenues. As a result, changes in the estimates occur. *The economic management of this changing cost environment requires periodic analysis of projects and fixed assets.* Technical managers, technicians, design and process engineers must monitor costs periodically to ensure that the processes produce the lowest cost products and services.

What-if analysis and continuous monitoring of investments and projects minimizes financial "surprises" and increases financial success.

Example: Purchasing New Equipment at Libing and Jones Co. Kathy O'Leary, process engineer at Libing and Jones, is considering the purchase of new equipment. The existing equipment cost $100,000 four years ago, and Kathy estimates that it has six years of life remaining with no salvage value at that time. The desired ROI is 15%, annual operating costs are $5,000/year, current book value is $40,000, and market value is approximately $60,000.

Kathy has been meeting with a supplier that has developed and manufactured similar equipment that can be purchased for $125,000. Its life is estimated to be 12 years, annual operating costs are $3,000, and salvage is $8,000 at the end of 12 years. All other costs, including labor and material costs, would be the same for the new equipment as for the existing equipment. The TVM diagrams for the two alternatives are shown in Figure 10–3.

For the existing alternative using the nonowner viewpoint,

$$\begin{aligned}
A_T &= A_{MV} + A_O - A_S \\
&= \$60,000(.2642) + \$5,000 - 0 \\
&= \$15,852 + \$5,000 - 0 \\
&= \$20,852 \text{ per year}
\end{aligned}$$

This assumes that the physical life of the existing equipment is six more years, and the operating and salvage values are as given. It is helpful at this point to determine if the economic life is less than or equal to the physical life. Table 10–4 shows the calculations for lives other than the six years calculated above.

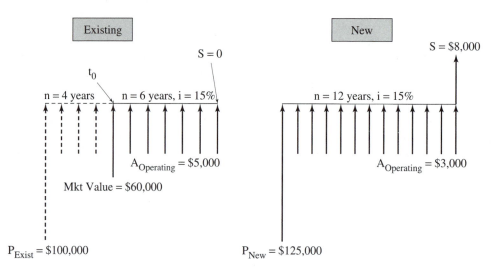

Figure 10–3   Libing and Jones Purchase of New Equipment

Table 10–4 Calculating Annual Costs of Existing Equipment with Revised Operating and Salvage Costs

Life Years	A_{MV} $/Year	A_{Oper} $/Year	A_{Sal} $/Year	A_{Total} $/Year
1	$69,000	$5,000	−$50,000	$24,000
2	$36,907	$5,000	−$18,605	$23,302
3	$26,279	$5,000	−$8,639	$22,640
4	$21,016	$5,000	−$4,005	$22,011
5	$17,899	$5,000	−$1,483	$21,416
6	$15,852	$5,000	0	$20,852

The annual cost for the new equipment is calculated in the usual manner:

$$A_T = A_P + A_O - A_S$$
$$= \$125,000(.1845) + \$3,000 - \$8,000(.0345)$$
$$= \$25,786 \text{ per year}$$

Based on only the above calculations, the cost of the existing equipment is $20,852/year and the cost of the new equipment is $25,786/year. On this basis, the decision to keep the old equipment is the best solution, *assuming that the existing equipment lasts six years as estimated.*

At some point in the future, it is likely that the existing equipment will wear out and/or incur greater costs. The existing equipment should be retained so long as it is the most economical and provides the required service to Libing and Jones.

Example: Economic Life and What-If Analysis for Libing and Jones. Kathy should ask, "Are there any costs that could change in the future during the life of either alternative that could change the conclusion to keep the old equipment?"

For the existing equipment, the physical life, the salvage value, or the annual operating costs could change within the next six years. The first step is to calculate the economic life of the existing equipment. The salvage values of the existing equipment are:

End of Year	Residual Salvage Value
1	$50,000
2	$40,000
3	$30,000
4	$20,000
5	$10,000
6	0

Manual methods or a spreadsheet can be used to set up a table that will calculate the annual costs of the existing equipment based on the above assumptions. The total annual cost can be determined by $A_{Total} = A_{MV} + A_{Oper} - A_{Sal}$.

The lowest A_T of the existing equipment, the economic life cost, is for a life of six years, which is the same as the estimate of the physical life. Since these costs are less than the new equipment's cost, the existing equipment can be retained.

Other what-if calculations can be made, and costs can be updated as they change.

> Replacement analysis is not a one-time event; it is a continuous process.

What About Trade-In?

Kathy knows that if the new equipment is purchased, the old equipment will be sold. Perhaps the seller of the new equipment would accept the old equipment as partial payment for the new, or a trade-in. If that is not possible, the existing equipment would be sold on the used equipment market. Can the trade-in value of the old equipment be used to reduce the cost of the new equipment? If so, this makes the new equipment less costly on an annual basis and makes the new alternative more attractive than keeping the old. There are two viewpoints on trade-in.

One viewpoint says that each investment must stand on its own and trade-in should not be considered. The belief is that when the old equipment is sold, that money can be used for other investments and should not be applied to the reduction in value of the new equipment. It is assumed that the revenue from the sale of the old equipment is placed in a 15% ROI investment. The new equipment "must stand on its own." This viewpoint is further strengthened by noting that when the old equipment was originally purchased, it had to "stand on its own" when compared to other alternatives of equipment purchasing, leasing, or other production methods.

The second viewpoint considers the equipment, material, labor, and overhead costs of the existing equipment. These costs are added to determine the total cost of producing the product or service and then to calculate a unit cost of the product. If labor or material costs can be reduced, that would yield lower unit costs. Likewise, the argument goes, the organization could spend additional dollars to improve the existing equipment by adding tooling, maintenance, or other equipment to lower the annual total costs and the unit costs of the product or service. Following this logic, it could be said that spending additional money to purchase a faster, higher quality, or lower cost new piece of equipment should be based on the net cost of that financial event. The *net cost* of the new equipment purchase is the market cost minus the value of the removed and sold old equipment if the trade-in is considered.

Both viewpoints are found in contemporary organizations.

Keeping Old Equipment as Backup

Often the old equipment that is being replaced by lower cost, more productive, and more efficient new equipment is kept as backup or spare equipment. The old, less efficient, and more costly equipment can be used for a short period if the new equipment is out of service or if output increases for a short period of time. Of course, if the output requirement increases permanently or the new equipment fails totally, then a completely new replacement analysis must be made.

Example: **Reducing Scrap Costs at I.D. Gooding Co.** Gooding is a medium-sized supplier to the automotive industry. Recently, Gooding's customers have been emphasizing product quality and continuous improvement of product designs and processes of their suppliers. Other customers are requesting that Gooding continuously improve design tolerances, reduce process scrap, and make quality improvements. One of Gooding's production processes has been studied by a process improvement team. The team has recommended a number of changes in the process that should result in lower rates of rejection and improved output per hour. The organization wants to determine whether the cost of these improvements is justified by the productivity and quality improvements.

The current scrap rate is an average of 1,000 units per week of the total output of 8,000 units/week for the 52-week year. The existing process was put in place five years ago at a cost of $300,000. It can be used for four more years, and its residual salvage value would be $5,000 if it is sold any time between now and then. The accounting department currently values the process at $46,000 and the estimated current market value of the process is approximately $25,000. The cost of the rejected parts, 1,000 units per week, is $1.10 per unit. Excluding the scrap costs, the other operating costs are $65,000 per year. A revised process recommended by the improvement team can be built for $86,000. This would reduce the operating costs to $55,000 per year, with a life of six years. The revised process would reduce the rejects to 500 units per week. Gooding Co. has a target of 15% as its return on investment.

The current cost of the scrap is:

$$(1,000 \text{ units/wk})(\$1.10/\text{unit})(52 \text{ wk/yr}) = \$57,200/\text{year}$$

The expected cost of the proposed process scrap is:

$$(500 \text{ units/wk})(\$1.10/\text{unit})(52 \text{ wk/yr}) = \$28,600/\text{year}$$

The TVM diagrams for the proposed improvements in Figure 10–4 show the market value of the current process, $25,000, using the nonowner viewpoint. The book value of $40,000 is neglected. For the remaining four years, residual salvage of $5,000 is estimated to be constant

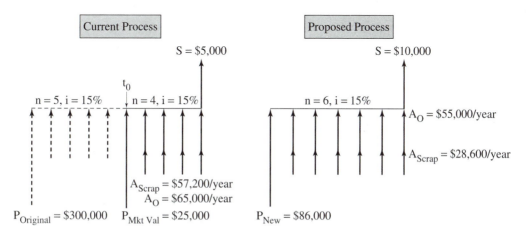

Figure 10–4   Gooding Co. Process Improvements

over the time period, and there are annual operating costs of $65,000 and annual scrap costs of $57,200. The initial cost of the new process would be $86,000 less the value of the old process equipment sold for its market value of $25,000. The proposed process would have a life of six years, annual operating costs of $55,000, scrap costs of $28,600 per year, and residual salvage value of $10,000. The TVM calculations are shown below. For the existing process, for four years of remaining life,

$$A_{Total} = A_{MV} + A_{Oper} + A_{Scrap} - A_S$$
$$= \$25,000(A/P, n = 4, i = 15\%) + \$65,000 + \$57,200 - \$5,000(A/F, n = 4, i = 15\%)$$
$$= \$25,000(.3503) + \$65,000 + \$57,200 - \$5,000(.2003)$$
$$= \$8,757 + \$65,000 + \$57,200 - \$1,001$$
$$A_{Total} = \$129,956/year$$

The calculations of the annual costs of the equipment based on lives of one, two, and three years are similar to the above calculation and can be summarized in Table 10–5.

The lowest cost and economic life of the existing equipment would be if the equipment were kept for four years.

For the proposed process:

$$A_{Total} = A_P + A_{Oper} + A_{Scrap} - A_S$$
$$= (\$86,000)(A/P, n = 6, i = 15\%) + \$55,000 + \$28,600 - \$10,000(A/F, n = 6, i = 15\%)$$
$$= \$86,000(.2642) + \$55,000 + \$28,600 - \$10,000(.1142)$$
$$= \$22,721 + \$55,000 + \$28,600 - \$1,142$$
$$A_{Total} = \$105,179/year$$

The proposed process is lower in cost than the existing process.

What-if analysis—For example, if the life is only four years, not six years, and the scrap rate improves to only 750 units per week, what is the annual cost of the new equipment? For the proposed process with revised scrap and life estimates,

$$A_{Total} = A_P + A_{Oper} + A_{Scrap} - A_S$$
$$= (\$86,000)(A/P, n = 4, i = 15\%) + \$55,000$$
$$\quad + (750 \text{ units per wk})(\$1.10/wk)(52 \text{ wk/yr}) - \$10,000(A/F, n = 4, i = 15\%)$$
$$= \$86,000(.3503) + \$55,000 + \$42,900 - \$10,000(.2003)$$
$$= \$30,126 + \$55,000 + \$42,900 - \$2,003$$
$$A_{Total} = \$126,023/year$$

Table 10–5 Economic Life Calculation of Existing Equipment at Gooding Co.

Life	$A_{Mkt Value}$	$A_{Operating}$	A_{Scrap}	$A_{Salvage}$	A_{Total}
1	$28,750	$65,000	$57,200	−$5,000	$145,950
2	$15,378	$65,000	$57,200	−$2,326	$135,252
3	$10,949	$65,000	$57,200	−$1,440	$131,709
4	$8,757	$65,000	$57,200	−$1,001	$129,956

Even with a smaller improvement in scrap costs and a shorter life, the revised process is less costly on an annual basis than the existing process. Other what-if calculations can also be made.

PERSONAL ASSETS AND THE NONOWNER VIEWPOINT

Replacement analysis applies to individuals as well as organizations. For example, an individual usually does not have the choice of purchasing one car without considering the sale of an existing car. A person might say, "I just spent $2,000 to overhaul my car. I can't afford to sell it!" Or, "This old car isn't worth very much as a trade-in. I guess I'll keep it." How should these economic issues be resolved? The techniques described in this chapter can be applied. Sunk costs are neglected and the nonowner viewpoint can be used. TVM calculations can be made. The operating costs for the old car and the new car can be included in the analysis. Finally, of course, nonfinancial considerations can be weighed.

> Economic analysis is good to use, especially when the numbers work out the way we want them to.

Another personal application of replacement analysis involves whether to sell a security and replace it with another security. Sunk costs are involved. The nonowner viewpoint can be applied to the analysis; price changes and dividends can be estimated. Finally, the purchase of a new home is another replacement analysis situation.

Replacement Analysis by Individuals

Example: Kathy O'Leary's New Car. Kathy O'Leary, process engineer at Libing and Jones Co., is considering purchasing a new car. In making a decision, she plans to apply the concepts of replacement analysis and TVM, which she uses on the job.

Her present car was purchased four years ago for $10,500. Its current value according to the standard used car valuation book is $2,900. Operating expenses during the past year have been $1,500, including maintenance, repairs, gas, insurance, and taxes. She thinks that she could drive this car two more years and then sell it for $800. The new car under consideration costs $16,500. She thinks that she could drive it for five years and it would then be worth $2,000. The operating expenses for the new car are estimated to be approximately $1,000. Kathy tries to earn a 10% return on all her investments. Diagrams for this decision are shown in Figure 10–5.

For the existing car, the sunk cost of $10,500 is not considered. The market value of the old car has been taken from the standard listing of used car prices. The actual market value may be more or less depending on the mileage of the car, its condition, and other factors. If the analysis shows that the annual costs of the old and new cars are similar, a more detailed estimate of the car's value might be necessary. Using the cost estimates, a calculation of the annual cost can be made using standard TVM techniques. For the old car:

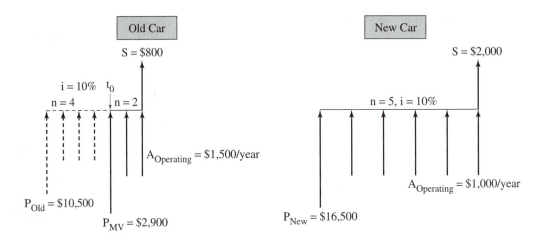

Figure 10–5   Kathy O'Leary's New Car

$$A_{Total} = A_{MV} + A_{Oper} - A_{Sal}$$
$$= P_{MV}(A/P, n = 2, i = 10\%) + A_{Oper} - S \ (A/F, n = 2, i = 10\%)$$
$$= \$2,900(.5762) + \$1,500 - \$800(.4762)$$
$$A_{Total} = \$1,671 + \$1,500 - \$381 = \$2,790 \text{ per year}$$

For the new car:

$$A_{Total} = A_{New-MV} + A_{Oper} - A_{Sal}$$
$$= \$16,500(A/P, n = 5, i = 10\%) + \$1,000 - \$2,000(A/F, n = 5, i = 10\%)$$
$$= \$16,500(.2638) + \$1,000 - \$2,000(.1638)$$
$$A_{Total} = \$4,353 + \$1,000 - \$328 = \$5,025 \text{ per year}$$

From the above calculations, Kathy sees that the new car is more expensive by $5,025 − $2,790 = $2,235 per year. An economic decision would be to keep the old car and place any unspent funds in investments that earn her target ROI of 10%. But perhaps nonfinancial considerations also enter into her decision. She may dislike the old car's color or may simply want a new car. Psychological wants often prevail over economic logic! In either case, once the calculations are made she has a sound economic starting point for her decision making.

Replacement Analysis by Companies

Example: Increasing Maintenance Costs at Circle City Construction. Rob Melton, owner of Circle City Construction, is considering replacing some of the company's earth-moving equipment. He thinks the repair costs are higher for the older equipment, and lower operating costs for the new equipment may make replacement economically justified.

The existing equipment was purchased three years ago for $175,000. Its current book value is $98,000 with a market value of $40,000. Its operating costs are currently $30,000 per year and will increase annually by $20,000. Rob estimates the equipment could be used for three more years and then be sold for $20,000 at the end of the third year, or $25,000 at the end of the second, and $30,000 at the end of the first. New equipment with a life of seven years and additional features could be purchased for $200,000. Maintenance and operating costs would be

$40,000 per year over the life of the equipment. The salvage value would be approximately $30,000 at the end of seven years. Rob has a target of 12% ROI on his investments.

The costs of the existing equipment increase over time. Therefore, the economic life of the existing equipment is an important factor in the decision process.

Figure 10–6 shows the TVM diagrams for the new and existing equipment. Let's calculate the costs of the existing equipment first. For 3 years,

$$
\begin{aligned}
A_{Total} &= A_{MV} + A_{Oper} - A_{Sal} \\
&= \$40,000(A/P, n = 3, i = 12\%) + [\$30,000 + \$20,000(A/G, n = 3, i = 12\%)] \\
&\quad - \$20,000(A/F, n = 3, i = 12\%) \\
&= \$16,654/\text{yr} + [\$30,000 + \$20,000(.9246)] - \$5,926/\text{yr} \\
&= \$16,654/\text{yr} + \$48,492 - \$5,926 \\
&= \$59,220/\text{year}
\end{aligned}
$$

For 2 years,

$$
\begin{aligned}
A_{Total} &= A_{MV} + A_{Oper} - A_{Sal} \\
&= \$40,000(A/P, n = 2, i = 12\%) + [\$30,000 + \$20,000(A/G, n = 2, i = 12\%)] \\
&\quad - \$25,000(A/F, n = 2, i = 12\%) \\
&= \$23,668 + [\$30,000 + \$20,000(.4717)] - \$11,792 \\
&= \$23,668 + \$39,434 - \$11,792 \\
&= \$51,310/\text{year}
\end{aligned}
$$

For 1 year,

$$
\begin{aligned}
A_{Total} &= A_{MV} + A_{Oper} - A_{Sal} \\
&= \$40,000(A/P, n = 1, i = 12\%) + [\$30,000 + 0] - \$30,000(A/F, n = 1, i = 12\%) \\
&= \$44,800/\text{yr} + \$30,000 - \$30,000/\text{yr} \\
&= \$44,800
\end{aligned}
$$

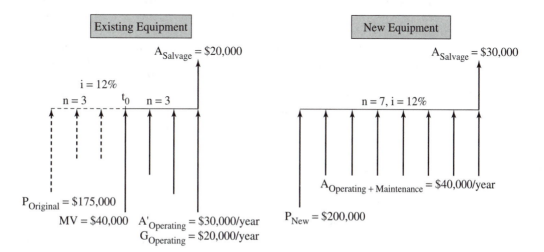

Figure 10–6   Circle City Construction's Increasing Maintenance Costs

The calculations indicate that the lowest cost for the existing equipment is for one year. The economic life of the existing equipment is one year. It should be replaced during the current year.

Next let's calculate the new equipment's costs:

$$A_{Total} = A_{New} + A_{Oper} - A_{Sal}$$
$$= (\$200,000)(A/P, n = 7, i = 12\%) + \$40,000/yr - \$30,000(A/F, n = 7, i = 12\%)$$
$$= \$43,820/yr + \$40,000 - \$2,973/yr$$
$$= \$80,847/yr$$

The economic life of the existing equipment is one year and the equipment should be replaced in this year if lower cost equipment can be found. However, the annual costs of the proposed equipment are *greater* than the cost per year of the existing equipment. Rob Melton should attempt either to find other new equipment with lower first costs, a longer life, or lower operating costs than the proposed equipment, or keep the existing equipment. Another option Rob can consider is to give the existing equipment a major overhaul. This might extend its life and perhaps reduce its annual costs. Leasing or outsourcing of products can be considered also. Rob must now search for other alternatives and continue with his replacement cost analysis.

This problem presents an unusual solution, which isn't usually encountered, of the existing equipment's annual cost increasing and the new equipment's cost still being higher. It occurs in the case of older office, retail, or manufacturing buildings. It is often less expensive to remodel the existing structure than to replace it. Each replacement situation must be analyzed within its own context to determine the optimum solution.

As always, what-if analysis should be conducted, and noneconomic advantages and disadvantages should be considered. Considerations such as safety, appearance, quality, customer preferences, aesthetics, environmental issues, and other elements cannot be weighed *only* in terms of dollars.

Example: Keep the Building or Move to a Shopping Center? Buy Lotz, Inc., a medium-sized retail store, purchased a building and land eight years ago for $300,000. The neighborhood in which Buy Lotz is located has deteriorated over the past few years, according to management. The building's current estimated market value is $140,000 with 10 years of life remaining, at which time the building would be torn down. The net value of the land after 10 years is estimated to be $35,000 after the tear-down. Maintenance of the building has averaged $11,500 per year, and the taxes and insurance have been $5,500 annually. Buy Lotz has the opportunity to move to a nearby shopping center, Shop 'til You Drop Mall, and rent comparable space for $3,500 per month. The mall owner would pay the taxes, building insurance, and maintenance costs. Buy Lotz's objective is to make 10% return on investments.

The TVM diagram for the building is shown in Figure 10–7. The calculation of the total annual cost of the building using the nonowner viewpoint is:

$$A_{Total} = A_{Mkt\ Value} + A_{Tax + Maint} - A_{Salvage}$$
$$= \$140,000(A/P, n = 10, i = 10\%) + \$17,000 - \$35,000(A/F, n = 10, i = 10\%)$$
$$= \$22,778 + \$17,000 - \$2,196 = \$37,582 \text{ per year}$$

The annual cost of the mall lease is (12 mo/yr)($3,500/mo) = $42,000 per year.

Considering cost factors only, the retail firm should stay in its present location. However, other factors can be considered. One question is whether sales would be the same at the new lo-

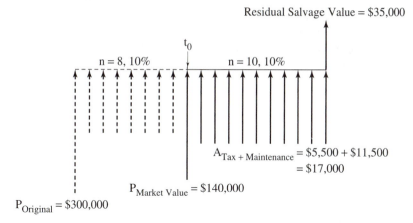

Figure 10–7   Buy Lotz What-If Calculations

cation. Will the current taxes and maintenance costs continue at the same rate, or will they increase over the next 10 years? Could the estimated salvage value change at the end of 10 years? Is the monthly lease cost of the mall location going to stay constant or increase? Making sensitivity or what-if calculations can be important in this case. Also, if the organization decides to remain in the current location, it should recalculate the cost of occupancy on a regular basis, at least annually. Some nonfinancial factors can be considered, such as the advantage of a free-standing location with the convenience of an on-site parking lot. Maintaining the same location might offer some advantages to regular customers. The neighborhood of the existing location and to the mall's retail environment can be compared.

Group Replacements

The discussion in this chapter has dealt with the case of replacing or keeping a single asset such as a building or piece of equipment. There are situations in which groups of assets are installed or replaced. Utility poles, fleets of trucks and vans, multiple units of identical production equipment, office furniture and equipment, lighting systems, and computer workstations are examples of when group or block replacement analysis can be effective. It may be more efficient to analyze such units as a group. Often, even though the lives and costs vary, the analysis and replacement can occur collectively.

Capacity and Replacement

Caution is stressed in regard to asset capacity expansion and replacement analysis. Sometimes equipment is justified based on its additional capacity, such as increased warehouse square footage, machine tool output, or computer speed or memory. By using the increased capacity, the analysis assumes that the capacity is needed or required. Often, analysis is made based on higher capacity that is underutilized. In these cases, the extra capacity may be a false and expensive advantage. Sometimes when the extra capacity is not considered, an opposite conclusion is drawn from the analysis. The argument can be made, "We are operating at maximum capacity. We need the increased capacity the new equipment will give us." Perhaps the new equipment will give too much capacity at too high a cost, and the more economical solution would be to use the existing equipment on an overtime and weekend basis or go to outside suppliers. Additional capacity can be profitable, but it must be correctly analyzed.

SUMMARY

This chapter has presented the basic concepts of replacement analysis, including economic life, nonowner viewpoint, sunk costs, market value determination, and continuous monitoring of replacement costs. Chapter topics include:

- Reasons for replacement, including deterioration, reduced performance, new service requirements, costs, obsolescence, abrupt failure, leasing, purchase of required services from outside supplier, product or services produced by the asset no longer required.
- Revised methods of drawing TVM diagrams.
- Nonowner viewpoint of using the current market value of the building or equipment as the current investment amount. Discussion of trade-in and book values.
- The ten steps of replacement analysis.
- Determination of market value using five methods; estimating future market or salvage values of the assets.
- Estimation of different types of lives, including the calculation of economic life.
- What-if or sensitivity analysis applied to replacement analysis.
- Fixed assets can be reanalyzed periodically with new estimates of life, salvage values, market values, and other relevant factors.
- Replacement analysis is driven by a desire to reduce costs; increase profits; improve productivity and product or service quality; expand capacity; reduce maintenance.
- Capital budgets and available funds must be considered. Even though it is economical to purchase new equipment, the funds may not be available. Perhaps, due to limited funds, the most economical alternative is to lease the asset.
- Future capacity requirements must be considered with replacement analysis.
- Noneconomic factors can be considered with replacement analysis.
- Continuous monitoring and what-if calculations are part of the replacement analysis and decision process.

QUESTIONS

1. State the reasons why fixed assets such as equipment and buildings may need to be replaced.
2. How is the time value of money diagram different for replacement calculations?
3. What are the steps of replacement analysis?
4. What is meant by sunk cost? How is it viewed in analyzing replacements?
5. What is meant by the nonowner viewpoint?

6. How is the market value of existing equipment or structure used in the analysis?
7. What are the methods of estimating market value?
8. What is the importance of book value in replacement analysis?
9. What is the best way to determine market value? What is the most common method of estimating market value?
10. What are the different types of building and equipment lives? Define each type and state why each is important.
11. Why is salvage value important in estimating economic life?
12. What is economic life? How does it differ from physical life?
13. What are the three types of economic life and physical life relationships that are typically encountered in analysis?
14. Why is sensitivity or what-if analysis important in replacement analysis?
15. How are different services lives of existing and new equipment considered in replacement calculations?
16. Why is the nonowner viewpoint difficult to accept?
17. How does minimum cost apply to replacement analysis?
18. What are the different points of view with respect to trade-in value being included in the analysis?
19. Where can replacement analysis concepts apply to personal finances?
20. Why should replacement analysis be a periodic activity for all assets?

21. As a process engineer, you are considering replacing some manufacturing equipment. A friend in the accounting department mentions that the equipment should not be sold since it is still being depreciated. What are your comments?
22. In general, determine the advantages and disadvantages to buying new equipment if the following happens:
    a. Salvage value of the old equipment increases
    b. Trade-in value increases
    c. New equipment life increases
    d. Book value is reduced
    e. Operating expense of the old equipment increases
    f. Desired or target ROI increases
    g. Market value of the old equipment increases
23. What is the basis for the reasoning that salvage value can sometimes be neglected? Why is this especially true when the salvage value is far into the future? (Hint: Consider the *A/F* factor.)
24. What is the effect of technological change on the following:
    a. Current market value of the existing equipment.
    b. Future salvage value.
    c. Cost of new equipment that is technically superior to the old equipment.
25. What are some noneconomic factors that may have bearing on the final decision whether to replace the existing equipment?
26. When can group analysis be applied to assets?
27. What are some cautions concerning the replacement of assets justified by increased capacity?

PROBLEMS

1. An existing piece of manufacturing equipment has been owned for eight years. It was purchased for $80,000. Its current book value according to the accounting department is $21,000; the current market value estimate is $25,000; and it has a trade-in value of $28,000, according to an equipment supplier. The equipment is expected to have three years of productive life remaining with operating costs of $1,500 per year and a residual salvage of $8,000 at the end of the three-year period. If 15% is the desired ROI, determine the following:
    a. Draw the TVM diagram for this alternative.
    b. Calculate the annual cost based on the initial equipment cost eight years ago.

    c. Calculate the annual cost based on the nonowner viewpoint and compare this answer with the one for part b.
    d. Is it appropriate to use the book value or trade-in value to calculate the annual costs?
2. A warehouse building was purchased 15 years ago for $400,000. Its current book value is $150,000 according to the accounting department; the current market value is $250,000; and it has an estimated 10 years of remaining life, at which time it is estimated to be worth $100,000. The operating costs, taxes, and other annual expenses are estimated to be $4,000 per year. If the owner has a target ROI of 12%, determine the following:

a. Draw the TVM diagram.
b. Calculate the annual cost based on the building's cost 15 years ago.
c. Calculate the annual cost using the nonowner viewpoint and compare this answer with the one for part b.
d. Is it appropriate to use the book value to calculate the annual cost?

3. An existing manufacturing building is 20,000 square feet. The organization is trying to determine the value of the structure. Three years ago, a similar building in the same industrial park sold for $450,000. It has 15,000 square feet. Property values in the area have been increasing at about 4% per year.
   a. Determine the current selling price per square foot of the building that sold three years ago.
   b. Estimate the current market value of the 20,000 sq. ft. building.
   c. Is the answer found in part b an exact estimate of value?

4. A machine tool is no longer required by an organization, which is trying to determine the tool's current value. It was purchased for $100,000 five years ago. It has a capacity of 80 units per hour. A sister company recently sold a smaller, similar unit for $40,000. That unit had an output of only 60 units per hour, and it was one year older than the equipment in question. ROI = 10%.
   a. On the basis of outputs, what is the value of the existing equipment?
   b. What effect does the fact that the current equipment in question is one year younger have on the value found in part a?
   c. Is the answer found in parts a and b an exact answer?

5. Using the income method, if a building is rented for retail space at $2,300 per month and the rate of return is 8%, what is the current value of the building?

6. A computer leasing company is able to lease a computer system for $10,000 per year. If the value of money is currently 7%, what is the approximate value of the equipment if the system was sold? Use the income method to determine the current value.

7. Current market value of equipment is estimated to be $45,500. Its cost of removal is estimated to re-

quire two workers for two days at a rate of $10 per hour and eight hours per day, and rental of material handling equipment to move the equipment of $500 per day for two days. What is the net salvage value?

8. A chemical process is estimated to have a total of $86,000 worth of salvageable materials, including pumps, scrap metal, electronic controls, and related equipment. The cost to disassemble the equipment and to deliver it to the scrap dealer will require three workers for three eight-hour days at a rate of $12 per hour, and crane and truck rental at a total of $1,100. What is the net salvage?

9. Manufacturing equipment was purchased seven years ago for $30,000; the current market value is estimated to be $15,000. If the physical life of the equipment is expected to be four more years and have no value at that time, what is the estimate of year-end salvage values for years 8, 9, 10, and 11 using linear extrapolation? Is it reasonable to assume that the future salvage values will decline in a linear manner?

10. Some robots were purchased three years ago for $225,000 each. They are expected to have three more years of useful life and have no value at that time. Their current market value is estimated to be $150,000. If linear extrapolation is used, estimate the value at the end of years 4, 5, and 6. Is using linear extrapolation a reasonable technique for estimating salvage value?

11. Using the nonowner viewpoint, draw diagrams and determine the annual cost of the following existing equipment.

First cost	$346,000
Current age	5 years
Remaining life	4 years
Current book value	$180,000
Current market value	$150,000
Salvage value at end of fourth year	10% of initial cost
Desired ROI	10%
Annual operating costs	$15,500 per year

12. Using the nonowner viewpoint, draw diagrams and determine the annual cost of the following existing equipment.

First cost	$49,000
Current age	4 years
Remaining life	2 years
Current book value	$20,000
Current market value	$15.000
Salvage value at end	
of sixth year	8% of initial cost
Desired ROI	12%
Annual operating costs	$900 per year

13. Existing equipment is estimated to have four years of life remaining. Its current market value is $50,000. Salvage values for the remaining years of life are estimated to be:

End of Year	Residual Salvage Value
1	$30,000
2	$24,000
3	$15,000
4	$6,000

Annual operating costs are estimated to be $1,500 per year. The desired ROI is 15%. Draw a TVM diagram. Calculate the annual cost for the equipment lasting 1, 2, 3, and 4 years. Set up a table showing components of the total annual cost. Graph the results. Use manual calculations or a spreadsheet. What is the economic life of the equipment?

14. Existing equipment is estimated to have five years of life remaining. Its current market value is $150,000. Salvage values for the remaining years of life are estimated to be:

End of Year	Residual Salvage Value
1	$100,000
2	$80,000
3	$50,000
4	$20,000
5	None

Annual operating costs are estimated to be $10,500 per year. The desired ROI is 10%. Calculate the annual cost for the equipment lasting 1, 2, 3, 4, and 5 years. Draw a TVM diagram. Set up a table showing components of the total annual cost. Graph the results. Use manual calculations or a spreadsheet. What is the economic life of the equipment?

15. Referring to problem 13, what if the Annual Costs were $1,500 for the first year of the remaining life, but then they increase by $5,000 each year for the remainder of the equipment's life? Draw the TVM diagram. Set up a table showing components of the total cost, and graph the results. A spreadsheet may be used. What is the economic life of the equipment? Is this an economic life case 1, 2, or 3?

16. An old building has a current net market value of $100,000. The estimated remaining physical life is 10 years. The residual value at the end of this year is $90,000; it is expected to decline by $10,000 per year for the remainder of the building's life. The operating costs including taxes and insurance are $22,000 for this year and are expected to increase by $5,000 per year for each year of the building's remaining life. Draw the TVM diagram. Set up a table of annual costs to determine the total annual costs. Graph the results and determine the economic life. ROI is 15%. Is this a Case 1, 2, or 3 example of economic life?

17. Alternative A has a first cost of $50,000, life of seven years, salvage value of $5,000 at that time, and annual operating costs of $3,000 per year. Alternative B has a first cost of $65,000, operating costs of $6,000 per year, salvage of $6,000, and a life of nine years. If the desired return on investment is 20%, select the best alternative.

18. A current piece of equipment was purchased three years ago for $55,000. Its current book value according to the accounting system is $25,000. Current market value of the equipment is estimated to be $28,000. It has an estimated life of four remaining years. The salvage values for the remaining four years are $30,000 for the end of this year, $22,000 for the end of next year, $15,000 for the end of the third year, and a final salvage value of $5,000 for the fourth year. Operating costs for each of the four remaining years are estimated to be $1,200 per year. The firm has a desired ROI of 12% for its investments. A lease has been offered to the firm that would cover the first cost and the nonoperating expenses of the equipment for $9,200 per year for each of the four years. Should the company lease or keep the existing equipment? What nonfinancial factors should be considered?

19. A machine was purchased four years ago for $100,000. It will last six more years. The depreciation method is straight line. ROI is 12%; salvage

value is $10,000. Annual operating costs are $4,000 per year; market value is $55,000. Determine the annual cost, first without using the nonowner viewpoint, and then using the nonowner viewpoint.

20. A production machine tool can be purchased for $200,000. Its life is estimated to be seven years. Salvage is $20,000 for the seventh year and can be assumed to decline over the equipment's life in a linear manner. Annual operating costs including material, labor, and overhead are $10,000 per year. The existing equipment was purchased four years ago for $150,000. Its current value is estimated to be $50,000 and it should have four more years of service. Salvage would be $8,000 at the end of the fourth year, $15,000 at the end of the third year, $25,000 at the end of the second year, and $35,000 at the end of the current year. The organization requires a 15% return on investment. Determine if the new equipment should be purchased. If the current equipment's operating costs are $10,000 for the current year, but will increase by $7,000 each year over the remainder of the equipment's life, should the new equipment be purchased?

21. A machine was purchased three years ago for $24,000. Its present value is $12,000. Operating expenses are $1,000 per year, with a salvage value of $2,000. Another used machine could be purchased for $5,000 with operating expenses of $3,500 per year and no salvage. The existing machine has six years of life remaining and the used machine would have four years of life remaining. If ROI is 15%, determine the best alternative.

22. An office currently has 25 personal computers with 10 printers. The computers cost $3,100 each and the printers were $400 each; they were purchased two years ago. The market value of the computers is estimated to be $600 each today and of the printers, $200 each today. The present equipment will be useful for four more years and have no salvage value at that time. Operating expenses are expected to be $200 per year for each computer and $150 per year for each printer. A new network system is being considered that would have 25 terminals with a cost of $2,500 each; five printers would be purchased for $1,500 each. The life of the new system is expected to be six years, with a salvage value of $300 for the terminals and $400 each for the printers at the end of that time. Operating expense is expected to be $6,000 per year for the total system. If

the firm desires a 20% return on invested capital, determine the best alternative. Assume that the salvage value of any of the components decreases linearly over the equipment's physical life.

23. An individual purchased 100 shares of a security two years ago for $52 per share including commissions. The stock is now worth $43.50 per share. For the past two years the firm has paid a $2.50 per share dividend, which is expected to continue. If this stock is sold, the investor is considering purchasing another 100 shares for $60 per share that pays a $3 per share dividend. A recent brokerage house research report estimates that due to the economy and current earnings per share estimates the potential price of the securities within the next year is $4.80 per share for the existing security and $6.50 per share for the new stock. Determine the best alternative for the investor, assuming the price estimates of the securities are correct on a pretax basis. What other factors could be considered?

24. A new piece of manufacturing equipment can be purchased for $150,000. Its life would be six years, at which time it could be sold for 15% of its first cost. The operating expenses would be $10,000 per year excluding maintenance. The maintenance would be $5,000 for the first year and increase by $1,000 each year during the equipment's life. Material and labor costs using the new equipment are $5.50 per unit; 5,000 units per year are produced and sold. The existing equipment cost $110,000 two years ago. It has three years of useful life remaining, at which time it would be sold for 10% of its first cost. The current book value is $60,000 and the estimated market value is $52,000. Operating costs are expected to continue at $12,000 per year; maintenance costs will be $7,000 next year and increase by $1,500 each year thereafter. Material and labor expenses are $6.25 per unit and are expected to continue at that amount in the future. If the desired ROI is 15%, determine the best alternative. Assume that salvage values of both the existing and new equipment decline at a linear rate.

25. A student owns a car that is four years old and cost $12,000. Its current estimated market value is $4,000 and is expected to be constant for the next four years. Operating expenses are expected to be $1,400 next year and increase by $500 each year for the next four years. The replacement truck that the student is considering costs $17,000, has operating

costs of $1,000 per year, a life of seven years, and salvage at that time of $3,000. An ROI of 12% is desired on all investments. Determine if the present car should be replaced. Is it valid to compare a car with a truck?

26. Existing equipment in a medical clinic was purchased five years ago for $400,000. It has three years of life remaining. Operating costs are expected to be $40,000 next year and increase by $18,000 each year. Current market value is $150,000 and is expected to remain the same for the next three years. New equipment would have a life of nine years, a current cost of $550,000, and annual operating costs of $30,000 each year for its life. Salvage value of the new equipment would be $90,000 in nine years. The clinic desires an ROI of 10%. Should the new equipment be purchased?

27. A retail store is currently leasing its space at a monthly cost of $5,000 per month. It has an opportunity to purchase a building for $600,000, which includes the building and the parking lot. The building would have maintenance and taxes of $7,000 per year. The store plans on occupying the building for 12 years, at which time the structure and land would be worth $350,000. The other operating costs would be the same for the new building as for the leased location. If the firm desires a 12% ROI, what is the best course of action? What noneconomic factors should be considered? If more information is available, describe the what-if analysis you would do.

28. An existing piece of equipment is currently generating 15% scrap. The value of the scrap including material, labor and overhead costs is $10.50 per unit. The current rate of output from the process is 1,000 units per month of both good and bad product. The existing equipment was purchased five years ago for $230,000. Current operating costs are $10,000 per year. Its current market value is $135,000. It has three more years of life remaining. A quality improvement team has estimated that the scrap rate

could be reduced to 4% if new tooling and major overhaul work could be performed and some of the major components could be replaced. The improvements would also reduce the annual operating costs from $10,000 down to $6,000 per year. If the organization requires 15% return on its investments, what is the maximum that could be spent on the equipment to reduce the scrap rate?

29. The current output from a process is 5,000 units per week with a total unit cost of $45 each and a selling price of $54.00 per unit. The existing equipment cost $125,000 three years ago and it is estimated to have five years remaining. It would have no salvage value at that time. Its current operating costs are $25,000 per year including maintenance, overhead, material, and labor. A new piece of equipment for the process could be purchased for $150,000 with a 10-year life and no salvage value at that time. The comparable operating cost of the new equipment would be $20,000 per year, and the output from the new equipment would be 6,000 units per week. If the firm desires an ROI of 10%, should the new equipment be purchased if all of the additional output can be sold at the current price? What if the additional output would have to be sold at a discount of $5 per unit? Should the new equipment be purchased? The current value of the old equipment is $75,000.

30. An existing delivery van for a small manufacturing company cost $15,000 four years ago, and it is estimated to have a current value of $4,000 and two years of life remaining, with a value of $500 at that time. Its operating costs are expected to continue at $3,000 per year. A new van can be purchased for $18,000 that is expected to last seven years and have a salvage/market value of $2,000 at that time. Its operating costs are expected to be $1,800 per year. Using the ROI/rate of return approach, determine the ROI at which the alternatives are equal. Use a spreadsheet for calculations and graph results.

DISCUSSION CASES

Regan Electronics

Mike Simpson, a process engineering technician at Regan Electronics, is analyzing existing electronics manufacturing equipment. He is making recommendations to the firm's upper management to replace the existing equipment with new equipment. The head of the accounting department, Jim Nelson, questions Mike's analysis about using the current market value of the equipment rather than the original cost of the equipment. Jim comments, "This equipment is four years old and we are depreciating it over a 10-year period. The original cost was $150,000 and it has a current value after accumulated depreciation of $90,000. With six more years to go, we must consider its original cost of $150,000, not its estimated market value of $75,000. It is impossible to accurately estimate the true market value of the equipment unless we actually sell it."

Jim Nelson continues, "We should use the original value of the equipment over its useful life as defined by the depreciation schedule since these are the figures we use on the Income Statement, which determine the taxes we pay. The new equipment that is being recommended is more expensive than the original equipment's cost by $50,000. When the accounting department was doing the new equipment purchase analysis, we always used the discounted cash flow method and brought all the costs to the present worth; I don't know why we changed from that method. It is true that the new equipment is estimated to have lower operating costs and produce a lower scrap rate, but we are always upgrading equipment around here before it is worn out. A very expensive way to do business, I think!"

1. If you are Mike, what is your response concerning the use of current market value to make a decision?

2. What should Mike answer about the use of the depreciated value as the estimate of the old equipment's value?

3. What should Mike say about the discounted cash flow/present worth question?

4. How should Mike respond to the use of depreciation for tax calculations?

5. What else should be considered in this decision?

Don Malard's Investments

Don Malard of Sandor Corp. is looking over his regular investments and his IRA investments. He finds one security that he purchased for $44 per share approximately three years ago. It has a current price of $31 per share. He has looked at the firm's financial statements, read a number of negative articles in the financial publications, and has talked with some of his friends who work for the company. All reports seem to be negative for the near future. For a while, Don thought that he should keep the security hoping that it would "come back." He now feels that since the current market has been on a moderate upswing, he would like to sell these securities and invest in others that have more potential. The other security that he is considering has a price of $19 per share and he estimates that it could go to $30 per share within two years. He realizes that there are no "sure things," but he has selected this security from many other potential investments, and he has analyzed the security.

1. What should Don do?

2. How is this situation similar to the equipment replacement problems he has encountered at Sandor?

3. Are there any potential tax considerations?

4. What other things should be considered?

Williams Sales Representatives

Kurt Williams has a small manufacturers representatives sales group that represents some major machine tool manufacturers. In their sales presentations, they always include time value of money calculations to show the buyer the costs of the existing equipment that is currently owned and a comparison of the new equipment that is being proposed. They include the target return on investment, operating costs, salvage values, and other costs of the old and the new equipment. In a recent sales presentation, a customer questioned the calculation of the cost of the new equipment by subtracting the current market value of the existing equipment from the cost of the new equipment.

The customer said that the new equipment "should stand on its own, and trade-in should not be considered." Kurt said that they typically help a customer find a buyer for the used equipment and that the value of the existing equipment was sometimes large and should be considered as a reduction in the first cost of the new equipment. He further said that the manufacturing process can be thought of as a "profit center" and if labor can be reduced from the process, or improvements can be made to reduce unit costs, these costs should be analyzed and taken into account when calculating the future unit cost of the product. Therefore, any savings or revenue from selling used equipment can be thought of as reducing the process cost too. The customer, who had bought many items from Williams in the past, said that the money from the sale of old equipment when new equipment was purchased went into the general fund for other investments and should not be considered as revenue for the existing process. Kurt Williams left the meeting not sure of how to handle the customer's questions.

1. Write a letter for Kurt to the customer that answers their questions.

2. Which approach is correct? Should Kurt consider trade-in?

3. What other factors should Kurt consider?

Chapter 11 Taxes

> The Congress shall have power to lay and collect taxes on incomes, from whatever source derived, without apportionment among the several States, and without regard to any census or enumeration.
>
> —Amendment XVI, U.S. Constitution

KEY TERMS

Capital Gains Tax Tax on the sale of assets and investments that are used as investments, not production and delivery of products and services.

Flat Tax A tax that has a constant rate over all taxable amounts; also called a fixed percentage tax rate.

Leverage Borrowing rather than using internal funds. Funds are usually borrowed at an interest rate that is lower than the ROI of the investment.

Limited Liability Corporation (LLC) A type of corporation permitted in some states that limits the liability of the owners and also permits income to flow to the owners/partners, who are taxed as individuals.

Net Operating Cash Flow Net income (net profit after tax and before dividends) plus depreciation plus amortization plus depletion plus write-offs plus deferred taxes, less purchases of buildings, equipment, and other fixed assets for the period.

Progressive Tax A tax that has increasing rates so that as the taxable amount increases, the tax rate increases.

Subchapter S Corporation A corporation that gives the owners the option of being taxed as individuals.

Total Cash Flow The cash into and out of the cash account during an accounting period. A summary of the sources and uses of the cash for a period. There are three components: operating, investing, and financing cash flow.

Types of Taxes Income and profits, assets and wealth, sales, value added.

LEARNING CONCEPTS

- Basic calculation of federal taxes for the corporation and the individual using the tax tables.
- Types of taxes for individuals and corporations.
- Effects of depreciation and other noncash expenses on the Income Statement and tax.
- Tax exemptions, deductions, subchapter S and limited liability corporations.

- Flat tax.
- Depreciation, operating cash flow, total cash flow, and taxes.
- Steps to calculate ROI based on operating cash flow and after-tax profit.
- Capital gains taxes on the sale of assets.
- Economic life, net operating cash flow, and taxes.
- Cash Flow statement and its components: operating, investing, and financing.
- Calculating operating cash flow and after-tax ROI.
- Application of leverage to funding investments with external funds.

INTRODUCTION

In the U.S., few are shy when it comes to talking about taxes. Everyone has an opinion about taxes and what should be done about them. Our task is to reduce the topic of taxes to the economic analysis of alternatives, calculating after-tax ROI, and expanding our economic analysis model by incorporating taxes into the calculations.

There are many types of taxes on corporations and individuals, including income, sales, property, wealth, and estate taxes. Some taxes are progressive, but most are fixed percentage or flat taxes.

Tax rates and calculations for individuals, partnerships, corporations, and special corporations are considered in this chapter. We discuss the flat income tax and double taxing of shareholders. The effect of taxes on the ROI of organizations and the relationship between depreciation, operating cash flow, and taxes is presented.

Three types of cash flow, operating, investing, and financing, are reviewed. After-tax operating and investing cash flows are the basis for calculating after-tax ROI. To make the after-tax ROI determination, TVM must be combined with accounting, depreciation, and total cash flow concepts. This chapter reviews the Accounting Equation, depreciation, and TVM techniques to arrive at after-tax economic analysis, and discusses the tax requirement when assets are sold. Concepts of leverage and borrowing to finance the purchase of equipment and buildings are covered as well.

Prior to this chapter, our economic analysis was based on *before-tax* calculations. The development of a comprehensive model to analyze investments of fixed assets and projects using all the previous analytical techniques combined with tax considerations is presented here.

The Income Statement and Balance Sheet were discussed as important statements that "keep financial score" and give a financial picture of the organization or individual. In this chapter, the Cash Flow Statement is reviewed as an additional important financial statement. The Total Cash Flow Statement is closely related to taxes and the after-tax analysis of equipment, buildings and other fixed assets. By combining TVM, total cash flow, and financial statement knowledge, after-tax ROI can be determined.

Because tax laws are the result of both local and national legislation in the U.S., they are complex and changeable. In this chapter, we will look at the overall effect of taxes on the organization and on economic analysis of fixed assets. Because of the changing nature of the tax regulations, no detailed discussion of specific regulations, deductions, or exemptions can be made.

Throughout the text, we have assumed that there is little difference in economic analysis when applied to the individual, proprietorship, partnership, or corporation. This is generally true in practice; however, tax requirements vary for different types of organizations and legal entities. This chapter differentiates between the application of taxes to the individual and the corporation.

By combining analysis tools such as time value of money, breakeven, minimum cost, and replacement with the tax topics presented in this chapter, the individual, manager, technician, or engineer now has a complete set of tools to apply to economic decisions.

> In this world nothing is certain but death and taxes.
>
> —Benjamin Franklin

TYPES OF TAXES

The types of taxes that corporations and individuals must pay include:

1. Income tax on revenues and profits.
2. Tax on assets and wealth.
3. Tax on sales.
4. Tax on value added to products and services.

In the U.S., we do not have extensive value-added taxes. A value-added tax is the tax on the difference in the cost of raw materials and the product's value after manufacturing. This tax has been proposed as an alternative or addition to the present tax system at the state and federal level; however, there is currently limited legislative support for it. Perhaps in the future, the value-added tax will become as "popular" in the U.S. as it is in some other countries.

Property tax and other wealth-based taxes, for the purposes of this text, are included in operating costs along with material, labor, and overhead expenses. Any sales taxes are also usually included in the cost of the purchase of equipment and materials when making an economic analysis.

The primary source of revenue for government entities is taxes on the income and profits of individuals and corporations. Profits and net income are functions of accounting system conventions such as depreciation. Taxes become a function of the accounting system as well. The discussion here will cover the concepts of the Accounting Equation, depreciation, and the effect of taxes on operating cash flow. From the determination of operating cash flow, the after-tax ROI can be calculated.

Historically, income, profit, and wealth taxes have their roots in Roman, Greek, and early Egyptian economies. Various taxes in early U.S. history were based on sales, wealth, property, and sometimes, to a limited degree, income. During World War I, the U.S. government began imposing income taxes on a "temporary basis" to fund the war.

Taxes on Income and Profit

Individual and corporation taxes are based on income less deductions and exemptions, which are based on the tax regulations. Individuals can have exemptions and deductions for dependents, home mortgage interest, donations to charities, etc. For the corporation, costs are deducted from revenues, and the resulting profit or income before tax on the Income Statement is the basis of corporate tax calculations. The taxable income for individuals or corporations is then multiplied by a percentage figure to determine the amount of tax. For both the individual and the corporation, *the percentage multiplier is not constant or fixed, but increases* as the taxable income increases. This is a **progressive tax.** The tables below show the progressive tax rates for the calculation of federal tax for individuals:

Single Individuals*

Taxable Income	Tax Rate and Amount
$0 to $24,650	15% of taxable income
$24,650 to $59,750	$3,697.50 + 28% of amount over $24,650
$59,750 to $124,650	$13,525.50 + 31% of amount over $59,750
$124,650 to $271,050	$33,644.50 + 36% of amount over $124,650
$271,050 and over	$86,348.50 + 39.5% of amount over $271,050

Married Filing Jointly

Taxable Income	Tax Rate and Amount
$0 to $41,200	15%
$41,200 to $99,600	$6,180.00 + 28% of amount over $41,200
$99,600 to $151,750	$22,532 + 31% of amount over $99,6000
$151,750 to $271,050	$38,698.50 + 36% of amount over $151,750
$271,050 and over	$81,646.50 + 39.6% of amount over $271,050

*These tables are based on the 1993 Tax Code and the IRS Corporation Tax Tables for 1997.

For example, if an individual has $31,000 in taxable income (the gross income less exemptions and deductions), the tax is calculated as shown below:

Tax on $24,650 = (.15)($24,650)	=	$3,697.50
Tax on ($31,000 − $24,650)	=	(.28)($6,350) = $1,778.00
Total tax	=	$5,475.50

The individual's *average effective tax rate* would be $5,475.50/$31,000 = 17.7%.

A married couple submitting a joint return with taxable income of $58,500 would owe taxes of:

Tax on $41,200 = (.15)($41,200)	=	$6,180
Tax on ($58,500 − $41,200)	=	(.28)($17,300) = $4,844
Total tax	=	$11,024

The couple's *average effective tax rate* would be $11,024/$58,500 = 18.8%.

Below is a table showing the progressive tax rates for corporations:

Corporation Rates (IRS Schedule)*

15% on the first $50,000 of net income

25% on the next $25,000 of net income

34% on the next $25,000 of net income

39% on the next $235,000 of net income

34% on the next $9,665,000 of net income

35% on the next $5,000,000 of net income

38% on the next $3,333,333 of net income

35% on all net income over $18,333,333

*These tables are based on the 1993 Tax Code and the IRS Corporation Tax Tables for 1997.

For corporations a surtax of 5% is added to the base rate of 34% for income between $100,000 and $335,000. For example, if a corporation has taxable net income or profit before tax (revenue − costs) of $250,000, the tax is calculated as shown below:

Tax on first $50,000	=	(.15)($50,000) = $7,500
Tax on next $25,000	=	(.25)($25,000) = $6,250
Tax on next $25,000	=	(.34)($25,000) = $8,500
Tax on next $150,000	=	(.39)($150,000) = $58,500
		$80,750

The corporation's *average effective tax rate* is $80,750/$250,000 = 32.3%.

The above examples are for federal taxes. There are, of course, similar additional state, county, and city taxes. Tax tables change frequently. Consult current IRS regulations for current federal rates.

Taxes on Proprietorships and Partnerships

A proprietorship isn't a taxable entity, but the organization's costs and income pass to the individual. The individual includes the amounts on his or her tax return.

A partnership, like a proprietorship, is not a taxable entity, but the income and costs pass to the partners for tax purposes. A partnership of two or more individuals divides the income and costs of the organization among the individual partners, and the partners include their share of the revenues and costs in their individual tax calculations.

Exemptions and Deductions

One of the most difficult tasks in determining taxes for the individual and the corporation is calculation of the taxable income. A detailed and complete discussion of deductions and exemptions cannot be included here. Internal Revenue Service publications, tax instructions, and tax advisors are the best source for all of these rules and regulations. The complexity of these regulations is one of the major reasons that a flat tax is offered as a simpler alternative.

In some ways, the determination of the taxable income for the corporation is more direct and simple than for the individual. The corporation's Income Statement is the basis for the calculation of the net income or taxable profit. Generally, the organization's revenue is reduced by the costs for the period, and the residual net income or profit before tax is the taxable income. There may be some exceptions to this, such as if the organization has a loss from previous years or other special situations.

Other Taxes and a Combined Tax Rate

In addition to the federal tax, the corporation and the individual reside in a state, county, or city that is likely to have taxes on income as well. The above individual and corporate income tax examples can be extended to include all of the other taxes paid on income.

For many medium-sized and larger corporations, the federal income tax rate is approximately 35% plus state and local income taxes. Adding the federal, state, local and other income taxes to the federal rate of 35% usually gives a total rate of 40% to 50% or greater. For many calculations in this chapter, a combined rate of 40% or 50% will be used, representing the combination of a number of income tax percentages.

Taxes on Dividends Paid to Shareholders—Double Taxation

We often hear, "The shareholder of a corporation pays taxes twice on dividends." This refers to the corporation paying taxes on its net income then distributing part of its after-tax earnings to shareholders, and shareholders then paying individual taxes on those dividends.

For example, if the corporation's net income was $250,000, the tax was $80,750. The difference, profit after tax or net income, is $250,000 − $80,750 = $169,250. These funds can be distributed to the shareholders or they can be retained by the organization and reinvested. Assume that $100,000 is paid out to the shareholders and the remaining $69,250 is retained in the corporation. The shareholders who receive the dividends must report this dividend income on their individual tax returns. If each shareholder pays taxes at an average rate of 20%, taxes are (.20)($100,000) = $20,000 for the shareholders. The total taxes paid on the earnings are the $80,750 paid by the corporation + $20,000 paid by the shareholders = $100,750 total taxes paid.

As a percentage, the effective tax rate is $100,750/$250,000 = 40.3% of the organization's net income. There are some special types of corporations and tax systems that can avoid or minimize this double tax, such as subchapter S and limited liability corporations.

Subchapter S Corporation

A corporation may qualify as a **type S corporation.** The corporation may elect to have the profits pass through to the owners without paying a corporate tax; the shareholders pay their proportionate share of the organization's profits, similar to a proprietorship. There are some very specific requirements to qualify as a subchapter S corporation for tax purposes. The IRS tax code and a tax advisor are the best sources for specific information about subchapter S taxes. If the requirements are met, a subchapter S corporation may have significant tax advantages for the shareholders.

Limited Liability Corporation (LLC)

Another special type of corporation is the **limited liability corporation (LLC),** which is permitted in some states. There are special provisions in the tax code that permit some corporations closely held by families or a few shareholders to file and pay taxes as though they were an S corporation. Check with a tax specialist or attorney for details on the LLC.

> "It was as true," said Mr. Barkis, "as taxes is. And nothing's truer than them."
> —Charles Dickens, *David Copperfield*

FLAT TAX

Taxes are often the subject of political debate. Recent discussion has focused on a federal **flat tax** for individuals. The belief is that the complex rules for deductions and exemptions could be eliminated. Gross income would be multiplied by a fixed percentage to obtain the amount of tax owed. The proponents of the plan say that the IRS tax reporting form could be the size of a postcard and result in approximately the same total taxes paid to the federal government. This would be a welcome relief for those who do their own taxes and attempt to understand all the complex deductions, exemptions, and calculations.

The flat tax is a fixed percentage tax. The tax tables shown above for individuals and corporations indicate an increasing tax percentage or progressive rate. The flat tax or fixed percentage is *constant*, not increasing, for all incomes above a no-tax minimum amount. Various proposals have suggested no individual tax up to approximately $26,000 of gross income and a rate of 16% to 19% on gross incomes above this amount. This eliminates the complex deductions and exemptions, while still maintaining approximately the same revenues to the government. It is likely that Congress will continue to discuss and change the tax laws as they have done in the past. Hopefully, future regulations will be simplified and a flat tax will be considered.

The main points of most federal flat or fixed percentage tax proposals are:

1. A constant tax rate in the range of 16% to 19% based on gross income rather than an increasing percentage (not progressive).
2. Limited or no special exemptions or deductions.
3. No tax paid by individuals below a certain income level ($26,000 to $35,000 are the suggested amounts).
4. Simple calculations by the taxpayer.
5. Approximately the same total amount of funds available for the federal government.

The Flat Tax Is Not New

Most sales taxes, property taxes, estate taxes, and some other special taxes are fixed percentage rate taxes without any progressive rates or special exemptions or deductions. And, when the federal income tax was started in the early 1900s, it too was a flat or fixed percentage tax. The current progressive tax is relatively new. Taxes existed as far back as three thousand years ago or more, and the majority of taxes have almost always been a fixed percentage or flat rate type of tax.

TAXES AND ROI

There are many consequences of taxes on return on investment. Taxes have been left out of our analysis until now in order to focus on the basic concepts of economic analysis. Now we are ready to include taxes as an important element in the analysis. When corporations or individuals state that they have "an investment with a return of 12%," they are usually thinking in terms of pretax ROI. Given that total income taxes are in the range of 30% to 40% or more for individuals and corporations, after-tax ROI may decrease by a proportional amount. And labor-based taxes such as unemployment tax, FICA, and related taxes can be included in materials, labor, and other operating costs.

There are also additional state and local income and property taxes. When adding all the taxes together, it is not unusual for the after-tax ROI to be one-half of the pretax ROI. Taxes are a significant and necessary part of any economic analysis of projects and investments.

Approximating After-Tax ROI

As an approximation, before and after tax ROIs have the following relationship:

$$ROI_{After\ Tax} = ROI_{Before\ Tax}\ (1 - t)$$

where t is the effective or total tax rate.

For example, if an organization has a federal income tax rate of 35%, and the state, city, county, property, labor-related (FICA, etc.), and other taxes are 10%, then the total or effective tax rate is 45% of the net income. If the organization's pretax rate of ROI is 12%, then the $ROI_{After\ Tax}$ is:

$$ROI_{After\ Tax} = ROI_{Before\ Tax}\ (1 - t) = 12\%\ (1 - .45) = (0.12)(0.55) = 6.6\%$$

This approximation may not always be a good one. In the actual case, the $ROI_{After\ Tax}$ should be calculated on the basis of operating cash flow after tax. Sometimes when the depreciation method and the total depreciation amount give a large depreciation for a particular period, the pretax and $ROI_{After\ Tax}$ may differ from the above approximation. As with all approximations, it should be applied carefully. When the approximation is not valid and an exact after-tax ROI is required, using the operating cash flow method for calculating the after-tax ROI is appropriate. This is sometimes called the after-tax internal rate of return as well as the after-tax return on investment.

These techniques can be modified to give after-tax ROI results. The concepts of TVM, depreciation, the Accounting Equation, operating cash flow, and related analysis are combined to determine the after-tax ROI.

Tax Rates and Total Revenue

In a study for the National Bureau of Economic Research, Cummins, Hossett, and Hubbard found that the average U.S. corporate tax rate fell from 1955 to 1965, and the average corporate return on equity (ROE) increased from 8% to 12% during that period—a 50% increase in return. As the tax rate went back up from 1965 to 1980, the return fell from an average of 12% to 3.5%. Then, from 1980 to 1995, when corporate tax rates fell again, the returns increased again to the 11%–12% range. When ROE is higher, the organization has a higher net income and more funds are available to distribute to shareholders and to reinvest into plant and equipment. The relationship between taxes and return is supported by the popular belief that lower taxes are good for the individual, the organization, and the overall economy ("Fair-Haired No More?" *Barron's*, 1 July 1996, p. 44).

> The organization's ROI is very sensitive to tax rates. ROI is inversely proportional to the prevailing tax rate.

Corporate profits are distributed to the shareholders or reinvested as retained earnings to purchase equipment, inventories, R&D, and for other investments. These investments, in turn, provide revenues, jobs, and profits for the suppliers of the equipment, inventory, and other purchases. All of this increased economic activity leads to more profits for organizations. These increased profits, even when taxed at a lower rate, provide more revenues to the federal government. Conversely, recent experiences show that when the federal tax rates are increased, corporate revenues and total profits decline, and federal tax revenues also decline. The argument is made that higher tax rates should yield higher federal revenues, but recent experiences of the past 40 years do not support that argument. Lower tax rates encourage economic activity, which results in higher ROIs, more jobs, and greater investment into equipment and buildings, and thus increased tax revenues to the government.

> Low tax rates encourage economic activity, higher profits, and increased tax revenues for the government.

Example: Effects of Depreciation at Adams Co. In order to understand the effect of taxes on profits, we will review depreciation. Depreciation, you will recall, is the reduction in value of a fixed asset over time. If a building purchased by Adams Co. for $600,000 is depreciated over a 30-year period using the straight-line method with no salvage value, there is a reduction in value of $600,000/30 years = $20,000 per year.

Recall that every entry in the Accounting Equation *must be offset* by an equal and opposite entry. When the building is purchased, the cash account is reduced and the asset, the building, is increased by the price of $600,000. The annual depreciation entry reduces the asset, the building, by $20,000 each year, and each year depreciation as an expense of $20,000 is entered under the Income Statement cost accounts. Diagrams of the entries into the Accounting Equation are shown in Figures 11–1 and 11–2.

When a cost is entered on the Income Statement for a given revenue, the gross profit decreases. A decrease in profit or net income results in reduced taxes and a lower tax payment for that period. The tax payment savings results in more cash available for the organization to distribute to shareholders and/or to reinvest.

Assets = Liabilities + Owners' Equity + Revenue − Costs

Cash − $600,000

Equipment + $600,000

When equipment or buildings are purchased, the cash account is *decreased* and the equipment account is *increased* keeping the accounting equation in balance.

Figure 11–1   Adams Co.—Purchasing a Building and the Accounting Equation

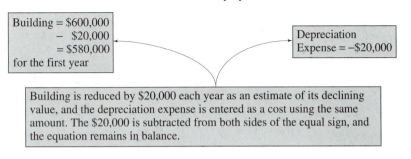

Assets = Liabilities + Owners' Equity + Revenue − Costs

Building = $600,000
        − $20,000
        = $580,000
for the first year

Depreciation
Expense = −$20,000

Building is reduced by $20,000 each year as an estimate of its declining value, and the depreciation expense is entered as a cost using the same amount. The $20,000 is subtracted from both sides of the equal sign, and the equation remains in balance.

Figure 11–2   Adams Co.—Accounting Equation and Annual Depreciation

The income statement with and without depreciation—The following example of the Adams Co. Income Statement demonstrates depreciation, taxes, and operating cash flow. Although it's not realistic to assume that depreciation can be either used or neglected as an expense, this may help to show the effect of depreciation on taxes and operating cash flow.

Adams Co.
Income Statement for Year Ended 19xx

	Without Depreciation	**With Depreciation**
Sales	$200,000	$200,000
Cost of Goods Sold	150,000	150,000
Gross Profit	$50,000	$50,000
Depreciation	none	20,000
Profit Before Tax	$50,000	$30,000
Tax @ 40%	20,000	12,000
Profit After Tax	$30,000	$18,000

The Income Statement without depreciation has the higher profit after tax of $30,000. If this is the case, why should depreciation be applied? Aren't high profits desirable? (Recall that depreciation expense is the result of a *decrease* in an asset value on the Balance Sheet.) Depreciation is a method of spreading the cost of the asset over time. Note also that the inclusion of depreciation results in the organization sending lower tax payments to the IRS. Only $12,000 rather than $20,000 is sent to the IRS. This savings of $8,000 in tax payments is a result of the depreciation being included as a noncash expense on the Income Statement.

Looking only at the Income Statement, the organization and its owners would likely prefer $30,000 in profits rather than only $18,000. But to understand the true tax advantage of depreciation, consider the effect of the "with depreciation" column on the cash account. In the cash account, the organization did not send a check for $20,000 depreciation expense to anyone. So it actually has $50,000 from operations less tax payments of $12,000 remaining in the cash account. Adams Co. pays its tax bill of $12,000, leaving $50,000 − $12,000 = $38,000 cash in the checking account for reinvestment or to distribute to the owners.

To summarize, the two tax effects of depreciation are:

1. A reduction in the amount of tax paid, in this case $12,000 rather than $20,000.

2. More cash available than only the net income as shown on the Income Statement. Depreciation expense is a noncash expenditure, and the amount stays in the cash account. In this case, the $20,000 depreciation expense is not disbursed and remains in the cash account

REVIEW OF CASH FLOW

It is not possible to discuss taxes without first understanding depreciation and operating and investment cash flow.

In the first section of the book, total cash flow and depreciation were introduced along with financial statements and depreciation. It may be helpful to go back and review that discussion now. Although the TVM diagram does graphically show "flows of cash," the term **total cash flow** has a very specific meaning in economic analysis. Before going further with after-tax analysis, we can briefly review cash flow and its relationship to taxes and the ROI$_{\text{AFTER TAX}}$ calculation.

> Depreciation is the result of assets declining in value rather than an action by the IRS. The IRS permits the depreciation method to be used. It is a misunderstanding to think that the purpose of depreciation is to provide tax benefits.

Total cash flow is composed of *operating, investing,* and *financing* cash flows. These three cash flows can be monitored by observing the cash in and out of the cash account, or they can be determined indirectly from the financial statements.

> To understand after-tax ROI calculations, accounting entries, total cash flow, and TVM must be understood and applied.

The cash that comes into or leaves the cash account can be divided into three categories.

1. Funds are received from customers as they purchase the goods and services the organization produces. Funds are disbursed to suppliers for materials, to workers for wages and salaries, for overhead expenses, and for interest expense and income. This type of cash flow is *operating cash flow*.

2. Funds may be disbursed from the cash account for investment into land, buildings, or equipment, or investments into subsidiary organizations in the form of loans or stock purchases. Funds may also be received from the sale or repayment of these investments. This second category of cash flow is *investing cash flow*.

3. Funds may flow into the cash account by borrowing money from a financial institution or from selling stock or bonds in the organization. Funds are disbursed by paying off loans, repurchasing stock from shareholders, and paying dividends to shareholders. This third category of cash flow is *financing cash flow*.

Three categories of cash flow:
1. Operating
2. Investing
3. Financing

Here, we are normally concerned with the first two of these cash flows, operating and investing. Analysis of financing cash flows, the third category, typically is the responsibility of the organization's accounting or finance department along with upper management.

The Income Statement is a summary of the revenues and costs from the operations of producing and selling services and products. Both cash and noncash revenues and costs are shown on the Income Statement. Noncash expenses on the income statement include depreciation, amortization, depletion, write-offs, and taxes that are owed but not yet paid. Depreciation and other noncash expenses on the Income Statement are noncash flows and are the result of a decrease in an asset or an increase in a liability, not a cash disbursement. Depreciation expense is usually the largest and most important of the noncash expenses.

To determine operating cash flow from the Income Statement, profit after tax (PAT) or net income is added to the depreciation. Other noncash expenses can be neglected unless they are significantly large amounts. As an alternative to observing the cash flowing into and out of the cash account, the Income Statement can be used to calculate operating cash flow.

The second category of total cash flow, investing cash flow, has a number of components, including investment into equipment, buildings, and land and investments into subsidiaries and securities. It is this investment that is of concern to manufacturing engineers, design engineers, production managers, and technicians. By comparing the investment in these fixed assets with returns from the operating cash flow, we can determine if the investment is warranted on an after-tax basis.

Depreciation and Operating Cash Flow

Example: Operating Cash Flow at Adams Co. Returning to the Adams Co. Income Statements, it would appear that the preferred result is the Income Statement that produces the higher profit after tax of $30,000 rather than $18,000. If this is so, why is the concept of depreciation thought of as a benefit to the organization and the shareholders? The answer to this question is found in an understanding of operating cash flow.

The depreciation and other noncash expenses are not cash disbursements, because they results from a noncash event such as reducing the asset value of the building. No one receives a check for payment of the depreciation expense. Therefore, the cash available to the organization, when depreciation is included, is the profit after tax *plus* the deducted depreciation amount. In the "With Depreciation" example, the profit or net income of $18,000 added to the depreciation expense of $20,000 yields a *total of $38,000* in cash that is available to the organization.

This flow of cash would be more apparent if we could see the Adams Co. checkbook for the year. Since that is usually not available, we can calculate operating cash flow indirectly from the Income Statement.

The calculations for the operating cash flow "With" and "Without Depreciation" of Adams Co. are shown below:

$$\text{Operating Cash Flow}_{I/S \text{ Without Depreciation}} = \text{PAT} + \text{Depreciation} = \$30,000 + \$0 = \$30,000$$
$$\text{Operating Cash Flow}_{I/S \text{ With Depreciation}} = \text{PAT} + \text{Depreciation} = \$18,000 + \$20,000 = \$38,000$$

We can see that the "With Depreciation" alternative results in $8,000 more cash in the Cash account than the "Without" alternative. This is the tax advantage of depreciation. Not only is the depreciation entry used to reduce the value of an asset, it also causes higher operating cash flow for the period. Because of the noncash depreciation expense of $20,000, there is $8,000 ($38,000 − $30,000) more cash available to be distributed to the shareholders or for reinvestment by the organization. For many external financial statement readers and for internal ROI calculations, the success of organizations and projects is best judged by looking at the operating cash flow rather than *only* the net income or profit after tax.

Alternate Method of Calculating Operating Cash Flow

There is another method of calculating the operating cash flow from the financial statements. In the previous abbreviated Income Statement of Adams Co. ("With Depreciation"), it can be shown that operating cash flow also is equal to gross profit minus taxes.

From the Income Statement,

$$\text{Gross Profit} - \text{Depreciation} = \text{Taxable Income}$$
$$\text{Taxable Income} - \text{Tax} = \text{Profit After Tax}$$

Operating Cash Flow is defined as:

$$\text{Operating Cash Flow} = \text{Profit After Tax} + \text{Depreciation (neglecting the other noncash expenses)}$$

Substituting for profit after tax from the above equations:

$$\text{Operating Cash Flow} = (\text{Taxable Income} - \text{Tax}) + \text{Depreciation}$$

Substituting for taxable income from the above equations:

$$\text{Operating Cash Flow} = (\text{Gross Profit} - \text{Depreciation}) - \text{Tax} + \text{Depreciation}$$

Cancelling depreciation:

$$\text{Cash Flow} = \text{Gross Profit} - \text{Tax}$$

Operating Cash Flow = PAT + Depreciation
or
Operating Cash Flow = Gross Profit − Tax

Applying this alternate method of calculation to the Adams Co. case:

$$\text{Operating Cash Flow} = \text{Gross Profit} - \text{Tax} = \$50,000 - \$12,000 = \$38,000$$

This is the same result as previously calculated using the other method.

This method of cash flow calculation is mathematically correct, but it doesn't indicate the actual sources of the flows of cash. This is a useful shortcut that can be used to identify after-tax cash flow from individual processes or products if the details of the Income Statement values are unknown. Both definitions of operating cash flow assume that other noncash flows such as depletion, write-offs, unpaid taxes, or amortization from operations are small and neglected.

In analyzing buildings, manufacturing equipment, and other investments, taxes must be considered as a cost of doing business using those assets and cash flow's effect on the tax payments must also be included in the analysis. The engineer, technician, or manager conducting economic analysis of the investments must understand and be able to apply the concepts of taxes, depreciation, and operating cash flow to investments in order to arrive at a correct solution to the equipment or project decision.

ROI AND CASH FLOW

Recall that the general definition of the annual pretax ROI is profit before tax (PBT) divided by the total investment. An extension to an after-tax basis, by including operating cash flow, is annual after-tax ROI.

$$\text{ROI}_{\text{Pretax}} = \frac{\text{PBT}}{\text{Investment}}$$

$$\text{ROI}_{\text{After Tax}} = \frac{\text{Annual Operating Cash Flow}}{\text{Investment}}$$

Effects of Depreciation Methods on After-Tax ROI

The organization and individual may be able to choose the most appropriate depreciation method for a particular asset. From a time value of money and economic analysis viewpoint, the most rapid reduction in asset value or "write-off" is the preferred method. A method such as MACRS, which has a high depreciation initially with lower depreciation later in the asset's life, will yield a higher ROI on the investment than, say, a straight-line method. The reason for this advantage is that the larger cash flows occur earlier in the asset's life and thus give a better ROI, since funds "flow back" to the cash account sooner than with the slower depreciation methods. All depreciation methods, of course, give the same *total* depreciation expense over the life of the equipment, even though each method gives variable annual depreciation amounts. The next example, purchasing a van, demonstrates the calculation of depreciation, operating cash flow, and after-tax ROI using MACRS and TVM techniques.

Example: Buying a Delivery Van at Center City Flower Shop. Mary Meek, the owner of Center City Flower Shop, is considering purchasing a new van for deliveries. She estimates that the new van would save $8,000 in operating expenses per year over the existing delivery van. The van costs $22,000 initially and will have a useful and economic life equal to the accounting life of five years. A 12% after-tax ROI is the target for the shop's investments. The applicable depreciation system is the Modified Accelerated Cost Recovery System (MACRS). This system uses a five-year life for light trucks.

The five-year MACRS schedule for property purchased using the half-year convention is shown below:

Year	Percentage of Initial Cost
1	20.00%
2	32.00%
3	19.20%
4	11.52%
5	11.52%
6	5.76%

Note that the schedule extends into the sixth year because of the half-year convention. The annual depreciation over the life of the delivery van is shown below.

Year	Depreciation Amount (rounded to nearest $1)
1	(.2000)($22,000) = $4,400
2	(.3200)($22,000) = $7,040
3	(.1920)($22,000) = $4,224
4	(.1152)($22,000) = $2,534
5	(.1152)($22,000) = $2,534
6	(.0576)($22,000) = $1,268 (+ $1 to make total of $22,000)

$$\text{Total} = \$22,000$$

The abbreviated Income Statements for the savings and investment in the new van are shown below:

Changes in Income Statement Accounts as a Result of the Investment in the New Van						
Year	1	2	3	4	5	6
Savings (Increase in Profit Before Tax)	$8,000	$8,000	$8,000	$8,000	$8,000	$8,000
Less: Depreciation of $22,000 Investment	4,400	7,040	4,224	2,534	2,534	1,268
Increase in Taxable Income	3,600	960	3,776	5,466	5,466	6,732
Increased Tax @ 40%	1,440	384	1,510	2,186	2,186	2,692
Increase in Profit After Tax	$2,160	$576	$2,266	$3,280	$3,280	$4,040
Operating Cash Flow	$6,560	$7,616	$6,490	$5,814	$5,814	$5,308

(Note: Operating Cash Flow = Depreciation + Profit After Tax)

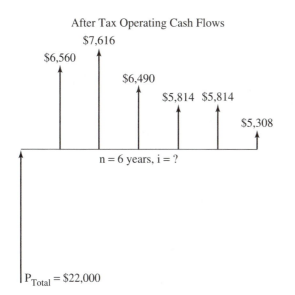

Figure 11–3   Center City Flower Shop's New Van

In the above statement, operating cash flow is the increase in profit after tax from the Income Statement plus depreciation expense, which is a noncash expenditure, recorded on the Income Statement because the van declines in value during each year. Neglecting other noncash flows, the above table shows the flows of funds into the cash account from the investment on an *after-tax basis*.

The operating cash flow for each of the six years represents the funds that are available to the business for reinvestment or for the owners. These funds *would not be available if the van is not purchased*. The tax "advantage" is due to the fact that the van is not a 100% expense when it is purchased. Instead, the purchase is "expensed" over a period of six years using the annual depreciation entry. In order to obtain the annual cash flow, the organization had to invest the initial $22,000. The objective is to earn 12% after tax on all investments, so annual cash flows after taxes can be brought back to the present worth using 12% and compared to the $22,000 to see if the investment is justified. The TVM diagram is shown in Figure 11–3.

The present worth of all operating cash flows amounts is calculated at ROIs of 12%, 15%, and 20% as shown below:

End of Year	Annual After-Tax Cash Flow Amount	Present Worth 12%	15%	20%	25%
1	$6,560	$5,857	$5,704	$5,467	$5,248
2	$7,616	$6,071	$5,759	$5,289	$4,874
3	$6,490	$4,619	$4,267	$3,756	$3,323
4	$5,814	$3,695	$3,324	$2,804	$2,381
5	$5,814	$3,299	$2,891	$2,337	$1,905
6	$5,308	$2,689	$2,295	$1,778	$1,391
Total Present Worth =		$26,230	$24,240	$21,431	$19,122

(These present worth calculations are based on the standard method of converting a future value to a present value using tables and a calculator.)

Since the present value of the future after-tax operating cash flows using 12% is greater than the original investment of $22,000, the investment has an after-tax ROI greater than 12%. The actual after-tax ROI of this investment is between 15% and 20%. The investment is justified based on the target ROI of 12%.

The after-tax calculations are more complex than the before-tax calculations because depreciation and tax rates are used. The after-tax ROI, however, gives a more complete picture of the funds available to the business and the acceptability of the investment. Certainly, as the size of the investment increases, complete after-tax analysis and the effect on finances become more important. A manufacturing engineer, designer, technician, project manager, production manager, and others responsible for spending the organization's money should have an understanding of the after-tax consequences of investments. The more financial knowledge become is available throughout the organization, the more effective financial decisions will be. Too often, the after-tax analysis of fixed asset investments is left to the accounting and finance departments because the design, process engineering, or manufacturing departments are not knowledgeable about TVM applications, depreciation, operating cash flows, and taxes. If personnel close to the technical decision can also perform economic analysis, the solution will be more complete.

Steps for Calculating After-Tax ROI

1. Determine the savings from making the investment over the old method or additional income from the new investment.

2. Calculate the depreciation using the appropriate IRS schedule and calculations.

3. Determine the increase/decrease in taxable income resulting from the investment.

4. Calculate the increased/decreased tax paid annually.

5. Determine the increase/decrease in profit after tax as a result of the investment.

6. Calculate the annual operating cash flow increase/decrease resulting from the investment.

7. Calculate the total present worth of the annual cash flows if service lives are equal. Otherwise, annual costs should be used.

8. Compare the total present worth of the cash flows with the original investment.

9. Once the above steps have been performed, find the actual after-tax ROI by using trial methods.

10. Consider noneconomic factors.

Example: Stevens Corp.—Make or Buy and After-Tax ROI. Bill Johnson at Stevens Corp. has estimated that if a machine is purchased to produce a component that is currently bought from an outside vendor, $12,000 per year savings in purchasing and operating costs can

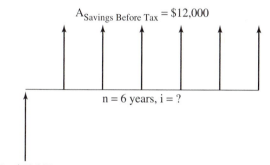

$A_{\text{Savings Before Tax}} = \$12,000$

$n = 6$ years, $i = ?$

P = $60,000

Figure 11–4   Before-Tax ROI for New Equipment at Stevens Corp.

be realized. The equipment would cost $60,000 initially and have no salvage value after six years. Straight-line depreciation is used. Stevens' effective tax rate is 40%. Bill wants to determine the ROI on both a before-tax and an after-tax basis. The TVM diagram for the investment is shown in Figure 11–4.

Before-tax ROI—This solution is a trial-type solution similar to previous examples.

$$\text{Try } 10\%: P = A(P/A, n = 6, i = 10\%) = \$12,000(4.3553) = \$52,263$$
$$\text{Try } 8\%: P = A(P/A, n = 6, i = 8\%) = \$12,000(4.6229) = \$55,474$$
$$\text{Try } 6\%: P = A(P/A, n = 6, i = 6\%) = \$12,000(4.9173) = \$59,008$$
$$\text{Try } 5\%: P = A(P/A, n = 6, i = 5\%) = \$12,000(5.0757) = \$60,908$$

The pretax ROI is between 5% and 6% (the exact answer is 5.47%).

After-tax ROI—If a straight-line depreciation method is elected, the equipment is depreciated as follows:

$$\text{Depreciation} = \frac{\text{First Cost} - \text{Salvage}}{\text{Life}} = \frac{\$60,000 - 0}{6 \text{ years}} = \$10,000 \text{ per year}$$

As a result of the investment of $60,000 in new equipment, the organization saves an additional $12,000 per year (pretax) than it had before the investment, so it must pay tax on the additional profit generated from the savings. The result is that operating cash flow after the tax of $11,200 is available annually for reinvestment into the organization or for distribution to shareholders as dividends.

A partial Income Statement for the process is presented below to determine the cash flow for the investment.

Annual Savings	$12,000
Depreciation	10,000
Profit Before Tax	$2,000
Increase in Tax @ 40%	800
Increase in Profit After Tax	$1,200

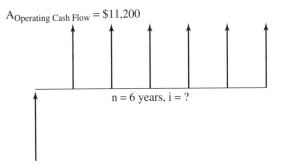

$A_{\text{Operating Cash Flow}} = \$11,200$

$n = 6$ years, $i = ?$

$P = \$60,000$

Figure 11–5   After-Tax ROI for New Equipment at Stevens Corp.

Operating Cash Flow = Depreciation + Profit After Tax = \$10,000 + \$1,200 = \$11,200/yr
Or, using the alternate method, the operating cash flow can be calculated as

$$\text{Savings} - \text{Tax} = \$12,000 - \$800 = \$11,200/\text{yr}$$

The TVM diagram for the after-tax ROI calculation is shown in Figure 11–5. The solution for the after-tax ROI using a trial method is:

$$\text{Try } 8\%: P = A(P/A, n = 6, i = 8\%) = \$11,200(4.6229) = \$51,776$$
$$\text{Try } 6\%: P = A(P/A, n = 6, i = 6\%) = \$11,200(4.9173) = \$55,074$$
$$\text{Try } 4\%: P = A(P/A, n = 6, i = 4\%) = \$11,200(5.2421) = \$58,711$$
$$\text{Try } 3\%: P = A(P/A, n = 6, i = 3\%) = \$11,200(5.4172) = \$60,673$$

The after-tax ROI is between 3% and 4%. The exact answer is 3.34% and can be determined by:

$$A = P\,(A/P, n = 6 \text{ yr}, i = ?)$$
$$A/P = \$11,200/\$60,000 = .1867$$

Using the tables or a calculator i can be determined:

$$i = 3.34\% \text{ after tax}$$

Using the approximation method,

$$\text{ROI}_{\text{After Tax}} = \text{ROI}_{\text{Before Tax}}\,(1 - t) = 5.47\%(.60) = 3.28\%$$

The pretax ROI is 5.47% and the after-tax ROI is 3.34%. Stevens Corp. can look at the after-tax ROI to determine if this ROI is satisfactory before making the decision to purchase the equipment. In this case, there is a positive operating cash flow resulting from the investment; however, it results in a very low ROI before and after tax. If this investment can be postponed or eliminated, perhaps it should be, because of the low ROI. Safety equipment and pollution control equipment are examples of investments that cannot be postponed. In nonelective cases such as these, the only feasible solution is finding equipment with the lowest costs. Often, investments must be made regardless of the return. In this case, Bill Johnson may decide to continue purchasing the component from the supplier due to the low ROI.

Example: Michaelson Electronics Manufacturing—Target After-Tax ROI. The previous example demonstrated how depreciation and operating cash flow are used to calculate the

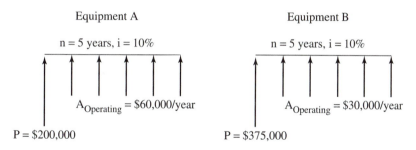

Figure 11–6   New Machines at Michaelson Electronics

after-tax ROI. An alternative calculation is to start with the organization's target or desired ROI to determine if the initial investment in the equipment is justified by the after-tax savings.

A process engineer at Michaelson Electronics is considering the purchase of one of two pieces of electronic manufacturing equipment. Equipment A has first costs totalling $200,000 and operating expenses of $60,000 per year. Equipment B has a first cost of $375,000 with annual operating costs of $30,000. A straight-line depreciation method is used for five years with no salvage value. Currently, Michaelson is purchasing the components from a supplier. The current annual cost to the organization for the supplier's services is $150,000. Michaelson's objective is a target value of 10% for after-tax ROI, and the tax rate is 40% of the profit before tax. A method similar to the analysis used in the Center City Flower Shop example can be applied for each alternative at Michaelson. Figure 11–6 shows the TVM diagrams for the equipment.

For Equipment A and B, the straight-line depreciation amounts are:

$$\text{Depreciation of A} = \frac{\$200,000}{5} = \$40,000 \text{ per year}$$

$$\text{Depreciation of B} = \frac{375,000}{5} = \$75,000 \text{ per year}$$

Partial Income Statements and calculations for equipment A and B are shown below. The Income Statements are the same for each year since straight-line depreciation is used.

	Equipment A	Equipment B
Savings from not using supplier	$150,000	$150,000
Increase in Operating Costs	60,000	30,000
Net Savings in Operating Costs	$90,000	$120,000
Depreciation	$40,000	$75,000
Net Increase in Profits Before Tax	$50,000	$45,000
Tax on Increase in Profits Before Tax @ 40%	$20,000	$18,000
Increase in Profit After Tax due to investment in equipment	$30,000	$27,000
Increase in Operating Cash Flow (Depreciation + Increase in PAT)	$70,000	$102,000

Or the operating cash flow can be calculated by subtracting the tax increase from the net savings:

$$\$90,000 - \$20,000 = \$70,000 \text{ for equipment A}$$
$$\$120,000 - \$18,000 = \$102,000 \text{ for equipment B}$$

The partial Income Statements demonstrate that both equipment A and B can save money for the organization and increase the operating cash flow. To obtain the increase in profits and cash flow, investments in the equipment must be made initially for $200,000 or $375,000 for equipment A and B, respectively.

To determine whether the investment is warranted on an after-tax basis, the annual increase in operating cash flow is brought to a present value at the organization's desired after-tax ROI of 10%. This amount is then compared to the investment required. An alternative calculation would be to bring the initial investment, as a present value, to an annual amount and compare the annual amount to the annual operating cash flow increase. Since the service lives are equal, either method will provide a reliable answer.

For equipment A,

$$P_{\text{Cash Flow}} = A_{\text{Cash Flow}} \ (P/A, n = 5, i = 10\%) = \$70,000(3.7908) = \$265,355$$

For equipment B,

$$P_{\text{Cash Flow}} = A_{\text{Cash Flow}} \ (P/A, n = 5, i = 10\%) = \$102,000(3.7908) = \$386,660$$

Because the present value of both A and B are greater than each equipment's initial cost, either alternative will exceed the 10% after-tax ROI target. Which of the two alternatives should be selected? The method used here to determine the best alternative is to calculate the ROI each alternative produces and selecting the one with the higher ROI.

For equipment A,

$$P = A_{\text{Cash Flow}} \ (P/A, n = 5, i = ?)$$
$$\$200,000 = \$70,000 \ (P/A, n = 5, i = ?)$$
$$i = \text{approximately } 22\%$$

For equipment B,

$$P = A_{\text{Cash Flow}} \ (P/A, n = 5, i = ?)$$
$$\$375,000 = \$102,000 \ (P/A, n = 5, i = ?)$$
$$i = \text{approximately } 11\%$$

Both alternatives exceed the desired 10% after-tax ROI. Equipment A, however, has a much higher ROI than equipment B. So equipment A should be selected.

Note that the two alternatives have the same lives. With identical lives, it is acceptable to make the comparison using present value techniques in calculating the after-tax ROI. When the service lives of the two alternatives are unequal, the annual amount or IRR technique is used.

Example: Process Selection at Circuit Board Inc. A component, CB-883, can be produced using one of two processes at Circuit Board Inc. The two processes can be analyzed using the operating cash flow savings and ROI techniques. One process is labor intensive and the other is equipment intensive. The process that uses more equipment has a higher initial cost due to the investment in the equipment, and it also has a higher depreciation and tax advantage. This advantage can be calculated using the concept of operating cash flow.

Shown below are the Income Statements for the two processes.

Proposed CB-883 Annual Income Statements for Circuit Board Inc.

	Process A Labor Intensive	Process B Equipment Intensive
Sales	$500,000	$500,000
Cost of Goods Sold	400,000	300,000
Gross Profit	$100,000	$200,000
Depreciation	60,000	100,000
Profit Before Tax or Net Income	$ 40,000	$100,000
Tax @ 40%	16,000	40,000
Profit After Tax	$ 24,000	$ 60,000
Dividends Paid	10,000	10,000
Retained Earnings	$ 14,000	$50,000
Cash Flow (PAT + Depreciation)	$ 84,000	$160,000

Difference in operating cash flow using Process B:

$160,000 – $84,000 = $76,000 (almost a 2 to 1 difference)

Examining the above Income Statements of the two processes, we see that the equipment-intensive process B has an operating cash flow of $76,000 more than process A. Since the depreciation deduction on the Income Statement is a noncash deduction, there are more funds available for reinvestment or distribution to the shareholders using process B.

A shortcut approach to calculating the cash flow is to subtract the tax paid from the gross profit before depreciation and tax. In the above example for processes A and B:

Operating Cash Flow = Gross Profit – Tax

	Process A	Process B
Gross Profit (before depreciation and tax)	$100,000	$200,000
Less: Tax for the period	16,000	40,000
Operating Cash Flow per year	$ 84,000	$160,000

This is the same operating cash flow calculated using the previous method.

Although process B has the greater operating cash flow, it must be compared to the investment to determine if the investment is justified based on after-tax ROI. If the initial investment in process A is $300,000 and in process B is $500,000, and each has a life of five years, then the TVM diagrams shown in Figure 11–7 can be drawn.

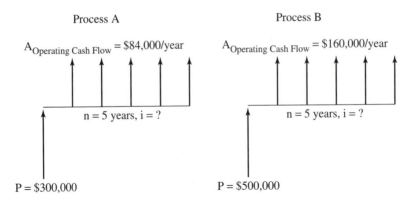

Figure 11–7   Process Selection at Circuit Board Inc.

The diagrams show the annual cash flow after tax, initial investment, and life of the investment. For process A,

$$P = \$300,000 = A_{\text{Oper Cash Flow}} \ (P/A, n = 5, i = ?) = \$84,000(n = 5, i = ?)$$
$$i = \text{approximately } 12\%$$

For process B,

$$P = \$500,000 = A_{\text{Oper Cash Flow}} \ (P/A, n = 5, i = ?) = \$160,000(P/A, n = 5, i = ?)$$
$$i = \text{approximately } 18\%$$

The equipment-intensive alternative, process B, gives the higher after-tax ROI. To obtain this higher ROI, an investment of $500,000 must be made. The lower ROI alternative, process A, requires an investment of $300,000. In order to take advantage of the higher ROI, Circuit Board Inc. must initially invest an *additional* $200,000. This may or may not be possible, based on the organization's amount of available internal funds or its external borrowing capability.

TAX CONSEQUENCES OF SELLING BUILDINGS OR EQUIPMENT

The discussion to this point has been on the acquisition of fixed assets such as buildings and equipment and the financial analysis of the acquisitions. Now we are ready to discuss the tax consequences when fixed assets are disposed of or sold. The asset may or may not be fully depreciated when sold.

Often, when new equipment is purchased, the older equipment is kept as a backup or transferred to another department or process. In these cases, the retention of the old equipment does not immediately result in a tax consequence. When new equipment replaces the old equipment, which is sold to another user or for salvage value, there may be tax and operating cash flow consequences. If the building or equipment is part of the operations that produce the services or products that are sold by the organization, the profit or loss from the sale is usually added into the other operating profits. However, if the asset is primarily used as an investment, rather than for operations, then the profit on the sale may be taxed at a **capital gains** tax rate.

The tax rate and procedures for capital gains are different from the normal income tax rate. Operating equipment that has been depreciated, inventory, financial receivables, and real property used in the organization are generally considered to be operating assets and are taxed at the income rates. Such assets as land, securities, antiques, personal property, investments in subsidiaries, and related assets are normally taxed at capital gains rates when sold. A tax advisor can help to determine if the profits on the sold asset will be taxed at the regular or capital gains rate.

As of this writing, capital gains on profits for assets held less than one year are taxed at the regular income tax rate. Property held more than 12 months is a long-term capital gain and is taxed depending on whether it is held between 12 and 18 months or for more than 18 months. The asset held 12 to 18 months has a tax rate of 28% or 15%, depending on the seller's tax bracket. The asset sold after 18 months or more is taxed at 10% if the individual is in the 15% income tax bracket, and at 20% if the individual is in the 28% bracket or higher. Under current regulations, the rates are to be lowered further in 2001. Special rates apply to real estate. A tax advisor can assist with the application of the new regulations.

Selling Partially Depreciated Assets

Example: Block Corp.—Selling Equipment. When an asset that has been partially depreciated is sold, there are usually tax consequences. For example, assume that Block Corp. sells some equipment for $35,000 that has a current book value of $30,000. Recall the effect of an asset's sale on the Accounting Equation: The cash account increases as the money is received from the buyer. This is an increase in an asset. The balancing entry of this asset increase would be the increase in a revenue, such as "income from sale of equipment" or just "sale of equipment."

Also, since the equipment is gone, the equipment account must be decreased by $30,000 to show the account at zero (no equipment, no book value). The balancing part of this entry is under costs on the Income Statement, "cost of equipment sold." The summary of these four entries into the Accounting Equation is shown in Figure 11–8.

Entries on the Income Statement accounts of a revenue of + $35,000 and cost of equipment sold of −$30,000 results in a "profit" of $5,000. This profit will be added to the other net income on the Income Statement and the resulting net income before tax will be higher. This will cause the resulting tax to be *greater* than it would have been without the sale. If the tax rate for Block Corp. is 45%, Block would pay (45%)($5,000) = $2,250 more in tax as a result of the

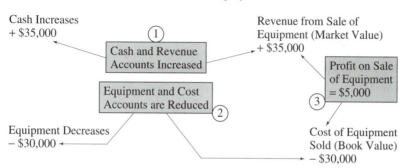

Figure 11–8   Block Corp.—Accounting Equation Entries When Equipment Is Sold

sale. If the sale had resulted in a loss, this loss would reduce the company's taxes. Whether the sale of an asset is taxed at the regular rate or at the capital gains rate can be determined by a tax advisor.

Depreciation and Taxes Are Not "Reversible"

Care must be taken to understand depreciation's effect on taxes. The concept that high depreciation results in lower pretax profits and thus lower taxes can cause misconceptions such as, "Let's buy the most expensive equipment since it has the highest depreciation and will permit us to pay the lowest taxes." This "reverse analysis" would lead to the purchase of the highest cost assets so that lower taxes would be incurred, and the investment would likely have a low ROI due to its high cost. With this thinking, the organization would pay lower taxes. However, it would have acquired high-cost equipment, which would mean that products and services would have to be sold at a higher price. High prices might cause fewer units to be sold, and lower revenue would result, which would lead to even lower profit and ROI. This is an example of backward thinking.

Minimizing Taxes?

Let's buy more expensive equipment. It will have higher depreciation. That will cause higher costs, which causes lower taxes!

(The higher equipment costs cause higher product unit costs, which cause lower sales. Profits decrease, and taxes really go down!)

The general rule for tax purposes is to select the depreciation method that permits the most rapid reduction in the asset's value. This permits the most rapid "recovery" or the fastest increase in operating cash flow to the organization. Financially, it is the most attractive method because of the effects of TVM. This assumes, of course, that the method selected for the particular asset is permitted by IRS regulations.

Investor: Is there anything we can do to eliminate the tax problem of my investments?
Stockbroker: Sure. No problem.
Investor: Really? How can that be done?
Stockbroker: Simple. Let's invest in stocks that have poor prospects, that will go down in price and not pay dividends. No profits, no tax!

OPERATING CASH FLOW AND ECONOMIC LIFE

The accounting or tax life of fixed assets is usually specified by the Internal Revenue Service and/or the in-house accounting system. As we have seen, when the asset is sold or disposed of and the selling price is different from the calculated book value, there are tax consequences.

There are three ways to compare economic life and accounting or tax life:

1. Economic life is *equal to* tax life
2. Economic life is *longer than* tax life
3. Economic life is *shorter than* tax life

When the economic life of the asset is equal to the tax life, the depreciation time period and the life of the equipment are the same. This would be an unusual coincidence. At the end of the life of the asset, it would be sold for replacement or salvage. The book value of the asset would become the cost of equipment sold and the sale revenue would be shown on the Income Statement. The difference would be the profit or loss from the sale of the asset. If the sale price is equal to the book value (another coincidence), then the net profit from the sale is zero and there would be no tax consequence.

In the second case, when the economic life is greater than the accounting or tax life of the equipment or buildings, the asset will continue to perform the required services, but the Balance Sheet will continue to show constant book value. In this case, no depreciation expense will be applied after the end of the accounting or tax life. Operating cash flow, in this case, will be increased by the amount of service, product, and profit that the asset delivers. There would be no correction or "add-back" of depreciation since depreciation expense is not on the Income Statement for this later period.

Finally, when the economic life is less than the accounting life of the asset, it will be sold or disposed of before all the value of the asset on the Balance Sheet has been depreciated. The remaining value of the asset would be an expense on the Income Statement and the revenue, if any, would be on the Income Statement. This would cause either a positive or negative profit and would affect taxes and operating cash flow.

TOTAL CASH FLOW

Limited cash, not low ROI, is often the cause of not acquiring or replacing equipment, buildings, and new projects. Many potentially high ROI investments in equipment and buildings are not made because of limited funds. For this reason, it is very important that managers, technicians, and engineers who do economic analysis of equipment and building alternatives understand the concepts of the three types of cash flow.

For most organizations, profits from operations are the most important source of funds. The firm's primary function is to satisfy customers by selling products or services, and the successful organization is rewarded with profits if it does the job right.

From the previous discussion in this chapter, we know that depreciation and other noncash expenses have a large impact on the operating cash flows of the organization. Operating cash flow has a major effect on the decision to purchase new equipment, buildings, or other long-term assets. Other components of total cash flows in and out of the cash account are also important in economic analysis. Operating cash flow is usually the largest component, but there are also cash flows in and out from investments by the organization and from financing the sale of securities and borrowing. (See Chapter 3 for a detailed review of total cash flow.)

1. **Cash flows from operations:** Operating cash flow results from inflow and outflow of cash from producing and selling services and products. The primary component of this type of

cash flow is net income or profit after tax. Also included here is depreciation expense and other noncash deductions on the Income Statement.

2. **Cash flows from investing:** This component includes the investments into and the sale of long-term fixed assets, including equipment and buildings. Also included are the sale and purchase of securities, or investing in subsidiaries, real estate, or other investments.

3. **Cash flows from financing:** This includes debt and equity transactions. It covers the sale and repurchase of securities and loans from financial institutions or the sale of bonds by the organization.

Cash Flow Statement

In addition to the Income Statement and the Balance Sheet, accounting departments publish the organization's Cash Flow Statement showing the sources and uses of the organization's cash during the period. This statement is normally included in the annual report. The Cash Flow Statement is a window into the cash account of the organization. Internal managers, engineers, and technicians who are responsible for investing in fixed assets should understand the organization's cash flow. Below is a typical Cash Flow Statement.

Cash Flows for Stuart Corp. for Year Ending 19xx		
Operating Cash Flows		
Profits After Tax	$70,000	
Changes in Current Assets	10,000	
Changes in Current Liabilities	(5,000)	
Depreciation	$50,000	
Net Cash Flow From Operations		$125,000
Investing Cash Flows		
Investment in Equipment	($45,000)	
Investment in Buildings	($40,000)	
Investment in Securities	($10,000)	
Net Cash Flow From Investing		($95,000)
Financial Cash Flows		
Funds from Bank Loan	$50,000	
Increase in Common Stock Sold	$15,000	
Dividends Paid	($ 5,000)	
Net Cash Flow From Financing		$60,000
Total Net Cash Flow Increase		$90,000

The statement above shows cash inflows as positive cash receipts and negative outflows in parentheses. The total cash flow amount on the above statement indicates that the organization has $90,000 cash available for payment to the shareholders and/or for reinvestment during the period. The primary *source* of the funds is profits after tax, which was $70,000. Another major *source* of operating cash flows is the $50,000 depreciation. The organization invested or *used* a total of $95,000 cash into equipment, buildings, and securities during the period. The financial cash flow section of the statement indicates that the organization had a *source* of $60,000 from loans and stock sale, less dividend payments to shareholders.

Projects, Investments, and Total Cash Flow

For the rapidly growing firm, such as a small state-of-the-art technology organization, there are often many more profitable investments and projects than there are funds available. It can be very discouraging to technicians, engineers, and managers whose high ROI projects and equipment investments cannot be funded due to limited total cash flow. The total cash flow analysis combined with TVM are very important tools in determining how the limited available funds should be apportioned to the investments and projects. Technical personnel who are responsible for equipment, buildings, R&D projects, and other assets should be familiar with the Cash Flow Statement as well as economic analysis. The accounting and financial departments can *generate* the cash flow statement, but its use and application become the responsibility of others in the organization.

For the mature and stable organization, total cash flow is often less of a factor. Investments into new projects and to replace existing buildings and equipment are often done at a steady rate, and funds that are received flow out at an even rate. The net effect is a steady-state flow of funds into and out of the cash account for this type of organization.

> The art of taxation consists in so plucking the goose as to obtain the largest amount of feathers with the least amount of hissing.
> —Jean-Baptiste Colbert

As a normal practice, financial cash flows are managed by the senior management or financial management rather than by the technical staff of the organization. Frequently, during the course of the budgeting process, it is discovered that many good potential investments could be made if additional financial cash flow is available from security sales and/or borrowing.

Sometimes management decides that external funds should not be obtained even though it appears there are many good internal investments capable of producing high ROIs. External factors such as high interest rates for borrowed funds, high bond interest rates, the state of the stock market, and the organization's profitability can all affect the organization's opportunities to acquire funds at a reasonable cost. Conversely, when external funds can be obtained at a low interest rate, more projects and investments can be funded. For this reason, the interest rates that banks charge and the interest rates set by the Federal Reserve Open Market Committee are watched very carefully. Changes in interest rates can have a major impact on organizations' project funding, which in turn affects the overall economy.

NET OPERATING CASH FLOW

Net operating cash flow is a further refinement of total cash flow analysis. Net operating cash flow relates depreciation and the equipment purchased to replace the declining assets. For a steady-state condition, the equipment that is wearing out is being replaced at the same rate. Therefore, the difference between depreciation and new equipment purchases would be zero.

> All earnings are not created equal. Net cash flow can be called the "owner's earnings," which are net profits plus depreciation minus capital expenditures to maintain the business.
> —Warren Buffett

An investor in a firm's stock is well served by including net operating cash flow in the analysis of the Balance Sheet, Income Statement, and other reports. Warren Buffett, one of the most successful U.S. investors, relies on the analysis of current and future net operating cash flow to determine the future funds available for payment to investors or for reinvestment into the organization. He considers all future estimated net operating cash flows. These are brought to a present value and compared to the current market value of all shares outstanding (the number of shares outstanding times their market price). It is this critical analysis of the present value of the net cash flow, along with other analyses, that determines whether Buffett decides to invest in an organization's stock.

This is the same cash flow concept applied by managers and technical staff to internal projects and investments. The techniques of determining an organization's internal project ROI using cash flow techniques are the same as for the valuation of the total organization.

LEVERAGE

Sources of funds are not limited to profits. Borrowing and the sale of securities are also important sources of funds for the organization. This is the third component of total cash flow: investing cash flow. **Leverage** is the purchase of buildings, equipment, and other assets using borrowed funds. Both individuals and corporations can borrow funds; however, only corporations can obtain funds by selling securities. If projects and investments have ROIs higher than the cost of the borrowed funds—the interest rate—then it may be desirable to obtain external funds for internal investment in the firm.

Some financial transactions have interest costs that can be used as a tax deduction. They include an individual purchasing a home and deducting the interest from taxable income; an organization borrowing funds to purchase equipment and paying tax-deductible interest on the loan; raising funds by the sale of bonds and deducting the interest paid to the bondholders; and other borrowing that permits the cost of the loan interest to be deducted before the tax is calculated. Bond interest is a pretax expense. This interest expense lowers the profits before tax, which in turn reduces the amount of tax owed. Stock dividend payments, however, are *not* a pretax expense, but are paid after tax, which requires both the corporation and the shareholder receiving the dividends to pay tax on the funds.

The cash flows from investing and financing have an important effect on the acquisition and replacement of buildings, equipment, and new projects. Borrowed funds and funds from the sale of stocks and bonds may be able to provide the means to finance the purchase and replacement of operating equipment when internal funds are not adequate. Of course, too much borrowing may require large interest payments to the lenders, which would have a negative effect on the total cash flow amounts.

Leverage and Borrowing by Individuals

When an individual borrows part of the purchase price of a home, the interest paid on the mortgage loan can be deducted as a pretax expense under current tax regulations. For example, if a home is purchased for $125,000, $25,000 is the down payment, and the remaining $100,000 is borrowed at 8%, then annual interest for the first year of the loan is (8%)($100,000) = $8,000. This interest expense can reduce the individual's taxable income. If the individual or family is in the 28% tax rate group, the tax savings is (28%)($8,000) = $2,240 for the first year. This amount is not paid to the IRS for the first year of the loan. As future payments are made to the lender of both interest and principal, the annual interest amount declines and future tax deductions decrease.

> The tax laws encourage home ownership. Sometimes it seems that all the grass cutting, painting, repairs, and other home upkeep duties give us a reason to question this "tax advantage."

The tax laws are similar for corporations with respect to loan interest expenditures. If equipment is purchased with borrowed funds, the interest on the loan can be used to reduce the taxable net income or profit before tax. The amount of tax that must be paid is reduced.

Example: Leverage at McDougal Corp. McDougal Corp. is a small manufacturing company in the electronic components supply market. It is considering expansion and requires some new equipment to update its manufacturing processes. The new equipment costs $200,000 and has an ROI of 25% before tax. John McDougal, the CEO, is trying to determine if the purchase should be made with internal funds currently invested in money market accounts or borrowed funds. He met with his bank loan officer and got a commitment from the bank of $200,000 on a five-year interest-only note at 10%.

Below is the Income Statement showing revenues and expenses for the new equipment at McDougal with and without the use of borrowed funds. The annual interest of the loan would be $200,000 × 10% = $20,000 per year for each of the five years. The principal is repaid in full at the end of the five-year period.

Projected Income Statement for McDougal

	Internal Funds	Borrowed Funds
Sales	$1,000,000	$1,000,000
Cost of Goods Sold	600,000	600,000
Gross Profit	$400,000	$400,000
Administrative and Marketing Expense	200,000	200,000
Depreciation Expense	40,000	40,000
Interest Expense	none	20,000
Profit Before Tax	$160,000	$140,000
Tax at 40%	64,000	56,000
Profit After Tax	$96,000	$84,000
Cash Flow from Operations	$136,000	$124,000

The internal funds alternative on the proposed Income Statement has an operating cash flow of $136,000 and the borrowed funds alternative, $124,000. Looking at the funds available for investment or distribution to shareholders, it would appear the internal funds alternative would be the preferred method. However, recall that if John McDougal borrows the funds for this purchase, *he still has $200,000 in money market funds* that could be used for other investments. Thus, $200,000 + $124,000 = $324,000 in funds could be available for additional investments during the next year. If this is a rapidly expanding organization, it will need all the funds for expansion it can obtain during the early years of its growth. The ROI of this expansion could be determined using TVM and the after-tax ROI is certainly higher (25%) than the 10% cost of the borrowed funds. If the ROI were not higher than the interest rate, it would be questionable whether the funds should be borrowed.

By using leverage—borrowing funds to finance investments with a higher ROI than the interest on the borrowed amount—McDougal is able to invest in equipment and expand the firm faster than if he had relied only on internally generated funds.

Leverage

Some use a general rule: If an investment has an ROI greater than the cost of the borrowed capital, the funds should be borrowed or obtained from the sale of securities.

OTHER CONSIDERATIONS

In investment analysis, it is important to consider after-tax ROI, total cash flow, and taxes. However, there are always other nontax and nonfinancial considerations that must be taken into account. Personal preferences, experience with past investments, employee safety, customer satisfaction, product and process quality considerations, and other nonfinancial factors are important considerations that may not be easily quantified. Tax regulations must be considered for equipment and other asset analysis, but there must also be room for qualitative factors in the decision-making process. The skilled manager, engineer, or technician uses not only analytical skills, but also experience and judgment to decide on investments into projects and assets.

The science of the time value of money and finance must be carefully combined with the art of experience and judgment to arrive at successful financial decisions.

Revenue and Cost Bias

As a final comment, it should be mentioned that after-tax analysis can only be as good as the data entered in the calculations. There may be a tendency to be overly optimistic about the revenues a project is expected to generate, and to understate the anticipated costs. The final after-tax decision must be based on the most accurate estimates.

SUMMARY

This chapter has presented the following concepts:

- Types of taxes and tax calculations for individuals and corporations.
- Tax exemptions and deductions.
- Taxes on dividends paid to shareholders and special types of corporations, including subchapter S and limited liability corporations.
- The flat tax.
- After-tax ROI calculation based on operating and investing cash flows.
- The Accounting Equation, purchase of equipment, depreciation entries into the Accounting Equation, and total cash flow.
- Calculation of operating cash flow for the organization, for individual projects and investments, and after-tax ROI.
- Steps of calculating after-tax ROI.
- Tax consequences of selling buildings, equipment, and other assets for less or more than their book value.
- Replacement analysis and operating cash flow.
- The Cash Flow Statement.
- Net operating cash flow, funding projects, leverage and loans by corporations and individuals, nontax considerations, and biased estimates of revenues and costs.

With the conclusion of the discussion of taxes, this section has covered analysis tools and the selection of projects and investments. The addition of after-tax analysis provides a set of tools that include Accounting Equation concepts, financial statements, time value of money, analysis and selection techniques, breakeven, minimum costs, and assets replacement analysis.

There are additional economic considerations. The next section covers the economics of quality, a relatively new topic in the economic analysis tool kit. Continuous financial improvement techniques are presented for monitoring and improving financial results. Economic analysis and improvement is a continuous process.

> I find all of the money a considerable burden.
>
> —John Paul Getty, Jr., 1985

QUESTIONS

1. List the different types of taxes and the corresponding amounts or percentages an individual pays.
2. List the different types of taxes and the approximate amount or percentage a corporation pays.
3. How are taxes calculated for the following:

   a. Partnership
   b. Proprietorship
   c. Corporation

4. What is meant by "double taxing" the corporation's profits?

5. What is one of the primary advantages of a subchapter S corporation?

6. What is an LLC?

7. What are the advantages and disadvantages of a "flat" or fixed percentage income tax rate?

8. Explain what is meant by the "tax advantage of depreciation."

9. What is an approximate method of estimating after-tax ROI from pretax ROI?

10. What are the Accounting Equation entries for the following:
   a. Purchasing a building.
   b. Depreciating equipment.
   c. Selling equipment.

11. What is meant by the statement, "Depreciation is a noncash expense."

12. What is the primary reason for a depreciation entry being made on the Income Statement?

13. What are the three categories of cash flow?

14. What is total cash flow as applied to a single project or investment?

15. Why is it important to consider total cash flow when making an economic analysis of building or equipment purchases?

16. State two ways that operating cash flow can be calculated if the cash account is not available.

17. Why should engineers, technicians, and managers understand operating cash flow?

18. Why is depreciation called an "add-back" to net income?

19. What are the steps in determining after-tax ROI?

20. What effect do different depreciation methods have on operating cash flow? What is the effect on after-tax ROI?

21. What is the tax implication of selling an asset that has been depreciated and that has a book value different from the selling price?

22. State the components of the three different categories on a typical Cash Flow Statement.

23. Why is it important for technicians, engineers, and managers to understand Cash Flow Statements?

24. What is net operating cash flow? How is it different from operating cash flow? How is it used?

25. What is meant by "the tax advantage of borrowing money"?

26. What is leverage? How can it affect total cash flow?

27. What is the "tax advantage" of funding projects with borrowed funds?

28. What are some nontax considerations when performing economic analysis on an after-tax basis?

29. What is meant by the statement, "Depreciation and taxes are not reversible"?

30. What is the relationship between operating cash flow and economic life?

31. Why are the Cash Flow Statement and net operating cash flow important to project managers, engineers, technicians, and shareholders?

32. What is meant by "revenue and cost estimating bias"?

33. How is buying a home a tax advantage for the individual?

34. How can lower tax rates contribute to higher tax revenues to the government?

PROBLEMS

1. A single individual has taxable income of $28,500. What is the federal income tax? What is the average effective tax rate?

2. A single person earns a taxable income of $23,750. What is the average effective tax rate?

3. A married couple, Jim and Karen, earn taxable incomes of $56,000. What is the federal tax amount and the average effective tax rate?

4. A corporation has net income before tax of $60,000. What is the federal tax amount and the average effective tax rate?

5. A corporation earns a net income before tax of $1,200,000. What is the federal tax amount and the average effective tax rate?

6. If a corporation has $500,000 in net income or profit before tax, what is the amount of tax owed?

7. If pretax ROI is 12.5%, and the combined total tax rate for the organization is 45%, what is the approximate after-tax ROI?

8. If a corporation has a combined effective tax rate of 42%, net income before tax of $1,600,000, and pays out 50% of its after-tax profits to shareholders, who have an average tax rate of 22%, what is the total tax amount and percentage paid by the corporation and stockholders?

9. The total average tax rate is 38% and the before-tax ROI is 16%. What is the estimate of the after-tax ROI?

10. A firm's revenue is $400,000 for the year, cost of goods sold is $300,000, depreciation is $50,000 for the year, and the average tax rate is 45%. Set up Income Statements with and without the depreciation expense included, and show the difference in profit after tax. Calculate the operating cash flow.

11. It is estimated that a new piece of equipment will save $30,000 per year in operating costs. The initial cost of the equipment is $80,000. It would have a 5-year life with no salvage value. The organization has a desired target of 10% after-tax ROI. Using the MACRS schedule of depreciation, calculate the annual operating cash flows and determine if the equipment meets the required ROI. What is the actual after-tax ROI of the investment? Tax = 40%.

12. In problem 11, what are the operating cash flow amounts and the after-tax ROI if straight-line depreciation is elected?

13. Show the Accounting Equation entries for the purchase of a building that costs $250,000, with a life of 19 years and salvage value of $30,000, using straight-line depreciation.

14. A chemical process is controlled with an outdated computer system. A new system of computers, software, and control devices can be installed for $150,000. It would have a life of 10 years and no salvage value. The new control system is expected to save $35,000 per year. Using a 10-year MACRS depreciation method, calculate the before-tax and the after-tax operating cash flow and the after-tax ROI. The tax rate is 40%. Use either manual or spreadsheet calculation.

15. New equipment would cost $175,000, and it is expected to save $35,000 per year. It has a 10-year life and a salvage value of $20,000. The target after-tax ROI is 8%. Straight-line depreciation is used. A effective combined tax rate of 45% exists. Does the investment meet the target ROI? What is the actual after-tax ROI?

16. Harrison Co. is currently spending $100,000 per year with a supplier that provides a necessary component. It is estimated that the firm could save $40,000 of the $100,000 per year in operating expenses if they made the component themselves. However, they would have to purchase equipment costing $125,000 and lasting 5 years with no salvage value. If straight-line depreciation is used, calculate the actual pretax ROI and after-tax ROI on the investment. Tax = 45%.

17. Three different methods of cleaning an electronic component are being considered. Each has an initial cost, savings per year, and salvage value. The desired after-tax ROI for the organization is 8%. Using straight-line depreciation and a combined tax rate of 40% determine the best alternative. The life of all the systems is expected to be six years.

Method	X	Y	Z
Initial Cost	$40,000	$50,000	$35,000
Annual Savings per Year	$5,000	$3,500	$10,000
Salvage	$8,000	$5,000	$3,000

18. An organization has an effective combined tax rate of 35%. It invests $74,000 initially into a new product that is predicted to have the following gross income per year before depreciation and tax. Depreciation is straight-line over five years with zero salvage. Determine the after-tax ROI for the new product. Tax = 40%.

Year	Gross Income and Savings
1	$25,000
2	$36,000
3	$30,000
4	$28,000
5	$17,000

19. New equipment costs $123,000 initially and has a salvage value of $20,000 after 10 years. Operating costs are expected to be $15,000 annually. The effective tax rate is 38%. The straight-line depreciation method is used. Savings are $30,000 per year. Should the investment be made?

20. A small test furnace is being purchased for the Reliability Testing Lab. There are two options under consideration. The Premier model has a first cost of $20,000, operating costs of $5,000 per year, and no salvage after five years. The Empire model costs $15,000 initially and has $7,000 per year operating costs and no salvage value after five years. The target after-tax ROI is 10%. Using straight-line depreciation and a tax rate of 40%, determine the best selection. Each saves $8,000/year.

21. A make or buy decision is being considered for a part that has been purchased from suppliers in the

past. The savings are estimated to be $15,000 per year. The equipment to produce the part would cost $55,000 and have a life of seven years and no salvage value at that time. The organization wants to earn 10% after tax on its investments. Determine the operating cash flow of the investment and if it meets the desired after-tax ROI if straight-line depreciation is used and the effective tax rate is 50%.

22. Two new products are being considered, Ping and Pong. Ping is estimated to have a revenue of $1,000,000 per year, cost of goods sold of $780,000 per year, depreciation of $80,000 per year, and a tax rate of 43%. Pong has an estimated revenue of $1,100,000 per year, cost of goods sold of $650,000, depreciation of $200,000 per year, and a tax rate of 43%. The initial investment for Ping is $400,000, and for Pong, $1,000,000. Both products and their processes have a life of five years and use straight-line depreciation with no salvage value. Using after-tax cash flow analysis, determine the after-tax ROI for each product.

23. A building has a book value of $48,000. It is sold for $55,000. If the tax rate is 30%, what are the entries into the Accounting Equation? What if the building is sold for $42,000? What are the required taxes?

24. A nondestructive test system can be installed for $50,000. It has a 5-year life with no salvage. Its operating costs would be $18,000 per year. Straight-line depreciation is used. The organization has a target of 8% after-tax ROI. An alternate solution is to send the components to be tested to an outside testing firm. The cost of the testing service is expected to be $35,000 per year. Determine if the purchase of the in-house system is recommended. Taxes are 45%. What is the actual after-tax ROI of the internal system?

25. Equipment that has a book value of $38,000 is sold for $42,000. If the tax rate is 28%, what is the tax consequence of the sale? What if the equipment could be sold for $48,000?

26. Harmon Manufacturing Co. just sold an old warehouse. It purchased the building 15 years ago for $100,000. The annual operating and maintenance costs, insurance, and taxes were $5,000 per year. The building was depreciated over a 30-year life using the straight-line technique and has a $10,000 salvage value The organization's tax rate has aver-

aged 45% of net income. The selling price of the warehouse is $230,000. What is the current book value of the building? If the firm must pay a tax on the gain when the building is sold, what is the after-tax ROI?

27. Equipment was purchased three years ago for $200,000 and is estimated to last four more years. Straight-line depreciation is used. Operating costs are expected to continue at $90,000 per year, and the salvage will be $10,000 at the end of four years. The current book value of the equipment is $105,000 and the market value is estimated to be $90,000. New equipment can be purchased for $250,000 with a life of eight years and a $20,000 salvage value. Operating costs would be $30,000. The tax rate is 48%. If the gross profit, before depreciation and taxes, is $200,000, determine the operating cash flows, and determine if the alternatives meet the desired after-tax ROI of 10%. Determine the actual after-tax ROI for the existing and the new equipment.

28. A small manufacturing firm is considering replacing its material handling system. A new system can be purchased for $250,000 that has a 10-year life and salvage of $20,000 at the end of 10 years. The annual operating costs are estimated to be $7,000 per year. The existing system was purchased five years ago for $200,000, and it is estimated to have a life of four more years and no salvage value. The current market value is $100,000 and the operating costs are expected to continue at $10,000 per year. Straight-line depreciation is used for the equipment. If the organization desires an after-tax ROI of 6% and taxes are 40%, determine if the new equipment should be purchased.

29. Prepare a Cash Flow Statement for Becom Co. given the following data. Present the statement in standard format using three categories. Parentheses represent a negative or cash outflow.

Depreciation	$40,000
Investment in Equipment	($50,000)
Changes in Current Assets	$12,000
Common Stock Sold	$20,000
Investment in Buildings	($55,000)
Profit After Tax	$85,000
Changes in Current Liabilities	($4,000)
Funds from Bank Loan	$40,000

30. It is estimated that a firm will have $50,000 per year in profit after tax into the future. The current long-term interest rate is 6%. The firm's stock is currently selling for $14.50 per share and there are 100,000 shares outstanding. Compare the current market value and the present value of the future operating cash flow of the organization.

31. A home costs $120,000. A potential buyer, using a down payment of $25,000 for 30 years at 8%, considering the tax advantage of purchasing the home. What is the advantage for the first year of ownership? The effective tax rate is 22%.

32. A married couple, Susie and Bob, have purchased a house that will be rented. The selling price of the house is $90,000 and a $20,000 down payment is made. It is estimated that the house has a value of $75,000 and the land is valued at $15,000. Only the house value is depreciated. The loan is for 30 years at a 9% fixed rate. Taxes, insurance, and maintenance are estimated to be $3,500 per year. Susie and Bob have an average tax rate of 28% on their net income. Straight-line depreciation with a 30-year life is used for the house. They plan on renting the house for $925 per month. If the house appreciates at a rate of 6% per year, and it is sold in 10 years, determine the following:

a. The monthly mortgage payment.

b. The total costs and revenues per year for the house.

c. The proposed Income Statement for the investment.

d. The estimated selling price of the house in 10 years.

e. The after-tax cash flow for the investment.

f. The tax owed at the 20% capital gain rate when the house is sold.

g. The after-tax ROI for the investment.

What other factors should Susie and Bob consider?

DISCUSSION CASES

Technical Software Co.

Jim Roberts and Bill Forbes have started a small software company. The software that they have developed has been presented to a number of potential buyers and has been well received. Jim and Bill are concerned about funding their new company's projects. Currently they work in Jim's basement, but they know that within a few months they will outgrow that space and add more employees. Also, they anticipate purchasing new computer equipment within two months. They plan to market their software nationally by advertising in trade publications, using extensive direct mail with sample software, and having booth space at regional and national trade shows.

They have talked with local banks concerning the possibility of short-term commercial loans. They have also considered obtaining second mortgages on their homes as a source of funds. Also, both have retirement funds from previous employers that they could liquidate to invest into their business. They have been approached by two investors who have offered to invest in the new organization in exchange for common stock. Although both Jim and Bill are optimistic about the prospects for their company, they are concerned that they might not have the funds to make the company successful if they only rely on the profits from the sales of the software.

1. How can Jim and Bill estimate how much capital they will need in the next 12 to 24 months? Suggest a format to estimate the amount of funds and the timing of the funds.

2. What are the different possible sources of funds for the company?

3. Identify the short- and long-run advantages and disadvantages of the possible sources of the funds from question 2.

4. How could cash flow and ROI for the next two years be estimated, including income, costs, depreciation, and taxes?

Trumble Restaurant Equipment Co.

Bill Peterson is a manufacturing engineer at Trumble. He is analyzing existing equipment to determine its efficiency and costs, and to compare it to newer equipment that is being considered for purchase. In the past all such analysis was done on a pretax basis. The information was then given to the finance department to determine the best alternative on an after-tax basis. Recently, the organization has been training its employees on its financial procedures. Topics included understanding the accounting process, cost accounting, financial statements, operating cash flow, taxes, and related topics. Bill has been asked by his boss, the plant manager, to make all of his recommendations directly to him, including all the operating cash flow and tax calculations that were made by the finance department in the past. Bill has all of the necessary information on his laptop computer, but he is not sure how to make the calculations for taxes, cash flow, and after-tax ROI.

1. How will Bill's analysis change with the new procedures?
2. How can his spreadsheet be modified to include operating cash flow, taxes, and after-tax ROI calculations?
3. Is this new analysis better without the finance department being part of the decision? Should they be included? What information can they add?
4. What other things should Bill consider?

Harrison Construction

Harrison Construction is a large civil engineering and construction company. They have approximately 185 different items of construction equipment including trucks, cranes, earthmoving equipment, specialized on-site mechanical equipment such as hoists, electrical generators, compressors, and related equipment. Linda Harrison, the daughter of the owner, has been assigned the task of determining a standard method of evaluating the purchase, replacement, and sale of equipment. Her father says that they spend a lot of time analyzing whether to purchase or lease equipment. He wants her to create a spreadsheet model that will take into account the costs of existing equipment, life, salvage, operating expense, taxes, depreciation, trade-in value, lease costs, interest on loans, return on investment, disposal costs, gain and loss taxes on sale of equipment, and all the other factors necessary to consider when making the keep, buy, or lease decisions.

1. Is it possible to have a comprehensive spreadsheet program that will include all the necessary information?
2. What assumptions and decisions will have to be made in order to implement the program?
3. Develop a spreadsheet that will provide the necessary calculations for Harrison.

Wynsum Corp. and Samules Corp.

Samules Corp. has been approached by one of its competitors, Wynsum Corp., with an invitation to buy the Wynsum firm. Wynsum has indicated that considering the past financial history, the firm's patents, and the projections for future profits, the company is worth at least the total market value of the outstanding stock, which is controlled by the Wynsum family that started the company. Currently there are 100,000 shares outstanding that are valued at $50 per share. The price of $5,000,000 ($50 × 100,000 shares) has been suggested for the value of the corporation.

Alan Samules, along with his company's engineers, managers, and accountants, has looked at the Wynsum financial statements, and technical analysis of Wynsum has been made by Samules engineers and technicians. Alan is considering Wynsum's current and future operating cash flow as an indicator of value. He feels that by assuming that the current operating cash flow (profit after tax plus depreciation) of $250,000 will continue into the future is a better way to value the company. He would use a pretax ROI that is equal to the current interest rate on long-term government bonds, which is $6\frac{1}{8}\%$, as a conservative estimate of future ROI. By considering the ROI and the operating cash flow per year, Alan determined the present value of the firm by dividing the estimated future operating cash flow of $250,000 by the $6\frac{1}{8}\%$ rate, to obtain $4,081,632 or approximately $4,000,000. Alan is considering offering Wynsum $3,500,000 for the firm. Wynsum strongly believes that since employees and family have recently purchased shares in the company at $50 per share, the value of the company is best determined by the total market value of the stock.

1. What should Alan Samules do?
2. Which is the best way to value the company, by market value or by present worth of the future cash flows?
3. Is the $6\frac{1}{8}\%$ rate a reasonable ROI for Alan to use? What if he assumes a higher rate?
4. What other factors should be considered?

New Office Building for Darden

Darden is a medium-sized engineering consulting firm. They are currently in rented space at three different locations. They want to combine all the locations into a single facility because they believe that will improve efficiency and will save rent. The building they are considering is a ten-year-old three-story building in a good location. The total square footage of the building is approximately 30,000 square feet, which is 10,000 square feet more than the current rented space.

The asking price for the new building is $1,250,000. The senior management is considering whether to obtain a mortgage to buy the building. They would be required to have a 30% down payment, 9% interest, 15-year conventional mortgage loan. Darden has been successful and has been placing profits into mutual fund investments over the past eight years; the value of the account is currently $800,000. The average ROI on the mutual fund investments has been about 12% pretax per year over the past eight years. The management is trying to decide if they should use the mutual funds to purchase the office building or use a combination of down payment with conventional mortgage financing. Some feel that the best approach is not to borrow funds in case there is a decrease in consulting contracts, and to use the firm's accumulated money in the mutual funds to purchase the building. Others believe that since the interest on the loan is tax deductible and would save taxes, which are currently at 35%, over the next 15 years, the loan alternative should be used.

1. What are the total cash flow considerations of the two alternatives?
2. What are the financial advantages of each alternative?
3. What nonfinancial factors should be considered by Darden?

Barbara and Hal's Duplex

Barbara and Hal currently live in an apartment and pay rent of $700 per month. They are considering purchasing a duplex home so they could live on one side of the duplex and rent the other half for $600 per month. The purchase price would be $150,000, and initial fix-up expenses would be $10,000. Taxes and insurance for the property would be $3,000 per year. The mortgage would be a 15-year fixed-rate loan at 9% based on a down payment of 25% of the purchase price. Their effective tax rate is 22%. They plan on living in the home for five years and then selling the property. They anticipate that the property will increase in value at 5% per year.

1. Estimate the annual operating cash flows for the ownership period.
2. What is the estimated ROI for the investment?
3. What other financial items should be considered?
4. What nonfinancial factors should be considered?

Taxes at Darrst Co.

Darrst Co., a retail sporting goods store, has begun to manufacture some sporting goods equipment on a small scale to see if they can sell the equipment in their stores. If successful, they plan on expanding by selling to other sporting goods retailers using manufacturer's representatives.

Margo, the owner and manager of the organization, has the firm's taxes prepared by a tax accountant. Now that the company is expanding into manufacturing, it will be purchasing equipment for manufacturing rather than only purchasing inventory for resale. She is trying to understand the tax implications of the new manufacturing activity. The accountant has said that there will be tax advantages for the depreciation of the new equipment. Margo understands that depreciation will be a cost on the Income Statement of the company, but doesn't see how this is an advantage.

She plans on purchasing the equipment using a line of credit that the firm has with a bank. The loan would be interest-only with the principal due at the end of five years, or sooner if Margo desires. Margo understands that the interest from the loan is a tax deduction also, but she isn't sure if that is an advantage.

As a successful retail company owner, she understands basic accounting and investment analysis, but she isn't sure how to do the financial analysis of a manufacturing company. With a retail store, she says, "You try to sell the purchased inventory for more than you pay for it, pay the labor and store costs, and hope that there is a profit left over at the end of the month." Now that she is involved with manufacturing equipment, depreciation, and taxes, Margo is having difficulty with her financial planning.

1. What is the difference between the financial planning of a retail company and of a manufacturing company?
2. How can Margo analyze the financial needs of the new manufacturing department?
3. Explain the concept of "tax advantage of the equipment purchase."
4. How is projected operating cash flow analysis different for a retail store and a manufacturing firm?

5. Would it be advisable for Margo to keep the financial planning and the financial results of the retail and the manufacturing departments separate or combined?

The Millionaires Club (F)

The Millionaires, a student investment club, does its own financial analysis of the firms in which it invests. They are considering the calculation of operating cash flow and after-tax income as an investment criterion since they read an article about a famous investor's technique of using net operating cash flow. Up to now they have used all the usual financial measures to evaluate the companies, but now they want to add net operating cash flow and tax analysis to their calculations.

They are considering estimating future net operating cash flows of a company and then calculating the present value of the future amounts. The present value would be compared to the market value of the outstanding stock. The market value is calculated by multiplying the market price times the number of shares outstanding. If the present value of the future value of the cash flows is less than the current total market value of the organization's stock, then the stock price may be thought of as a "bargain" price.

Some members of the group feel that the techniques they are using now, such as earnings per share, dividends, current price, market price trends, and related factors, are adequate. They feel that net operating cash flow and tax calculations are not applicable and too complicated for their analysis.

1. What are the problems of applying the calculations that are proposed?
2. Is the group right in their opinion that net operating cash flow or taxes are not applicable?
3. What are the advantages of considering net operating cash flow of the club's investments?
4. Is this approach conservative or will it encourage investing in many more firms?
5. What else should the club consider with their analysis?

IV CONTINUOUS FINANCIAL IMPROVEMENT

Chapter 12 Economics of Quality

> All good things are cheap, all bad are very dear.
> —Henry David Thoreau

KEY TERMS

Appraisal Cost The cost of inspecting, testing, supplier audits, gauge control, and inspection equipment.

Baldrige Award The Malcom Baldrige National Quality Award is awarded annually to manufacturing and service organizations in the U.S. The award recognizes both quality systems and economic performance.

Continuous Improvement A quality concept of not just meeting specifications, but making improvements in products and processes continuously. It emphasizes ongoing improvement of designs, processes, materials, and procedures, resulting in lower costs as well as improved quality.

Deming's Chain Reaction Improving quality reduces scrap, which reduces the product's or service's unit cost, and improves productivity by increasing the percentage of good output, which in turn reduces the selling price and improves quality to the customer, which expands the market and increases sales and the number of jobs.

External Failure Cost The cost of warranty, field failure, complaint analysis, product liability, and related costs of quality after the customer receives the product or service.

Goalpost Quality A good product or service is defined as anything between the upper and lower tolerances, and is not dependent on nearness to a nominal specification.

Internal Failure Cost The cost of scrap, rework, troubleshooting, and other defects found before shipping to a customer.

ISO 9000 An international quality system standard for quality documentation and other quality standards for products and processes. Currently, quality cost standards are not included, but they are being considered for future versions.

Pareto Distribution The presentation of data, usually in bar chart form, with the largest frequency cause shown first and the remaining causes in descending order. The reduction of quality costs can begin by determining the largest cost and its root cause, and then correcting the cause and reducing the cost. Then the process is repeated for next largest cost, and so on.

Prevention Cost The cost of preventing poor quality. It includes quality planning, design, and training costs.

QS 9000 A standard developed by U.S. automobile producers to monitor quality systems. It includes ISO 9000 requirements plus automobile-industry and customer-specific standards. It requires that producers have a quality cost system.

Return on Quality The measure of a quality characteristic, such as customer complaints or scrap costs, compared to the investment to attain the quality.

Taguchi Experiment A version of product and process experiments that can reduce the number of trials and costs with improved results. It relies on team approaches to process and product improvement rather than including all possible mathematical combinations of variables as does the traditional experiment design.

Taguchi Loss Function The cost or dollar loss from the service or product's failure anytime during its service life with the customer. The costs of quality over the product's life increase as the product characteristics move further away from nominal or target specifications.

Total Quality Cost to Sales Ratio The sum of the four categories of quality costs divided by the sales for a period. It may be as high as 20% or more for organizations that have little or no formal quality program.

Total Quality Costs The sum of prevention, appraisal, internal, and external quality costs for a period of time. This amount includes manufacturing, nonmanufacturing, and administrative quality costs.

LEARNING CONCEPTS

- Background of quality economics.
- Quality and cost relationships.
- Quality management, culture, and philosophy.
- Four categories of quality costs.
- Cost of quality to sales ratio.
- Quality cost reports and graphical presentations of quality costs.
- The Pareto distribution as a guide to reducing quality costs.
- Minimum costs and the ideal cost of quality.

- Nonmanufacturing and manufacturing quality as contributors to total quality costs.
- Return on quality.
- Implementing and monitoring a quality cost improvement program.
- Demonstrating Deming's chain reaction, scrap improvement, reduced quality costs, and improved productivity.
- ROI calculations of quality improvements by combining the chain reaction and TVM calculations.
- Rework or scrap decisions based on economic analysis.
- Goalpost quality, Taguchi's loss function, and economic analysis.
- Steps in improving quality costs.

INTRODUCTION

This final section of the book discusses the fourth component of the financial cycle, continuous financial improvement. Managers, design engineers, manufacturing engineers, or technicians do not limit their activities only to understanding and analyzing financial statements, time value of money, breakeven, minimum costs, and replacement. They also monitor and improve the costs of designs, processes, materials, and methods on a continuous basis. Economic analysis is not a one-time event, but an ongoing activity. This chapter considers the continuous improvement of financial results using quality economics.

Since the beginning of the factory system, organizations have pursued financial analysis and improvement using traditional cost analysis and return on investment techniques. Recently, the focus has been on the cost of producing poor or defective products as well. The economics of quality has become an important component of economic analysis and financial improvement tools for the organization. The financial rewards of successfully reducing the costs of producing a quality service or product by reducing or eliminating defective products or services may be large. Many organizations are improving their financial results significantly by focusing on quality improvement. Quality economics techniques have rewarded producers with lower costs and higher profits, and customers with lower prices and higher quality—a winning situation for both supplier and customer. Today, the costs of quality for some organizations can be as high as 10% to 25% of the organization's sales if there is no quality program in place. A potential reduction of quality costs to less than 5% of sales can result in the savings and profits going directly to the net income and/or to reducing prices for customers.

By focusing on measuring and improving the costs of quality, organizations can reduce quality costs to a small percentage of sales. This requires the effort of operating employees, quality specialists, design engineers, process technicians, and senior management. Improvements are the result of long-term efforts, which may require the organization to modify its culture, beliefs, and values in order to place importance on long-term customer expectations and quality rather than only on short-term cost reduction and profits. The potential returns are great, but there is no easy quick fix.

This chapter presents tools that can be used to measure and reduce quality costs. These techniques include quality cost measurements, quality ratios, Deming's chain reaction, time value of money, minimum costs, and Taguchi's concepts of quality and costs.

The challenge is not just in understanding the calculations of the economics of quality costs, but in the diligence an organization must use to apply these techniques over a long period of time. This long-term application effort requires understanding the quality cost concepts, and often revising long-established beliefs concerning customers, quality, and financial improvements. Often, the key to the survival and success of the organization in an internationally competitive market is its quality improvement efforts.

> Quality is free. It's not a gift, but it's free. What costs money are the *unquality* things—all the actions that involve not doing jobs right the first time.
>
> —Philip Crosby

THE BACKGROUND OF QUALITY ECONOMICS

During the 1800s and the beginnings of the factory system, production was expanding rapidly and profits were growing. Economic analysis and financial improvements were based on increased mechanization, increased productivity, and specialization. There was little emphasis on quality or the costs of quality. Products were produced, then inspected to determine if they met customer and design specifications. Defective products were scrapped or reworked and good products were shipped. Quality *detection* was emphasized, not *quality improvement* or *prevention* of poor quality. This approach continued into the twentieth century.

In the 1920s, Walter Shewhart at ATT's Bell Laboratories developed the quality concept and principles of statistical process control (SPC). His effort was driven by the desire to improve quality and reduce costs. His book *Economic Control of Manufactured Products* (1931) summarized the in-house quality training sessions that he presented at Bell Laboratories and Western Electric Company. The book's title revealed the concepts of linking quality with economic control. These concepts were successfully applied at Western Electric. Other organizations became aware of Shewhart's techniques, but did not generally adopt the SPC concepts. In the 1940s, during World War II, the Department of Defense required many military suppliers to use SPC to control and monitor the processes used to produce military products. After the war, organizations returned to producing consumer goods and abandoned SPC techniques. Quality was again maintained by inspection and detection rather than prevention. Cost reduction was implemented by increased mechanization, automation, methods improvements, and specialization, not by linking costs and quality.

Quality Economics Milestones

1920s—Walter Shewhart develops statistical process control as more economical method than inspection and detection of product quality at Western Electric Co.

1940s—SPC used to monitor and control production of military product manufacture.

1950—Edwards Deming and others visit Japan to give seminars on improving quality. Deming develops the chain reaction theory relating quality and costs.

1951—Joseph Juran writes *Quality Control Handbook*. Includes cost of quality concepts.

1961—A. Feigenbaum includes cost of quality as a significant part of his Total Quality Control approach to quality.

1961—American Society of Quality Control establishes Quality Cost Committee.

1963—Mil Std Q-9858A quality systems standards established by U.S. Department of Defense for military contractors. It contained a general requirement, "Costs Related to Quality," for an internal quality cost system to monitor and improve costs of quality by suppliers. Q-9858 became the basis for NATO quality standards, which became the basis for British Quality Standards. All of these standards evolved into the ISO 9000 series of standards.

1981—NBC airs the special "If Japan Can, Why Can't We?", a documentary about Japan's economic recovery to become a world-class producer of quality and economical products. Deming is identified as a major influence on Japan's management, quality, and economic success.

1980s—**Total Quality Management (TQM)** becomes the new paradigm for organizations. It includes customer-driven quality, team-based quality activities, continuous improvements, poor quality prevention, statistical applications, supplier monitoring and auditing, management participation, significant culture change of the organization, and economics of quality.

1987—**Malcom Baldrige National Quality Award** is created in response to Japan's Deming Award. The criteria consider both quality and financial performance measures.

1990—**ISO 9000** Quality Standard begins to become important in U.S. Current standard does not contain quality costs provisions. Planned versions will likely incorporate quality cost systems.

1994—**QS 9000** Quality Standard created by U.S. automobile producers using ISO 9000 standards plus automotive industry standards. It contained quality cost system requirements.

Today—Quality philosophy and techniques from manufacturing organizations are being applied in medical organizations, education, government agencies, nonmanufacturing and service organizations. Quality economics is a significant part of the overall quality activities in these organizations.

The financial yardstick in organizations that are committed to quality improvement and total quality is a cost of quality measurement and monitoring system. Quality economics is an integral part of any cost reduction or financial improvement system. It is one of the many tools that quality improvement teams can use to analyze, select, monitor, and improve processes, projects, and investments.

Cost of Poor Quality—Where the Money Is

After a notorious bank robber was captured, he was asked by a naive reporter, "Why do you continue to rob banks?"

His answer was, "Because that's where the money is."

Many organizations find that by focusing on the economics of quality, they can reduce costs and increase profits and ROI significantly.

Why is cost and economic analysis an important part of quality efforts today? Because that is where large savings can be discovered. An organization whose quality effort is limited to inspection of completed products may be spending as much as 10% to 25% or more of its sales dollars on poor quality. An organization that has a comprehensive quality effort in place for a period of time may be able to reduce the costs of quality to under 5% or even 2–3% of the sales revenue. The potential savings from quality activities is significant. By improving quality, financial rewards can be achieved as well as improved products and services.

WHAT IS QUALITY?

The concept of quality encompasses practically every aspect of a product's design, manufacturing process, delivery, and use by the customer. Quality may mean different things to different organizations and individuals.

What Is Quality?

Quality is fitness for use.
—J. Juran

Quality is conformance to requirements.
—P. Crosby

Quality is the total composite product and service characteristics of marketing, engineering, manufacturing and maintenance through which the product and service in use will meet the expectations of the customer.
—A. Feigenbaum

Quality is the totality of features and characteristics of a product or service that bear on its ability to satisfy a given need.
—ISO 9000

The quality of a product is determined by the minimum loss imparted to society from the time the product is shipped.
—G. Taguchi

In technical usage, the word quality relates the features and characteristics of a product or service to the ability of that product or service to satisfy stated or implied needs.
—ANSI/ASQ A-3 Quality

QUALITY ECONOMICS AND ORGANIZATION

The language of quality is statistics, but the language of management is money. If quality concepts are converted into money language, management understands the problems, alternatives, solutions, and results rather than trying to understand statistical concepts. The 1980s and '90s have seen shifts of management thinking from traditional techniques with roots in the 1800s to contemporary techniques including total quality. Today, the quality paradigm has permeated almost every organization. Quality is not justified only because, "It is the right thing to do." Quality efforts and expenditures are also analyzed, monitored, and measured in economic terms as are other equipment, product, and facilities decisions.

When economic analysis and quality techniques are combined, organizations can increase "partnerships" with suppliers, increase auditing of internal and supplier quality systems, reduce product unit costs, focus on preventing the cost rather than the detection of poor quality, increase customer satisfaction, increase productivity, improve office processes, reduce inventories, and improve profits and return on investment. All of these advantages yield significant financial benefits.

Quality Economics as a Competitive Strategy

In the late '70s and early '80s, the automobile and electronics industries were feeling the pressure from market share loss to Asian and European competitors who were producing low-cost and high-quality goods. Quality and management expert W. Edwards Deming's teaching and consulting in Japan during the 1960s and 1970s was "discovered" by U.S. organizations as a result of NBC's 1981 program, "If Japan Can, Why Can't We?" Ford Motor Co. brought in Deming as a consultant to advise about its quality. Ford began the long, difficult process of changing its culture by focusing on improved quality, supplier partnerships and audits, increased training, and other changes. Quality was at the heart of the new culture. Other organizations began to follow the quality path, and the culture and philosophy of American service and product providers change to meet the international competitors with improved quality and lower costs. The old concept that better quality was only available at a higher cost was replaced with the concept that by improving the quality of materials, processes, training, and management, costs could be reduced and profits increased. The economics of quality and the cost analysis of quality became a high priority.

Classification of Quality Costs

The first step in the economic analysis of quality is to define the cost categories of quality.

Prevention Costs. Prevention costs are the costs of activities designed to prevent defects in the product or service. The category includes:

Product and process design	Purchasing
Quality training	Market research
Customer surveys	Design reviews
Field trials	Supplier reviews and training
Operator training	Quality planning
Quality research	
Quality engineering	

For the organization that has very little quality effort, prevention costs are low, while for the organization that has a well-developed quality program, this cost category can be high. Even though this cost increases as more emphasis is placed on quality improvements and prevention, however, it is generally less costly to *prevent* poor quality rather than having to *correct* defective products after production. As prevention costs increase, other quality costs of appraisal, internal failure, and external failure typically decline. The result is that the **total cost of quality** is lower.

Appraisal Costs. **Appraisal costs** include the measurement and inspection of the service or product. Appraisal of quality for many organizations is the major quality effort. Rather than prevent poor quality, inspection determines quality after the product or service has been created. This could be called the inspection or sorting approach rather than the prevention approach to quality. This cost category includes:

Material, labor, and overhead for inspection

Inspection and test equipment

Quality software

Reliability and life testing

Inspection equipment and gauge repair and calibration

Supplier audits

Internal quality audits

Gauge control costs

Internal Failure Costs. The costs of scrap and rework of products and services that are found to be defective before they are delivered to the customer are part of the category of **internal failure costs.** For an organization that is dependent on inspection for its quality control, internal failure can be the largest cost component. This category includes:

Scrap	Rework
Failure due to design	Troubleshooting
Material failure	Reinspection after rework
Material review board	Storage and monitoring of rejected material
Disposal of defective product	Data collection and analysis
Redesign of product for quality	Downtime and lost production due to poor quality

This category of quality cost is the first one that most organizations think of and track when quality costs are mentioned. Frequently, in the early stages, the manufacturing department collects these costs because the accounting system is often not available to collect these data.

External Failure Costs. Included in the category of **external failure costs** are the costs of replacing defective products or services when the customer discovers the defects. This category includes:

Warranty cost	Returned goods
Recalls	Complaint investigations
Penalties	Field repairs
Customer claims	Product liability costs
Cost of lost future business	Legal costs

The most difficult cost to determine in this category is the cost of lost future business if the customer is dissatisfied and buys the product from a competitor in the future. Also, "bad news travels fast." Studies show that customers that buy poor quality services and products tell others, and more potential revenue is lost. We know that this occurs, but it is difficult to put a dollar value on this cost. This category of cost may be high for the organization that has minimal quality efforts. It is likely to be low and to approach zero for the organization that has a well-developed quality effort.

> What is the cost of low employee morale, the cost of unhappy customers, the cost of losing a good employee or customer, the cost of relying on 100% inspection, or the cost of high variation in the process? These costs are unknown and unknowable.
>
> —W. Edwards Deming

Why Measure Quality Costs?

There are a number of reasons for collecting, monitoring, and using quality costs:

- To prevent poor quality rather than detect it after it has been produced. Identifying costs focuses on the worst quality areas that provide the greatest opportunity for preventing poor quality and lowering costs.
- To reduce internal and external failures.
- To reduce appraisal costs by incorporating appraisal into the manufacturing process.
- To identify high-quality cost areas, determine root causes, correct and prevent future failures.
- To reduce product/service costs, which can amount to as much as 10–25% of sales.
- To improve customer satisfaction.
- To help quality management understand costs and make decisions.
- To isolate and identify problems that could be disguised when quality costs are included in other material, labor, and overhead costs.
- To justify expenses and investments for quality improvement.
- To obtain lower costs and higher productivity, which results in increasing market share.
- To monitor the progress and the evolution of the quality journey.

On an organization's financial statements, there are usually no cost of quality categories. These costs are embedded in other costs. Even when looking at the internal cost accounting statements, quality cost categories often cannot be found. As an organization becomes more conscious of quality and as its culture changes, it will likely want to "keep score" on how it is doing with its quality improvement system, and it will start using a quality cost system. Many organizations have some data on scrap and rework costs even if these data are incomplete. Fewer organizations have complete cost information on quality planning, inspection costs, and external failure costs.

Once the costs are collected regularly, efforts can focus on improving process and product quality. Measuring and collecting quality costs is often a result of increasing emphasis on quality by the marketing, technical, and manufacturing areas, and often does not originate in the financial or accounting areas. In some organizations, the manufacturing and quality departments are responsible for the collection and distribution of the quality cost data.

> In the U.S., about a third of what we do consists of redoing work previously "done."
>
> —J. M. Juran

Benefits of Measuring Quality Costs

What are some expected results from measuring quality costs? A survey by the firm Imberman, Imberman, and DeForest of 2,400 firms among 50 industries found that the average **ratio of the cost of failures to sales** was higher than the average of the net profit to sales ratio. A variety of industries were surveyed, including aluminum die casting, candy, printing, electronic components, appliances, industrial machinery, lawn equipment, plastics, and athletic equipment. *More money was spent on failures than was obtained in net profits.* Only internal and external failure costs were considered; prevention and appraisal costs were not included.

> Once the quality cost levels are determined, the opportunity for improvement may be obvious. It is not uncommon to find initial quality cost estimates ranging from 10% to 25% of sales. While direct comparison cannot be made, some manufacturing companies with extensive quality management/quality cost experience are showing total quality costs in the neighborhood of 2% to 4% of sales.
>
> —ASQC Quality Costs Committee, *Principles of Quality Costs,* 1986

An early 1987 Gallup survey found results similar to the above. Forty-four percent of the executives surveyed believed that their organization's percent of quality costs to sales was in the 0% to 4% range and that the average of all U.S. firms was in the 5% to 9% range. Later, results of analysis concluded that the actual percentages exceeded 20% of sales. Two-thirds of this cost is composed of failure costs; one-third includes prevention and appraisal costs of quality. Recent studies have confirmed that these ranges of percentages still exist in many organizations.

> ### Increased Profits from Quality
>
> Net income as a percentage of sales in many organizations is typically 5% to 10%. It is not unusual for the costs of quality to be in the range of 10% to 25% of sales.
>
> What if these costs of poor quality could be reduced or partially eliminated? What would this do to the organization's profits and net income?

Table 12–1   Quality Costs and Sales

Quality Cost Report
Technical Manufacturing Corp.
1st Quarter, 19xx

Prevention		
Quality Training	$80,000	
Design Reviews	$25,000	
Customer Survey	$13,000	
Supplier Training	$12,000	
Field Trials	$4,000	
Total		**$134,000**
Appraisal		
Inspection	$350,000	
Reliability Testing	$145,000	
ISO 9000 Audit	$40,000	
Gages Expense	$23,000	
Software	$5,000	
Total		**$563,000**
Internal Failures		
Scrap	$400,000	
Rework	$100,000	
Reinspection	$10,000	
Total		**$510,000**
External Failures		
Returns and Replacements	$150,000	
Warranty	$75,000	
Complaint Investigation	$30,000	
Recalls	$50,000	
Field Repairs	$70,000	
Total		**$375,000**
Total Quality Costs		**$1,582,000**
Sales for Period		$14,800,000
Total Quality Costs as a Percent of Sales		10.7%

Summary of Technical Manufacturing's Quality Costs for 1st Quarter

	$	Percentage
Prevention	134,000	8%
Appraisal	563,000	36%
Internal Failure	510,000	32%
External Failure	375,000	24%
Total	1,582,000	100%

Table 12–1   Quality Costs and Sales (continued)

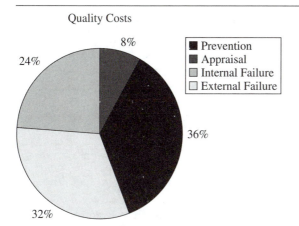

Each organization, as a result of its own quality efforts, has different quality costs and quality cost/sales ratios. Many organizations, particularly those with limited quality efforts, are surprised to discover the high costs of preventing, appraising, and producing defective products. "It is quite an attention-getting effort," as one director of manufacturing stated. "After pursuing one or two percentage point increases in many cost ratios and trying to increase ROI by a percentage point or two for many years and reducing unit costs by only a few pennies, finding the cost of quality is like finding a gold mine of potential savings." These quality savings are not limited to the manufacturing area. Many organizations discover significant savings in the office areas, including purchasing, accounting, marketing, and design engineering. The efforts to extract the savings from the quality cost gold mine are not easy, but may be very rewarding.

Efforts to reduce quality costs/sales ratios from approximately 20% or more to below 5% of sales can take years rather than just weeks or months. Is it worth it? Considering that 20% of the customer's dollar could be going toward quality efforts, it is a worthwhile activity. Of course the other benefits from improving quality are that the customers' expectations are met and exceeded, which promotes future sales and income for the organization. Customers are happier, costs are lower, and, as will be seen later, productivity increases when quality improves.

Quality Cost Reports

Reports from the monitoring of quality costs may be in a variety of formats. A typical report relating quality costs and sales is shown in Table 12–1.

The quality cost report for Technical Manufacturing Corp. shows the four categories of quality costs and components of each of the four costs. A percentage comparison of the total quality cost to the sales for the period, 10.7%, gives an overall measure of the magnitude of the costs of quality at Technical. Also shown are the four categories as a percent of total quality costs. A pie chart is useful for this comparison. The table and chart show the organization spends more than 10% of sales revenue on quality costs. Of this total quality cost amount, the largest portion ($563,000) is in the appraisal category, which is 36% of the total. Failure costs are 56% of the Total Quality Costs.

To improve quality costs, various strategies can be used. One approach is to focus on the largest of the quality costs, find its root cause, and correct it, which reduces total quality costs. Then select the next highest cost and continue the strategy of working on the largest component until all categories are reduced.

Another approach is to focus on the external failure costs first, since this is a customer-based cost and it is important to customer satisfaction. Even though this may not be the largest cost, when external failure root causes are determined, correction results in increased customer satisfaction. This is a customer-driven strategy with emphasis on customer-based quality problems.

Frequently, both of these approaches are implemented by quality improvement teams, which may include operating personnel, supervisors, quality technicians, design engineers, manufacturing engineers, vendors, and sales persons or purchasing staff.

Example: Scrap Costs at Technical Manufacturing Corp. The companywide quality cost report can be divided into departmental and individual process and product quality costs. For example, at Technical Manufacturing, the largest individual category on the report is the total scrap cost of $400,000. Scrap can be subdivided into the scrap of the individual departments. A summary of the department scrap costs for the quarter at Technical Manufacturing is shown below.

Technical Manufacturing Corp.

Department	Scrap Cost for 1st Quarter, 19xx
Machining	$95,000
Electronics	$60,000
Small components	$40,000
Assembly	$130,000
Casting	$75,000
	$400,000

Another useful way to display quality cost data is by using a **Pareto distribution,** which is a bar chart showing costs in descending order. (See Table 12–2.)

By presenting the scrap costs in a Pareto format, the quality engineers, managers, and technicians can see that the largest improvements can be made in the assembly department. This would be a likely place to start to find the root causes of the scrap cost problem. If improvements can be made to reduce future assembly department scrap costs below those of the other departments, then the next department, machining, would have the next largest scrap costs. This strategy would continually focus on the next-highest scrap cost area while holding the gains and improvements in the assembly department. In this way, team personnel and resources can be focused on the worst scrap area, improve/correct the cause, and then reduce the next worst problem. With limited money and personnel, the focus can be on those parts of the process that offer the highest savings and returns. This improvement process can continue until the scrap is no longer a major cost factor. Then efforts turn to another high-quality cost category such as inspection, which has costs of $350,000. This is a continuous activity of quality improvement and cost reduction.

Table 12–2   Scrap Costs by Department for First Quarter, 19xx at Technical Manufacturing

Assembly	$130,000
Machining	$95,000
Casting	$75,000
Electronics	$60,000
Small Components	$40,000
Total	$400,000

Pareto Distribution of Department Scrap Costs

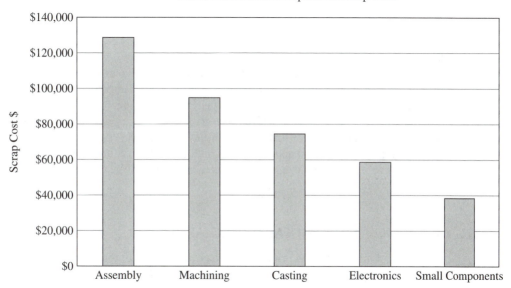

External Failure Costs at Technical Manufacturing. Another strategy is to attempt to reduce external failure costs. The basis for this approach is that the most important goal is customer satisfaction. Anything that detracts from that goal, such as failures after the product has been delivered to the customer, should be considered before other quality cost problems. If the root cause is found to be the most common problem in this cost category and it is corrected and improved, the quality costs should decrease and customer satisfaction should increase. The long-term goal would be to have zero costs of external failure.

MINIMUM COSTS AND IDEAL QUALITY

Discussing minimum costs, we saw that quality costs have a minimum cost relationship. The ideal relationship between the four components of quality cost, total quality cost, and the quality level of the organization's products is shown graphically in Figure 12–1. For a given quality

Figure 12–1   Costs of Quality

level and percent of good product, there are costs of prevention and appraisal and costs of internal and external failures. At poor levels of quality, organizations typically have low prevention and appraisal costs and high failure costs. As the quality improves, prevention and appraisal costs usually increase at a slower rate than the failure costs decrease.

Ideally, at the perfect quality level, there is 100% good product and the internal and external failure costs are zero. At higher levels of quality, the total cost of quality is primarily composed of the prevention and appraisal costs and the cost of failures is small. Ultimately it may be possible to reduce the prevention and appraisal costs further, which will decrease total quality costs.

QUALITY COSTS IN NONMANUFACTURING AREAS

Traditional thinking in some organizations has assumed that product quality is the only quality issue in the organization. However, many organizations find that significant costs are associated with nonmanufacturing quality functions. Errors in invoices, mistakes in purchase orders, omissions in quotations, mistakes in internal reports, errors in engineering design calculations, and accounting errors are examples of poor quality in the office. This is often referred to as *white collar quality*.

Ritz-Carlton and American Express, winners of the Baldrige Award, demonstrate the importance of measuring, monitoring, and reducing quality costs in the white collar area of the service industry. Thermo-King, a Westinghouse company, found that white-collar rework costs were greater than the rework and scrap costs in the manufacturing area. It also discovered that white collar quality problems caused a "multiplier effect" in the manufacturing areas. Mistakes made in the office area cause excess manufacturing inventories, increased manufacturing wait time, low value added, long deliveries, and increased downtime, which contributed to high quality costs in the manufacturing area. Correcting quality problems in the office can often result in improved quality and operations in the manufacturing area, as well as significant cost savings.

QUALITY COST RATIOS AND PERCENTAGES

Measuring quality costs can require a variety of ratios and percentages. The measures can be organizationwide, or at the department, process, or product level. Some commonly used ratios include:

$$\frac{\text{Total Quality Costs}}{\text{Sales}}$$

$$\frac{\text{Prevention Costs}}{\text{Total Quality Costs}}$$

$$\frac{\text{Appraisal Costs}}{\text{Total Manufacturing Costs}}$$

$$\frac{\text{Department Scrap} + \text{Rework}}{\text{Total Department Costs}}$$

$$\frac{\text{Purchased Components Appraisal Costs}}{\text{Total Purchasing Costs}}$$

These percentages can be customized for any organization's quality effort. As with all measures, whatever is measured is usually what is improved. Unproven quality measures should be applied prudently. The application of a few critical measures are generally more effective than many unnecessary ones.

Another possible quality ratio is analogous to return on investment. **Return on quality (ROQ)** is the ratio of a customer-based measure, such as customer satisfaction or reduced field failures, compared to the investment in a quality improvement for a period of time.

Quantitative values may include customer complaints, net sales resulting from quality improvements, responses on quality surveys, and other measures. The cost of quality investment can be measured in dollars. A ratio of complaints per sales dollar, complaints per quality cost dollar invested, for example, can show trends in complaint data over time.

Customer satisfaction can be a combination of many factors, including survey results, increased sales, decreased product returns, and other external quality cost measures. This sum or combined value can be compared to the cost of the investment in quality for a period of time. This ratio is not as easy to derive from values as other ratios, but it can be a guide for quality personnel and management to use in measuring the result of investments in quality. It can be applied to individual products and processes as well as to the organization.

$$\text{Return on Quality} = \frac{\text{Customer Satisfaction}}{\text{Quality Investment}}$$

IMPLEMENTING A QUALITY COST IMPROVEMENT PROGRAM

Because the potential for financial benefits is great, a quality cost and improvement program may be implemented to pursue both increased profits and improved quality simultaneously. By starting with the monitoring of costs due to scrap, rework, and inspection, a department or organization can begin to measure its quality costs. These costs do not have to be reported by the accounting department; they can be estimated by the manufacturing department. Often the accounting department is reluctant or unable to provide the data in a timely manner. This should not be a deterrent to the organization that wants to improve its quality and reduce its costs. Once the costs of scrap, rework, and inspection are monitored, additional data from customer-located failures and quality-planning costs can be included and then improved. The cost of a quality system can grow over time. Although the tasks of data collection, measuring, improving, failing, and then trying again may be difficult for improvement teams, the benefits are usually worth the efforts.

Quality costs are long-term activities that cover years, not months. Typically, it took the organization a long time to get into the poor quality condition, and it will take time to correct the damage. There is no quick fix, but many long-term potential benefits. Also, remember Deming's admonition that 85% of the quality problems are caused within the management system and only 15% at the operator level. For quality costs to show significant improvement, management must provide the leadership and resources for improvement of the system.

> Defects are not free. Somebody makes them, and gets paid for making them.
> —W. Edwards Deming

Measuring Quality Inputs and Outputs

For a traditional organization, the emphasis is on the quality results: appraisal, inspection, and detection of defects, rather than prevention. "Make the product and then inspect it to see if it meets the specification" is the method. This is measurement of the *output of the process*. The contemporary organization should emphasize the *prevention of poor quality* by spending more effort and money in the planning and improvement of products and processes. By monitoring and improving the inputs into the process, the quality of the output is improved. Generally, inputs to the process include materials, personnel, equipment and facilities, methods and techniques, and environmental inputs. The change in focus from monitoring outputs to inputs requires a significant change in the leadership attitudes, beliefs, and values from "bottom-line thinking" or "management by the numbers" to process-based management. Once the change occurs, quality and quality costs can often improve rapidly.

Quality Cost Improvements

After a specific quality cost problem is identified, the possible causes can be determined, the most likely cause is selected, and a solution is tried. If no improvement occurs, the team can go back to other possible causes and try again. It is valuable to use team members from many areas of the company. A cross-functional team has many different skills to use in solving the problem.

The steps for cross-functional quality improvement are:

1. Form a team of individuals related to the problem from throughout the organization.
2. Collect quality cost and financial data.
3. Use brainstorming techniques to identify possible root causes of the problem.
4. Group them into related causes and identify the most likely causes.
5. Correct them or make recommendations for correction.

6. Implement changes.

7. Collect quality cost data to see if the solution worked.

8. Repeat the process if more improvement is required.

If the solution is acceptable, disband the team and move on to other quality cost problems.

As the firm's quality system matures and quality improvements begin to take hold, the quality costs to sales ratio can eventually drop to below 5%. Experience indicates that it may take four to six years or longer for this reduction to occur, even with concentrated efforts toward quality improvement. A typical graph showing the four quality cost categories over two years for Technical Manufacturing Corp. is shown with Table 12–3.

As Technical Manufacturing Corp. begins to emphasize quality, the prevention costs increase. Appraisal costs may initially increase as more emphasis is placed on inspection and quality. The benefits of these increased costs are reductions in internal and external failure costs. Continued emphasis on quality reduces total quality costs. Monitoring the quality costs continues. This provides feedback to managers, engineers, and other personnel involved in the continuous quality effort.

DEMING'S CHAIN REACTION

W. Edwards Deming is well known for his quality management philosophy and his Fourteen Points, and for his belief that 85% of quality problems are in the system and can only be changed by management, with the other 15% the responsibility of the operator of a process. Another of his important concepts has to do with the relationship between costs and quality, what he called the **chain reaction**. He believed that cost reduction and productivity improvement programs by themselves are "doomed to fail" unless other changes occur in the organization. Experience often supports his viewpoint. Any improvements in productivity and decreases in costs are usually short lived unless other changes are made in the organization's culture and methods. A major root cause of low productivity and high costs can be found in the quality of products, processes, and services, he believed.

Traditional management has pursued cost reduction and increased productivity. Many times these efforts are unsuccessful or short-lived. Employees say, "Oh, here comes another cost reduction program, here's a new technique to increase productivity—this won't last long. Management will be off in another direction soon. It's just a fad." And many times the employees are right. Deming stated that the reason why a cost reduction or productivity plan did not work was that the process and quality had to be considered first, and then costs and productivity would improve.

Deming believed that the first steps are to improve quality through changing the organization's culture, using **continuous quality improvement** techniques based on statistical process control principles. Then, as the quality improves, scrap and rework costs are reduced, giving higher output for the same input, resulting in increased productivity. Costs decrease as scrap and rework decrease. These reduced costs can be passed on as reduced prices to the customer, higher wages, increased investment in equipment, and/or increased dividends to shareholders. Customers then have a higher quality product at the same or lower price, which causes sales to expand and the organization's market share to increase. Finally, the firm expands, hires more employees, and the chain reaction is repeated. (See Figure 12–2.)

This chain reaction is an important part of Deming's philosophy. Often companies find his recommendations too radical. However, once the chain reaction is understood and put into context,

Table 12–3   Quarterly Quality Costs for Technical Manufacturing Corp.

	1	2	3	4	5	6	7	8
Prevention	$134,000	$145,000	$160,000	$180,000	$185,000	$200,000	$190,000	$195,000
Appraisal	563,000	590,000	600,000	570,000	555,000	545,000	535,000	520,000
Internal Failure	510,000	480,000	460,000	450,000	435,000	380,000	360,000	350,000
External Failure	375,000	345,000	305,000	280,000	250,000	230,000	180,000	160,000
Total Costs	$1,582,000	$1,560,000	$1,525,000	$1,480,000	$1,425,000	$1,355,000	$1,265,000	$1,225,000

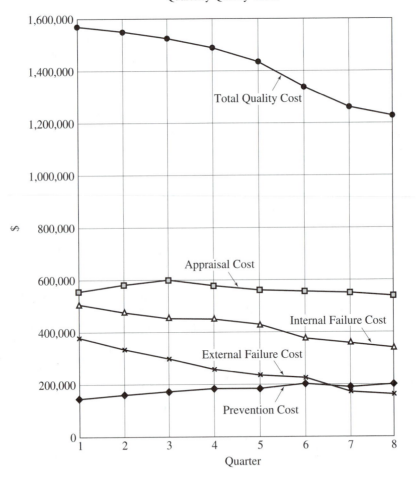

the philosophy becomes more understandable. The examples that follow are based on Deming's chain reaction.

The question is often asked, "How much does it cost to improve quality? Could the cost of improving quality be greater than the benefits?" Generally, as the quality becomes a high priority and part of the organization's culture, the cost of prevention or appraisal may go up in order to attain the lower internal and external failure costs. However, the usual result is that the

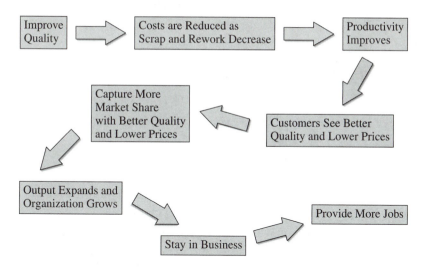

Figure 12–2   Deming's Chain Reaction

costs of the failures decrease faster than the increase in prevention and appraisal costs. This is especially true for the organization with high total costs of quality that is just beginning to expand its quality measurement and improvement activities.

This is a dynamic relationship. As more effort is put into quality, some costs increase and others are reduced. Once failure costs are low and the organizational culture has changed, prevention and appraisal costs can be lowered. This can result from increased quality from suppliers, team approaches to quality improvements, or management emphasis on the culture of quality, as well as from specific quality improvement techniques such as SPC or design of experiments.

Examples of Quality Improvement Techniques

Technical Manufacturing Corporation and the Chain Reaction. At Technical Manufacturing, the largest cost component on the quarterly quality cost report is $400,000 for scrap (Table 12–1). Of this amount $130,000 is found in the assembly department. The quality improvement team in the assembly department has been working on reducing the existing scrap rate of 14%. They have found that causes of the assembly scrap cost include defective material coming from internal suppliers, inadequate training of existing and new assemblers, worn fixtures, errors in the assembly instructions, inaccurate test equipment, and lack of timely supervision. The production rate in the assembly department is 20 units per hour, and the unit cost of the assemblies is $8 per unit. The material flow chart of the assembly department is shown in Figure 12–3.

The assembly department is on a schedule to produce 100,000 units per quarter. With its existing current scrap rate of 14% of the incoming material, it must work on a total of 116,279 units per quarter to obtain the required 100,000 good units. Calculation of this incoming quantity is shown below.

Incoming units = Good outgoing units + Bad outgoing units
Incoming units = Good outgoing units + (14%)(Incoming units)
Incoming units (1−.14) = 100,000 Good outgoing units
Incoming units = 100,000/.86 = 116,279 units required

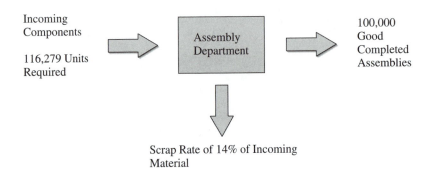

Figure 12–3   Technical Manufacturing Corp. Before Quality Improvement

The extra production of 16,279 units per quarter or 65,116 units per year is required because of scrap. If the quarterly value of the scrap produced is $130,000 (16,279 units × $8/unit), any reduction in the scrap rate in the assembly department would reduce these costs.

The quality improvement team has been able to suggest improvements that will reduce the scrap rate from the present 14% to 8% (see Figure 12–4). The revised required incoming units are calculated in a similar manner as above:

$$100,000/(1 - .08) = 100,000/.92 = 108,696 \text{ units required}$$

If the improvements are implemented and the scrap rate can be reduced to 8,696 units per quarter, the savings will be

$$116,279 - 108,696 = 7,583 \text{ units per quarter}$$

At a unit cost of $8, the savings in scrap cost is

$$(7,583 \text{ units per quarter})(\$8.00 \text{ per unit}) = \$60,664 \text{ per quarter}$$

which is $242,656 per year. This savings could be kept as increased profits for Technical Manufacturing; used to reduce prices to customers; and/or reinvested into the organization as increased wages or new equipment.

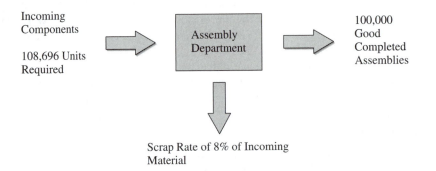

Figure 12–4   Technical Manufacturing Corp. After Quality Improvement

If the production rate is 20 units per hour, the total hours required to produce the 116,279 units originally was:

116,279 total units/20 units per hour = 5,814 hours per quarter

To determine the productivity based on the number of good units produced,

Productivity = 100,000 good units output/5,814 hours input = 17.2 good units per hour

After the improvement, the incoming units are reduced to 108,696 per quarter, and the total number of hours to produce this quantity is:

108,696 total units/20 units per hour = 5,435 hours per quarter

The improved productivity based on good units produced is:

100,000 good units output/5,435 hours input = 18.4 good units per hour

This is almost a 7% increase in productivity—(18.4 − 17.2 units per hour)/17.2. The hours saved each quarter permit the assemblers to work on other products. Or a greater quantity of the current product could be produced in the same time if there is a market for it. In either case, less raw material and labor are used to produce the 100,000 units. *By improving quality, costs are reduced and productivity is increased.*

This Technical Manufacturing example demonstrates Deming's chain reaction. Is it really possible to reduce scrap rates from 14% to 8%? It is, if concentrated effort is made by the improvement team and if the team has the support of senior management. Organizations are doing it regularly. As the team becomes more skilled at quality and process improvement, even further gains are possible with additional cost reduction and productivity benefits. When this type of quality improvement occurs throughout the company, not just in the assembly area, even larger savings and benefits are possible.

Quality Improvements at Sunrise Medical Products Corp. For one product at Sunrise, the VS-10, 10,000 finished units are required per year. The scrap rates and material quantities are shown on the material flowchart in Figure 12–5.

Department D is the last step in the process for producing the VS-10. If 10,000 good finished components are required and department D has a scrap rate of 9% of the incoming material, the required incoming material for department D is calculated as:

Total incoming product = Good product out + Scrapped product out
Total incoming product = 10,000 Good units out + (9%)(Incoming product)
Incoming product − (.09)(Incoming product) = 10,000 Good outgoing product
Incoming product (1 − .09) = 10,000 Good outgoing product

$$\text{Incoming product} = \frac{10,000 \text{ Good product}}{1 - .09} = \frac{10,000}{.91} = 10,989 \text{ Incoming product for D}$$

If 10,989 incoming units are required for department D, this must also be the *output* quantity of department C. The incoming product of department C can be found by dividing the required output by 1 minus the scrap rate, similar to the department D calculation.

$$\frac{10,989 \text{ Outgoing units C}}{1 - \text{Scrap rate of C}} = \frac{10,000 \text{ Units}}{.84} = 13,082 \text{ Units into dept. C}$$

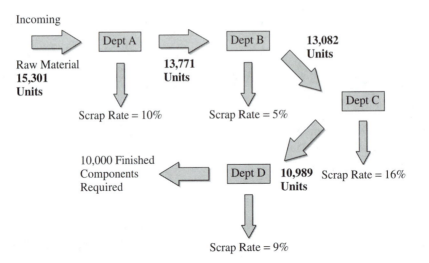

Figure 12–5   Sunrise Medical Corp.—Material Flowchart Before Quality Improvements in Product VS-10

The input to department C is the output from department B, and the input to department B is calculated in a similar manner,

$$\frac{13,082 \text{ Outgoing units B}}{(1 - .05)} = 13,771 \text{ Units into dept. B}$$

Continuing with department A,

$$\frac{13,771 \text{ Outgoing units A}}{(1 - .05)} = \frac{13,771}{.95} = 15,301 \text{ Units raw material into A}$$

From the above calculations, in order to produce 10,000 good finished products, raw material for 15,301 units must be ordered from suppliers. This is more than 150% of the required 10,000 good units output. In addition, labor must be expended in each of the four departments to produce defective units and units that will be scrapped in downstream departments. For example, department B must work on 13,771 units to produce 13,082 for department C, which in turn will scrap 16% of its production as it supplies 10,989 units to department D.

Given this current scrap situation for product VS-10 at Sunrise, and after much work, the quality improvement team is able to reduce scrap by the following amounts.

Product VS-10 Department Scrap Rates

Department	Incoming Material Units	Current Scrap Rate	Current Scrap Cost per Year	Improved Scrap Rate
A	15,301	10%	$35,000	4%
B	13,771	5%	$18,000	3%
C	13,082	16%	$56,000	7%
D	10,989	9%	$20,000	4%

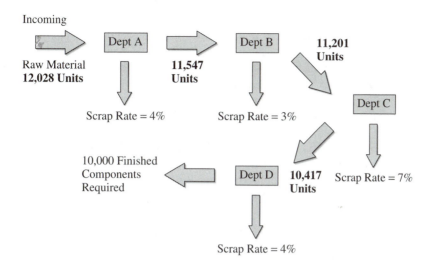

Figure 12–6   Sunrise Medical Corp.—Scrap Rates and Quantities After Quality Improvements in Product VS-10

The reduced scrap rates in the four departments will result in lower material and labor costs.

The revised flowchart for the VS-10 product (shown in Figure 12–6) shows the improved scrap rates and the resulting reduced material requirements. The estimated improved scrap costs, assuming that the new scrap costs are reduced in an amount proportional to the material requirement reduction, are shown in the table below.

Reduced Scrap Costs for VS-10 Component as a Result of Improved Scrap Rates

Dept.	Original Incoming Dept. Quantity— Units (2)	Improved Incoming Dept. Quantity— Units (3)	Original Scrap Cost (4)	Estimated New Scrap Cost— Col. 4 times (Col. 3/Col. 2) (5)
A	15,501	12,028	$35,000	$27,158
B	13,771	11,547	$18,000	$15,093
C	13,082	11,201	$56,000	$47,948
D	10,989	10,417	$20,000	$18,959
		Total	$129,000 per Quarter	$109,158 per Quarter

The improved scrap rates result in an estimated scrap cost savings of $19,842 ($129,000 – $109,158) per quarter, which is almost $80,000 per year.

Sunrise demonstrates the concept of Deming's chain reaction. Improvements in quality and scrap reduction result in lower costs and improved productivity for the manufacturing process.

Productivity increases for each department could be calculated similar to the method used for Technical Manufacturing. Additional improved customer satisfaction for both internal and external customers is also part of the chain reaction benefits. Benefits to the customer are more difficult to measure, but as suggested by Deming, increased customer satisfaction would result in increased sales, market share, output, and expansion of the organization. *Quality is the beginning of the chain reaction* for reducing costs and improving productivity. Quality improvements are only possible with some expenditure of money to implement them. Economic justification of quality improvement expenditures is demonstrated in the following example.

Foxrun Corp.—Justifying Quality Improvements. Linda Farrel, quality engineer at Foxrun, and the quality improvement team are working on a process that is currently producing 15% scrap at a cost of $1.50 per unit. The annual production requirement is 11,500 units per year. After extensive work over two months by a quality improvement team (spending 221 labor hours at a cost of $9.00 per hour), the scrap rate is reduced to 10%. Linda wants to determine the return on investment. It is anticipated that the product and process will have a life of three more years.

The current cost of scrap is:

$$\text{Raw material needed} = \frac{\text{Output}}{1 - \text{Scrap rate}} = \frac{11,500}{.85} = 13,529 \text{ Units of raw material}$$

The revised raw material requirement after scrap reduction:

$$\text{New raw material requirement} = \frac{11,500}{.90} = 12,778 \text{ Units}$$

Savings in raw material:

$$(13,529 - 12,778 \text{ Units})(\$1.50/\text{Unit}) = \$1,126 \text{ per year}$$

The savings of $1,126 per year is a result of the initial investment of 221 hours of the quality improvement team's effort.

The cost of the improvement team is:

$$(\$9.00/\text{hour})(221 \text{ hours/yr}) = \$1,989$$

The TVM diagram is shown in Figure 12–7.

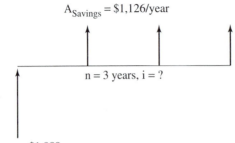

$A_{\text{Savings}} = \$1,126/\text{year}$

$n = 3 \text{ years, } i = ?$

$P_{\text{Team Cost}} = \$1,989$

Figure 12–7   Foxrun Corp.'s Quality Improvements

Solving for the unknown i,

$$P = A(P/A, n = 3 \text{ yr}, i = ?)$$
$$\$1,989 = \$1,126 \ (n = 3, i = ?)$$
$$P/A = 1.7664 \text{ at unknown } i \text{ for 3 years}$$
$$i \text{ is approximately } 32\%$$

If Foxrun's target ROI is less than 32%, this project is acceptable. If Foxrun's combined tax rate is 48%, the after-tax ROI can be approximated as equal to

$$\text{ROI}_{\text{Before Tax}}(1 - t) = (.32)(1 - .48) = 17\%$$

Process Improvements at Foxrun Corp. Foxrun has been conducting a Taguchi-type experiment (discussed later in the chapter) to improve another of its processes by decreasing the variation in the incoming material and thus lowering the outgoing product's scrap rate. The quality improvement team, composed of Linda Farrel, the quality engineer, equipment operators, process engineers, product designers, quality technicians, and purchasing personnel, run an experiment and find that the material from one supplier results in a lower scrap rate. The cost of running the experiment is 110 labor hours at $12 per hour and lost production during the experiment of 10 equipment hours at $300 per production hour. Even though the other suppliers are meeting the specifications, this supplier has a lower variation in their product and the yield from the machine is higher when this material is used. But the preferred material is $1.80 per unit more expensive than the current material. The current material cost is $39.50 per unit. The required annual output is 4,550 units per year. The scrap rate by using the new supplier decreases from 7% to 2%. If the firm desires a 12% ROI and the process has six more years to run, Foxrun needs to determine if the change in suppliers is warranted.

Current scrap cost:

$$\text{Current incoming raw material requirement} = \frac{4,550 \text{ Units}}{(1 - .07)} = 4,892 \text{ Units}$$

$$\text{Revised incoming raw material requirement} = \frac{4,550 \text{ Units}}{(1 - .02)} = 4,643 \text{ Units}$$

Cost of existing raw material:

$$(\$39.50/\text{unit})(4,892 \text{ units}) = \$193,234 \text{ per year}$$

Cost of new raw material:

$$(\$39.50 + \$1.80) \text{ per unit } (4,643 \text{ units}) = \$191,756 \text{ per year}$$

Process savings from using new material:

$$\$193,234 - \$191,756 = \$1,478 \text{ per year}$$

Team cost of the experiment to improve the process:

$$\text{Team cost} + \text{Lost time cost} = (\$12.00/\text{hour})(110 \text{ hours}) + (\$300/\text{hour})(10 \text{ hours})$$
$$= \$1,320 + \$3,000 = \$4,320$$

The TVM diagram for the project is shown in Figure 12–8.

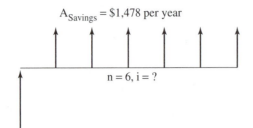

Experiment Cost = P = $4,320

Figure 12–8   Foxrun Corp.'s Process Improvement

$$P = A(P/A, n = 6, i = ?)$$
$$\$4,320 = \$1,478 \ (P/A, n = 6, i = ?)$$
$$P/A = \$4,320/\$1,478 = 2.9229$$

Solving for the unknown i results in an $\text{ROI}_{\text{Before Tax}}$ of approximately 25%. If Foxrun has a desired target ROI of 12%, this project is acceptable.

REWORK OR SCRAP?

Calculations of ROI from quality improvements are based on the reduction of scrap, which in turn reduces material, labor, and overhead costs. There is another decision to be made when scrap is produced: either sell the scrap for its salvage value or rework the defective product into a good product. Typically, defective products are collected and then at the end of the week or month, a decision is made to sell or rework them. These rework decisions are usually made by the manufacturing, quality, or manufacturing engineering personnel, rather than the cost accounting personnel. How can this economic decision to rework or scrap the product be made?

No Salvage Value

A product sells for $11 per unit and has a material cost of $2/unit, a labor cost of $3/unit, and overhead costs of $5/unit for a total cost of $10/unit. There is $1 per unit profit, and salvage value is zero. What should be the *maximum amount* to spend on the defective units to rework and correct the defect so the product can be sold for $11?

From Table 12–4, we see that if the product is disposed of *without any rework,* the loss is −$10 of the costs invested up to this point. This is shown in the "No Rework" column.

Rework cost + Original cost = Total cost of reworked unit

What if the rework costs are −$5? Now the total costs are −$10 − $5 = −$15, the revenue is +$11, and the loss is −$4. This loss of −$4 is preferable to a loss of −$10 when there is no rework.

What if the rework costs are −$10? Now the cost is −$10 − $10 = −$20 and the selling price is +$11, for a loss of −$20 + $11 = −$9. This is still better than the no rework alternative of a loss of −$10.

Table 12–4 Rework or Scrap?

No Scrap Value Unit Cost	No Rework (Dispose)	$5 Rework	$8 Rework	$10 Rework	$11 Rework	$15 Rework
Material	$2	$2	$2	$2	$2	$2
Labor	$3	$3	$3	$3	$3	$3
Overhead	$5	$5	$5	$5	$5	$5
Total Cost	$10	$10	$10	$10	$10	$10
Rework Cost	None	$5	$8	$10	$11	$15
Total Cost after Rework	$10	$15	$18	$20	$21	$25
Selling Price	Dispose	$11	$11	$11	$11	$11
Profit (Loss) after Rework and Selling for $11	($10)	($4)	($7)	($9)	($10)	($14)
	***				***	

*** Same loss for a product disposed with no rework expenditure as for a product that has $11 of rework costs.

Up to the point where the rework cost is as high as the selling price of the product, it is economical to invest additional rework dollars.

What if the rework costs are −$11 (equal to the selling price)? This would give a total cost of −$10 − $11 = −$21 and a loss of −$21 + $11 = −$10. This is the same net cost as when the defective product is thrown away at a cost of −$10 per unit.

> As a general guide, it is economical to spend rework costs up to the amount equal to the selling price of the product if there is no salvage value.

Salvage Value Is Not Zero

What if there is a salvage value of the defective unit? Continuing the above example, what if the defective unit could be sold to the scrap dealer for $3 per unit?

In this case, when the defective unit has a +$3 scrap value, the sale of the defective unit without rework would yield a revenue of +$3 and a cost of −$10 or +$3 − $10 = −$7 net loss when the defective unit is sold for its salvage value. Now, what is the *maximum to spend on rework* when the part has a salvage value of $3 per unit?

It can be shown that if a maximum of the selling price of a good product less the salvage value is spent on rework, it is economical to rework the unit.

Selling price of good product − salvage value of defective unit = maximum to spend on rework
+$11 − $3 = +$8

This is the maximum amount, $8, to spend on rework if the unit has a $3 scrap value.

It is not economical to rework the units if the rework cost exceeds $8. Prove this to yourself by making a table similar to Table 12–4.

> As a general guide, it is economical to spend additional rework costs up to the amount equal to the difference between the selling price of the good product less the salvage value of a defective unit.

Example: Greenrich Corp. Electronic components are manufactured by Greenrich. It collects defective units and once a week decides if the units can be reworked or sold for salvage value. Two major products are manufactured, A-1920 and B-4678. The cost of A-1920 is $56.50 and its selling price is $65.00 per unit. It has no salvage value as a defective unit. B-4678 has a cost of $178, a selling price of $210, and a scrap value of $45 for a defective unit. Christine, the manufacturing engineer, is deciding whether to rework the defective units or sell them for scrap. She examines each defective unit, estimates its rework cost, and if she thinks the unit can be reworked, she tags it with the estimated cost to rework. Some units cannot be reworked. The nonreworkable B-4678 units are sold for $45 each.

To determine the maximum amount to spend for rework on the remaining units, Christine makes the following calculations.

A-1920 reworkable units,

Maximum to spend on rework = Selling price − Salvage value = $65.00 − $0 = $65.00

B-4678 reworkable units,

Maximum to spend on rework = Selling price − Salvage value = $210.00 − $45.00 = $165.00

By comparing the estimated rework cost on the tags on the defective units with the above calculations, Christine can decide which units to sell for scrap and which to rework.

Rework costs can be even higher than the original cost of manufacturing the product when the troubleshooting time and costs are high. The above analysis is useful in making the decision whether to rework or scrap a defective component. Of course, the estimates of the cost to rework are not always the same as the actual costs, but estimating accuracy improves with experience and knowledge of the processes.

> Quality comes not from inspection, but from improvement of the process.
> —Mary Walton

Noneconomic Factors in the Scrap/Rework Decision

There are noneconomic factors that influence the rework/scrap decision.

Capacity. When the equipment is working at maximum capacity, an interruption to repair and rework an item is less likely to be made. In this case, the scrap will be accumulated until production volumes are lower, or the scrap will be disposed of to make room for the good products. When business is good, less rework is performed. An opposite case occurs when the organization is working at lower than maximum capacity. In this case, it is likely that rework is performed even when it is not economical to do so, in order that personnel can be kept busy rather than laid off.

Raw Material Availability. Another case of noneconomic factors influencing the rework decision is when the raw material takes a long time to acquire. An order may require 100 units, for example, but only 98 units are finished because two units were defective. If it takes a long time to get more raw material and the delivery date is soon, it is likely that the decision to re-work the two defective units will be made even though it is not economical, in order to meet the delivery schedule.

THE TAGUCHI LOSS FUNCTION AND GOALPOST QUALITY

G. Taguchi has introduced quality concepts and techniques that have influenced the analysis of quality costs in many organizations. He has defined quality to include the service the product delivers to the customer *over the life of the product*. This extends the costs and responsibility of the producer into the future until the time when the customer no longer uses the product.

A traditional view of specifications and tolerances is that if the product is between the low-est and highest permitted specifications, the product is acceptable. Anything outside of these specifications is defective product and is rejected. Using the football analogy of kicking a field goal, anything between the goalposts, the lower and upper specification limits, is a good three-point field goal, or **goalpost quality.** (See Figure 12–9.)

However, Taguchi states that quality is best when it is at the *nominal value of the specifi-cation*. Given that the producer's design department or the customer has an optimum or best value for the specification requirement, Taguchi believes that the further away the product is from this best nominal value, even though it is within the permitted specification range, the lower the quality of the product and the higher the future costs to the user of the product. Vari-ation costs money to the producer and to the customer over the life of the product. Taguchi sug-gests that the costs vary as a second-order parabolic function—the **Taguchi loss function**—as deviation from the nominal value increases.

> The quality of a product is determined by the minimum loss imparted to society from the time the product is shipped.
>
> —G. Taguchi

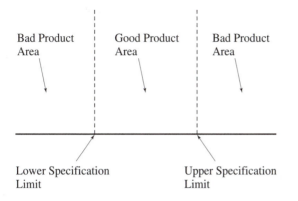

Figure 12–9   Goalpost Quality

Taguchi's definition of quality cost is "the minimum loss to society from the time the product is shipped to the customer." This contrasts with conventional definitions of quality that refer to meeting specifications. His definition comes closer to the satisfaction of customer expectations. If we extend the idea of customer satisfaction over the life of the product, we approach Taguchi's definition of quality. Rather than satisfying the customer and meeting their specifications only at the time of delivery, Taguchi's view of quality is that it extends over the useful life of the product. Using this concept, not only are the original costs relevant to the producer, but also the costs that the customer incurs as a result of using the product. This concept of quality and loss is similar to the concept of time value of money and life-cycle costs used by some organizations.

Taguchi's Quality Costs

For example, a manufacturer sold plastic sheet material to agricultural users. The plastic was to be used for covering the frames placed over crops, similar to a greenhouse, to protect the crops from harsh weather. The producer made the thinnest plastic sheet permitted by the specifications of a national trade association. This thin material, when used, would not withstand strong winds or mishandling. It would tear and would have to be replaced. The manufacturer was not at fault, since the thickness was within trade association specifications (goalpost quality). But, for the customers, there were additional costs due to:

1. The cost of replacing the plastic.
2. The loss and cost of the unprotected crops.
3. The loss to consumers, who paid higher prices because of the reduced supply of good produce.

Using Taguchi's definition of quality, the costs to the customer of replacing the plastic, losing crops, and losing unhappy consumers who bought the produce would have to be included in the cost of quality of the plastic sheet product, not just the selling price that the farmers paid for the plastic sheet. A competitor could capitalize on this by producing a thicker plastic, charging more, but still having lower total costs. Future buyers of the plastic sheet would likely buy the thicker, higher quality plastic even at an additional first cost if the other costs could be eliminated or reduced.

Taguchi would conclude that the ideal plastic has an optimum thickness—thick enough to withstand the damage by winds and misuse, but thin enough to be affordable. This ideal thickness would be the target or nominal thickness for the producer. Any deviation from the ideal nominal would increase the cost to the user, either as a higher first cost (too thick) or higher costs over the service life of the plastic (too thin).

> A product may only spend brief minutes or hours in the producer's organization, but may spend years providing service to the customer.

Loss Function Graph

Figure 12–10 shows a Taguchi loss graph with a parabola cost function touching the X-axis at the nominal value and extending upward from the nominal in both directions. The vertical axis

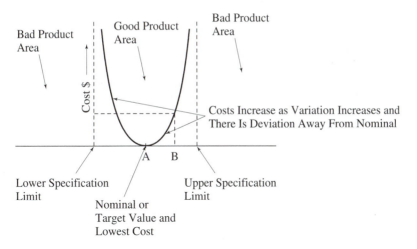

Figure 12–10   Taguchi Tolerances and Cost of Variation

is in dollars. A product with value *B* has a higher cost than a product of value *A* even though both are inside the specification limits as "good" products.

When the concept of tolerance limits defining good and bad products is combined with Taguchi's loss function, there is increasing cost when the product characteristic is not at the nominal value. And, even though the product is good by specification standards, it still has lower quality with higher costs than if the nominal had been delivered. Variation away from the target or nominal results in higher costs over the product's life, according to this concept.

Example: Transmission Quality. For another example that demonstrates this concept of nominal, consider the U.S. automobile manufacturer that required additional transmissions for one of its models. Because the manufacturer was producing at maximum capacity in its transmission facility, it had to look for an outside vendor for the transmissions. A Japanese manufacturer was found and was asked to produce the transmissions using the same drawings and specifications that the U.S. producer used.

When the transmissions had been in service for some time, a difference between the U.S. producer's and the Japanese supplier's transmissions was discovered. Even though both were producing the transmissions within the specifications, the supplier's products were consistently closer to nominal than were those of the U.S. manufacturer. The failure and repair rate of the supplier's transmissions were lower than for the U.S. producer. The supplier had set up their processes to produce as close as possible to nominal. This was more costly initially, but would deliver a more dependable product over its total life. This resulted in fewer field failures, repairs, and replacements of the transmissions, giving a lower total cost to the customer.

> Quality can be improved by reducing variation of product output even though the output is within specifications.
> —G. Taguchi

Other Contributions by Taguchi

In addition to the quality loss function, there are two other important quality and cost concepts developed by Taguchi: shortcuts in experiment design, and the application of signal to noise analysis to the development of robust products and processes.

Shortcuts in the traditional design of experiments used by designers, process engineers, and quality teams have been developed by Taguchi. These techniques permit significant improvements in the design and process by using a lower cost, shorter time approach, rather than the lengthy and costly traditional experimental techniques. Although using these abbreviated experiments may result in a less than optimal design or process, improvements in the range of 70% to 90% of the optimum, at 5% to 15% of the cost of full experiments, are possible, resulting in improved products and processes.

The third contribution by Taguchi has been the application of signal to noise analysis (a type of variation analysis) of a product's or process' capability of functioning under extreme or unusual conditions. This is the concept of optimizing robust design of processes and products. Robust processes and products are those that produce consistent output and results under all conditions of input variation; they can be determined by experiment. Increased robustness is a quality improvement that can result in lower failure costs.

Combining the three techniques of loss function, less costly quality improvement experiments, and more robust processes and products can lead to lower costs, higher product quality, and more efficient processes. Taguchi's approaches are important contributions to the continuous improvement of quality and the reduction of the cost of products and services to customers.

SUCCESS AND FAILURE OF QUALITY COSTS EFFORTS

There are some common, recurring events in both successful and unsuccessful quality improvement and quality cost programs. These are not absolute "dos and don'ts." Any one occurrence will not guarantee success or failure, but combined, they may have an effect on the outcome of an organization's quality cost efforts. They are summarized here as suggestions that may assist with cost reduction and quality improvement implementation.

Implementing Quality Costs

Steps to Successful Quality Cost Implementation

1. Measure quality costs to establish a current benchmark. Divide costs into their components.

2. Educate senior management on the current cost situation and potential benefits. Obtain their support and commitment.

3. Select a few trial projects in office and manufacturing areas to demonstrate the concepts of the program. They can be the easy, obvious problems.

4. Train operating personnel and teams in quality cost identification, concepts, and improvement techniques.

5. Establish an internal cost of quality reporting system.

6. Analyze quality costs using time value of money and other economic analysis tools, costs reports, and scrap reduction techniques.

7. Connect the quality cost efforts to the organization's strategic plans.

Causes of Potential Failures of Quality Cost Efforts

1. Efforts are staff and consultant driven, rather than by operating personnel.

2. Large, diffuse efforts use shotgun rather than rifle techniques to target projects.

3. No training of management and operating personnel.

4. No white-collar quality cost efforts.

5. No estimates of costs of lost customers.

6. Accounting system reports are too slow, too complicated, and/or not focused on critical areas.

7. Search for a "quick fix."

8. Efforts are not connected to a strategic plan.

SUMMARY

The economics of quality is an important part of economical design, continuous reduction of costs, and manufacture of products for contemporary organizations.

The pursuit of quality is not a fad. It requires a change of the organization's values and beliefs, which may have existed for more than a century. By moving from an authoritarian to a customer-driven philosophy, the organization is able to modify its culture and belief system. By meeting and exceeding customers' expectations, and by continuously improving the quality of the services and products delivered, costs are reduced, productivity is increased, customers' satisfaction is improved, profits are increased, markets expand, and the organization grows.

Deming tells us that by improving quality, waste and scrap are reduced, which reduces the cost of service and products. Productivity is increased as a result of a higher proportion of good output. Customers share in the reduced costs and improved quality. Then market share increases, the organization expands, and more jobs are created. Quality economics and this chain reaction permit customers to receive higher quality and lower prices from reduced internal costs. It is a customer-directed system with benefits to customers, employees, and shareholders.

The economics of quality combines measurement of quality costs, cost accounting, Deming's chain reaction, TVM analysis, Taguchi's loss concept, and financial statements. Quality economics has large potential savings, returns on investment, and increases in net income. Internal financial rewards can be high, and external benefits to customers are increased.

The most difficult part of the application of quality economics is that the organization must change its culture, pursue quality, and change its financial results horizon from a few months or quarters to a number of years.

Specific items presented in this chapter:

- Philosophical differences between quality and customer-driven organizations and traditional organizations.
- The four classifications of quality costs.
- Measuring quality costs, quality cost reports, and Pareto graphs.
- Minimum costs and quality.
- Scrap cost reduction by quality improvement teams resulting in lower costs and higher productivity.
- Measuring quality costs using ratios and percentages and return on quality.
- Deming's chain reaction.
- Calculating quality improvement ROI using TVM techniques.
- Rework and scrap decisions.
- Taguchi's loss function, goalpost quality, and reducing quality costs by reducing variation.
- Steps to successful quality cost reduction and potential causes of failures.

Quality economics uses all the parts of the financial cycle. Historical costs are collected and reported, time value of money analysis of the quality costs is made, the economic tools are applied, and finally the quality projects and investments are measured and improved during their life. The cycle is repeated continuously. Customers receive lower prices and higher quality, market share increases, and the organization grows. What begins as quality improvement results in expanded markets, higher profits, improved ROI, more satisfied internal and external customers, and creation of a strategic competitive advantage. The combination of contemporary quality improvement techniques with economic tools can have significant benefits to the organization.

QUESTIONS

1. When did interest in quality costs begin?
2. What are some milestones of the development of costs of quality?
3. What are some definitions of quality? How are they different?
4. What are the basic classifications of quality costs? Give examples of each.
5. State some reasons why an organization should measure quality costs.
6. What are some benefits of collecting and monitoring quality costs?
7. What are typical values of the cost of quality to sales ratio?

8. What are the contents of a typical cost of quality report?

9. As the organization begins to monitor and improve quality costs, what can be expected to occur to the magnitudes of the four quality costs over time?

10. Sketch an example of a Pareto distribution of quality costs. How can this distribution be used to reduce quality costs?

11. What is the relationship of quality costs and minimum costs?

12. What are some defects that occur in the office, accounting, purchasing, engineering design, and other nonmanufacturing areas that can contribute to costs of quality?

13. State some typical cost of quality ratios and percentage measurements that can be used.

14. What is return on quality? How can it be used?

15. How can a program to reduce the costs of quality be implemented?

16. What is the Deming chain reaction? How does the concept relate to productivity and profits? How does it relate to the cost of quality?

17. How is the incoming material requirement calculated if the required output of good product and the scrap rate is known?

18. How can reducing scrap decrease unit costs and increase productivity?

19. How can time value of money analysis be used to analyze quality improvements?

20. What is a decision rule that can be used to determine how much to spend to rework a defective product?

21. What are some nonfinancial considerations in the scrap/rework decision?

22. How are TVM and quality costs combined to measure the ROI of a quality investment?

23. What is meant by goalpost quality?

24. What has Taguchi contributed to improve the quality of processes and products and to the study of quality economics?

25. What is the Taguchi definition of quality?

26. What are some examples of how reduced variation can contribute to reduced costs over the life of the product or service?

27. What are some suggestions for successful quality cost implementation?

28. What are some possible causes of failures of quality cost systems?

PROBLEMS

1. Identify the following costs as prevention, appraisal, internal, or external quality costs.

Quality training	Troubleshooting
Inspectors' wages	Field trials
Replacement of field failures	Recalls
Customer surveys of quality	Quality engineering
Quality software	Supplier audits
Rework costs	Internal quality audits
Lost sales due to	for management
dissatisfied customers	Returned goods

2. If an organization's sales for a year are $40,000,000 and the total of the four categories of cost of quality is $6,450,000 for the year, determine the cost of quality as a percent of sales.

3. The cost of quality to sales ratio for an organization is 22%. What is your conclusion concerning the length of time that the organization has been committed to a quality improvement effort and the emphasis it has placed on the effort?

4. Given the following costs of quality, rearrange the costs into four categories. Determine the costs and percentage of each of the four cost categories. Calculate cost as a percentage of sales if sales for the quarter are $1,500,400.

Supplier Training	$5,000
Scrap Costs	$34,900
Audits	$3,000
Customer Returns of	
Defective Product	$29,000
Inspection Cost	$51,000
Field Repairs	$17,000
SPC Employee Training	$9,000
Reliability Tests	$6,000
Rework Cost	$7,500

5. Arrange the following costs into the four quality cost categories. Then present a Pareto distribution of the cost categories, and suggest which cost area should be reduced first.

Warranty Costs	$1,200
Inspection	5,000
Quality Training	800
Field Replacement	1,000
Scrap	2,500
Return Allowances	500
Rework Labor Costs	2,200
Supplier Audit Expenses	1,500
Invoice Error Costs	400
Supervision of Inspection	1,500
Reliability Testing	1,800
Quality Manual Preparation	$1,000

6. Using manual methods or spreadsheet software, make a pie chart of the four costs of quality from the results of problem 4. Present a Pareto distribution of the data.

7. Using manual or spreadsheet methods, generate a pie chart of the four quality cost categories of problem 5.

8. Given the costs and causes of defects in the assembly department, construct a Pareto distribution.

Cause of Defect	Third Quarter Costs
Mismatched components	$12,500
Scratches	$5,900
Broken switches	$9,000
Defective supplier material	$8,400
Wrong color	$2,050
Other	$2,300

9. Using the data and analysis of problem 4, construct ratios that you consider important measures.

10. From the analysis and data of problem 5, construct ratios that could be used to monitor the quality cost in the future to show the progress of quality cost improvements.

11. Construct a graph that shows costs versus time of the following information.

Month	1	2	3	4	5	6	7	8
Quality Cost (in thousands of dollars)								
Prevention	5.1	5.4	5.8	6.1	6.8	7.1	8.5	10.5
Appraisal	15	18	16	17	14	15	13	12
Internal Failure	22	23	23	21	20	18	17	15
External Failure	30	32	31	28	26	28	25	23

State any conclusions from the data or graph.

12. If 5,000 units of good product are required per year and the process scrap rate is 11%, what is the required order quantity of the raw material? If the scrap rate is reduced to 7% and the cost of scrap is $27 per unit, what is the savings in raw material from reducing the scrap? If it takes 7.6 minutes to produce a unit of product, what is the output productivity improvement from reducing the scrap rate?

13. A quality improvement team is able to reduce the scrap rate in a process from 15% to 9%. The required number of good units from the process is 30,000 per year. The value of a part is $4.52 per unit. The average time to produce a unit is 1.45 minutes. Determine the original scrap loss in units, the improved scrap loss in units, the original productivity, and the improved productivity.

14. A quality improvement team, after spending 324 labor hours at an average cost of $11 per hour, revises the methods and the tooling for a particular process. The changed process reduces the scrap rate from 8% to 5%. The unit cost of the raw material is $23.60 and the annual required output is 1,680 units per year. The product is expected to be sold for four more years. What is the ROI of this quality improvement effort? What will occur on the quality cost report as a result of this improvement activity?

15. A new supplier's product is $.65 per unit more expensive than the current supplier's. The firm is selling 12,000 units per year of this product and expects this rate to continue for six more years. By using the new supplier's material, the organization estimates that the savings due to scrap will be a reduction from 7.6% down to 5.2%. The cost of scrap is $41.55 per unit. Should the firm buy from the new supplier?

16. A product sells for $357, has costs of $340 to produce, and has a salvage value of $45. What is the maximum to spend for reworking the defective product?

17. A component has no salvage value and costs $430.50 per unit to produce. The selling price of the product is $475 per unit. What is the maximum to spend on reworking defective units? What might cause the organization to spend more on reworking the units? To spend less? What if the product had a salvage value of $200 per unit?

18. A process has an output requirement of 10,000 units per month. The historical scrap rate has been 6% of the incoming material.
   a. Determine the incoming material requirement.
   b. If the part value is $7.50 per unit, what is the cost of scrap in the process?
   c. If direct labor is 0.2 hours per unit, what is the productivity?
   d. If the process scrap rate can be improved and reduced to 3%, what is the cost of scrap? What is the new productivity?

19. A process can be improved and the scrap rate reduced from 6% to 3%. It is estimated that this will save approximately $2,500 per year. What is the maximum to spend on the initial improvement costs such as equipment and labor hours for team improvement work? The desired ROI is 10% and the estimated life of the process is eight years.

DISCUSSION CASES

Barnes Power Equipment Corp.

Barnes has been producing transformers and other power distribution equipment for more than 75 years. It is a family-owned company producing equipment used by power companies to distribute power in residential neighborhoods and for commercial/industrial applications. The company's transformer line currently has 15 basic models with a number of different configurations for each model design. Some of the transformers are ordered from stock and shipped within two weeks of receiving the order from the power company. Special designs require between four and six weeks to manufacture and ship. The company manufactures approximately 30 other power distribution products, but they are best known for their transformers.

Barnes has been receiving numerous customer requests for documentation of their quality system, and as a result has begun a quality effort. As part of the documentation effort, Harrison Barnes, the president, would like to be able to measure the cost of implementing the quality system and justifying all quality improvement expenditures, rather than spending money on expensive quality projects that are not needed. He feels that quality is important, but the costs of implementing the quality system may be too high. He feels that Barnes has produced a quality product at a reasonable price for years, and that all of the recent interest in quality is just another fad. He believes that the current cost of poor product quality is small, in the area of less than 5% of the total revenue of the company. Because of this low cost of quality, he does not want to spend a lot of money improving quality that is already very good.

Stuart George, his son-in-law, is the manager of manufacturing. He is a electrical engineer and worked at the company during summer vacations while in school. He has held various positions with the company during the past 15 years. Currently, he has his secretary keep track of the scrap and rework that each of the eight departments generate each month. These are estimates from the supervisors or process engineers in each of the departments. Stuart has asked the cost accounting department to help keep track of the scrap and rework numbers, but they say they are busy with other projects and never seem to get around to his request.

Stuart is also starting to keep track of returns of transformers from the field that are shipped back after not passing their electrical tests upon installation. He is also concerned that the current inspection costs are a lot higher than only the wage costs of the three inspectors in the manufacturing area. Some customers are requiring Barnes to submit a quality survey to them every six months in order to be on the approved suppliers list. These reports have become a lot of work for Stuart. He has also heard of some power companies visiting their suppliers and performing audits of the supplier's design department, manufacturing area, and quality system. Stuart thinks it is only a matter of time until this will happen to Barnes.

Trade journals have had recent articles about foreign suppliers having superior quality power distribution components at a lower cost. Stuart has discussed this with the sales manager, and the sales manager says he is "hearing more talk of quality" every time he visits customers. Stuart is worried that his father-in-law is not serious about quality and that this could be a problem in the future, but he doesn't know what to do about it.

1. What can Stuart do to make Harrison Barnes more interested in the quality issue?

2. Is this quality issue, as Harrison states, "just another fad"?

3. Is it likely that the cost of quality is less than 5% of sales at Barnes? How could this be proved or disproved?

4. How can the quality cost information be collected more efficiently? Must the cost accounting department be the department to collect and distribute the quality cost data?

5. How can the quality issue be less of a nuisance and more of a technical and marketing strategic weapon for Barnes to increase their competitive position in the face of foreign competition?

6. What other factors should Stuart consider?

Circle City Electronics

Ben Maxwell, the chief engineer of Circle City, has just returned from a three-day conference on quality in the electronics industry. The conference was also attended by the president of Circle City, Bill Vernon; the manufacturing manager, Janice Tagaret; and the quality manager, Jim Handman. The conference was one of many presented by an industry trade association. As they left the airport after returning home, Ben said, "I think there is a lot of information for us to think about from the conference. Let's plan on getting together in a couple of days to discuss what we should do, if anything, to implement the quality issues presented at the conference." That was two days ago, and Ben has called a two-hour meeting this afternoon.

Everyone was there except Bill; his secretary said he was on the phone with a European customer. Bill arrived late and the meeting began. "That call I had just now was from Hans at German Electronics. He was unhappy. They had just completed some life tests of their assemblies and found that the cause of 40% of the failures were due to *our* components! It seems like we are all spending more and more of our time lately dealing with quality. It hits us from all sides. I'd like your thoughts about the things we picked up at the conference—what should we do?"

Ben said, "I don't know about the rest of you, but I think we are okay on quality. We design to the customer's and industry specs. Sometimes we even put on tighter specs than the customer wants. I thought that the quality conference was okay, but we don't have problems like some of those other producers."

Janice responded, "We are having more trouble holding the tolerances due to the age of the equipment. We have gone to more and more inspection to make sure that we only ship good parts. I went to the session on the Deming approach to quality at the conference. One of the things that was discussed was that it is cheaper to prevent bad quality rather than rework or scrap defective units before they are shipped. Our approach has always been to make the components and then inspect them and either ship or repair them. Our scrap rates are somewhere around 10–15% in some areas."

Jim interrupted, "Janice, it's over 20% in some areas when things are not working well! We can rework some of the defective units, but sometimes we just sell them to the scrap dealer for pennies on the dollar. I think our quality and scrap problem is larger than we think. I've heard that some companies spend as much as 25% of their sales revenue on bad quality costs. We should have zero scrap!"

"I've heard that percentage figure too," Bill added, "I think that quality is a competitive problem. Prices and deliveries are important to customers, but quality is really becoming a competitive issue!"

Ben spoke again. "We don't even know if we have a problem. We get a few customer complaints, but no more than we have ever had in the past. Some of our new customers complain more about quality than they do about price. That's the way customers are."

The meeting continued for another 40 minutes with discussion about the topics presented at the conference and about Circle City's quality. There was little agreement about the key issues or what should be done. Finally, Bill said, "I am going to have to cut this meeting shorter than planned; I have to get ready for some customers that are arriving. Let's do this. Between now and the staff meeting next week, keep track of the quality problems and the time you spend on quality problems. I think we will all be surprised at the time we spend on quality. And, Jim, could you get together estimates of the amount of money we spend on scrap and rework. You will probably have to estimate it: I don't think the cost accounting guys do much data collection about quality. And, the rest of us, let's try to estimate the costs in our area that are related to quality to see if we can get a handle on the dollar size of the problem. Thanks for coming."

1. Is Bill taking a good approach?
2. How can Ben's comment about not "knowing if there is a problem" be dealt with?
3. Is quality something that should be handled only by Jim since he is the quality manager?
4. What is the customer side of quality that Bill talks about? How should the customer's quality problems be combined with the other quality issues at Circle City?
5. What other things should the group consider?
6. Is Jim right about having zero scrap? Is this possible?

Appliance Repairs Inc.

Jim Roberts, the president of Appliance Repairs Inc., a small firm of 65 employees, has led the organization in Total Quality Management effort for the past 10 months. It seems to be going slower than he has anticipated, but the results have been generally good. He wants to incorporate a cost of quality measurement system in order to monitor the results of the quality program.

The firm is located in Centerville and has seven offices within a 75-mile radius of the city. Each office serves an area of homeowners and businesses. The company prides itself on fast repair service, reasonable fees, and quality service. The company is growing rapidly.

Roberts is anxious to increase the speed of quality improvement at the company, and he hopes that keeping track of quality costs will help with that goal. The company doesn't manufacture any products; it only provides a service to customers, and Roberts is not sure how cost of quality and the four categories will fit into a service organization. Currently, return visits are sometimes required to repair customers' appliances, and there have been some complaints about slow response to emergency calls. Roberts hopes that the quality cost system will be a positive motivator for employees to become even more conscious of quality improvement. He is not sure how to start.

1. What advice could you give to Roberts?
2. What quality cost categories would you suggest be setup?
3. How is a quality cost system different in a service firm and a manufacturing company?
4. Comment on Roberts's desire for the TQM program to move faster, and the prospects of a quality cost system providing positive motivation to employees.
5. How could cost of quality measures be combined with other financial measures as management tools? Suggest an implementation plan.

Quality Plastics Corp.

A medium-sized plastic injection molding company, Quality Plastics is finding that more customers are requiring it to monitor and improve its process quality. The quality manager, Jane Mitchler, is trying to implement a Total Quality Management effort. She is currently working with the accounting department to set up cost of quality accounts in each of the four categories. Also, she wants to study the total cost of a product to a customer. She thinks that a failure or repair of a Quality Plastics product while it is in service should be considered in the cost of quality system. She is not sure how to proceed with this idea, however.

Customers return defective units and they also send units in for repair even after the units are out of warranty. Sometimes the customers repair or replace units that have failed in the field without telling Quality Plastics. Jane feels that the company's reputation for good products could be improved if they could study the causes of field failure and the incidence of repairs and replacement during the customer's use. Bob Morrison, the plant manager and Jane's boss, feels that if they can control the quality of the product in their plant, Quality Plastics has met their responsibility to the customer. "Build product to the specs and only ship the good ones!" he is fond of saying.

1. Do you agree with Jane's or Bob's viewpoint? Why?
2. How can a quality cost system monitor both internal quality of production and external quality after the product is sold? What data and information should be collected to implement Jane's system?
3. How does the concept of the Taguchi loss function apply to this problem? Is Taguchi's viewpoint theoretical or does it have some practical application to the situation at Quality Plastics?

Chapter 13 Continuous Financial Improvement

> Every day in every way, I'm getting better and better.
>
> —Emile Coue

KEY TERMS

Balanced Metrics A term that includes the financial, nonfinancial, and Baldrige criteria measurements.

Brainstorming Collecting many unedited ideas from a group, representing different viewpoints, leading from one idea to another.

Breakthrough Improvement Sudden improvements in products, services, and processes that result in financial benefits.

Cash Flow Return on Investment (CFROI) The present value of future cash flow generated by a project or investment, divided by the investment to generate the cash flow.

Continuous Financial Improvement Making small frequent improvements in processes, services, and products that result in incremental financial benefits.

Culture The values, beliefs, and attitudes of the organization.

Economic Value Added (EVA) Return on investment after the cost of all capital is subtracted.

Fishbone Diagram Also called cause and effect or Ishikawa diagram. It collects the possible causes of a problem and arranges them in related groups to determine relationships.

Horizontal Analysis Comparing cost and financial data between periods to determine trends and changes in financial results, to identify problems, and to lead to improvement.

Kaizen Japanese term for the philosophy of making small continuous improvements.

Nonfinancial Audit Studying products and processes to ensure that they meet a standard. The audit may be done internally, by a customer or a supplier, or by an independent third party. ISO 9000 and Baldrige Award audits often identify areas whose improvement results in financial improvements even though the areas are indirectly related to finances.

Return on Net Assets Before Interest and Taxes (RONABIT) Earnings before interest and taxes divided by net assets, permitting a focus on operating returns.

Return on Net Assets (RONA) Similar to ROI and ROA, except the denominator is net assets (assets less accumulated depreciation).

Root Cause Analysis Tracing the surface or apparent cause to its source so that permanent correction as well as correction of related problems can be achieved.

Run Chart A graph of a measurement's value to see the variation and level of the measurement over time.

Sources and Uses of Funds A horizontal analysis technique that identifies increases and decreases in financial statement accounts over a period of time.

Statistical Process Control (SPC) Monitoring outputs and inputs to a process by sampling data, calculating the mean and variation in the sample data, and comparing the mean and variation to the process limits, which are statistically determined to include 99% or more of the common or system variation. SPC charts also indicate special caused variation. SPC identifies variation, but does not indicate its source or root cause.

Variances Differences between the planned budget amounts and the actual amounts, which can be positive or negative.

Vertical Analysis Collecting and analyzing financial data within a single period using percentages and ratios to measure financial results that lead to improvement.

LEARNING CONCEPTS

- Eight prerequisites to continuous improvement.
- Continuous Financial Improvements Cycle.
- Basic improvement tools and techniques, including training, brainstorming, fishbone diagram, run chart, and flowchart; and their application to continuous financial improvements.
- Contemporary measurements including ROI variations, time-process improvements, value added, quality cost improvements, nonfinancial audits, profit sharing, and statistical process control.
- Traditional analysis of financial statements using ratios, percentages, horizontal and vertical techniques.

INTRODUCTION

This chapter presents the final component of the Financial Cycle (Figure 13–1), continuous financial improvement (CFI). After the collection of historical financial data, the selection

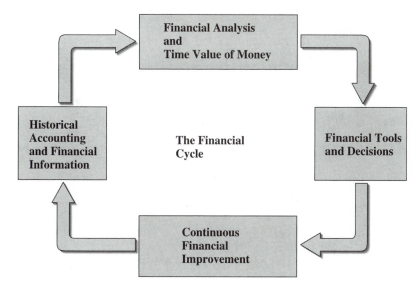

Figure 13–1   The Financial Cycle

and economic analysis of projects and investments, and the application of financial tools, the task of improving the financial results is implemented. It is not enough for engineers, managers, accountants, and others in the organization just to analyze investments at the beginning. Projects must be continuously monitored, measured, and improved. The cycle is *continuous*. Operating and technical personnel are in an ideal position to contribute significant financial improvements.

Successful continuous financial improvement (CFI) combines a philosophy based on the organization's culture and an application of techniques and tools that work on specific problems to improve financial results. Continuous financial improvements may be easy and natural for some organizations, or difficult and slow for others. Obviously, each organization has different problems to solve and improvements to make, but there are some universal concepts. Successful financial improvements are based on:

- Connecting historical financial data with engineering economy and financial improvements.
- Creating an organizational environment that includes prerequisites for continuous financial improvement.
- Collecting financial data and measuring results.
- Applying financial improvement tools.

The ability to collect, measure, and analyze financial data are important skills for engineers, managers, and others in the organization. The accounting system generates historical financial records; however, others must know how to use the reports, data, and statements for improving profits and lowering costs. Monitoring and improving financial results is the work of

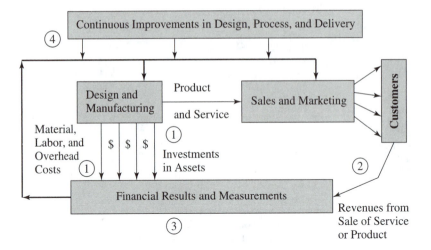

Figure 13–2   Measurement and Continuous Financial Improvement

individuals and teams throughout the organization. This continuous improvement of investments and projects over their life often requires more time and effort than the initial selection and analysis.

The **measurement and continuous improvement cycle** (Figure 13–2) shows:

1. The organization spends material, labor, and overhead costs and makes investments into projects and assets.

2. After sale of the services and products to customers, money flows into the organization.

3. The financial information is collected by the accounting system and presented in financial statements and other reports.

4. The organization, particularly technical personnel, uses the financial measurements and analysis to target improvements in manufacturing processes, products, designs, and projects.

News Release

New York—Ajax Corp. reported record earnings for the third quarter today. Net income for the quarter ending September 30 was $5,500,000 on sales of $96,000,000, which is $.25 per share. A year ago, the quarterly earnings were $.19 per share. A management representative commented that this increase of more than 30% was due to increased sales, reduced costs, and the accumulation of many small financial improvements for the period. Management looks forward to continued financial improvement and increased investment in facilities for the coming quarter.

FINANCIAL GOALS

A fundamental task of the organization is to define its philosophy of financial measurement and improvement. It is driven by owners/shareholders who expect continuously increasing earnings per share, sales, dividends, and net income; by employees who want higher wages and benefits; and by customers who want lower prices. The goal becomes meeting these money expectations. Is the "bottom line" profit the goal? Is the firm focused exclusively on profits for the short term—next quarter or the current year? Are the financial results the primary goal of the organization or are the financial results the rewards for doing everything else right, such as serving customers?

W. Edwards Deming, management consultant, has stated that, "Emphasis on short-term profits is one of the 'deadly diseases' that can kill the organization." He recalls a company executive telling him, "In our company the quarterly profit-and-loss statement drives us. No one gives a hoot about long-term profits. Toward the end of the quarter we will take all kinds of action to make our numbers, never mind the future. I know we once loaded a customer up with products he couldn't really use, but we showed good end of the year numbers. We lost that customer, but no one cared. We made our bonus for that year" (W. Latzko and D. Saunders, *Four Days with Dr. Deming* [New York: Addison-Wesley, 1995]).

Does the Financial Chicken or Egg Come First?

Are profits and other financial measures *goals* of the organization? Or are they *results and outcomes* from doing everything else right?

Prerequisites to Financial Improvements

Financial improvements work better when prerequisites are in place. Although these prerequisites do not always precede improvements in every organization, experience shows that improvements of all types, including financial ones, are easier, more permanent, and more effective when the prerequisites are included with the improvement efforts. Improvement efforts must be:

1. Supported by top management
2. Customer directed
3. Based on the organization's culture
4. Continuous
5. Implemented by teams
6. Connected to strategic plans
7. Directed at processes, not just outcomes
8. Appropriate for measuring financial results

Senior Management Support of and Participation in Financial Improvement

There is little disagreement that top management wants financial success and improvement. Any conflict, if it occurs, typically concerns the methods used to improve finances. Traditional financial improvement techniques have emphasized cost reduction, increased productivity, downsizing, plant closing, and working harder and faster to make and sell more services and products. These techniques work, but there are others. Contemporary management techniques pursue quality improvements, increased return on investment of fixed assets, improved customer satisfaction, increased market share, improved products and processes, faster delivery and response times, and many other small continuous financial improvements.

Financial improvement requires genuine involvement, investment of resources, and support by upper management. It cannot be, "Sure, continuous financial improvement is a great idea. How much will it cost to implement? I'll approve the expenditure and you can get started. Let me know in six months how it's working." Management must participate and lead the effort for it to be successful.

For some, these changes are as different as football is from ballet. There is still the need for some aggressive, rough-and-tumble sales increases and cost cutting, but there is also a need for long-term quality changes that will satisfy customers, planned improved manufacturing processes to reduce cycle time, improved assets utilization, product design based on customer expectations, and small continuous improvements that will result in large accumulated financial rewards. These are planned, longer term, choreographed movements—different trails to the same financial mountain peak.

Edwards Deming also suggested that 85% or more of the causes of an organization's problems are a result of management policy and are embedded in the organization's system. Only 15% of the problems are due to "local" causes. Local causes include the operator, the equipment, material, and other nearby causes. This 85/15 rule applies to financial problems as well. Upper management leadership and participation is the vital 85% of the improvement of financial results.

Customer Expectations and Satisfaction and Financial Improvement

The customer is a moving target. The competition is tracking the customer with increased vigilance. To be successful, the organization can no longer offer marginally adequate products and service, but must deliver products that are designed and manufactured to meet and exceed customers' expectations in a timely manner. The customer has more choices, is more knowledgeable, has higher quality standards, expects after-purchase service, evaluates purchases economically, and wants continuous improvements in service and products. In the past customers, perhaps, were easygoing, loyal, and had lower expectations. The contemporary customer is fickle, demanding, critical, and intelligent. If financial results and improvements are to occur, they must be earned by improved quality, faster service and delivery, meeting and exceeding customer expectations, lower prices, and support of the customer's needs after the product or service is purchased. The organization needs bifocals to continuously monitor the customer *outside* the organization and to *internally* pursue small improvements in designs and processes that affect the customer.

Watching the customer continuously and making small corrections and improvements can minimize the financial surprises from lost customers and lower sales. Mistakes in understanding the customer's needs may result in the organization's warehouses being full of unsold buggy whips, 8-track tape players, spare parts for Edsels, and decisions to discontinue Classic Coke. Continuously monitoring and responding to the small changes in customer expectations can sustain and improve financial results significantly.

> The organization needs bifocals to simultaneously see the external customer and the internal process improvements so that financial improvements occur.

The Organization's Culture and Financial Improvement

> It's very good for an idea to be commonplace. The important thing is that a new idea should develop out of what is already there so that it soon becomes an old acquaintance.
>
> —Penelope Fitzgerald

The organization's **culture**—beliefs, attitudes, and values—is the foundation on which all the financial measures and improvements are made. It is a dynamic process. Individuals give the organization its cultural structure and, in turn, individuals are continuously learning values, beliefs, and attitudes from the organization. Management, technical personnel, staff, suppliers, and customers all participate in the culture values and beliefs of the organization. These values and beliefs are transmitted by memos, decisions, actions, stories, gossip, interviews, meetings, and in many other ways. Continuous financial improvement seeds planted in an accepting culture can grow well. Seeds that fall on the rocky soil of abrupt cost cutting, short-term financial fixes, and management by numbers are less likely to germinate.

What Is Culture?

Some questions that can help define the organization's values, attitudes, beliefs, and culture are:

- What are the rules and policies of the organization? Are they written, flexible, inflexible, or informal?
- How is the success of people, projects, products, and departments defined, measured, and rewarded?
- What is normal in the organization? What are exceptions to normal?
- What are the customs? The unwritten rules?
- How does training occur? What behaviors does it teach?
- Is the organization structure formal, informal, mixed, dynamic, open, rigid, or flexible?
- How does management behave—formally? Informally?

- Does everyone think and talk about customer satisfaction? About quality? About money?

- Do meetings begin with a review of the profits of the project, product, or plant; or are quality and customer complaints discussed first?

- What do things look like in the organization? Are desks neat or messy? Is the manufacturing area clean or dirty? Is dress informal or formal? Generally are there more smiling faces than frowns?

- What are the stories, legends, and myths? Are they about customers, employees, managers, suppliers, successes, failures, winners, losers?

- Who are the heroes of the organization? Why are they famous, respected, and loved?

- What are the ceremonies or events? What and who is recognized and rewarded?

- Is management by example, by fear, by openness, by consensus, by team, by a single individual, or by another technique?

- Are there traits of cooperation, trust, fairness, honesty, friendliness, or are there turf battles, distrust of everyone, unequal treatment of products and people, dishonesty, and unfriendliness?

- Is there a changing or static environment? If there is change, is it abrupt and sudden, or is it planned, discussed, and gradual?

- Is there balance or imbalance between the goals and expectations of customers, employees, shareholders, managements, community, and suppliers?

- What proportion of the time is spent on goals and what is spent on processes to accomplish the goals?

- What are the formal and informal communications?

- Is there open discussion of the organization's values, beliefs, and culture? Of the individual's values and beliefs? Or are these "out-of-bounds" topics?

The answers to these and other questions can begin to define the values, beliefs, attitudes, and culture of individuals and organization. This culture can have a significant impact on the success of financial improvement projects and investments.

For example, consider Harley-Davidson, maker of heavyweight motorcycles, which went from the edge of Chapter 11 bankruptcy in 1981 to sales of $1.5 billion, a return on equity of 27% for the owners, and 56% of the heavyweight motorcycle market in 1994. How did change happen? Buying the company from AMF at the eleventh hour before bankruptcy, thirteen top managers changed its culture. It was a long, slow process, but it was successful. Once suffering from poor quality, poor management-union relationships, low productivity, and negative cash flow under AMF, Harley can now sell every motorcycle it produces, and it has retail customer waiting lists of 6 to 18 months. Market share would be higher, but capacity is at the maximum.

Employees carry a laminated card that describes the culture issues and characteristics that saved the company: quality, participation, productivity, flexibility, and cash flow.

The organization, full of tough, no-nonsense bikers, believes in values like truth between workers and managers; fairness with the unions, including cooperation and outsourcing practices; respect for opinions, which may mean questioning of decisions by everyone; curiosity, which is at the heart of training and retraining everyone; and considering all employees as stakeholders connected to the organization. These values and beliefs not only help improve the financial results of the organization, they can save the company from bankruptcy, as Harley-Davidson has shown (B. Filipczak, "The Soul of the Hog," *Training Magazine*, February 1996, Lakewood Publications).

> The organization's culture determines the speed and success of continuous financial improvement.

Balanced Financial Rewards from the Organization's Culture. The organization's culture, values, and beliefs can also determine the financial relationship between the shareholders, employees, customers, suppliers, and management. What is done with the money from the improved financial results? Are there increases in management salaries, wages, profit sharing, employee bonuses, lower prices for customers, increased dividends for shareholders? Is there a balanced combination of these beneficiaries?

Financial improvements that reward only one or two of these groups will be perceived differently than if all share in the benefits of the improvement. There is no perfect balance or correct answer to the problem, but the distribution of money from financial improvements becomes part of the organization's culture. These values affect how individuals and groups work on future financial improvements. Often employees' improvement efforts are related to the degree to which they share in the rewards. Profit sharing, bonuses, and employee stock ownership are strong motivators for financial improvements. If employees see 100% of the financial rewards from their improvement efforts go to increasing shareholder value or senior management bonuses, they may be reluctant to pursue future financial improvements. Customers that see a supplier's continually increasing net income, while prices increase, may find other suppliers. Similarly, shareholders that see little improvement in the value of their stock over time will unlikely be anxious to fund future expansion by purchasing new shares or bonds.

American Cultural Values. A critical prerequisite to financial improvements in the organization is to understand its culture. Certain cultural values are universal in American organizations, according to two researchers who have studied both corporate and American cultures.

In their book *The Stuff Americans Are Made Of*, Josh Hammond and James Morrison identify seven uniquely American cultural forces found in U.S. organizations:

1. Insistence on choice

2. Impossible dreams

3. Big and more

4. Now!

5. Oops!

6. Improvise

7. What's new?

(J. Hammond and J. Morrison, *The Stuff Americans Are Made Of* [New York: Macmillan, 1996])

Understanding these cultural forces may lead to more successful implementation of financial improvement efforts in the organization.

Insistence on choice—We love big menus and cafeterias. We choose different jobs from our parents and live in different parts of the country. Too many choices may be bad for us, but we'll take the risk. If the government or the boss tells us to do something, we may do it, but with reluctance. We would rather have a choice.

Financial improvement teams work well when they can choose projects, when there are alternate solutions to a problem that must be analyzed and selected. They work best if the members have chosen to be on the team in the first place. If the organization wants to improve the way it delivers products to the customer, let the customer help decide, within bounds, how they would like to have it delivered. They will place more orders. Do shareholders want more dividends or lower dividends, with the organization reinvesting earnings and improving the share value? Let shareholders choose.

Impossible dreams—We dream about winning the lottery. We dream about a new house, a new car, and retirement. We dream about a better way to make the product and about selling more products. The bigger the dream, the more we like it. Every new company began with a dream. All new products and patents began with a dream. They began with something that didn't exist, and perhaps was thought not to be able to exist. We want our leaders to talk about visions and dreams. We don't ask that they be possible dreams, just The Dream Team, the dream vacation, the dream employee, the dream boss, the dream home—all become possible as we move from the present and known to the dream in small steps. There are often many setbacks, but the dream endures and eventually we arrive at the American dream.

To pursue continuous financial improvements, cost reduction, profit improvement, scrap reduction, increased sales, and lower supplier prices, the individuals or team must have a vision or dream. If this "dream force" is ignored by a cost reduction or financial improvement team, it may not succeed. If this dream force can be harnessed, the team may work toward the goal with enough excitement to achieve the impossible. Americans love to see the underdog win; the more impossible the dream, the more we pursue it.

> I could never convince the financiers that Disneyland was feasible, because dreams offer too little collateral.
> —Walt Disney

Big and more—Bigger homes, cars, profit-sharing plans, stock prices, offices, TV screens, pizzas, plants, shopping centers, and schools are what we like. "Less is more" is not a philosophy that receives much attention in U.S. culture. Customers want more product for less money, faster delivery, and more quality. Employees want more money, more time off, better fringe benefits, more equipment, and better bosses. Is big and more really better? We Americans think so.

Financially, we want more dividends, higher earnings per share, more sales, more productivity, more inventory, more equipment, higher ROI, more wages, more cash flow, and more money in all forms. This is a major driving force of almost all financial improvements. If it is recognized and built into the financial improvement system, the participants will understand the goals and methods of "big and more" quickly.

Now!—This American cultural force is related to "hurry" and "faster." We don't like to wait at the supermarket check-out counter, at the bank, on the freeway, at the stoplight, or anywhere else. We want the quick-fix and the instant cure, microwave dinners, and instant pudding. We like UPS, Federal Express, and the Web better than the post office, because they are faster. Fast cars, deliveries, promotions, results, feedback and anything else speedy is important to us. Henry Ford reduced production time from more than 12 hours per car in 1911 to less than 15 seconds per car in 1925. Ford offered black cars after he first made them green and red. He had a big backlog, almost 13 years of orders, and it was important to get the cars to the customers fast (the *now* force). He used a single color, black paint, because it dried faster. He decided that the customer would rather have a black car *faster* than a *choice* of color. Sometimes the cultural forces of "choice" and "now" are in conflict. Did "now" make a difference in Ford's financial results? Surely.

Financially, we want sales to grow faster, cost reductions now, and shorter cycle times. We should realize that if we can improve speed, improved financial results will occur. Larry Bossidy, CEO of AlliedSignal Inc., made speed a corporate value. He trained people in using speed and minimized non-value-added work so work flow would be faster. He wanted conversations to be short and fast, called limited time or "elevator speeches." At Allied, sales increased faster than expected, cash flow turned around faster than planned, and earnings per share went from negative to positive in less than a year. Speed has financial benefits. Speed also is part of the American culture, so the successful organization can build on that cultural value to become more financially successful sooner. Some of the improvement tools presented here consider speed, product flow rate, and inventory velocity and their financial advantages.

Oops!—We often would rather have action than a plan. This cultural force causes us to start before we are ready, and make mistakes rather than plan. Do it again rather than do it right the first time, fix it while under way, try again and make it work. If we are going to use this cultural force, we can expect some failures in the financial improvement effort. Spending time placing blame about failures will only take away from the effort. If "mistake making" is part of our culture, focus on *what* caused the failure, not *who*, and improvements will be more effective. When we talk about planning, and doing it right the first time, the cultural force of "oops" is often stronger than the resolve to "plan and do it right."

> Why do we never have time to make it right the first time, but always have time to do it over?
>
> —Sign in an electronics company

We expect to make mistakes. We even like them because it gives us something to improve on. According to Hammond and Morrison, the Danish toy company that makes Legos discovered that German children and American children use Legos differently. The German child studies the instruction booklet, builds the first project according to the instructions, and then moves on to the next project until all the projects are completed. The American kid opens the box, tosses the instruction manual aside, begins to assemble something different from the instructions, and then makes something else. These American kids grow up to be part of organizations and make financial improvements the same way they built their Lego models.

Action is important to us, often more important than initial success. We can get it right next time and maybe even read the instruction manual. If we are to include this "oops" force, cost reduction teams should move fast and not spend all their time planning or avoiding mistakes. This can get us into trouble, too. Doing something, even if it is wrong, can be expensive. Small, quick financial improvements may fit our cultural values easier than large, well-planned, slow efforts that become boring before they show results.

> If you're not making a few mistakes, you're not doing enough to earn your paycheck. Just be sure the mistakes are small ones.
> —Boss to newly hired engineer

Improvise—Jazz music is an American invention. It is improvisation. A classical symphony leaves little to chance; all the notes are there, and the instructions on how to play them.

We think on our feet, solve problems on the fly, play it by ear, wing it, and don't always follow the rules. The financial improvement team that has the freedom to try new things, break the rules, move ahead not always knowing where they are going, and improvise will likely succeed. The organization must understand this cultural value and let them improvise. Breakthroughs or small continuous improvements are not always planned or developed "by the book."

In the traditional organization, the CEO makes decisions, sets priorities and policy, rewards success, and generally is the center of expertise and control. He or she is the symphony conductor. If the rest of the organization has the American trait of improvisation and wants to play jazz, but is required to adhere strictly to a "classical score" with an inflexible conductor, little financial improvement may occur.

What's new?—We like new things. We tear down the old and build something new in its place. We throw things away before they are worn out even though they could be repaired. Good is equal to new sometimes. We get trapped into thinking that a new fad is good. We discover that it doesn't work and throw it away, and try another new fad that may work.

Financially, an investment or project may be still financially profitable. But it's old, and we want something newer. This "new" force causes us to change things within the organization when they should only be maintained and improved. Searching for what is different and new can cause us to replace standard financial improvement techniques with techniques that may or may not be better. New for new's sake can be a trap; we need to keep our eyes open.

Financial improvement and the seven cultural forces—We should *choose* the financial improvement projects that offer the most ROI. Individuals like to *choose* or not to work on improvement projects. We can *dream* and pursue difficult and impossible financial improvements. We should understand the difference between a fad and a valid *new concept* that will have a financial reward. It's okay to *fail* and make *mistakes* or *oops* in the improvement efforts, and it's also okay to under-plan the project so long as we can *fix it* along the way. Planning is okay, but let's *improvise* and *get going now. Bigger* profits are better than smaller. *Faster* profits and returns are better than slower profits. *High* ROI is better than low ROI. Improvement team members, management, customers, and suppliers all understand that *faster results* and deliveries are better and generally will make *more* money. These same improvement ideas may not work as well in other countries as they do here in the U.S. These are the ways Americans like to do things.

Other American Cultural Values. There are other values, attitudes, and beliefs that are part of the American culture, such as our desire for freedom of all kinds, the Puritan work ethic, religious beliefs and values, respect for private property, and emphasis on the family. An organization may notice different values and beliefs in offices and facilities located in different parts of the country. The U.S. is still a regional heterogeneous mix, rather than completely homogenized.

An obvious subculture can be found in some unionized organizations. The values and beliefs of management and the union membership may be different, traditionally or because of their job responsibilities. The labor contract may solidify these differences rather than reduce them. One of the major results of the quality emphasis in many organizations over the last few years has been to discover that both union and management believe many of the same things about quality and its overall importance to the organization. Often the same is true for financial improvement values if the benefits are shared. Subcultures within the organization may be geographical, departmental, or by some other division. Financial improvements can be positively or negatively affected by the subculture differences.

Heroes—A characteristic of a culture is the heroes that it creates, part reality, part myth. Children's heroes are cartoon characters, fairytale heroes, cowboys, spacemen, athletes, TV actors, and movie stars. They can do anything, solve everything; they may be real or imaginary. Washington could not tell a lie; Lincoln walked miles to return a borrowed book; Jackie Robinson broke the race barrier in baseball; Babe Ruth hit only home runs; Patton and Eisenhower became military legends; the Beatles and Elvis started a music revolution; Edison was an inventive genius; Jobs and Wozniak invented the personal computer; Bill Gates dropped out of Harvard and became the richest person in the U.S. These are some of our heroes, creators of legends, people to follow and respect, part of the culture.

Heroes are cultural symbols in the organization. There are super salespeople, creative engineers in the design department, talented researchers in the lab, retired employees who are remembered, senior managers who can perform miracles, operating employees who are legends, software wizards who can tame the computer. There may be heroes outside the organization: competitors, suppliers, and customers. Formal and informal recognition can create heroes.

The financial improvement teams can be heroes. If the organization's culture places value on heroes that create financial improvements, others will follow.

Continuous Financial Improvement and Kaizen

> Nature does not make jumps.
>
> —Carl Linnaeus

The Japanese term for small continuous improvements is **kaizen**. It is part philosophy and part technique. It comes from beliefs and values that place importance on small improvements continuously linked together over a period of time producing large results. It is the opposite of sudden, major improvements that may be based on equipment, technology, and other break-throughs. Breakthroughs are great when they occur, and an environment can be established that encourages breakthroughs, but they do not happen continuously. As one project engineer stated, "Continuous small improvements are what you do *between* trying to make the major break-throughs happen."

Financial improvements are similar. Quarterly earnings per share often improve by many small increases in sales, reductions in costs, improvements in quality, acquisition of new cus-tomers, small decreases in scrap, procedure improvements, and many other small events. **Breakthroughs** are bonuses and are welcome, but the small **continuous improvements** also increase financial benefits.

> The message of the kaizen strategy is that not a day should go by without some kind of improvement being made somewhere in the company.
>
> —M. Imai

Holding the Breakthrough Gains. Sometimes, after major financial breakthroughs, the organization becomes complacent and the hard-fought gains are allowed to slip, perhaps not to their original level, but to a lower level. This is not uncommon in breakthrough-type cultures. There is a search for the sudden, cure-all event, and then a letdown after it occurs, with back-sliding.

As Figure 13–3 shows, sometimes the small gains made continuously can result in large improvements. *Holding the gains is part of the improvement process.* Focusing continuously on the improvement makes backsliding less possible. Involving everyone in the improvement process, making and holding gains, and achieving small continuous improvements results in significant financial benefits.

> Continuously focusing on the small financial improvements helps prevent backsliding.

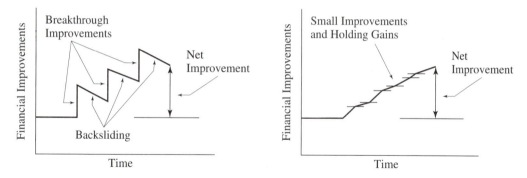

Figure 13–3   Breakthroughs and Kaizen Improvements

Is the typical American organization the culture in which to practice small continuous improvements? Some organizations have the philosophy of "watching the pennies so that the dollars will take care of themselves." But others prefer the 40-yard touchdown pass (the breakthrough) in the last two minutes of the football game rather than making small yardage gains all afternoon. A balance between the two is possible. It depends on the organization's culture. It may be easier to work within the existing culture before trying to introduce a new philosophy.

At United Technologies, CEO George David talks about the makeover and culture change at his organization. He has not approached the organization's financial problems with traditional cost-cutting techniques. Instead, he implemented kaizen, small continuous improvement concepts at the operating level of the organization, modeled after United's Nippon Otis Division efforts in Japan. By involving employees in kaizen, root cause analysis, insistence on quality in all steps in the production process, and cycle time analysis, United has made impressive gains.

A three-month engine assembly process at United's Pratt and Whitney Division was reduced to 10 days. The time from order to delivery of replacement engines has been cut from two years to seven months. At the Carrier Division's plant, output has increased by over 80% since 1991, while using one-third less floor space. Inventories have been reduced throughout the organization, resulting in increased cash flow, reduced storage space, and increased ROI. By involving employees in the problems and their solution, along with management leadership, major financial improvements have resulted. Earnings per share in 1992 showed a loss of $0.05 per share, but in 1995 were a positive $5.70 per share. This increase in earnings per share was on the same companywide revenues of approximately $22 billion per year each year during the four-year improvement period. Small continuous improvements of cycle time reduction, inventory reduction, team involvement, kaizen, quality, and root cause analysis all contributed to the financial turnaround (J. R. Laing, "Going Up," *Barron's*, 25 November 1996).

> Think naught a trifle, though it small appear. Small sands make the
> mountain, moments make the year, and trifles life.
>
> —Edward Young

Teams and Financial Improvements

When we discuss American culture and American organizations, we often visualize the lone heroic individual solving problems and overcoming obstacles to reach success. We think of a single Conestoga wagon heading West, John Wayne rescuing a town from outlaws, or Luke Skywalker saving the galaxy. In organization terms, it's the project engineer saving the project from technical or financial disaster, or the lone salesperson making the critical sale. If this image is true, is there any place for teams in the contemporary organization?

The factory system of the 1800s, which continued into the 1900s, can be generalized as the era of mechanization and Taylor's scientific management. Taylor's approach was, "Workers do the work, and the expert, all-knowing managers plan and control the work." Couple this with autocratic, almost military-disciplined organizations. Management had little room for empowered teams, consensus management, or self-directed work teams.

Why, then, do teams seem to be effective in the discovery and implementation of financial improvements?

- Processes are more complex and technical than in the past. These processes require experienced and detailed knowledge in order to improve them. The boss doesn't have total knowledge. Knowledge and skill are at the operations level with design and process engineers, supervisors, and others close to the problem. Unlike Taylor's scientific management, the manager does not know all the answers.

- The improvement of products and processes involves many different parts of the organization. Changes in one part of the process may result in other parts being less efficient. This "suboptimization" in one area may result in a worse overall process rather than an improved process. A cross-functional team minimizes this problem.

- The customer, both internal and external, is often known by the front-line workers. They understand the customer's needs. Financial improvements should be customer directed. Middle and senior management may not have all the information about the customer's wants. Engineers and technicians can be included in customer-based decisions to technically improve products and processes, which in turn will improve finances.

- Small, incremental improvements can be made throughout the organization. The coordinated effort of different individuals can be combined into teams to make larger and continuous improvements.

- Management can provide the leadership to make the team effort part of the strategic plan. Teams are not leaderless or random. They can be focused on specific financial improvements that are customer directed and fit into the overall strategy of the organization.

- The teams can take the form that is best for their functioning. Not every team in the organization is the same. A team may be advisory only. It may have the responsibility of arriving at partial solutions, with management or other technical parts of the organization having the remainder of the responsibility. Teams may be established for a specific task or have ongoing responsibility. They may be completely automated and self-directed with little assistance from technical or management personnel. They can be created to fit the purpose.

> A team trained in financial techniques that has management support and participation can focus on continuous improvements that are connected to the organization's strategic plan. Customer-directed and team-based efforts can be a powerful force to improve products and processes that will result in financial successes.

Within an improvement team, there is still individual responsibility and effort. Teams need leaders that can balance individual contributions with the team effort. Teams need training if their efforts are to be successful. Training should be directed toward two areas. First, teams usually require assistance in managing themselves. Second, teams need financial data and knowledge of financial tool application.

The Strategic Plan and Financial Improvement

Upper management, individuals, and financial improvement teams can target specific projects for improvement. Targets can be connected to the organization's strategic plan, which is connected to customers' needs. The plan states the mission and goals of the organization. Is market share more important than net income? Is shareholder return targeted for increasing? Is new product expansion and investment into R&D a key strategy? Are the plant and equipment old and in need of replacement? These and other targets can become part of the overall strategy to meet customers' needs more effectively.

The strategic goals and plans of the organization are usually based on meeting the external customers' expectations, but may also be directed at internal customers. From these customer expectations, the internal strategy plan, including financial improvement, is developed and guided.

The specific techniques should also recognize the culture and values of the organization. If an improvement technique is in conflict with the organization's culture, the improvements may not be as successful. Training can help change cultural behavior, but one of the strongest cultural changes can be for senior management to modify their values and beliefs and communicate these to the rest of the organization. Then, if the actions of senior management confirm that the change has really occurred (walking the talk), the organization's culture may follow. The degree to which the revised culture is aligned with the external customers' expectations is often one of the soundest bases for CFI. In many cases, change may be a slow process even when an ideal situation exists.

Processes versus Outcomes and Financial Improvements

> It is better to travel well than to arrive.
> —Tibetan saying

"The project is behind schedule and over budget. Drastic action is needed," says the project manager. "More overtime, cut the extra costs, let's get back on schedule now!" Or, "Sales are down!" says the sales manager. "Let's have a sales meeting and fire up the salespeople so they will sell more! Let's introduce new products and increase advertising expenditures." These and other techniques are the normal and often correct responses to some projects, but there may be other techniques that can help the project or increase sales.

Individual process improvements can indirectly help increase sales. Better designs can help increase customer satisfaction. Improving quality, accepting customer suggestions for design improvements, responding to customer complaints faster, decreasing the time to deliver the services or product, training, having senior management visit customers in the field, and other process-based improvements also may have positive effects on sales. These activities are directed at improving the processes that provide the service to the customer. Some improvement techniques target the outcome (advertising, sales meetings, new products), and some techniques improve the process, which in turn will improve sales because they are customer-based improvements. Continuous improvement often focuses on the *process* rather than the *outcomes* of the process.

> We should work on our processes, not the outcome of our processes.
> —W. Edwards Deming

Measurement of Financial Improvements

The map of the financial improvement journey shows unknown territories. Measuring the progress of the journey is important. Traditional financial measures look at historical accounting information and financial statements to see where the organization has been. Measures are needed to evaluate current and planned improvements to be sure that progress is being made toward the organization's strategic goals, targets, and customers rather than being only an interesting side trip.

Contemporary financial measures consider customer-driven improvements such as the cost of product failure in the field, the time to respond to a customer's complaint, the current project ROI, the time to set up and change a machine from producing one product to another, the cost of purchasing versus making the product, the cost of holding inventory, the cost of a flexible manufacturing system that will provide products in a just-in-time manner, and the time value of money analysis of equipment. Contemporary tools of continuous financial improvement are closely linked to the measurement techniques. Historical financial measurements from financial statements and cost accounting may be inadequate to monitor fast-moving projects, the ROI of facility improvements, quality economics, and financial improvements based on cycle time reductions.

> When you can measure what you are speaking about, and express it in numbers, you know something about it; but when you cannot measure it, when you cannot express it in numbers, your knowledge is of a meager and unsatisfactory kind.
> —Lord Kelvin, 1891

Questions to Ask Before Beginning the Continuous Financial Improvement Journey

- Is the financial improvement project supported by senior management?

- Will senior and middle management participate in the financial improvement effort?

- Is the financial improvement consistent with the existing culture or can the cultural behavior be modified to make the improvement consistent?

- Are there benefits to the external customer if this financial improvement is realized? Are there any internal customer benefits?

- Will the financial improvement be based on team analysis and implementation?

- Is this improvement related to other small continuous financial improvements?

- Is the financial improvement connected to the strategic goals?

- Is the financial improvement process or outcome directed?

- Are the measurement systems in place to know when improvements have been achieved?

Positive answers to these questions are a good start to the financial improvement journey.

STEPS FOR CONTINUOUS FINANCIAL IMPROVEMENT

Although each organization and its problems are different, some universal steps can be considered by the financial improvement team.

Continuous Financial Improvement Steps

1. Form the team.
2. Train the team in both team and financial techniques.
3. Ensure that the prerequisites to continuous financial improvement are in place.
4. Select the most critical internal process, customer-driven problem, or strategy problem.
5. Target the improvements to be made.
6. Define the selected process using process flowchart techniques.
7. Collect and display data using run chart, SPC, or other techniques.
8. Search for common and special causes of problems.
9. Use brainstorming and the fishbone diagram to determine potential root causes.
10. Apply financial improvement tools and analysis including TVM, breakeven, minimum cost, and replacement analysis.
11. Make corrections in process variables.
12. Continue to measure process inputs and results to determine if improvements occur.
13. Hold onto gains and prevent backsliding.
14. Do it all over again and further improve financial results.

FINANCIAL IMPROVEMENT TOOLS

Financial improvement techniques include three types of improvement tools and some additional techniques.

Basic Tools

1. Financial improvement training
2. Brainstorming, root cause, and cause and effect analysis
3. Run chart
4. Process flowchart

Contemporary Tools

5. ROI techniques
6. Statistical thinking
7. Statistical process control
8. Time process changes
9. Quality economic analysis
10. Total Quality Management
11. Audits
12. Sharing financial rewards

Traditional Financial Measurements

13. Vertical and horizontal statement analysis
14. Budget variance analysis

Basic Tools

Some basic techniques are common to many improvement team efforts: brainstorming, cause and effect analysis, root cause techniques, simple run charts, and process flowcharts. These basic techniques combined with contemporary economic analysis tools are methods to collect and display financial information in a simple manner that can lead to improvements.

Financial Improvement Training. As engineers, technicians, project managers, and others in the organization focus on improving the finances of their projects, departments, and products, they can develop skills of measuring and improving financial results. Training of financial improvement teams typical includes the following:

- Learning how to function as a team.
- Understanding basic tools: brainstorming, cause and effect analysis, root cause analysis, basic statistical techniques.
- Training in basic financial statements and simple Accounting Equation and cost accounting techniques.

- Learning analysis techniques: time value of money, breakeven, minimum costs, replacement analysis, and quality economics.

- Applying contemporary financial improvement tools and techniques to specific improvement problems.

Brainstorming, Root Cause, and Cause and Effect Analysis. An effective method of collecting financial improvement ideas from a group is to do **brainstorming.** Everyone gets involved. The team defines and identifies the key problem. Each team member gives ideas concerning the causes of the problem, without discussion. One idea may give birth to an additional idea, and the final collection of ideas is larger and better than if each person contributed separately. Discussion of the merit of the ideas comes later. When brainstorming is successful, it demonstrates that "many heads are better than one."

Brainstorming

Brainstorming is a technique that initially accepts all ideas as equal possibilities. A rule is not to criticize the suggestions until all the ideas have been collected so that everyone participates without bias or criticism. After all of the possibilities have been collected, they can be grouped, revised, or eliminated if there is consensus among the members. The team can then focus on what they believe to be the causes of the problem. Then these causes can be corrected, and more root causes can be identified.

Upstream or **root causes** of the problem can be located in many different parts of the organization. The causes may be identified by operating and technical personnel, or they may be policy or culture based and have roots in past management decisions. It is likely that no single individual or department knows all the root causes. Their complex interrelationships often require many different viewpoints for identification.

Fishbone Diagram

The financial benefits of brainstorming and developing a fishbone diagram are that many different root causes of the problem are revealed, which may lead to simultaneous solutions and multiple financial benefits.

The collection of causes can be summarized in a cause and effect or **fishbone diagram**, as shown in Figure 13–4. On the right is the "head" of the fish, which states the problem or effect, improving ROI. To the left are the major causes, or "bones" of the fish: reduce assets, increase sales, increase profits, increase productivity, improve quality, and decrease costs. Subcauses, the "minor bones," are shown with arrows into the major causes. These bones or causes have usually been identified by brainstorming techniques.

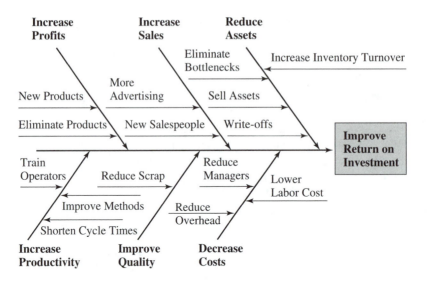

Figure 13–4   Fishbone Diagram for ROI Improvement

The next step is to identify the dominant causes in the problem. This can be done by polling the team members based on their experience and knowledge. Another method is to create an experiment that contains many of the causes as inputs into the process outcome. These causes can be given different values during the experiment and changed under controlled conditions to determine which factors improve the outcome and which do not. Both the polling and the experiment approach are effective methods. Next, measurement data are collected to see if the change was effective. By making small continuous changes, the improvements in ROI, or other desired outcome, can be continuously improved.

> The root causes and solutions to financial problems are not always obvious, and may be upstream in many parts of the organization.

Run Charts. Data can be presented in a simple **run chart** with the important characteristic plotted on the graph's vertical axis and time on the horizontal. (For an example, see Table 13–1 and the accompanying graph.) This simple time graph can show trends and variation. Often this can be enough information to determine if the financial improvement is working. For example, is scrap cost increasing, decreasing, or constant? Is the average product delivery time decreasing? Is the ROI of completed projects improving? If a more detailed analysis of variation, averages, and improvement data is needed, statistical process control charts can be constructed.

Process Flowcharts. An initial step for measurement and improvement techniques is to define the process using a **process flowchart.** Often a system of work steps or procedures to complete a task evolves over time and results in a process that is performed differently from its original design. Or, different individuals may perform the same process differently. Not only can

Table 13–1

Week	Scrap Percent
1	8.6
2	7.5
3	9
4	10.3
5	6.5
6	7.6
7	8.1
8	7.9
9	9.4
10	9.1
11	8.2
12	7.9
13	6.4
14	5.9
15	5.9
16	5.1
17	5.3
18	4.7

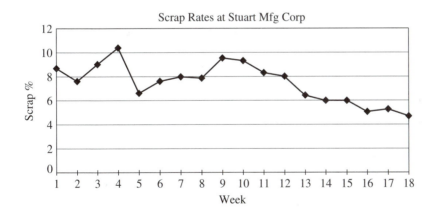

processes be simplified, they should be performed consistently. Process variation can be minimized by using consistent methods. By observation of the actual process, managers, operators, technicians, and engineers can identify delays, waiting time, bottlenecks, inefficient steps, and other undesirable components. It is important to record the actual process, not the intended or desired process.

The process flowchart can be simple. Most processes can be described on a single piece of paper. It can be handwritten by the team and does not necessarily need to be computer generated.

Using symbols, arrows, and words, the steps and the flow of information, people, or material can be shown. The initial process flowchart should show the actual, not the desired process. After the improvement team agrees that the process has been accurately recorded based on the actual steps in the process, they can begin to ask questions and use brainstorming and fishbone diagrams to simplify, combine, eliminate, or revise the process. It may be helpful to indicate times, distances, or other measures to give the magnitude of the delay, operation, or movement. Times may be from a few minutes for a form on someone's desk to be signed, to weeks for extensive calculations and analysis by engineers in the design department, wait time for a management decision, or inventory sitting in the warehouse.

Making flowcharts of the procedures and processes of the organization may seem like an overwhelming task. In fact, there may be only 100 to 125 processes that the organization performs with regularity. Of these, only a few are critical to customers' satisfaction and have financial impact. These are the key processes and products on which the team can focus. Figures 13–5 and 13–6 show examples of process flowcharts.

Ebert Corp.'s invoice payment improvement team should include Mary, Jane, and Karen. Figure 13–5 shows the team's process flowchart. The team members can ask questions, in brainstorming style, that may lead to an improved process. Typical questions the team may ask are:

Is it necessary for the signature step to take three to four days?

What is the total number of invoices and checks written per month?

What are the total hours spent on this process per month?

Why do invoices sit for as long as 30 days on Mary's desk?

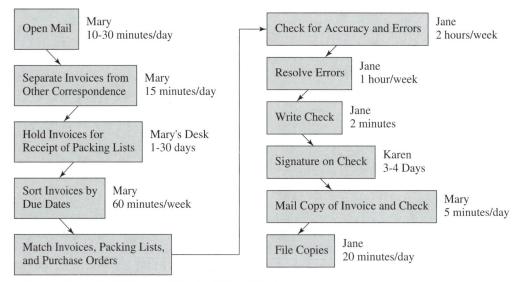

Mary or Jane—1 or 2 Times per Week—2 Hours/Week

Figure 13–5   Ebert Corp. Flowchart for Paying an Invoice

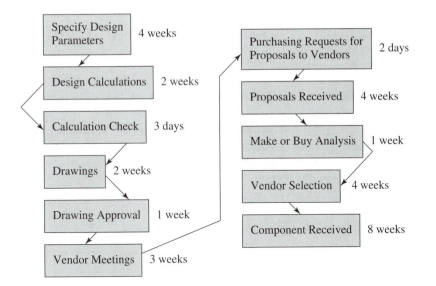

Figure 13–6   Ebert Corp. Flowchart for Design and Purchase of Component 12-XM3

Is there a way to reduce the time spent on error checking and resolution?

Are the steps in the best order? Is there a better order?

What is kept in the files? Is it necessary?

Can Mary sort invoices by due date when she opens the mail?

Is there a history of mistakes or errors made in issuing the checks? What was the cause?

Does the actual flow agree with how everyone thought the process occurred?

How can time delays be reduced?

How can work times be decreased?

Can any of the steps be replaced with a computer application?

Is it possible to reduce the number of people involved in the process?

Once all the questions have been asked, the team can organize the questions and possible causes of the problem into related groups on a fishbone diagram. The groups of questions can be studied to determine their relative importance. The most important questions are dealt with first, and then alternate solutions can be proposed. The team can agree on a revised procedure and try it for a period of time such as a month. At the end of the period, the team can review results, make revisions, and formally establish the new procedure. Participants can note any problems while they are working in the process and follow up with the team when it evaluates the process. The revised procedure should be written down and given to each participant. It is sometimes valuable to have the steps of the process posted where the process is performed, avoiding the possibility of the procedure being lost on someone's desk or filed and ignored. After the steps are revised again, the team can measure and summarize any benefits as a result of the changes such as fewer errors, shorter delays, faster cycle time, lower costs, and higher returns on investment. Any advantages and successes should be recorded and publicized to other teams, co-workers, and management. Success stories become important parts of the culture, and the techniques that were used in this process may be helpful to other improvement teams. Finally, the team should be formally disbanded so they can move to other improvement projects.

Although this example is a simple process, it demonstrates the team approach. All the techniques can be modified to fit the particular project as needed. The process flowchart is the basis for many financial improvement techniques. An example of a process flowchart for engineering design and manufacturing is shown in Figure 13–6.

Contemporary Tools

> If you keep doing what you've been doing, you're going to get what you've been getting.
>
> —American folk saying

Contemporary tools include time value of money, breakeven, minimum cost, and replacement analysis, which were presented in previous chapters, as well as other techniques that are discussed next.

ROI and Financial Improvements. We have seen the importance of ROI and TVM as analysis and selection tools for new investment and projects. These same ROI and TVM techniques are vital to monitoring and improving projects and investments as they are under way and after they have been completed.

Comparing the after-the-fact, actual ROI to the planned ROI can help improve the decision-making abilities of the organization. Financial improvement teams can trace the causes of poor performance or success and correct or improve them.

An improvement team can continuously monitor the progress of long-term projects over a period of months or years using ROI techniques. Actual costs, revenues, profits, asset investment, and financial benefits can be improved throughout the life of the investment to ensure that it is meeting the original ROI target. The project's economic life can be continually monitored to decide whether equipment or buildings should be retained or replaced.

> We spend hours, days, and even weeks of analysis, discussion, computer time, and meetings to analyze, select, and decide if an investment should be funded or not. We do very little during or at the end of the project to see if all of our estimates were correct and what *actual* return the project made or how to improve our future project analysis.
>
> —Chief engineer, Fortune 500 company

ROI Questions a Project Manager or Engineer Can Ask

- How does the original estimated ROI compare with the current actual ROI on the project?
- How do the original cost estimates and revenue estimates compare with the actual costs and revenues?
- How does this project's or investment's ROI compare with the ROI of other projects?
- What are the reasons for differences between the actual and estimated ROI of the project?
- For future projects, what techniques of analysis should be changed? Which should be eliminated?

The basic financial improvement tools can be incorporated into the analysis of actual ROIs. Brainstorming, fishbone diagrams, root cause analysis, run charts, and SPC can be used to monitor and improve actual and estimated ROIs.

Example: Stevens Corp.'s past ROIs—Two approaches are used at Stevens for monitoring past ROI. The first is to summarize and compare actual ROIs that the projects earned for a certain period. The second is to compare the actual ROI with the original estimated ROI.

Table 13–2 and the accompanying graph show that the majority (18 of 39 projects) of Stevens' actual ROIs are in the 11% to 20% range. There are some project ROIs above 30%, some below 10%, and even some that are negative. Table 13–3 shows that the average estimated ROI is 21% and the average actual ROI is 19%. The improvement team can look at both the high and low ROI projects to identify the root causes of the high and low values. If root causes can be determined and dealt with in future projects, the overall average ROI of Stevens' projects can be improved.

The ROI variance summary in Table 13–3 reveals that another improvement technique can be applied. Those investments with large negative variances, such as the warehouse project, can be examined for root causes to see if any of the causes can be eliminated in future projects. By reducing the variance, future projects will produce closer to the expected results. This lower variance improves future budgets. Fewer financial surprises help financial planning.

There can be many root causes of variation between the estimated and the actual ROI values. Changes in annual costs, revenues, lives of equipment, salvage values, operating expenses, and taxes are some of the surface causes. Overly optimistic or pessimistic estimates are often a root cause. With monitoring, the accuracy of the initial target ROI estimate can be improved. Continuous monitoring can aid in fine-tuning the ROI estimating process and improving future ROIs.

An important reason for continuously monitoring the investment's costs and revenues or savings is to monitor the economic life of the investment. As previously discussed, the equipment, process, or building can have a physical life that exceeds its economic life. The economic life of the investment can change as maintenance and operating costs, remaining physical life, and salvage values change. Even though the economic life is estimated initially, it is valuable to check its actual values during the life of the project.

Looking-backward ROI techniques similar to these can also be applied by an individual investor who selects a security or other investment for purchase based on its potential ROI. It can be monitored frequently for its actual ROI. Expectations about an investment's ROI can be initially enthusiastic, but later, the investment may not meet those high expectations.

Table 13–2   Distribution of Investment Projects: Actual ROI for Stevens Corp.—19xx

ROI %	Frequency
−10% to 0%	3
1% to 10%	8
11% to 20%	18
21% to 30%	5
31% to 40%	3
41% to 50%	2

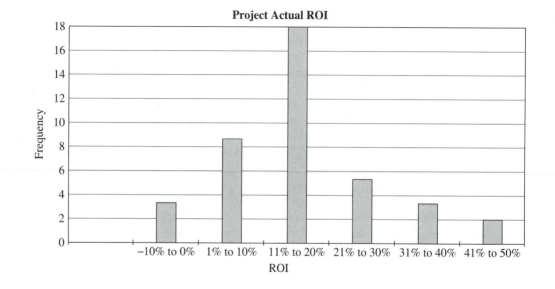

Table 13–3   Current Projects' Actual and Estimated ROI

Project	Estimated ROI	Actual ROI	Variance
A-43	20%	18%	−2%
A-567	20%	22%	2%
Burgess	15%	21%	6%
Walker	45%	38%	−7%
Warehouse	12%	2%	−10%
33-Deater	13%	10%	−3%
Average ROI	21%	19%	−2%

Return on Investment and Financial Improvements

ROI measurement can be applied in retrospect to projects and investments by individuals and organizations. Financial improvement teams can compare actual project ROIs with estimated ROIs. By determining the root causes of high and low ROI variance and correcting the causes, future financial results can be improved.

Variations on ROI—Organizations can use a number of ROI variations to monitor and measure projects, investments, and companywide performance.

Tight control of the company's assets also provides a lever for managing the business. Lower profitability may be acceptable when it's accompanied by lower asset usage, as the free cash flow generated may be the same. And management can usually have more of an effect on assets than on the income statement.

— CFO, computer equipment manufacturer

Currently, modified ROI-based financial measures are being used by many organizations. Some related nonfinancial measurements are also being applied. These nonfinancial measurements include customer satisfaction, employee satisfaction, and criteria such as Baldrige Award measurements. Indirectly, these measurements point to potential financial improvements. In addition, many organizations use cross-functional teams to measure and improve the financial results that were the responsibility of the chief financial officer in the past. Engineers, technicians, managers, and operating personnel are now part of the financial measurement and improvement team.

Example: ROI Measurement at Xerox—ROI based on net income divided by total assets fell from 20% in the 1970s to the low single digits in the 1980s.

Xerox is a "comeback story." With the invention of xerography in 1938, the company's first 45 years were successful. In the late 1970s and early 1980s, trouble started to surface as patents expired and intensive international competition began. Financial ratios including ROI reflected the company's situation. The ratio of indirect workers to direct workers in U.S. operations was twice that of the nearest Japanese competitor. New product development required twice as many workers; it took twice as long to develop a product; and it took almost three times as long to deliver a new product to the marketplace. Product defect levels were approximately seven times higher than those of competitors. The selling price of similar Japanese equipment in the U.S. was equal to the manufacturing cost of Xerox's product, and the competitors were making a profit. Market share declined. ROI was falling. Worldwide layoffs occurred, wages were frozen in 1983, declining ROI reduced employee profit sharing, and employee satisfaction with senior management decreased. Things were not good at Xerox in the early 1980s.

David Kearns was appointed CEO. In 1982, he began to see improvements in the Fuji Xerox operations in Japan that resulted from their quality management initiatives. Then some employees at U.S. Xerox operations approached Kearns about applying quality and financial improvement initiatives to the domestic operations. With outside consultants' help, Kearns and senior management began a plan for the organization called "Leadership Through Quality." The investment of time and work began to pay off. ROI measurements began to show the improvements that were taking hold at Xerox in the mid-1980s and that continue today. Profits on assets rose to above 10% and market share improved. By the fiftieth anniversary of xerography in October 1988, things were a lot better for Xerox in terms of ROI (R. Palermo and G. Watson, *A World of Quality* [Milwaukee: Quality Press, 1993]).

Return on net assets (RONA)—For **return on net assets** the denominator is total assets minus depreciation with net income for the numerator. Discussing financial measurement in his book *Beyond the Numbers* William Simon refers to National Semiconductor's CFO, Donald Macleod. Macleod's early goals at National were to move away from profit-and-loss, bottom line–oriented management to focus on return on net assets. By improving RONA, assets could be reduced and net income could be increased. This change to return on net assets gives managers a broader perspective and lets them manage assets as well as revenues and costs.

Cash flow return on investment (CFROI)—Later at National Semiconductor, Macleod and outside consultants developed a measurement system that became known as National Semiconductor Value Management. It is a combination of cash flows and ROI (present worth of future cash flow divided by assets). This was **cash flow return on investment** (CFROI) after tax. Some of National's businesses were returning less than the cost of capital. They didn't want teams to focus on an outmoded effort to "grow out of the problem"; they wanted them to move up in ROI over time, first improving the rate of return on their existing asset base, before increasing the investment rate.

Return on net assets before interest and taxes (RONABIT)—Another variation of return on net assets is **return on net assets before interest and taxes**, or RONABIT. This measure excludes interest and taxes, which are financial rather than operational in nature. Managers and engineers can use this to focus on the return that operating assets are providing to the organization. Simon, in *Beyond the Numbers*, reports that Sun Microsystems put in a bonus plan that was tied to RONABIT. Sun rewarded employees for not using cash. By changing to RONABIT, employees understood that cash was valuable. Hundreds of projects were performed to get cycle time down and $1 billion in cash on operations was generated. This was not just profit, it was reduced inventories and reduced receivables. Simon reported that Sun has RONABIT up to 50% to 60% in some areas—significant results.

Economic value added (EVA)—Another measure that is being used by a number of organizations is **economic value added** (EVA). It is net income after tax divided by total assets less the cost of the capital. The cost of capital may be based on the loan rate for borrowed funds, the dividend rate on preferred stock, or another capital cost. By implementing EVA as a tool, Simon reports, Harnischfeger Industries was able to reduce the capital employed in the business from $1.2 billion to less than $900 million, a reduction of $300 million in less than three years during a time when sales increased. It permitted them to measure both the income and cost sides of the business as well as the capital and balance sheet side of the business.

R. R. Donnelley evolved from a cash flow return on investment (CFROI) system to an economic value added system as well, Simon reports. One of the advantages that Donnelley attributed to EVA was that it was easy for the line operating people to understand this technique, compared to most measures. They modified the EVA measurement by including the costs of startup capitalization, FIFO inventory, leased assets, goodwill, asset disposal, and cash taxes. These are added to the capitalized asset base and the net income numerator of the EVA value.

Financial and nonfinancial measurements—At General Electric, presidents of the different areas look at both financial and nonfinancial measures on a regular basis. According to Simon, presidents of the companies at GE look at cash flow, return on total capital, and net income. For nonfinancial measures, they're looking at speed, quality, and customer and employee satisfaction. Dennis Dammerman, CFO of GE, believes that the ideal goal would be to worry just about the inputs. He looks at how to manage inputs to the process, and then how they influence the outputs, cash, earnings, and sales. He feels that it's a much more activist role for leaders of the organization to have an impact on the inputs to the system rather than only the results. Jack Welch, CEO at GE, expects the associates at GE to be well versed in the financial and nonfinancial measurements applied to their areas of responsibility.

Balanced metrics—At National Semiconductor, the approach is balanced metrics. **Balanced metrics** uses both financial and nonfinancial measures. The common element is that the measures are directly based on the Baldrige criteria. The divisions and manufacturing sites use the Malcolm Baldrige Quality Award application criteria to measure the process gaps and to drive their continuous improvements. The team approach wanted to used the Baldrige criteria as a basis for studying and improving their critical business issues. They wanted to look at the organization in terms of both financial and nonfinancial measures. Does it work? A study done in 1995 by the National Institute for Standards and Technology showed 14 Baldrige winners' stock values outperformed the Standard & Poors 500 stock average by more than four to one.

ROI is a critical tool for analysis, selection, measurement, and improvement by the project manager, CFO, manufacturing engineer, purchasing buyer, design engineer, and supervisor. Financial and nonfinancial measures can be used by everyone in the organization, not just the financial personnel. RONA, RONABIT, EVA, balanced metrics, and CFROI are some of the contemporary measures that are being used to improve projects, investments, and overall success. Financial personnel are becoming more knowledgeable about the products, people, and processes for which they keep the financial history. Engineers, technicians, managers, and operating personnel are monitoring both technical and financial decisions, measurements, and improvements.

> It is axiomatic that whatever is measured is what will be managed. Don't bother to measure it unless you are going to do something with the data.

Statistical Thinking and Financial Improvements. Continuous improvement is based on the recognition of variable results—sometimes up, sometimes down. The financial variation as well as hitting the financial targets can be improved. These twins, target and variation, must be continuously measured and improved.

Statistical Thinking and Financial Improvement

Two types of financial improvements can be considered: meeting the desired goal or *target*, and reducing *variation* of the results. Improvements rarely occur in a straight line without any up and down variation. Improvements usually take the form of hitting the target with smaller variation from the norm.

Applying statistical thinking to financial measurements can benefit projects. One characteristic of statistical thinking is understanding that measurement and improvement data are variable, not fixed over time. For example, if inventory turnover is 5.9 times a year, it is understood that, due to variation in the sales data and inventory values, the turnover could have been higher or lower and *still been in its normal range*. An organization could have a turnover target value of 6.0 times. If no statistical thinking is applied, a measurement of 5.9 could be interpreted as a failure, and any value higher than 6.0 as good. It is likely, however, that inventory has a normal range of turnovers, for example, from 5.5 to 6.5 times per year. These variations in outcomes must be accepted as normal if improvements are to be made. Variation can be traced to either random or special causes.

Random causes of variation—A *random* cause of variation is within the system. The small variation of each component combines with the variation of other components and results in the overall variation. This is normal or common variation. It can only be reduced or improved by changing and improving the system causes. Changing such things as purchasing policy, equipment maintenance policies, procedures, retaining, or designing can significantly reduce variation.

Special causes of variation—There may be external variation on the process that makes the total process variation unusually large. For example, the sales target of $500,000 per salesperson is not met because rising interest rates slowed customers' economic expansion, and so customers reduced their purchases for the year. This is not a random, system-based cause but a negative and external *special* cause. Or perhaps a competitor goes out of business; the market share is divided between fewer sellers and sales exceed the $500,000 target. This is a positive special cause. Construction equipment damaged at the job site and weather delays are other special causes that can prolong project completion and reduce ROI. In these cases, the special cause is external to the organization's normal patterns. This is not random or normal variation.

Random and special causes in financial measurements—How can we know if the cause of financial variation is special or common? W. Shewhart, a statistician for Bell Laboratory and Western Electric in the 1920s, proposed a statistical answer to this question. He suggested collecting random samples of product measurements and calculating the average value and a measure of variation, either the range of the data or its standard deviation. Typically, the *mean* and the *range* of the sample are used. Using mathematics of sampling distributions, Shewhart constructed one chart that showed the average of the samples and another chart that showed the ranges of the samples over time. Shewhart charts have been successfully applied to processes in product quality, product design, manufacturing, hospitals, government offices, hotels, educational systems, restaurants, banks, and other organizations for measurement and improvement of financial results.

Sometimes the run chart does not show enough information. The team can select a statistical process control chart (discussed in the next section) to monitor measurements and assist in their improvement. It is important to understand that the SPC charts and other tools are only tools; it is the *team* that determines root causes and improvements. Charts may indicate that something is out of control, but they do not tell the cause of the problem. The cause must be diagnosed by those familiar with the problem.

Statistical Thinking Points for Financial Improvement Teams

■ Use data rather than intuition to initiate financial improvements.

■ The process of collecting data to assist with an improvement does not have to be complex or even computerized. It can be a simple run chart. Decisions can be based on simple data collected by operating employees as well as technical personnel.

■ Understand that variation is inherent in all financial measurements and targets.

■ Training in understanding variation and its measurement and analysis should be part of financial improvement activities.

■ Processes have variation in their inputs and in the process itself, which causes variation in the outputs.

■ The majority of variation is common and is caused by systemwide policy. It can be measured, but it can be corrected only by changing and improving the system. This may represent as much as 85% of the variation in financial results.

■ Local or special variation is more easily identified and corrected than common variation. This represents approximately 15% of the variation in financial results.

■ Special variation is usually the result of operator errors, wrong procedures, improper raw materials, poor equipment, environmental factors, and other causes local to the product or process.

■ Special causes can be identified and usually corrected by the work group and operating employees. System changes may be required to make permanent financial improvements.

■ Processes can be improved by reducing variation as well as hitting the target.

■ Overcontrolling a process when only common or normal variation exists should be avoided. When no special causes exist, the process may become worse through overcontrol.

Statistical Process Control (SPC) and Financial Improvements. If the variation and averages of improvement data are not obvious using a run chart, or if the common and special causes cannot be identified, **statistical process control (SPC)** can be applied. Sample data are collected over time. The average and range of the samples are calculated and plotted on two charts, the average chart and the range chart. The three standard deviation limits from the centerlines are calculated from the sample data using control chart tables available in handbooks and shown on the charts. If the data are experiencing only normal, common variation, the average and range values are expected to be inside the three standard deviation limits. To improve the process, the inputs must be modified to give either smaller variation or nearness to the desired target of the output.

The SPC chart for monthly inventory turnover, shown in Figure 13–7, shows normal or common variation in the turnover except for one point on the X chart above the upper control limit. This is a likely indication of a special cause. The team can search for the root cause of this out-of-control point and correct the cause. Improving the common variation and changing the average value requires making improvements to the inventory system such as trying different procedures, purchasing policy, make or buy decisions, vendor selection, vendor production systems, and storage procedures.

Basic Steps of Statistical Process Control for Financial Improvements

1. Collect samples of data from the process.
2. Calculate the mean and range of each sample. Calculate the grand mean of the samples x double bar, and the grand mean of the sample ranges R bar.
3. Calculate the limits of the control charts using statistical tables.
4. Draw the mean and range charts, centerlines and limits, and plot the sample data.
5. See if the data are meeting the desired target value or goal. See if the data are within the control limits. If not, special causes may be present. Look for trends and other special cause patterns.
6. Use brainstorming, fishbone diagrams, and other tools to locate potential causes of variation or missed targets. Select likely causes and trace to the root cause. Correct and continue to plot data to see if improvement occurs.
7. Continue monitoring and repeat the steps to further improve the process.

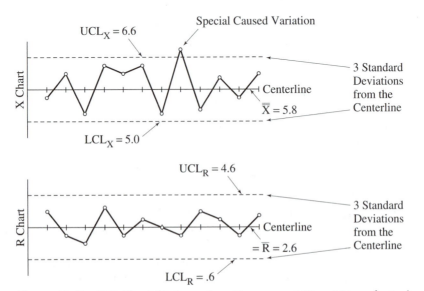

Figure 13–7   SPC Chart for Inventory Turnover at Stuart Manufacturing

---

Statistical Process Control

Statistical process charts are powerful tools for showing the occurrence of special and common variation. SPC identifies the existence of a problem, but does not identify the cause. The team can identify a likely cause, make the correction, and continue to use the SPC charts to see if the results improve.

Time, Value Added, and Financial Improvements. The processes that an organization performs, such as designing and assembling a product, paying an invoice, purchasing raw material, writing paychecks, delivering a product, performing engineering calculations, and inspecting a component, all require time. Time-based analysis and improvements of the processes in the office, manufacturing, and field can produce significant time reduction and in turn financial benefits. Generally, less time means lower costs, higher ROI, and higher customer satisfaction. As Ben Franklin said, "Time is money." Typical time reduction tools are:

- Value-added flow analysis
- Cycle time and delivery time decrease
- Just-in-time purchasing and manufacturing
- Wait time analysis
- Constraint and bottleneck reduction and elimination
- Capacity measures

Value added and time measurement and improvement—As the raw material is received, it begins its journey through the manufacturing area to become a finished product. Inventory turnover measures the average frequency or number of times that inventory material moves compared to the annual sales of the organization. It may be 6 to 10 times for a typical American manufacturing firm. With an inventory turnover of 8 times per year, it takes an average of 45 days (365/8) for the raw material to move from receipt to the customer in the form of the finished product. But the actual productive time for the product to be changed from its raw state may be only a few minutes or hours. If the time that the inventory is in the system can be shortened, then inventory can be decreased, total assets are lower, and ROI is increased. Measuring the time that inventory is in the system is an initial step to improvement. Abruptly cutting oversized inventory is one approach; however, this may result in the wrong items being reduced and lead to shortages and scheduling problems.

This technique is not limited to inventory reduction. Increasing the flow rates of product designs, purchase orders, reliability experiments, or accounting procedures will increase the value added per time, reduce delays, decrease response time, speed delivery, improve customer satisfaction, increase ROI, and lower costs.

Example: Computer Products Corp. and value-added flow rate for component 64-A— When increasing flow rates and reducing inventory, care must be taken not to starve the production system or the final customer. Just-in-time systems must be managed carefully. A related measure of inventory turnover and JIT is the value added to the product compared to the time that the product is in the system. The costs of material, labor, and overhead are added as the material moves through the system. By measuring the dollar amount of value added, and comparing it to the time that the material is in the system, the value-added flow rate can be determined.

Value-added flow rate calculations—The dollars added to the component's value in each step of the process are important measures, and they can be combined with time data. The cost-time relationship lets us measure delays and flow-rate costs. This is shown in Table 13–4 for Computer Products' component 64-A.

Column 1 identifies the department or step in the process.

Column 2 shows the value-added dollars to the product in the department, composed of material, labor, and overhead costs.

Column 3 shows the time that the product is in the department, including storage and processing time.

Column 4 is the actual processing time for the product.

Column 5 is the value added from column 2 divided by the average time in the department from column 3. A relatively low value indicates that the product is waiting a long time to receive small amounts of added value. Improvement can be made by moving the product more quickly.

Column 6 is the value added from column 2 divided by column 4, the actual processing time. Low values here indicate that the product is receiving little added value for long processing times. Improvement can be made by revising the process, equipment, methods, or materials.

Table 13–4   Computer Products Corp. Value/Time Data for Component 64-A

(1) Product Routing	(2) Material, Labor, and Overhead Value Added in Dollars	(3) Average Time in Storage or Department in Days	(4) Processing Time in Department in Hours	(5) Value Added per Department and Storage/Day (Col. 2/Col. 3)	(6) Value Added per Processing Hour (Col. 2/Col. 4)
Raw Storage	$10.00 Raw Material Cost	10	0.3	$1.00/Day	$33.33/Hour
Dept. 1	$2.30	5	0.5	$.46/Day	$4.60/Hour
Dept. 2	$8.10	4	1	$2.03/Day	$8.10/Hour
Dept. 3	$4.50	6	0.7	$.75/Day	$6.43/Hour
Dept. 4	$6.00	12	0.5	$.50/Day	$12.00/Hour
Dept. 5	$1.25	4	0.2	$.31/Day	$6.25/Hour
Assembly	$9.50	7	0.5	$1.36/Day	$19.00/Hour
Final Storage	—	14	—	—	—
Total	$41.65	62 days	3.7 Hours	—	—

Total process flow values can also be calculated and compared to values of other processes in the organization.

1. The ratio of total value added to total time in storage or the department is:

$$\$41.65/62 \text{ days} = \$.67 \text{ per day}$$

2. The ratio of total value added to total processing time is:

$$\$41.65/3.7 \text{ hours} = \$11.26 \text{ per hour}$$

3. The ratio of total time in storage or the department to total processing time is:

$$62 \text{ days}/3.7 \text{ hours} = 16.76 \text{ days of storage per hour of processing}$$

The first ratio measures the value added to the total time in the department or storage. An average of \$0.67 is added to the product each day it is in the system. A low value shows that costs are based primarily on waiting, not working. If "time is money," then the time that the product is waiting to be processed has some costs associated with it. The costs of waiting are usually overhead costs, such as storage costs, rather than variable costs of material and labor. This holding cost can be measured by determining the space, utilities, insurance, time value of money, and related holding costs of the organization. Many studies indicate that storage/holding costs can be in the range of 10 to 30% of the part value per year or \$25 to over \$100 per square foot per year. Reducing wait time permits inventory reduction and lower space requirements, both of which can result in higher ROI.

The second ratio shows the total value added in dollars compared to the total processing time. This is an indicator of the cost compared to the speed at which the processing costs are being applied to the product. In this case it is \$11.26 per hour. This can be a benchmark against which future improvements can be compared.

The third ratio shows the time in the system compared to the total processing time. In this case, for every hour the product is worked on, it spends almost 17 days in the facility.

These and similar value added/time ratios can be calculated easily by employees or improvement teams. The value-added amounts can be found from the internal cost accounting reports and other financial data. The average time in the department or in storage can be found by sampling techniques used on the material as it comes into and leaves the work area. Processing time can be determined from time studies, machine processing times, and other studies. The results can be monitored over time to see if changes to improve the process were effective in improving flow rates, value added, and inventory reduction, and result in increased ROI.

Bottlenecks and delays—By increasing the value added per time unit and decreasing the non-value-added activities such as wait or storage time, inventory can be decreased, costs reduced, and ROI improved. Ideally, the flow of the material through the steps in the process would be continuous without any delays or wait time. The only way this ideal situation can occur is for all of the operations in the process to be balanced, with the same amount of time per unit. If any step in the process takes longer than another and cannot be reduced, there will be wait time. This wait causes inventory to accumulate in front of the slower operations. These bottlenecks cause the material flow to be delayed. Value added per time decreases and the costs of delays, such as higher inventory, increase.

One typical approach that leads to imbalance is to run all equipment and work groups at their individual maximum output. This is based on the belief that idle workers or equipment is inefficient and should be avoided. If some of the work groups or equipment are faster or more efficient than other steps in the process, they will produce inventory that must wait while the less efficient or lower capacity equipment processes the material. This inventory buildup between steps in the process results in higher storage costs and lower throughput over time. It may be less costly to have idle equipment and/or work groups so that flow is more constant. Even a perfectly balanced process would have random variation in the output of each step.

Improvement in one step of the process should not cause problems in other parts of the process. This suboptimization can result in a less efficient and more costly total process even though one step has been improved.

No Quick Fix

The visitors were completing their tour of the manufacturing facilities, and the guide asked if there were any questions or other areas the guests would like to see. One of the visitors asked to see the warehouse where the raw, in-process, and finished goods were stored. The guide said that they did not have a warehouse. When the product was finished, it was shipped daily to the customers. "But what about the raw material storage?" the visitor asked.

"We call our vendors each afternoon and order tomorrow's material. At night, they come into the plant using keys that we gave them, and leave the next day's material near the machines that require it so that material is here when the work starts the next morning. And the processes have been balanced so labor or equipment does not overproduce. Limited in-process inventory is accumulated," the guide replied.

The guest exclaimed, "That's great. We should try that in our facility! Low inventory and low storage costs! I'll recommend we do this when I return to our plant. By the way, how many days did it take to perfect this concept?"

The guide answered, "We're still refining the system. We started 11 years ago."

Wait time—The delay or wait time can be used as a measure of process efficiency. Reducing the wait time of personnel, inventory, project completion, deliveries, engineering designs, customers, or equipment can result in reduced inventory, increased ROI, improved flow rate, improved productivity, increased utilization of equipment, reduced space, and lower costs.

Equipment utilization—The amount of time equipment is used is a financial measure. Equipment is not productive 100% of the time. It can be out of service because of breakdowns, normal maintenance, tooling changeover, waiting for material, or unavailable operators. By monitoring the utilization of the equipment, financial improvements can be made.

Two techniques have been successfully implemented with respect to equipment utilization: setup time reduction and total preventive maintenance (TPM). They are both designed to reduce unscheduled downtime caused by lengthy setups and equipment breakdowns.

In the chapter on minimum costs, we discussed reduction of setup times. Setup reduction can reduce equipment downtime, improve its efficiency, improve response times to customers, reduce inventory storage, and improve ROI.

Many organizations have reduced their unscheduled equipment downtime by revising maintenance programs. By improving awareness and training employees in preventive maintenance, downtime can be reduced. A sound preventive maintenance program helps maintain the quantity of the equipment's output, and it also has a positive effect on reducing downtime and total maintenance costs. Measuring downtime, finding the root causes of the problems, correcting the causes, and continuously improving the maintenance processes have a financial impact that is reflected in reduced maintenance costs and improved efficiency. For equipment-intensive organizations, monitoring and improving maintenance and downtime can lead to financial benefits.

Delivery and cycle times—A primary measure of employee and customer satisfaction and an important financial measurement is the time required to respond to an order, question, or complaint. Time lags or delays in response times can influence sales, costs, and future financial success. The customer, who measures both the cost of a product or service and the time required to receive it, will base future decisions on the delivery time as well as the price.

The growth of fast-food restaurants and drive-through services, the proliferation of speed-oriented services like Federal Express, and the fact that everyone would prefer to spend less time at the post office or dentist's office are examples of the impact of the speed factor on our culture. Organizations and individuals that meet time and delivery expectations are rewarded financially and with repeat business.

Most contemporary organizations have responded to the speed factor with initiatives such as:

- Reducing time from design to delivery by using concurrent engineering techniques.
- Responding to employee suggestions within 48 hours of receiving them.
- Measuring the time it takes to answer the phone on customer service "help lines" that are available 24 hours, 7 days a week.
- In restaurants, time-stamping when food orders are taken and when the order is delivered to the customer's table.
- For package delivery services, reducing domestic and international delivery time from weeks to hours.
- In inventory systems, reducing wait time and increasing JIT scheduling and flow rates.
- Making speed a corporate strategy.

Not only are these time and delivery efforts rewarded financially, time becomes a strategic force that differentiates the organization from its competitors. The measurement of response and delivery times is important if financial results are to be improved.

Process-Time Changes and Financial Improvement

Incorporating just-in-time purchasing and manufacturing and using cycle time analysis, value-added flow analysis, wait time analysis, and bottleneck reduction can result in financial gains. The benefits of successful process improvement are inventory reduction, shorter time to respond to internal and external customers, reduction of storage costs, increased inventory turnover, and improved quality. The flowchart is at the heart of all of these techniques.

Quality Economics and Financial Improvements. The economics of quality, which we discussed in an earlier chapter, is a vital financial improvement area that has large potential returns for many organizations.

Total Quality Management and Financial Improvements. Total Quality Management (TQM) is often implemented by organizations in financial trouble. They have falling market share, increased competition, loss of profits, and related problems. These problems are strong motivators for an organization to pursue the improvement of quality, change its culture, do things differently, introduce teams, and monitor and improve products and processes. The rewards for these efforts are multiple, including many financial benefits. Surveys of the Japanese Deming Prize for Quality (similar to the U.S. Baldrige Award) winners by N. Kano over the past 40 years show that the winners report direct benefits of increased profitability and cost reduction, and indirect benefits of reduced customer problems, company growth, defect reduction with process improvement, and increased employee development.

U.S. evidence of successful TQM implementation is similar. Total Quality efforts have important financial benefits. A study by the Strategic Planning Institute of 3,000 organizations found that those that had the highest perceived quality by their customers also had the highest ROIs (R. King, "Using Total Quality Management to Improve Bottom-Line Results," GOAL/QPC Annual Conference Proceedings, 1992).

Total Quality Management and Financial Improvements

By doing the things that are involved in a TQM effort, the organization not only improves the quality of services and products delivered to customers, but it also receives financial benefits. Typically, quality improves, scrap decreases, costs decrease, profits improve, market share improves, and customer satisfaction increases.

Nonfinancial Audits and Financial Improvement. Auditing of the financial and accounting system is a common technique to ensure that correct financial procedures have been followed. Nonfinancial auditing techniques can also benefit the organization, its products and processes, and provide for financial improvements.

Nonfinancial auditing and measurement occurs in manufacturing organizations, hospitals, educational institutions, and other organizations. These audits are directed at the organization's processes and products rather than the accounting system. The audits are performed by internal auditors, customers' auditors, or third-party auditors such as ISO 9000 and other registration organizations.

One of the early nonfinancial audits practiced in the U.S. was Military Standard 9858A, an audit of quality documentation and procedures by the Department of Defense to ensure that suppliers of military equipment were meeting DOD standards. From this and other standards, NATO created international standards for military equipment and similar standards were adopted by NATO members. Later, the NATO standard was modified to apply to nonmilitary products and became a basis for the current international standard of ISO 9000. The ISO 9000 Series standard is applied in almost every industry and economy in the world today.

Specific industries have their own audit systems also. The automotive industry in the U.S. has added its own requirements to the ISO 9000 standards, called QS 9000 (Quality System 9000). The pharmaceutical and medical equipment industry must meet federal food and drug standards. Aerospace firms have created an audit system that includes ISO 9000 standards as well as criteria unique to aerospace processes. Many other industries have their own audit procedures and organizations.

The Malcom Baldrige National Quality Award is a standard that can be applied to many organizations. It was created by the U.S. Department of Commerce to recognize successful American organizations. Prior to its creation, a number of U.S. firms had applied for and won the Deming Award, a Japanese quality award named after the American quality expert Edwards Deming. Although only a few companies win the Baldrige Award, thousands of organizations apply its standards internally.

The Baldrige Award and financial improvements—The Malcolm Baldrige National Quality Award has been awarded to small and large manufacturing firms and to nonmanufacturing organizations since 1987. There are seven **Baldrige Award criteria** of measurement.

Malcolm Baldrige National Quality Award Criteria

Category	Points
Leadership	110
Strategic planning	80
Customer and market focus	80
Information and analysis	80
Human resource development and management	100
Process management	100
Business results	450
Total points	1000

Source: Malcom Baldrige National Quality Award, 1997.

These seven categories are further divided into a total of 20 subcategories. Financial performance is a major subcategory in the business results area (130 of 450 points.) Not only are the current financial results measured, but also improvement of the financial results and trends. These financial measures represent 13% of the total point value in the Baldrige criteria.

The results from audits based on Baldrige criteria, ISO 9000, QS 9000, and other audit systems can be used as baselines for improvement. As one Baldrige examiner stated, "Success with financial performance measures confirm that the organization has done everything else right— such as customer satisfaction, management of processes, sound leadership, human resource development, and strategic planning. If these are done well, financial success follows."

Nonfinancial Auditing and Financial Improvements

Monitoring processes, procedures, quality, systems, and performance characteristics is costly, but contributes to the continuous financial improvement of the organization.

Sharing the Financial Improvement Benefits. One of the important reasons why individuals and teams should be able to understand financial statements, TVM, accounting concepts, and CFI as well as traditional economic analysis techniques is to see the impact of individual and team actions on the organization's financial success. Once the relationship between small financial improvements and the overall financial results is understood, teams and individuals become more financially aware. They are likely to perform their responsibilities with more enthusiasm if they participate in profit sharing, bonus plans, stock ownership, and other rewards.

Reward systems from improved financial results are being changed by many companies. Rather than the benefits going only to the shareholders and senior management, middle managers, project team members, operating employees, and others throughout the organization are sharing in the financial savings.

The average pay of factory workers at Lincoln Electric Co. of Cleveland, which has been a leader in employee financial incentives, was $61,000 (of which $19,000 was from profit sharing) in 1995. Not only does the company guarantee no layoffs, it also has a well-known profit-sharing system for both upper management and operating employees. That is why the average shop employee earned over 30 percent of his or her total income from a profit-sharing bonus.

By placing responsibility for decision making and improvements on individual employees and teams, Lincoln is able to reduce supervision costs and have fewer supervisors and middle managers, in the range of 1 supervisor per 100 employees. Some U.S. firms typically have a 1 to 20 ratio. In addition, over 60% of Lincoln employees own shares in the company. An advisory board of employee representatives meets with top management to discuss problems. Lincoln is well known for the advice, technical expertise, and other assistance it gives customers, which helps it maintain its market share and obtain higher prices than competitors. Even Lincoln's foreign plants in Europe and Australia are based on the incentive and profit-sharing system and stock ownership. Do profit sharing, stock ownership, and employee empowerment have financial benefits? Lincoln certainly thinks so! ("Highly Motivated," *Barron's*, 11 November 1996, p. 24; J. Lincoln, *Inventive Management* [Cleveland: Lincoln Electric, 1951]).

Rewards from financial improvements can be divided between shareholders, employees, managers, and customers (in the form of reduced prices). Organizationwide participation in the financial rewards from improvements provides an incentive to improve even more. Shared financial rewards become the basis for future financial improvements.

> Organizational values, beliefs, and culture can change quickly for the better when every employee has some financial ownership in the organization, and participates in the organization's financial success with bonuses or profit sharing.

Traditional Financial Measurements

Traditionally, the accounting and financial functions of the organization perform financial analysis using the financial statements and other reports. These organizationwide measures are used to monitor financial progress. What gets measured usually gets managed. The tradition of using financial statement–based ratios goes back to the 1920s when the DuPont Formula was developed. Return on investment from the shareholders' viewpoint was measured. It was named return on equity (ROE), and calculated as:

$$\text{Return on Equity} = \text{ROE} =$$

$$\left(\frac{\text{Net Income}}{\text{Sales}}\right)\left(\frac{\text{Sales}}{\text{Total Assets}}\right)\left(\frac{\text{Total Assets}}{\text{Common Share Equity}}\right) = \frac{\text{Net Income}}{\text{Common Share Equity}}$$

The goal when using this measurement was to monitor and manage each of the three components so that ROE is maximized. This was a shareholder-based measurement.

Today, nonfinancial employees and teams, in addition to the financial departments, are becoming more active in the financial monitoring and improvement of products, projects, and processes, which ultimately improves the organization's financial results. Improving net income, cash flows, and returns on investments is becoming everyone's job. This section provides a brief summary of the traditional financial measurements.

Vertical and Horizontal Measurements. **Horizontal analysis** is the analysis of Income Statements, Balance Sheets, other statements, and data on projects *between* periods of time such as months, quarters, or years. **Vertical analysis** is the analysis *within* a period such as a single quarter's Income Statement, Balance Sheet, or other financial reports. It was this type of financial analysis for which some of the original spreadsheet software was invented in the 1980s. Now, new spreadsheet software and powerful hardware make these calculations even easier.

Example: Vertical and horizontal analysis at Stuart Manufacturing—The horizontal and vertical techniques applied to the firm's financial statements can also be applied on the department level. Shown below are the assembly department's results for two years.

Stuart Manufacturing Corp. Assembly Department Costs			
	Year 1	**Year 2**	**Percentage Change**
Indirect Materials and Supplies	$12,000	$13,000	+8%
Direct Materials	110,000	125,000	+14
Indirect Labor	54,000	90,000	+67
Direct Labor	340,000	390,000	+15
Scrap Expense	25,000	31,000	+24
Rework Costs	$10,000	$13,000	+30
Production volume	109,000 Units	118,000 Units	+8

Production volume increased by 8% over the two-year period, but some of the department expenses increased by a higher percentage. By comparing the increase in direct labor to the increase in production volume, we see that the direct labor costs went up more than production volume. This indicates a *decrease* in productivity.

$$\text{Productivity} = \frac{\text{Output (units or dollars)}}{\text{Input (total labor dollars or hours)}}$$

	Year 1	**Year 2**
Productivity	109,000 units/$394,000	118,000/480,000
	= 0.28 units/total labor dollars	= 0.25 units/$

The drop in productivity is equal to more than a 10% (.03/.28 = 10.7%) decrease for the two-year period. This is a significant decrease. The department manager, manufacturing engineer, operating employees, and technical staff can investigate, find the cause of the problem, and correct it.

There are other calculations that can be made, such as:

$$\frac{\text{Scrap Expense}}{\text{Production Volume}} = \frac{\$25,000}{109,000 \text{ units}} = \$0.23 \text{ per unit}$$

For the second year, scrap expense is $0.26 per unit.

Other relationships between material and labor, direct and indirect costs, and volume can be calculated. Using this simple horizontal analysis and monitoring these ratios and symptoms, root causes can be detected and corrected and financial results can be improved.

Example: Ratio analysis at Stuart Manufacturing—One difficulty in comparing financial statements from different time periods or different organizations is that the dollar amounts are difficult to compare. Ratios and percentages, rather than dollar values, can be used. These traditional financial ratios measure past performance, and they can be useful starting points in the search for improvement targets. Typical ratios are calculated for Stuart Manufacturing below.

Stuart Manufacturing Corp.
Income Statement
January 1–December 31, 19xx

Sales	$200,000
CGS	120,000
Gross Profit	80,000
Expenses	40,000
Profit before Tax	40,000
Tax	20,000
Net Income	20,000
Dividends	10,000
Retained Earnings	$10,000

Stuart Manufacturing Corp. Balance Sheet December 31, 19xx

Assets		**Liabilities**	
Cash	$5,000	A/P	$10,000
Marketable Securities	5,000	N/P	15,000
Accts. Receivable	5,000	Common Stock 10,000 shares	100,000
Inventory	50,000	Retained Earn.	40,000
Plant & Equip.	100,000	Total Liabilities & Owners Equity	$165,000
Total Assets	$165,000		

Current market price of Stuart Manufacturing Corp. common stock: $20.00 per share.

Table 13–5 presents the year's ratios of Stuart Manufacturing from current financial statements. Similar ratios for Stuart over five years, for example, could be calculated and compared. These ratios could also be compared to industrywide or competitors' ratios and lead toward potential financial improvements.

Other ratios—Other percentages and ratios can be created as required. For example:

- Manufacturing engineers could be interested in inventory ratios comparing the amount of dollars invested into in-process inventory compared to total inventory value.

- Quality engineers would be interested in the ratio of scrap compared to good product in a particular process.

- The manager of a department would be interested in the amount of investment made into a process and its increased output or reduced scrap.

- Sales employees would be interested in the ratio of sales dollars in one geographic area compared to total sales.

- Manufacturing management would be interested in the ratio of indirect labor to direct labor dollars.

- The credit manager would be interested in a supplier's accounts payables compared to cash.

- The human resources department would be interested in the employees' benefits, training costs, and cost of on-the-job accidents compared to the total labor cost or hours worked.

Ratios and measurements can be created to meet the needs of the user.

Example: AP Corp. and horizontal analysis—Horizontal analysis is a useful tool for application at the company level as well as the department/process level. By comparing cost and income data over time, trends and problems can be identified. Shown below are financial statements for AP Corp. in condensed form.

AP Corp. (thousands of dollars)			
	Year 1	**Year 2**	**Percent Change**
Sales	$1,923	$2,130	+11%
Cost of Goods Sold	1,270	1,430	+13%
Selling, Admin., and R&D Exp.	267	261	−2%
Interest and Other Exp.	42	56	+33%
Total Expense	1,579	1,747	+11%
Income Before Taxes	344	383	+11%
Income Taxes	96	127	+32%
Net Income	$248	$256	+3%
Dividends Paid	23	25	+9%
Retained Earnings for Year	$225	$231	+3%

Sales increased by 11% between the two years, but net income increased by only 3%. Both CGS and interest and other expenses increased more than the sales change for an overall increase in total expenses of +11%. Taxes increased disproportionally as well.

Table 13–5   Summary of the Financial Ratios for Stuart Manufacturing Corp.

Ratio Formula/Calculation	Ratio Measure and Name	Ratio Value
Liquidity Ratios		
Current Ratio $= \dfrac{\text{Current Assets}}{\text{Current Liabilities}} = \dfrac{\$65,000}{\$25,000} = 2.6$	Current	2.6 times
Quick Ratio $= \dfrac{\text{Current Assets} - \text{Inventories}}{\text{Current Liabilities}} = \dfrac{\$15,000}{\$25,000} = 0.6$	Quick	0.6 times
Profitability Ratios		
Profit Ratio $= \dfrac{\text{Net Income}}{\text{Sales}} = \dfrac{\$20,000}{\$200,000} = 10\%$	Profit	10%
Earning Power $= \dfrac{\text{Profit Before Tax and Interest}}{\text{Total Assets}} = \dfrac{\$40,000}{\$165,000} = 24\%$	Earning Power	24%
Return on Investment of Assets $= \dfrac{\text{Net Income}}{\text{Total Assets}} = \dfrac{\$20,000}{\$165,000} = 12\%$	Return on Assets	12%
Return on Common Equity $= \dfrac{\text{Net Income}}{\text{Common Equity}} = \dfrac{\$20,000}{\$140,000} = 14\%$	Return on Common Equity	14%
Asset Ratios		
Inventory Turnover $= \dfrac{\text{Sales}}{\text{Inventory}} = \dfrac{\$200,000}{\$50,000} = 4$	Inventory Turnover	4 times per year
Days of Receivables Outstanding $= \dfrac{\text{Accounts Receivable}}{\text{Average Sales per Day}} = \dfrac{\$5,000}{\frac{\$200,000}{365\ \text{Days}}} = 9\ \text{Days}$	Days of Receivables	9 days
Revenue to Assets $= \dfrac{\text{Sales}}{\text{Total Assets}} = \dfrac{\$200,000}{\$165,000} = 1.2$	Revenue to Assets	1.2 times
Debt Ratios		
Debt Assets $= \dfrac{\text{Total Debt}}{\text{Total Assets}} = \dfrac{\$25,000}{\$165,000} = 15\%$	Debt to Assets	15%
Debt to Net Worth $= \dfrac{\text{Total Debt}}{\text{Net Worth}} = \dfrac{\$25,000}{\$140,000} = 18\%$	Debt to Net Worth	18%
Security Ratios		
Earnings per Share $= \dfrac{\text{Net Income}}{\text{Number of Shares}} = \dfrac{\$20,000}{10,000\ \text{Shares}} = \$2\ \text{per Share}$	Earnings per Share	$2 per share
Price Earnings $= \dfrac{\text{Market Price per Share}}{\text{Earnings per Share}} = \dfrac{\$20\ \text{per Share}}{\$2\ \text{per Share}} = 10$	Price Earnings	10 times
Book Value $= \dfrac{\text{Owners Equity}}{\text{Number of Shares}} = \dfrac{\$140,000}{10,000\ \text{Shares}} = \$14\ \text{per Share}$	Book Value	$14 per share
Payout $= \dfrac{\text{Dividends}}{\text{Net Income}} = \dfrac{\$10,000}{\$20,000} = 50\%$	Payout	50%
Yield $= \dfrac{\text{Dividend per Share}}{\text{Market Price}} = \dfrac{\frac{\$10,000}{10,000\ \text{Shares}}}{\$20.00\ \text{per Share}} = 5\%$	Yield	5%

Example: Percentage financial statements at Stuart Manufacturing—Another way to present financial data for comparison to other financial statements is to convert the Income Statement and Balance Sheet to Percentage Statements. Using the Stuart Manufacturing financial statements, the dollar values can be converted to percentages as shown below.

Stuart Manufacturing Corp.
Percentage Balance Sheet
December 31, 19xx

Cash	3%	A/P	6%
Marketable Securities	3%	N/P	9%
Accts. Receivable	3%	Common Stock 10,000 shares	61%
Inventory	30%	Retained Earn.	24%
Plant & Equip.	61%	Total Liabilities & Owners Equity	100%
Total Assets	100%		

Stuart Manufacturing Corp.
Percentage Income Statement
January 1–December 31, 19xx

Sales	100%
CGS	60%
Gross Profit	40%
Expenses	20%
Profit before Tax	20%
Tax	10%
Net Income	10%
Dividends	5%
Retained Earnings	5%

These percentage financial statements allow comparison to other organizations' statements, even though those firms may be larger or smaller than Stuart. Stuart's past financial statements can be converted to percentages and compared to the current statements.

Horizontal, vertical, and ratio analysis, like any financial measurements, are not ends in themselves. The goal is to analyze the financial results of the firm, product, or process. If a problem is discovered, an attempt to find the root cause of the problem and to correct it can be made.

> Measurement is not an end, but a means that should lead to improvement of financial results.

Budget Variance Measurement. When the actual costs are determined from the accounting system, they can be compared to the original budget values. These differences are budget **variances**. These variances can be studied to find root causes leading to improvements. Below is a departmental variance report. Variances are shown in dollars and percentages.

Variance Report for July
Department 250
Harrison Black, Supervisor
Maggie Johnson, Manufacturing Engineer

Category	Budget	Actual	Variance $	Variance %
Direct Material	$50,000	$47,000	−$3,000	−6.0%
Indirect Material	$10,000	$11,000	+$1,000	+10.0%
Direct Labor	$85,000	$89,000	+$4,000	+4.7%
Indirect Labor	$23,000	$21,000	−$2,000	−8.7%
Overhead	$150,000	$155,500	+$5,500	+3.7%
Total	$318,000	$323,500	+$5,500	+1.7%

For department 250, total department costs were 1.7% over budget. Indirect material was 10% over budget.

Next, the abbreviated Income Statement below shows results for the first quarter at Paper Corp. Results are better than predicted. Sales were 13% higher than estimated, and net income was 16.1% higher than projected. These are favorable variances, which isn't always the case. In either case, it is important to identify the causes of the variances so that the results can be corrected and incorporated into the next period's budget.

Budget Variance Report
Paper Corp.
Quarter 1, 19xx

	1st Quarter Budget	1st Quarter Actual	Variance $	Variance %
Sales	$230,000	$260,000	+$30,000	+13.0%
Cost of Goods Sold	$138,000	$154,000	+$16,000	+11.6%
Operating Profit	$92,000	$106,000	+$14,000	+15.2%
Expenses	$41,000	$44,000	+$3,000	+7.3%
Profit Before Tax	$51,000	$62,000	+$11,000	+21.6%
Taxes	$20,000	$26,000	+$6,000	+30.0%
Net Income	$31,000	$36,000	+$5,000	+16.1%
Dividends	$12,000	$14,000	+$2,000	+16.7%
Retained Earnings	$19,000	$22,000	+$3,000	+15.8%

Causes of variances lead to improvements—As with other financial measures, the budget variance measurement is only a first step in the process of improving financial performance. If root causes can be found and the over-budget variances can be analyzed, improvements can be made on a permanent basis and incorporated into future budgets for further improvement of financial results.

Variation is a part of all estimates. It is impossible to estimate future financial data with 100% certainty. In the budget and estimating process, variances can be reduced by continuously improving the estimating techniques.

A significant cause of variances is volume differences. Budget forecasts of sales are based on volume or number of units. When the number of units is greater or less than the estimated values, the revenues and costs also vary. Recalling breakeven concepts, total fixed costs do not generally change with output volume. For example, if fixed costs are $100,000 for a process and the budget volume was estimated at 100,000 units during the period, then the fixed cost per unit is expected to be $100,000/100,000 units = $1 fixed costs per unit. However, if the actual volume is only 90,000 units during the period with constant fixed costs, then the unit fixed cost is $100,000/90,000 units or $1.11 per unit. When actual volume is different from budgeted volume, total and units costs vary and variances occur.

Additional Financial Improvement Techniques

Concurrent Activities. Sequential activities of product design and then process engineering can be changed to be done concurrently, or concurrent engineering. This saves time and shortens cycle time from concept and customer request to delivery which reduces costs, obtains revenues sooner, and increases ROI. There may be many other sequential activities that can be converted to concurrent activities. A process flowchart is useful to discover these activities.

Project Management. This concept has been "rediscovered." It was developed formally in the 1960s and used in the design and construction of Polaris submarines, DuPont's chemical plants, NASA's Moon Project, and IBM's software development. Large projects using planning, scheduling, resource allocation, and cost analysis can be managed to meet the target dates and to optimize time and minimize cost. Program evaluation and review techniques, popular in the 1960s, are being used again. They are techniques that can minimize a project's costs as well as shorten the project's schedule.

Suggestion Systems. Toyota increased the number of suggestions from 463,000 to 860,000 per year within a five-year period. This was an increase from 10 suggestions per employee per year to almost 19 per employee per year. The acceptance and implementation percentage of suggestions was over 80%. Top management at Toyota participated actively in the suggestion process. Thousands of small product, process, and financial improvements were suggested. While many U.S. organizations have suggestion boxes throughout their facilities, the boxes at some companies, as one plant manager said, "collect more chewing gum wrappers than valuable suggestions." The organization's culture may not encourage making suggestions. However, a successful suggestion system, such as Toyota's, that grows gradually and has strong management participation can be a very powerful component of the financial improvement system (S. Shingo, *Study of Toyota Production System* [Tokyo: Productivity Press and Japan Management Association, 1989]).

Value Analysis and Value Engineering. These concepts have also been rediscovered. Popularized by General Electric and other organizations in the 1960s, their focus is on the design and assembly of products. Different materials, methods, designs, and assemblies are considered to reduce costs and increase functionality. Combining brainstorming, work simplification, methods improvement, design simplification, and other team techniques can result in better products at a lower cost.

Target and Kaizen Costing. By creating target costs for new products, the cost of the new product is continuously reduced as the product matures. This protects market share from competitors and ensures a growing market for the product. Kaizen costing combines continuous technical improvements in processes and products with cost reductions, increasing ROI, and lower prices to the customer.

FIREFIGHTING OR FINANCIAL IMPROVEMENT?

> The best servants of the people, like the best valets, must whisper unpleasant truths in the master's ear. It is the court fool, not the foolish courtier whom the king can least afford to lose.
>
> —Walter Lippmann

The techniques and tools presented here are basic and commonly used. As an organization applies them, it is likely to discover variations and additional approaches to financial improvements.

Must an organization be at the door of disaster before it uses financial improvement techniques? Even though a crisis increases the motivation to survive, many organizations can become more successful by improving slowly and continuously. This is the ideal case. Prevent problems by strengthening the improvement system so that financial problems do not have a chance to take hold—preventive medicine rather than surgery.

The culture of some organizations encourages "putting out fires" and moving from crisis to crisis. Other organizations rely on preventing problems and anticipating potential financial problems. The discussion here has been based on preventing, anticipating, and improving rather than "fighting fires and managing crises." These tools and techniques can be used in financial crisis situations, but they work best in an improvement and prevention environment.

> **Reacting versus Anticipating Financial Events**
>
> Some organizations spend time and money improving and managing their firefighting and crisis techniques. Others try to prevent and anticipate financial events, functioning in a culture based on making improvements before the emergency occurs.

SUMMARY

This chapter has presented the fourth and final part of the Financial Cycle. Important concepts of this chapter are:

- Eight prerequisites to continuous financial improvements.

- Continuous financial measurement and improvement tools and techniques.

- The prerequisites to CFI are not absolute, but when they are present, the results are faster and team functioning is smoother. Reluctant management, too specific expectations of the team's activities, hoping for breakthroughs to save the day, no rewards for small continuous improvements, and opposition to the existing culture are all potential land mines in the team's financial improvement journey. The tools and techniques work best in an environment that is compatible with the techniques.

- Personnel throughout the organization are becoming responsible for financial improvements. As the tasks of the organization become more complex, international competition more intense, acceleration of information flow increases, response times shorten, and higher quality of services and products is required, the successful organization of the twenty-first century must make continuous financial improvements. The traditional, autocratic, multilevel, centralized organization is giving way to the decentralized, team-based organization in which both employees and management make decisions.

- It is a race without a finish line. The world is catching up with capitalistic countries. The U.S. is a leader, but as we look over our shoulder, the others are catching up. The Asian firms are alert and eager. The Eastern Europeans are learning, watching, and copying the successful capitalists that they once opposed. Africa and the Middle East are resolving their differences so they can become international competitors in the economic race. The techniques and tools for improving and staying financially ahead are available to any individual or organization who wants to apply them. The concepts of financial improvement have been tested and found to be sturdy and useful. Small improvement steps will get the successful organization moving.

- If historical financial measurements and results are the primary goal and are used to drive the organization, it is like using the rearview mirror to drive a car. Extrapolating past results into the future is mathematically simple, but it doesn't consider the unseen obstacles and turns in the road ahead. Historical financial data are useful, but they may not be a good predictor of future results since the sales environment, the competition, interest rates, technical developments, and many other factors are continually changing. These are "moving targets," and it is not possible to simply extrapolate past financial data into the future to hit these moving targets, without continuous financial improvements by project managers, department managers, design engineers, process engineers, operators, and others throughout the organization.

> In the past, we steered by the wake; in the future, we'll steer by the new metrics.
>
> —Former CEO, computer systems company

- The measurement of outcomes is not the end of an organization's journey, but the beginning of continuous financial improvement. Measurements can show if the changes actually provide the desired results.

- Although the tools presented here appear straightforward and even simple, their application requires time, patience, and work to yield financial improvements. The organization or individual may be ready for these techniques, resulting in financial improvements, or more preparation may be necessary. Each individual and organization is unique and starts with different resources.

QUESTIONS

1. What is the measurement and continuous financial improvement cycle?
2. What are the eight prerequisites to continuous financial improvement?
3. In what ways can management demonstrate its support of continuous financial improvement teams?
4. Why are customer-driven improvements sometimes different from "boss-driven" improvements?
5. What is meant by the culture of an organization? How does it affect CFI?
6. Describe the culture of an organization of which you are a member.
7. What are the seven American cultural forces? Give examples.
8. What is kaizen? How does it relate to financial improvement?
9. What is the difference between breakthrough improvements and kaizen improvements?
10. Why are teams important in making financial improvements?
11. What is a strategic plan and how does it relate to continuous financial improvement?
12. What is meant by the statement, "Measure and improve the outcomes by focusing on the inputs to the process, not the outputs"?
13. What are the steps for continuous financial improvement?
14. What are the differences between training in team techniques and training in financial improvement techniques?
15. What are the basic, contemporary, and traditional tools of continuous financial improvement?
16. How can team brainstorming result in more and better ideas than individuals can produce?
17. Why is it important to arrange improvement ideas on a fishbone diagram?
18. What is meant by root or upstream causes?
19. How is a flowchart constructed?
20. Describe a run chart. How can it assist with improvements?
21. How can past ROIs of projects and investments be used to improve future ROI?
22. Identify the following measurements:
    a. DuPont Formula
    b. RONA
    c. CFROI
    d. RONABIT
    e. EVA
    f. Balanced metrics

23. What is meant by statistical thinking as applied to CFI?

24. What are the differences between random and special causes of financial measurements? What are some examples?

25. How can SPC be used to help a financial improvement team?

26. What is meant by the statement, "SPC alerts the user to a problem, not its solution"?

27. When would a run chart and an SPC chart be useful?

28. How can value added be used to measure and improve financial results?

29. What is the relationship between the time to deliver a service or product and financial benefits?

30. What is value-added flow rate?

31. How can reducing a bottleneck improve financial results?

32. How are wait time and equipment utilization related to financial improvements?

33. How can improving quality result in financial improvement?

34. What are examples of nonfinancial audits and how can they contribute to financial improvements?

35. Why is it important to consider how the benefits from financial improvements will be distributed and shared?

36. What are vertical and horizontal analysis?

37. What are some important historical ratios that can be calculated from the financial statements?

38. What is a Percentage Financial Statement? How is it helpful?

39. How can budget variances be used as a financial improvement tool?

40. What are some additional financial improvement tools?

41. Discuss these statements:
    a. "The goal of the organization should be to maximize profits and to do what is necessary to reach that goal."
    b. "Profits and financial results are a reward for doing everything else right including meeting the customer's expectations."
    c. "Using past financial data to determine the future is like using a rearview mirror to drive."
    d. "Firefighting is a different culture from planned continuous financial improvement."

PROBLEMS

1. In teams of four or five, select a financial problem, brainstorm possible causes of the problem, construct a fishbone diagram, collect data from the problem, identify the most likely causes, make revisions, and check to see if the changes improved the process.

2. Weekly scrap rates for a department are shown below. Calculate the average scrap rate and construct a run chart. Make any conclusions that seem appropriate.

Week	Scrap Rate
1	9.8%
2	7.5%
3	8.4%
4	7.9%
5	8.9%
6	6.4%
7	7.0%
8	7.8%
9	6.1%
10	6.8%
11	5.9%
12	6.1%

3. Sales for the past 15 months for a small company are show below. Plot a run chart. Can you make any observations from the chart? (Values in $1,000.)

Jan	$42
Feb	$47
Mar	$40
April	$46
May	$44
June	$50
July	$58
August	$43
Sept	$50
Oct	$49
Nov	$63
Dec	$57
Jan	$44
Feb	$51

4. Karron Co. has completed some projects during the past quarter. They have calculated the actual ROIs

of the recently completed projects. The table shows the original estimated ROIs and the actual ROIs.

Project/ Investment	Estimated ROI %	Actual ROI %
A	12	14
B	15	25
C	12	5
D	14	14
E	10	3
F	16	30
G	15	16
H	15	27
I	12	8
J	18	10
K	20	21
L	15	16
M	13	25
N	18	21
O	14	7
P	12	−10
Q	10	3
R	12	23
S	14	18

a. Calculate the average and range of the estimated ROIs and compare with the average and range of the actual ROIs.
b. Plot a histogram of both sets of data.
c. Calculate the variances of the ROIs.
d. State any conclusions

5. Obtain an annual report of a publicly held company and calculate the following:
   a. ROI
   b. RONA
   c. CFROI
   d. RONABIT
   e. EVA

6. As a team, construct a flowchart of one of the following processes:
   a. Making toast.
   b. Getting gas at a service station.
   c. Collating and stapling 10 pages of a report.
   d. Making a peanut butter and jelly sandwich.
   e. Any manufacturing process with which you are familiar.
   After making the chart, make any changes in the process and determine if the improvements result in less time than the original method.

7. Below are the results of a study at Stuart Manufacturing Corp. for one of their products showing material flow, value added, and the times.

Department	Dollar Value Added	Average Time in the Department	Processing Time in Department
Storage	$51.50 Raw Material Cost	18 Days	0.2 Hour
1	$48.90	6	0.8
2	$12.00	8	0.5
3	$67.00	12	1.3
4	$23.50	14	1.0
Assembly	$34.70	21	2.0

Calculate the following ratios for the product:
a. Value added in dollars per time in the department.
b. Value added in dollars per processing time.
c. Time in the department per hour of processing.
d. Total value added to total time in storage.
e. TVA to total process time.
f. Total time in departments to total process time.
If you were attempting to improve the flow, time, and value added of this product, which departments would require further analysis, study, and improvement?

8. Using vertical analysis for each year of the Flip Flop Corp. statements (Table 13–6) calculate the following (teams may be formed to analyze the problem):

Table 13–6   Financial Statements for Flip Flop Corp.—Year Ending December 31 (All data in thousands of dollars)

	Year 1	Year 2	Year 3	Year 4	Year 5
Sales	300	360	378	454	567
Cost of Goods Sold	170	200	223	254	315
Gross Profit	130	160	155	200	252
Expenses	60	75	85	110	139
Interest Expense	7	7	10	12	15
Depreciation	9	10	11	13	12
Net Income Before Tax	54	68	49	65	86
Tax	22	28	20	25	33
Net Income	32	40	29	40	53
Dividends	13	16	14	16	19
Retained Earnings	19	24	15	24	34
Assets					
Cash	40	44	35	35	46
Accounts Receivable	36	42	50	67	78
Investments	10	10	15	15	16
Inventory	65	91	98	120	150
Current Assets	151	187	198	237	290
Buildings and Equipment	130	138	140	145	165
Land	39	39	45	45	45
Fixed Assets	169	177	185	190	210
Total Assets	320	364	383	427	500
Liabilities					
Accounts Payable	40	50	54	62	90
Notes Payable	15	21	22	25	28
Current Liabilities	55	71	76	87	118
Bonds—9%, 2010	80	80	80	80	80
Long-Term Loans	50	54	53	62	70
Term Liabilities	130	134	133	142	150
Total Liabilities	185	205	209	229	268
Owners Equity					
Common Stock—100,000 shares outstanding	35	35	35	35	35
Retained Earnings	100	124	139	163	197
Owners Equity	135	159	174	198	232
Total Liabilities and Owners Equity	320	364	383	427	500
Market Price per Share	$4.50	$5.25	$7.50	$6.75	$8.25

a.
- Profit to sales
- Inventory turnover
- Current ratio
- Quick ratio
- Earning power
- Return on common equity
- Return on investment in assets
- Revenue to assets
- Days of receivables
- Debts to assets
- Earnings per share
- Price earnings
- Book value
- Cash flow
- Yield

b. Considering the percentages and ratios of Flip Flop, are any existing or potential problems indicated?

c. Using horizontal analysis, make the following calculations and graphs for the five years of Flip Flop financial statements:
- Sales growth over time.
- Changes in net income over time
- The CGS compared to sales over time.

- Inventory turnover.
- Cash flow over the five-year period.
- ROI over time.
- Share price, P/E ratio, and dividends on the same graph.

d. What was Flip Flop's investment policy in equipment and buildings for the five-year period?

e. Convert year 3 and year 4 financial statements to percentages and compare. Make conclusions or observations.

f. Construct an Income Statement budget for year 6, assuming a sales increase of 12% from year 5. Use the percentages of Income Statements from years 4 and 5 to project the year 6 budget.

9. Given the following departmental financial information for Satellite Components Corp., set up the data manually or using spreadsheet software. Using horizontal and vertical analysis, calculate:

Satellite Components Corp. Thrust Flow Monitor Component Department Costs					
Cost	Jan	Feb	March	April	May
Volume—Units	10,500	12,000	11,000	14,000	13,000
Direct Matls—$	400,000	510,000	476,000	650,000	600,000
Indirect Matls—$	15,000	30,000	25,000	41,000	39,000
Direct Labor—$	800,000	980,000	900,000	1,000,500	990,000
Indirect Labor—$	40,000	45,000	44,000	76,000	67,000
Rework—$	20,000	15,000	10,000	9,000	11,000
Scrap—$	5,000	8,000	10,000	15,000	12,000

a. Scrap as a percentage of total output from the department for five months.

b. Productivity for each month.

c. Graph of productivity over time in the department.

d. The percentage of direct labor to total labor costs for each month. Graph the results.

e. Ratio of direct labor to total units produced for the period.

What additional information would you like to have if you were the team leader of this department's cost improvement team? State any conclusions from the above analysis.

10. The following are data for a small department in Henry Products Co. that takes phone orders for mail order items. Calculate and plot the horizontal and vertical information indicated. Use a computer spreadsheet or manual methods.

Henry Products Co.

Cost	Week 1	Week 2	Week 3	Week 4	Week 5
Direct Labor	$10,000	12,000	14,000	14,000	11,000
Phone Calls Rec'd	2,000	2,400	3,100	2,800	2,500
Supplies—$	500	300	400	250	450
Sales/wk—$	40,000	53,000	78,000	65,000	51,000
Shipping, Invoice, and Order Errors	78	125	200	145	98

a. Calculate and plot the average value of an order for each week.

b. Determine the productivity in sales dollars per direct labor dollar.

c. Determine the productivity in number of calls per order dollars. Which is the better measure of productivity, your answer in part b or part c?

d. Determine the error ratio by comparing errors per sales dollars

e. If you were the team leader for cost improvements, what would you suggest?

11. The budget for year 1 is shown below for Flip Flop Corp. Prepare a variance report.

Budget for Flip Flop Corp.—Year 1

Sales	$310,000
Cost of Goods Sold	165,000
Gross Profit	145,000
Expenses	48,000
Interest	7,000
Depreciation	10,000
Net Income Before Tax	80,000
Tax	36,000
Net Income	44,000
Dividends	14,000
Retained Earnings	30,000

12. Below is the November budget and actual values for department 10 of Renolds Co. Calculate the variance percentages.

	Budget	Actual
Direct Materials	$156,000	$170,000
Indirect Materials	40,000	49,000
Direct Labor	270,000	290,000
Indirect Labor	60,000	54,000
Overhead	314,000	310,000

13. Obtain the annual report of a firm in which you are interested. Using data from two or three years of Income Statement and Balance Sheet information, make comparisons between the years, using horizontal and vertical analysis including ratios. What conclusions can be made?

> By right means, if you can, . . . make money.
>
> —Horace

DISCUSSION CASES

Bomar Manufacturing

Last year, Bomar Manufacturing implemented six major investments in equipment, revised processes, and buildings. The ROI for each project was estimated. The original estimates and the actual returns for the first year of each project are shown below.

Project ROI at Bomar Manufacturing

Project	Estimated ROI (%)	Actual ROI (%)
Warehouse Expansion	14	20
Replaced Equipment—Dept 33	22	20
Redesign of Component A-11	10	6
Heat Treating Equipment Revision/Scrap Reduction	15	3
Computer Network Revision	25	21
Sales Department New Cars	8	10

The six projects were funded based on estimates of their initial ROIs. Measuring project or investment ROI during the life of the project is valuable to determine if the project is meeting its original target. If it is meeting the target, the decision process is working. If the actual ROI is different from the estimated target ROI by a significant amount, then the estimating methods may have to be revised to better predict future project ROIs.

1. Which projects should be studied further?
2. What effect would the project's first cost have on the analysis?
3. What is the effect of the project or investment life on the analysis?
4. What other things should Bomar consider?

Lakeview Engineering

Vice president and chief engineer Dick Grapelli has just returned from a conference. One of the themes of the conference was continuous improvement in engineering consulting organizations. Dick attended a number of presentations and workshops about making improvements in services to clients and in internal processes to improve both customer relations and financial results. Much of the information was based on the successful application of improvement techniques in manufacturing and service companies in the past ten years. The consulting industry was behind the other industries, particularly manufacturing and other technical organizations.

Dick had asked 15 key individuals to a meeting to form a team and to discuss the improvements that he thought needed to be made at Lakeview. He had his own list of projects that he wanted them to study. The members of the team were from different parts of the organization. Dick asked both the chairman of the board and the president of Lakeview to attend the first few meetings to demonstrate the support of senior management, but they both declined. Each said that they thought it was a good idea to improve finances and they would approve any budget that Dick requested to do the job. The president asked how long the project would take before they could expect significant financial results.

Dick's agenda listed the 10 key financial ratios, financial goals, and projects that he thought were important for the improvement team to consider. He also had goals for each of the ratios that he thought were reasonable to achieve. He would tell the group that he would like to see the improvement results in one year, and they could make any changes they wanted so long as the goal was reached.

The meeting began and Dick made his 45-minute presentation with transparencies and handout information. At the end, he asked for questions and comments. There was silence until one of the managers asked, "How will this help us serve the clients better? Competition is getting tough. We're losing some customers!"

Dick said that if the financial goals could be met, they would not have to raise the prices of their services.

Another question from an older, experienced engineer was, "Two years ago we saw the strategic plan for the next five years and had a few meetings about that. Is this part of that plan?"

"Not exactly," Dick responded. "This is related to improving the financial results of the organization. The strategic plan looked at new services we were going to offer and the reorganization that we would be making to offer the new services."

Dick continued, "I know that in the past we have made most of the decisions at the main office. What we want to do is to try to have this group make some of the decisions that have been made by upper management in the past. Of course, they will have to approve our decisions, but I was just in a meeting this morning with the president and he told me our team has the ball. We'll keep him posted. I am anxious to begin training in the improvement tools such as brainstorming, fishbone diagrams, team development, and using statistical process control charts that I learned at the conference. We can probably start next week."

1. Comment on the eight prerequisites to making continuous financial improvements and Dick's plan.
2. Does the effort seem to be outcome or process directed?
3. Is the effort customer directed?
4. Is the effort connected to the strategic plan?
5. Are the activities considering the existing culture?
6. Is there strong management support and participation?
7. What are the chances for success by this effort?
8. Should they start training in use of the tools next week?
9. What should Dick do next?

Software Corp.

The team meeting was starting on time. It would only last 30 minutes, long enough to discuss some problems and to make assignments. Today's main topic was to look at some data that had been collected about the team's project, reducing the time from when an order was received until it was delivered to the customer. Joan and Bill had collected delivery times for the past six months and plotted a run chart. Everyone was looking at the charts projected on the wall.

Jim, a salesman, spoke. "Looks like some orders take less than a week and some take five or six weeks, which is too long compared to our competitors."

"Lots of variation," said Donna, from accounting.

"There were over 250 orders that we studied. We have identified the customer, the quantity, the dollar values, and made notes about any problems with the order. Some were small orders under $5,000 and some large orders over $60,000," Joan added.

"Maybe we should separate the large and small orders into two different charts. We might be looking at apples and oranges," Jim suggested.

Bill commented, "The important thing is to move on to the causes of the high delivery times and the causes of the wide variation. I think the next step is to create a fishbone diagram."

Joan said, "I think the next step is to look at the order process flow from the time we receive an order until it arrives at the customer. Then we can look for bottlenecks and delays in the process that might be common to the long delivery orders. Also, to find the reasons for the delays in these old orders is going to take a lot of detective work and too much time. Let's use these data as a baseline for future improvements, but look at current and future orders to see if our improvements really reduce delivery times."

Donna added, "I'm not a statistician, but we get five or six hundred orders a year. We can't look at all of them. Should we sample some of these past orders, current orders, and future orders so we can manage the workload? The sample will tell us if we are on the right track."

Jim said, "We've only got a few minutes left. Let's get copies of these data, and decide what each of us needs to do during the week for our meeting next week."

1. Is this project only improving delivery times or are there some financial benefits too?
2. Should large and small orders be separated as Jim suggests?
3. Is Bill right? Should a fishbone diagram be made?
4. What about Joan's suggestion to make a process flowchart?
5. Is Donna correct that there is "lots of variation"? How can she be sure?
6. Should the group look at samples or all of the orders?
7. What should be done before next week's meeting? What should be the topic for next week's meeting?

Valley Hospital

Jean Edwards is a nurse at Valley Hospital. She has been a surgery nurse for eight years, a recovery room nurse for six years, and the past year she became the liaison nurse in the human resources department for nurse training. She had been in meetings all morning. She started the day with the administrative staff at their weekly meeting. The topic was finances and the budget. Next she was with a group of the floor nurses to discuss new patient care procedures, and to discuss the upcoming visit by the hospital accreditation team. She was beginning to prepare for the internal reviews and training that would precede the accreditation visit.

This meeting was being held during lunch in the staff lunch area to discuss improvement techniques in the emergency room. Some of the ER nurses, assistants, clerks, and volunteers were there. She began the meeting with the question, "Do you think there are any benefits in studying the ER administrative procedures to see if patient care could be improved and if costs to the hospital and patient could be reduced?" She liked to begin with open-end questions to see if the group could become interested or if they were neutral to the ideas.

One of the more experienced nurses was first to speak. "I'm not interested in cutting costs as much as improving patient care. The administration is always cutting costs! The patients should come first."

Ben, a volunteer, asked, "How would we do this? The doctors make all the ER medical decisions. What would *we* do?"

Jean answered, "We can study and make recommendations for changes in the administrative procedures in the ER. We can form an improvement team, make studies, get feedback from patients, collect data, discuss solutions, and recommend changes. We could start with a survey of patients to have them evaluate the administrative procedures they receive. We can survey the ER doctors and nurses to see what they think of the administrative procedures. They sure complain enough about them! Then we can analyze the data and see if we can come up with any improvements."

Maggie, one of the younger ER receptionists, said, "Back to the cost-cutting idea. If the administration procedures are improved and simplified, they should also be less costly and if the patients are happier, they are going to come back and also recommend us to their friends. They think we have great medical care. It's our paperwork procedures and billing that patients complain about."

An admissions clerk, Roberta, asked, "What do we do? I've never been on an improvement team. I know that we have had them in other departments, but I don't know what they did or how they worked."

Jean answered, "If you all want to work on this I can show you how to collect the data, analyze it, and work together to get the job done. I've helped the recovery room group to improve procedures, the floor nurses to make improvements, and the pharmacy to work as an improvement team. I'll show you."

There were some nods around the table, and a few "okays" from some of the others. Jean continued, "One thing is important. This is a volunteer team. I don't want anyone to think that it's required. I'll send out a description to all the different shifts including the weekend staff about what we're trying to do. Maybe we'll even get some of the doctors and residents on the team! Thanks for coming. We'll probably have our first meeting in a week or ten days; I'll be in touch."

1. Is Jean off to a good start? What processes in the ER could be studied?

2. Is it correct to make the team voluntary?

3. Do the eight prerequisites to improvements apply in a hospital situation?

4. How can costs be reduced if the goal is to improve services to the patients and the medical staff?

5. What improvement tools are likely to be used by the team to study the ER area?

6. If you were Jean, what would you do at the next meeting?

7. What might be the complaints that patients and medical staff make about ER administrative procedures?

Dan Burk's Physical Exam

Dan had just completed his annual physical at Dr. Janss' office at Valley Hospital. The results were all good, except this time the doctor was very firm about Dan's weight. "Twenty-eight pounds overweight," he said. He wanted Dan to decrease his weight and he gave a long list of medical reasons why Dan should do that. Dan was concerned. He knew he was overweight, but how to reduce? He knew the latest fad diets didn't work. No quick fix here.

As he was driving back to the office, he began to think about the financial improvement team meeting that he had to lead in 45 minutes. They were making good, but slow, progress. The team had identified six critical processes of the 80 major processes in the company. They had decided to work on two of the six. Both were customer-related processes with potential for big financial improvements, and also some potential for revenue increases and cost reductions. They were through collecting data and were in the stage of analyzing the data. "We're on track and making steady progress," he thought.

Sitting in .his car before going into the building, he thought about his overweight problem. Would it be possible to apply some of the improvement techniques that he had used for financial improvements to his personal weight problem? Sure, they would work. The fishbone diagram: surely causes included lack of exercise and too many desserts! His wife had bought a new set of scales last year. He could measure his weight daily and post a run chart of his weight on the wall. A Pareto chart of the highest calorie foods would work. He could reduce the times he ate the number one item on the Pareto chart, and then work on the others. This would be a long-term project, but it might be fun to apply the improvement tools to the problem. He could even form a "Weight Improvement Team." His kids might have fun doing that. He could start today!

"Better get into the office and get ready for the meeting," Dan thought.

1. Will the improvement techniques that apply to financial improvement work for reducing personal weight?

2. What other tools should Dan consider?

3. How do the eight prerequisites to improvements apply to this problem?

4. Are there financial benefits to Dan, or only health benefits?

5. Is it important to involve his family or friends at work in the project?

6. What else should Dan consider?

Ellington Corp.

As a supplier to major computer manufacturers, Ellington wants to improve its performance and financial results. The CEO, Ann Jackson, has instituted a policy of training the work groups, middle managers, and senior management in understanding financial and nonfinancial measurements. Each department collects data along with the accounting reports about its performance. The data are displayed in chart form. The data and charts are posted on the "Wall of Measurements," a wall approximately 10 feet long on both sides of the hallway leading to the company cafeteria. Employees are encouraged to look at the data on their way to and from the cafeteria for meals and breaks.

The wall information includes budgets, variances, scrap reports, corporatewide financial ratios and percentages, time and delivery data of products to customers and to internal departments, supplier data, employee satisfaction information from surveys, quality information, project ROI information, value-added data, safety information, sales data, and other measurements.

Recently, Ann has noticed that the Wall is getting less attention than when it was first installed. She believes that it is very important to have this information available to all employees. She hopes that the work groups and middle management can use this information to improve their processes and products as well as the company's financial performance, but she has not seen many improvements lately.

1. What should Ann do next?

2. Is the Wall a good idea?

3. Why are employees not reading the Wall?

4. How can the information be used to make improvements?

Moon Valley Electronics

Jane is the manager of housekeeping and maintenance at a large company in the southwestern United States. The facilities include component manufacturing, assembly areas, offices, parking areas, lawns, and warehouses. Jane is well liked by the people that work for her as well as senior management. The only difficulty that she has is with the general manager, her boss. He is continually referring to the budget and variance report. He says that Jane should pay more attention to the budget and the reports that the cost accounting department prepares monthly. She is frequently off-budget and has high variances. He says that Jane does an excellent job of motivating her employees, and that the housekeeping and conditions of the plant are excellent. Many employees and visitors compliment the company on its housekeeping and physical facilities.

Jane says that the accounting department doesn't have any idea of the work her department does. She says that the budget and variance report is used as a device to punish managers. Her goal is to have the best maintenance and housekeeping department in the business. She keeps employee and management satisfaction data posted in the employee meeting room. She frequently reports to the employees about how well they are doing on maintaining the HVAC system, equipment repair, facilities management, and gives them feedback from the survey information that she collects. Jane has monthly training sessions for her employees and user satisfaction is always the first item considered. "All the memos about budgets and variances are written by people that never get out of the office and see what is going on," Jane says. "The users pay our salaries. They are more important than a few variances or overbudget items in the "bean counter" reports. Besides, the budgets aren't realistic anyway."

1. Is Jane right?
2. Should she be spending more time on reading the budget reports and trying to stay in budget?
3. How can this problem be resolved?
4. Is this conflict typical or is this an isolated situation?

Millionaires Club (G)

The club is six years old now. A few of the original founders are still members. Most past members have graduated and moved on. Some alumni of the club correspond using the club's Web page. They exchange investment ideas and stories. All of the investments are now done on the Web. Many of the graduated members have established personal investment portfolios for their retirement and regular investment accounts. Some past members work for small startup organizations that are just going public with their stock, and they alert the club to possible investments. The club has been successful with annual ROIs of 15%, 23%, 38%, 19%, 41%, and 22% for each of the past six years. The current portfolio, after all withdrawals, is just under $50,000. There are 24 members now and they all have invested their own funds into the club through the purchase of shares. The group no longer needs operating funds from the university. It finances its own activities and pays dividends to the members. Most members reinvest the dividends into the purchase of more stock. Things have gone well, but they want to do better.

The members have decided to do a study of the past six years and the current investments to determine the reasons for both the successful and unsuccessful investments. There is a special team of five members to study the improvement problem. They are searching for the causes of both good and poor results. The committee knows that the data are available, but isn't sure about the other steps they need to take. They are not sure how to proceed.

1. Should the team look at individual ROIs, the profit or loss of each investment, or some other criteria?
2. Are the successes of the investments a function of the members' ability to select and analyze potential investments, something that cannot be quantified?
3. How should the group proceed?

Epilogue

> Begin the journey from where you are now.

THE FINANCIAL JOURNEY

We have explored historical record-keeping of financial data, the Accounting Equation, depreciation, and financial statements. We traveled the analytical territory of time value of money and analysis of projects, equipment, and other assets. Our trip took us to the mountains and valleys of breakeven, minimum costs, replacement, and taxes. The final leg of the trip was the region of continuous financial improvements, measurements, and quality economics. Is the journey over, or are we ready to explore new financial territory?

Three themes stand out. One is that financial analysis and improvements are not just for accountants and corporate financial departments. Individuals, families, one-person organizations—everyone should participate in the financial journey. If financial success is sought, everyone is part of the team traveling on the financial analysis and improvement trip.

A second theme is that every technique and application requires both quantitative and qualitative analysis. The Accounting Equation must be balanced, the ROI calculated, and the analysis done; but people must be trained, management must be supportive, and teams must target improvements with combined efforts. The financial journey is a mixture of analysis and judgment; techniques and cooperation; human understanding and calculations.

A third theme is that economic analysis, selection, monitoring, and improvement are a complex combination of technical, economic, and financial factors that work best when a team considers the project or investment.

TRADITIONAL VERSUS WORLD CLASS FINANCIAL PERFORMANCE

The following table compares traditional approaches to economic analysis and decision making, philosophies, and activities to contemporary approaches and total financial management. These are not absolutes, just suggested starting points. These general concepts can be useful in mapping the location of the organization on its financial journey. Most organizations are somewhere between traditional and world class financial performance. Some are applying the techniques more successfully than others. The movement from the traditional to the world class is a continuous process. Once the journey is begun, it is difficult to turn back. As the culture changes and evolves, the organization's values and beliefs change, and it becomes difficult to return to the old beliefs, values, and culture. The journey continues forward.

Comparison of Traditional and World Class Financial Performance

Characteristic	Traditional	World Class
Financial Performance	Average financial success. Acceptable profits, ROI, and financial ratios. Meeting traditional financial statement measures.	Best in industry financial performance. Leader in field and financial innovator. Superior financial results. Financial rewards for shareholders, employees, managers.
Organization	Vertical financial decision making. Traditional line and staff employees. Many-level pyramid. Static.	Emphasis on horizontal and team-based economic analysis for financial decision making. Few levels. Inverted pyramid with customer at top. Dynamic, evolving, and improving.
Financial Tools	Cost accounting focus. Cost-based pricing, not market strategy–based. Payback analysis, no TVM. Cost/benefit analysis. Bottom-line driven. Intuitive decisions. Nonstatistical, fixed targets.	Customer-driven cost decision tools. Market share cost decisions. TVM techniques including breakeven, minimum cost, replacement analysis. Quality, loss function, and variation reduction techniques. Fact-based analysis and decisions. Statistical thinking and goals.
Management	Centralized, authoritarian. Central financial control. Management by financial numbers. Profits first. Limited support of and participation in financial improvement activities.	Team approach to financial decisions. Consensus financial analysis. Financially empowered employees at all levels. Managers as facilitators. Leadership and vision-based managers. Customer-driven profits. Active management participation and leadership in financial improvements.
Financial Culture	Bottom-line oriented. Numbers driven. Narrow traditional reward system. Focused on outcome and historical financial results. Unaware of organizational culture. Centralized financial decisions. No financial understanding or decisions in nonfinancial part of organization. Domestic focus.	Management support and participation of employee decisions. Broad financial strategy. Customer driven. Continuous improvements. Team based. Balanced reward system. Focused on process. Aware of organizational culture. Financial improvement teams throughout the organization. Financial responsibility and decisions throughout the organization. Global focus.

Characteristic	Traditional	World Class
Quality Economics	No quality costs understanding. Cost of quality = 20–25% of sales. No quality cost system. Limited scrap accounting. "Good enough" quality thinking. Detection of defects mode.	Understanding of quality costs throughout organization. Cost of quality < 5%. Full quality cost measurement and improvement system. Continuous quality improvement. Prevention of defects mode.
Cost Management	Cost reduction. Cost-cutting techniques. Downsizing. Layoff strategies. Crisis or firefighting emphasis. Short-term profit focus only.	Costs reduced by team approach. Continuous improvements in quality, process, value added, design improvements, cycle time reduction, inventory management and concurrent engineering. Deming chain reaction. Target and kaizen costing. Fire-prevention emphasis. Balanced focus between short- and long-term costs and profits.
Financial Growth	Emphasis on external mergers and take-overs to improve financial growth.	Emphasis on internal development and improvement for growth.
Auditing	Only financial audits.	Continuous audits of processes, products, designs, and team results using contemporary measures including ISO 9000, Baldrige criteria, and other industry audit criteria. Frequent internal, customer, and third-party audits Nonfinancial audits connected to financial improvements.
Financial Planning	Time horizon = weeks or months. Firefighting.	Time horizon = years. Prevention.
Ownership and Financial Rewards	Primarily shareholder and upper management rewards from financial improvements. Owners are owners, employees are employees philosophy. Emphasis on short-term financial results. External owners. Focused on share price.	Large portion of financial benefits shared based on organization-wide performance among shareholders, employees, customers, and management. Long-term financial emphasis. Get rich slowly. Extensive employee ownership. Focus on total ROI.
Economic Decisions	Centralized. Historical accounting–based. Suboptimization. Overhead allocation emphasis. "Rearview mirror" decisions.	Customer driven. Decentralized. Team-based financial decisions. Improvement and valued-added based.

(continues)

Comparison of Traditional and World Class Financial Performance *(continued)*

Characteristic	*Traditional*	*World Class*
Economic Decisions *(continued)*	Driven by earnings per share. Income Statement and Balance Sheet based. Unit cost emphasis. Intuitive. Everyone defers to financial department for money decisions.	Organizationwide optimization. Activity-based cost allocation. ROI, TVM, COQ, market share, cycle time, and process cost emphasis. Analysis using cash flow and TVM techniques. Fact-based coupled with experience.
Customer Focus	Acceptable "required" quality for customers. Profit maximization driven. Only external customers considered. Customers are only providers of revenues. Arms-length, adversary relationships with customers.	Partnership with customers to help reduce their costs and improve quality, profits, and performance for customer's customers. Customer-based employee rewards. Both internal and external customers considered. Focus on customer's customer.
Suppliers	Multiple suppliers with lowest cost as the primarily goal. Adversary relationship.	Single-source suppliers. Cooperation with supplier as a partner in quality, costs, delivery times, and improvements. Partnership relationship.
Value Added and Value Engineering	Development and design based on specialization and organization hierarchy. No value analysis or value engineering.	Organization based on value-added functions. Target costing. VA and VE throughout the organization.
Technical, Staff, Middle Management, and Operating Employees	Employees provide labor only and are financial costs.	Employees are financial decision makers as well as providers of labor. All employees are providers of improvements in process costs, product costs, and ROI. Internal customer philosophy. Employees critical to financial decisions and performance.
Financial Measurement	Financial statement and cost accounting based measures. Historical scorekeeping approach. Outcome and bottom-line numbers driven. Traditional financial ratios.	Measuring customer-based financial data. Process-based financial measurements based on customer expectations. Financial improvement approach. ROI-based measures. Cash flow measurements. Utilizing external measurements such as ISO 9000 and Baldrige financial criteria. Comparison with industry measures. Everyone involved in financial measurement—team measurement. Publicize successes widely.

Characteristic	Traditional	World Class
Financial Reward System	Time-based pay emphasis. Management bonus and stock options based on revenue, costs, and profit outcomes. Emphasis on shareholders and upper management rewards.	Performance-based pay. Organization-wide, performance-based wages and salaries. All employees participate in company-wide bonus, incentives, and profit sharing. Rewards based on ROI, RONABIT, EVA, CFROI, and related measures. Large portion of benefits for all employees based on organization's financial performance.
International Financial Direction	Limited to domestic focus.	Global focus on economic analysis, measures, financing, manufacturing, and markets.
Financial Improvements	Breakthrough based. Downsizing, cost cutting. Absolute targets and goals of improvements.	Continuous improvements from all parts of the organization. Team based. Customer directed. Variation- and statistical-based improvements.
Personal Financial Management	Organization has limited impact on employees' financial knowledge. Limited personal budgeting. No TVM application. Limited retirement planning. Little financial knowledge. "Leave money decision to the experts"—tax accountants, stockbrokers.	Organization interested in financial knowledge of employees. Family as a business focus. Extensive retirement planning and budgeting. Continuously learning. Application of TVM, tax analysis, breakeven, accounting, investments, economic analysis, and improvements to family. Financial interaction between organization and individual.
Financial Improvement Teams	None or limited financial improvement teams. Costs are primarily engineering and manufacturing based.	Well-trained financial improvement teams throughout organization. Financial improvement is both white and blue collar based.
Financial Training	Limited. Only financial personnel understand or have access to financial data and information. No financial training outside of finance department.	All employees receive financial training for job. Employees receive training about personal finances. Continuous training in financial improvement techniques and information.

WORLD CLASS TOTAL FINANCIAL MANAGEMENT

The term *world class total financial mangement* implies that the organization is the best in the world, not just domestically. It may seem obvious that the market for services and products is now worldwide, but some organizations still view themselves only as domestic businesses. Future financial success is likely to depend on the degree to which an organization can satisfy international markets as well. The application of economic analysis and financial improvement techniques is translated into other cultures and modified to fit those cultures.

Living and Learning Organization

International competition, an emphasis on customers and quality, and knowledge as a resource will continue to shape companies' financial success. Arie de Geus, who formerly headed group planning for Royal Dutch/Shell Company, has some suggestions that point to the financial future of international organizations. In corporate planning work at Shell over the past 20 years, he concluded that the successful organization can be considered a living, organic entity that learns as it grows. If it doesn't learn to survive in new and changing environments, as do humans, it dies a premature death (Arie de Geus, *The Living Company* [Boston: Longview Publishing, 1997]).

Surprisingly, de Geus and others at Shell found that the average life expectancy of a multinational corporation (Fortune 500 or equivalent) was only 40 to 50 years. Approximately one-third of the 1970 Fortune 500 companies had disappeared by 1983: merged, acquired, broken into parts, or died. Some organizations live for centuries, and in the United Kingdom there is even an association for companies that have lasted more than 300 years, the Tercentenarians Club. At Shell, which began in the 1890s, de Geus wondered why some organizations die prematurely and others live for centuries. He found only 40 large companies, excluding small and family organizations, that were older than Shell. Of these, 27 were studied in detail, from which interesting reasons for their long life were discovered.

Four key factors stood out as characteristics of these long-lived organizations. They were conservative in financing, sensitive to their environment, had a strong sense of identity, and were tolerant.

1. *Conservative in financing.* These companies used their capital with care and practiced conservative economic analysis. By having cash in hand, they were able to do things that would not otherwise be possible if they depended on equity or loans. This has given them a competitive edge. Control of one's future depends on the ability to enter markets and create products without outside interference.

2. *Sensitive to environment.* As political activities, wars, depressions, and technological innovations took place, the older organizations were always keeping their feelers out. Information and communications were vital to them and they were able to take economic advantage of the events going on around them. These long-lived firms continually learned from their surroundings.

3. *Strong sense of identity.* Although widely diversified throughout many economic systems, long-lived organizations had a sense of togetherness, economic comradeship, and community feelings among employees, customers, and even suppliers. Disconnected divisions knew they were part of the bigger organization. There was a sense of connection over time. Older managers were respected and acknowledged as new ones took over. Unless a crisis was banging at the door, senior management's primary responsibility was the health of the overall organization. The organization had an individual personality.

4. *Tolerant.* Long-lived, large organizations avoided centralized decision making and control. They were tolerant with experimenters and eccentricities. They were able to build relationships and alliances with outside organizations as well. This tolerance stretched their knowledge and possibilities.

What de Geus and other Shell planners *did not* find in the long-lived organizations was also interesting.

- Return on investment to shareholders did not seem to be related to longevity. Profitability was a symptom of corporate health, not a cause, and not a predictor or determinant of success.

- Historical and accounting information was vital to the managers and engineers of the firms, but it did not predict future deteriorating health. By the time any organization identified trouble on the financial statements, it was too late to prevent the problem.

- Longevity did not seem to have anything to do with material assets.

Another conclusion from their work is that if these characteristics of long-lived organizations lead not just to short-term success, but to long-term survival, then other organizations should copy them. If a business survives for a long periof of time in changing environments, maintains its personality, has well-functioning components that establish relationships with other entities, and conservatively finances its efforts, it must be adapting and learning. "Adapting and learning" sound like human characteristics. If so, the organization resembles a living, organic, learning entity, de Geus and others at Shell concluded. Perhaps the next era of economic analysis and financial management will focus on techniques to nurture a living, learning organization, providing for long-term survival and success rather than premature death in 40 to 50 years. That would be advantageous to owners, employees, customers, suppliers, and society in general. The tools and techniques presented throughout this book are consistent with de Geus' conclusions, particularly the application of conservative financial and economic analysis based on long-term measures and decentralized economic decisions.

The traditional mechanical concept of the organization—to hire labor to work on the raw materials, using equipment as inputs, so that the product can be sold profitably as an output to the customer—takes different economic, management, engineering, and organizational skills than the living organization concept. If the organization is a machine, it can be managed mechanically. If it is a living, learning, organic entity, it can be managed as a living organsim. It's too soon to tell if the philosophy that de Geus identified has potential in the next evolution of financial mangement. We will have to stay tuned.

Arie de Geus studied world-class companies between 100 and 700 years old in his book, *The Living Company.* He discovered these organizations shared four common traits. They were:

1. Conservative in financing and economic analysis.
2. Sensitivity to the world around them.
3. Aware of their identity.
4. Tolerant of new ideas.

THE PERSONAL FINANCIAL JOURNEY

In the early days of an organization's formation, perhaps only one or two people serve in several roles as employee, owner, and manager. Their individual finances are mixed with the organization's. Many of today's large global organizations began as one- and two-person businesses. Their economic analysis and financial management mixed family and organization money. Part of the philosophy of total financial mangement is to rediscover the relationship of an individual's and the organization's finances.

Most individuals manage their personal finances well. They receive a paycheck for their services, they pay bills, save money for retirement, invest in mutual funds, and borrow money for purchasing a home. If these individual skills of financial management can be assisted and applied to the organization, the organization can improve its financial results.

> Good economics begins at home. The word *economics* is rooted in the Greek word *oikos*, meaning house or household.

Training employees in both organizational and personal financial skills can benefit the family and the organization. Training individuals in how to calculate a mortgage payment, prepare personal financial statements, create a budget, obtain a loan, invest in a mutual fund, and manage a retirement plan can have a positive effect on the organization since employees understand that the same financial techniques can be applied to the organization. Conversely, training employees in improving organizational financial results by practicing cycle time reduction, increasing ROI, making TVM calculations, learning cost accounting basics, investing in stock, and other financial techniques has a positive effect on families' ability to manage their finances successfully. The financial concepts and tools of the individual, family, and organization are closely interrelated. A world class organization using total financial management techniques understands and applies these interrelated dynamic financial skills.

The journey continues. May we all enjoy many successful financial journeys.

Appendix A Answers to Odd-Numbered Problems

These answers have been calculated using tables, financial and programmed calculators, and spreadsheet software. Slightly different answers to some problems are possible due to differences in rounding and the number of decimal places used in the tables, calculators, and software.

CHAPTER 2

5. retained earnings = $530; total assets = $5,530; ending cash = $3,630
7. retained earnings = $160; total assets = $7,810; ending cash = $1,140
9. retained earnings = $450; total assets = $7,830; ending cash = $2,600

CHAPTER 3

1. a. capitalize—long life, high value
   b. expense—low value, used rapidly
   c. capitalize—value, life
   d. capitalize—value
   e. expense—low value
   f. expense—low value, short life
   g. expense—low cost
   h. expense—low value, small repair
3. a. cash = −$600,000; bldg = +$600,000
   b. depreciation = $20,000/year
   c. BV_{14} = $320,000; BV_{25} = $100,000
5. purchase entry—cash = −$50,000, equipment = +$50,000
   annual entry—equipment = −$6,429/year; depreciation expense = −$6,429/year
   end of life—cash = +$10,000, sale of equipment = +$10,000, equipment = −$5,000, cost of equipment sold = −$5,000 (profit on final transaction = $5,000)
7. initial entry in Accounting Equation—cash = −$50,000, equipment = +$50,000
   annual entry—equipment = −$6,429, depreciation expense = −$6,429/year
   end of 5th year Accounting Equation entries—cash = +$2,000, sale of equipment = +$2,000, equipment = −$17,855, cost of equipment sold = −$17,855 (loss of $15,855 on final transaction)

9. SOYD = 55
   a. first year depreciation = $20,545
      D_2 = $18,491
      D_3 = $16,436
      D_4 = $14,382
      D_5 = $12,327
      D_6 = $10,273
      D_7 = $8,218
      D_8 = $6,164
      D_9 = $4,109
      D_{10} = $2,055
   b. BV_6 = $32,546
   c. end of 6th year entries—cash = +$50,000, revenue from equipment sold = +$50,000, equipment = −$32,546, cost of equipment sold = −$32,546
11. D_1 = $22,500
    D_2 = $15,750
    D_3 = $11,025
    D_4 = $7,718
    D_5 = $5,402
    end of 5th year entries—cash = +$2,000, sale of equipment = +$2,000, equipment = −$12,605, cost of equipment sold = −$12,605 (loss of −$10,605 on final entry)
13. a. use 7-year MACRS life for office equipment
    D_1 = $10,718

$D_2 = \$18,368$
$D_3 = \$13,118$
$D_4 = \$9,367$
$D_5 = \$6,697$
$D_6 = \$6,690$
$D_7 = \$6,697$
$D_8 = \$3,345$
b. $10,714

	MACRS	Straight Line
c. BV_1	$64,282	$64,286
BV_2	$45,914	$53,572
BV_3	$32,796	$42,858
BV_4	$23,429	$32,144
BV_5	$16,732	$21,430
BV_6	$10,042	$10,716
BV_7	$3,345 (rounded)	$0
BV_8	$0	

15. a. $33,929
    b. BV = $12,500

c., d. Year	Dep/year	BV
1	$35,725	$214,275

2	$61,225	$153,050
3	$43,725	$109,325
4	$31,225	$78,100
5	$22,325	$55,775
6	$22,300	$33,475
7	$22,325	$11,150
8	$11,150	$0

f. $148,213; cash = +$200,000, sales = +$200,000, equipment = −$148,213, cost of equipment sold = −$148,213; tax on $51,787
$109,325; cash = +$200,000, sales = +$200,000, equipment = −$109,325, cost of equipment sold = −$109,325; tax on $90,675

17. $960; $2,141; $1,728; $1,152; $1,584
19. $600,000
21. cash = −$5,150, inv = +$5,150
cash = +$1,675, sales = +$1,675
inv = −$1,287.50, CGS = −$1,287.50
23. $110,000

CHAPTER 4

1. 7.97%
3. $4,441; $3,080
5. $13,788; $9,788
7. $29,575
9. $175
11. $3,780; $780
13. $102,087
15. $28,394; $47,845; $65,435; $135,854; $228,923
17. 11.77%
19. 10.97 years
21. 6 years
23. 24.49%

25. 6.90%; 6.32%; 7%; 6.18% (Bank Z is best)
27. 1.25% per month, nominal = 15%, effective = 16.08%
29. 10.83%
31. 6.5×10^{10}, 1.18×10^{11} (as of 1998)
33. 9.11%
35. $7,551; $3,777
37. 18% (as of 1998)
39. −5.68%
41. 4.34%; yes
43. 4.88%

CHAPTER 5

1. $212,793
3. 1.38 cents
5. $29,419; $1,518,490
7. $38,382
9. $845,965; $1,163,652; $2,027,408
11. $285,849; $254,310
13. 4%, 8.9%
15. $36,135 and ROI = 32%; $29,954 and ROI = 28%
17. $2,076/month; 9.38%
19. 10.6%

21. $918/year; $6,448
23. $1,565; $7,260
25. $265,065; $329,885
27. $120,000
29. $66,250
31. $981/year
33. a. yes
    b. 15.5%
35. $1,601; $8,541; $15,371; $2,881
37. $107,171; $86,210

CHAPTER 6

1. $47,308/year
3. $16,508/year
5. a. $100/year
   b. $123/year
   c. $23
   d. small effect
   e. $135/year
   f. $105/year
   g. discount price and credit card
7. a. $5,865/year
   b. $5,702
   c. $6,340
   d. for $n = 4$, $6,986/year
9. $10,370/year
11. buy = $72,796/year; lease = $77,757/year (lease has better initial cash flow)
13. $6,019
15. $37,500
17. $n = 5$, $16.35/unit, $n = 7$, $15.49/unit

19. a. $5,525/year
    b.
Years	8%	10%
4	$6,500	$6,825
5	$5,420	$5,739
6	$4,702	$5,022
    c. See table; reduce unit price? Difficult to quantify customer benefits—ask customer, should improve market share.
21. a. $725.00/month (approx.)
    b. $719.60/month (exact)
    c. $856.26/month
    d. $177,056
    e. $72,127
23. a. $14,183.94 per quarter
    b. $20,207
    c. $54,008.59
25. for 10%, $143,177; for 11%, $164,066
27. $76,803; $25,200
29. 8.5%
31. 4.18% + 3.06% = 7.24%

CHAPTER 7

1. $976
3. $127,568
5. 17.2 years
7. $250,000
9. 5%
11. $123,270/year
13. $730 versus $978
15. P_{Rev} greater by $38,890; A_{Rev} by $10,258
17. 10.83%, 14.86%
19. 50%
21. a. $114,346; $144,061
    b. $283,116; $356,691
    c. $23,018/year; $29,000/year
    d. 18% < ROI < 19%

23. A—$26,528; $30,543; $33,668
    P—$88,926; $102,385; $112,860
25. project's IRR—13%, 22%, 21%, 20%; B, C, D, A
27. $13,514/year vs. $14,448/year
29. 2.3 years, 3.0 years
31. IRR rank—G, D, B, I, E, A and H, F, C
    Risk rank—D, B and G, A, E and I, C and F, H
    P rank—D, H, B, A, F, E, G, C, I
33. Aggressive choice is Proj 46-MM and 30-I with high ROIs; they consume $57,000; $7,000 can be borrowed at 8.5%; risks are high, but projects are very important to customer.

CHAPTER 8

1. a. −$25,000 (loss)
   b. +$26,000
   c. 13,333 units/year
3. 75,342 miles/year; 3.77 years
5. $3,640; $2,780; $1,700 (if her time = $0)
7. a. $50,000/year; 35,000/year
   b. 2,500 units/year, 48 more per week

9. 45,958 miles/year
11. 33,700 hours/year (not possible—use A)
13. $446,400; $414,000 (best; use mechanical)
15. $394/year vs. $409/year
17. $24.04/unit

CHAPTER 9

1. 2,236 units/order, 11 times/year
3. a. 1,697 units/order, 30 times/year
   c. 2,191 units/order, 23 times/year
   d. 2,078 units/order, 36 times/year, 1,200 units/ order, 21 times/year
5. 1,200 units/order, 42 times/year
7. 822 oz., every 2 months
9. a. 371 units/order, 15 times/year

b. $55,746/year
c. 3 times
d. $55,365—OK to take discount quantity
11. a. 2,318 units/order, 9 times/year
    b. 1,728 units/order, monthly
    c. 864 units versus 1,159 average units in inventory
13. OK to use monthly repair

CHAPTER 10

1. b. $16,457/year
   c. $10,146/year
   d. use market value, not trade-in
3. a. $30/sq. ft.
   b. $33.75/sq. ft.
   c. $675,000
   d. Must be used with cost and income methods also
5. $345,000
7. $44,180
9.

Year	Salvage Value
7	$15,000
8	$11,250
9	$7,500
10	$3,750
11	$0

   (works best when n is short)
11. $55,366

13. $n = 1$, $29,000/year
    $n = 2$, $21,093$
    $n = 3$, $19,079$
    $n = 4$, $17,812$ (economic life = 4 years)
15. $n = 1$, $29,000$
    $n = 2$, $23,419$
    $n = 3$, $23,614$
    $n = 4$, $24,443$
    (type 3 economic life—search for replacement after year 2)
17. $16,484/year, $21,837/year
19. $16,145/year, $21,128/year
21. $3,944/year, $5,251/year
23. yields = 5.75% and 5%; both have PE ratios of 9 times
25. $2,560/year, $4,428/year
27. $5,000 versus $7,446 per month
29. $0.17, $0.14 per unit; buy new

CHAPTER 11

1. $4,775.50; 16.76%
3. $10,324; 18.44%
5. $408,000, 34%
7. 6.88%
9. 9.92%
11. $24,400; $28,240; $24,144; $21,686; $21,686; $19,843
    $P_{Cash Flow} > $80,000$ at 10%; $ROI_{AT} > 10\%$
13. equipment = −$11,579/year; depreciation expense = −$11,579/year; cash = −$250,000; equipment = +$250,000
15. oper CF = $26,225; $ROI_{AT} > 8\%$
17. less than 8%; less than 8%; greater than 8% (select z)
19. ROI_{AT} = approx. 3%

21. ROI_{AT} approx. 10.4%; oper CF = $11,428/year
23. cash = +$55,000, sales = +$55,000, bldg = −$48,000, cost of bldg sold = −$48,000, tax = $2,100, cash and sales = +$42,000, equipment and cost = −$48,000, tax = −$1,800
25. cash = +$42,000, sales = +$48,000, bldg = −$38,000, cost of bldg sold = −$38,000, tax = $1,120, cash and sales = +$48,000, equipment and cost = −$38,000, tax = $2,800
27. both $ROI_{AT} >> 10\%$
29. OCF = +$133,000, ICF = −$105,000, FCF = +$60,000
31. $1,672 savings

CHAPTER 12

1. | P | A | IF | EF |
    |---|---|----|----|
    | Qual Trng | Insp Wages | Rewk cost | Replacement of field fail |
    | Qual Engr | Cust survey | Troubleshoot | Lost sales |
    | | Qual software | | Recalls |
    | | Field trial | | Returns |
    | | Supplier audits | | |
    | | Internal mgmt audits | | |

3. Very high percentage, little quality emphasis or time, great amount of potential rewards from quality improvement efforts

5. | P | | A | | IF | | EF | |
    |---|---|---|---|----|----|----|----|
    | Trng | $800 | Insp | $5,000 | Scrap | $2,500 | Warnt | $1,200 |
    | Qual Manual | $1,000 | Audit | $1,500 | Rewk | $2,200 | Field Rep | $1,000 |
    | | _____ | Super | $1,500 | Inv Error | $400 | Allow | $500 |
    | | $1,800 | Relia | $1,800 | | | | |
    | | | | | | $5,100 | | $2,700 |
    | | | | $9,800 | | | | |

7. 9%, 51%, 26%, 15%

9. quality costs/sales; total failure/total quality costs; inspection/internal failure

13. a. Q = 35,294 units; loss = 5,294 units
    b. Q = 32,967; loss = 2,967 units
    c. 35.17 good units/hour
    d. 37.66 good units/hour; +7.08%

15. a. 12,987 units/year
    b. 12,658/year
    c. savings = $5,442/year

17. max = $475/unit
    max = $275/unit

19. max P = $13,337

CHAPTER 13

3. high variation—month to month; upward trend

7.
	a. $/day	b. $/process time	c. day/hr
Storage	2.86	257	90
1	8.65	61	7.5
2	1.50	24	16
3	5.58	52	9.2
4	1.68	24	14
Assmb	1.65	17	10.5

    d. $13.20/day
    e. $40.97/hour
    f. 13.6 days/hour

9.
J	F	M	A	M
a. $.47/unit	.66	.91	1.07	.92
b. .012 unit/$	.012	.012	.013	.012
d. 95.2%	95.6%	95.3%	92.9%	93.7%
e. 76.19	81.66	81.82	71.46	76.15

11.
	Variance
Sales	−3%
CGS	+3%
Gross P	−10%
Exp	+25%
Int	0%
Dep	−10%
PBT	−32%
Tax	−38%
Net Inc	−27%
Div	7%
RE	−37%

    Expenses 25% higher than budget. Net income 27% less than budget. Sales 3% less than budgeted.

Appendix B Interest Tables

These tables are provided as an alternative to using the formulas. The tables show the most commonly used interest rates. For rates not shown, the formulas can be used.

4% Interest Factors

n	F/P	P/F	F/A	A/F	P/A	A/P	A/G
1	1.040	0.9615	1.000	1.0000	0.9615	1.0400	0.0000
2	1.082	0.9246	2.040	0.4902	1.8861	0.5302	0.4902
3	1.125	0.8890	3.122	0.3204	2.7751	0.3604	0.9739
4	1.170	0.8548	4.246	0.2355	3.6299	0.2755	1.4510
5	1.217	0.8219	5.416	0.1846	4.4518	0.2246	1.9216
6	1.265	0.7903	6.633	0.1508	5.2421	0.1908	2.3857
7	1.316	0.7599	7.898	0.1266	6.0021	0.1666	2.8433
8	1.369	0.7307	9.214	0.1085	6.7328	0.1485	3.2944
9	1.423	0.7026	10.583	0.0945	7.4353	0.1345	3.7391
10	1.480	0.6756	12.006	0.0833	8.1109	0.1233	4.1773
11	1.539	0.6496	13.486	0.0742	8.7605	0.1142	4.6090
12	1.601	0.6246	15.026	0.0666	9.3851	0.1066	5.0344
13	1.665	0.6006	16.627	0.0602	9.9857	0.1002	5.4533
14	1.732	0.5775	18.292	0.0547	10.5631	0.0947	5.8659
15	1.801	0.5553	20.024	0.0500	11.1184	0.0900	6.2721
16	1.873	0.5339	21.825	0.0458	11.6523	0.0858	6.6720
17	1.948	0.5134	23.698	0.0422	12.1657	0.0822	7.0656
18	2.026	0.4936	25.645	0.0390	12.6593	0.0790	7.4530
19	2.107	0.4747	27.671	0.0361	13.1339	0.0761	7.8342
20	2.191	0.4564	29.778	0.0336	13.5903	0.0736	8.2091
21	2.279	0.4388	31.969	0.0313	14.0292	0.0713	8.5780
22	2.370	0.4220	34.248	0.0292	14.4511	0.0692	8.9407
23	2.465	0.4057	36.618	0.0273	14.8569	0.0673	9.2973
24	2.563	0.3901	39.083	0.0256	15.2470	0.0656	9.6479
25	2.666	0.3751	41.646	0.0240	15.6221	0.0640	9.9925
26	2.772	0.3607	44.312	0.0226	15.9828	0.0626	10.3312
27	2.883	0.3468	47.084	0.0212	16.3296	0.0612	10.6640
28	2.999	0.3335	49.968	0.0200	16.6631	0.0600	10.9909
29	3.119	0.3207	52.966	0.0189	16.9837	0.0589	11.3121
30	3.243	0.3083	56.085	0.0178	17.2920	0.0578	11.6274
31	3.373	0.2965	59.328	0.0169	17.5885	0.0569	11.9371
32	3.508	0.2851	62.701	0.0160	17.8736	0.0560	12.2411
33	3.648	0.2741	66.210	0.0151	18.1477	0.0551	12.5396
34	3.794	0.2636	69.858	0.0143	18.4112	0.0543	12.8325
35	3.946	0.2534	73.652	0.0136	18.6646	0.0536	13.1199
40	4.801	0.2083	95.026	0.0105	19.7928	0.0505	14.4765
45	5.841	0.1712	121.029	0.0083	20.7200	0.0483	15.7047
50	7.107	0.1407	152.667	0.0066	21.4822	0.0466	16.8123
55	8.646	0.1157	191.159	0.0052	22.1086	0.0452	17.8070
60	10.520	0.0951	237.991	0.0042	22.6235	0.0442	18.6972
65	12.799	0.0781	294.968	0.0034	23.0467	0.0434	19.4909
70	15.572	0.0642	364.290	0.0028	23.3945	0.0428	20.1961
75	18.945	0.0528	448.631	0.0022	23.6804	0.0422	20.8206
80	23.050	0.0434	551.245	0.0018	23.9154	0.0418	21.3719
85	28.044	0.0357	676.090	0.0015	24.1085	0.0415	21.8569
90	34.119	0.0293	817.983	0.0012	24.2673	0.0412	22.2826
95	41.511	0.0241	1012.785	0.0010	24.3978	0.0410	22.6550
100	50.505	0.0198	1237.624	0.0008	24.5050	0.0408	22.9800

5% Interest Factors

n	F/P	P/F	F/A	A/F	P/A	A/P	A/G
1	1.050	0.9524	1.000	1.0000	0.9524	1.0500	0.0000
2	1.103	0.9070	2.050	0.4878	1.8594	0.5378	0.4878
3	1.158	0.8638	3.153	0.3172	2.7233	0.3672	0.9675
4	1.216	0.8227	4.310	0.2320	3.5460	0.2820	1.4391
5	1.276	0.7835	5.526	0.1810	4.3295	0.2310	1.9025
6	1.340	0.7462	6.802	0.1470	5.0757	0.1970	2.3579
7	1.407	0.7107	8.142	0.1228	5.7864	0.1728	2.8052
8	1.477	0.6768	9.549	0.1047	6.4632	0.1547	3.2445
9	1.551	0.6446	11.027	0.0907	7.1078	0.1407	3.6758
10	1.629	0.6139	12.587	0.0795	7.7217	0.1295	4.0991
11	1.710	0.5847	14.207	0.0704	8.3064	0.1204	4.5145
12	1.796	0.5568	15.917	0.0628	8.8633	0.1128	4.9219
13	1.866	0.5303	17.713	0.0565	9.3936	0.1065	5.3215
14	1.980	0.5051	19.599	0.0510	9.8987	0.1010	5.7133
15	2.079	0.4810	21.579	0.0464	10.3797	0.0964	6.0973
16	2.183	0.4581	23.658	0.0423	10.8378	0.0923	6.4736
17	2.292	0.4363	25.840	0.0387	11.2741	0.0887	6.8423
18	2.407	0.4155	28.132	0.0356	11.6896	0.0856	7.2034
19	2.527	0.3957	30.539	0.0328	12.0853	0.0828	7.5569
20	2.653	0.3769	33.066	0.0303	12.4622	0.0803	7.9030
21	2.786	0.3590	35.719	0.0280	12.8212	0.0780	8.2416
22	2.925	0.3419	38.505	0.0260	13.1630	0.0760	8.5730
23	3.072	0.3256	41.430	0.0241	13.4886	0.0741	8.8971
24	3.225	0.3101	44.502	0.0225	13.7987	0.0725	9.2140
25	3.386	0.2953	47.727	0.0210	14.0940	0.0710	9.5238
26	3.556	0.2813	51.113	0.0196	14.3752	0.0696	9.8266
27	3.733	0.2679	54.669	0.0183	14.6430	0.0683	10.1224
28	3.920	0.2551	58.403	0.0171	14.8981	0.0671	10.4114
29	4.116	0.2430	62.323	0.0161	15.1411	0.0661	10.6936
30	4.322	0.2314	66.439	0.0151	15.3725	0.0651	10.9691
31	4.538	0.2204	70.761	0.0141	15.5928	0.0641	11.2381
32	4.765	0.2099	75.299	0.0133	15.8027	0.0633	11.5005
33	5.003	0.1999	80.064	0.0125	16.0026	0.0625	11.7566
34	5.253	0.1904	85.067	0.0118	16.1929	0.0618	12.0063
35	5.516	0.1813	90.320	0.0111	16.3742	0.0611	12.2498
40	7.040	0.1421	120.800	0.0083	17.1591	0.0583	13.3775
45	8.985	0.1113	159.700	0.0063	17.7741	0.0563	14.3644
50	11.467	0.0872	209.348	0.0048	18.2559	0.0548	15.2233
55	14.636	0.0683	272.713	0.0037	18.6335	0.0537	15.9665
60	18.679	0.0535	353.584	0.0028	18.9293	0.0528	16.6062
65	23.840	0.0420	456.798	0.0022	19.1611	0.0522	17.1541
70	30.426	0.0329	588.529	0.0017	19.3427	0.0517	17.6212
75	38.833	0.0258	756.654	0.0013	19.4850	0.0513	18.0176
80	49.561	0.0202	971.229	0.0010	19.5965	0.0510	18.3526
85	63.254	0.0158	1245.087	0.0008	19.6838	0.0508	18.6346
90	80.730	0.0124	1594.607	0.0006	19.7523	0.0506	18.8712
95	103.035	0.0097	2040.694	0.0005	19.8059	0.0505	19.0689
100	131.501	0.0076	2610.025	0.0004	19.8479	0.0504	19.2337

6% Interest Factors

n	F/P	P/F	F/A	A/F	P/A	A/P	A/G
1	1.060	0.9434	1.000	1.0000	0.9434	1.0600	0.0000
2	1.124	0.8900	2.060	0.4854	1.8334	0.5454	0.4854
3	1.191	0.8396	3.184	0.3141	2.6730	0.3741	0.9612
4	1.262	0.7921	4.375	0.2286	3.4651	0.2886	1.4272
5	1.338	0.7473	5.637	0.1774	4.2124	0.2374	1.8836
6	1.419	0.7050	6.975	0.1434	4.9173	0.2034	2.3304
7	1.504	0.6651	8.394	0.1191	5.5824	0.1791	2.7676
8	1.594	0.6274	9.897	0.1010	6.2098	0.1610	3.1952
9	1.689	0.5919	11.491	0.0870	6.8017	0.1470	3.6133
10	1.791	0.5584	13.181	0.0759	7.3601	0.1359	4.0220
11	1.898	0.5268	14.972	0.0668	7.8869	0.1268	4.4213
12	2.012	0.4970	16.870	0.0593	8.3839	0.1193	4.8113
13	2.133	0.4688	18.882	0.0530	8.8527	0.1130	5.1920
14	2.261	0.4423	21.015	0.0476	9.2950	0.1076	5.5635
15	2.397	0.4173	23.276	0.0430	9.7123	0.1030	5.9260
16	2.540	0.3937	25.673	0.0390	10.1059	0.0990	6.2794
17	2.693	0.3714	28.213	0.0355	10.4773	0.0955	6.6240
18	2.854	0.3504	30.906	0.0324	10.8276	0.0924	6.9597
19	3.026	0.3305	33.760	0.0296	11.1581	0.0896	7.2867
20	3.207	0.3118	36.786	0.0272	11.4699	0.0872	7.6052
21	3.400	0.2942	39.993	0.0250	11.7641	0.0850	7.9151
22	3.604	0.2775	43.392	0.0231	12.0416	0.0831	8.2166
23	3.820	0.2618	46.996	0.0213	12.3034	0.0813	8.5099
24	4.049	0.2470	50.816	0.0197	12.5504	0.0797	8.7951
25	4.292	0.2330	54.865	0.0182	12.7834	0.0782	9.0722
26	4.549	0.2198	59.156	0.0169	13.0032	0.0769	9.3415
27	4.822	0.2074	63.706	0.0157	13.2105	0.0757	9.6030
28	5.112	0.1956	68.528	0.0146	13.4062	0.0746	9.8568
29	5.418	0.1846	73.640	0.0136	13.5907	0.0736	10.1032
30	5.744	0.1741	79.058	0.0127	13.7648	0.0727	10.3422
31	6.088	0.1643	84.802	0.0118	13.9291	0.0718	10.5740
32	6.453	0.1550	90.890	0.0110	14.0841	0.0710	10.7988
33	6.841	0.1462	97.343	0.0103	14.2302	0.0703	11.0166
34	7.251	0.1379	104.184	0.0096	14.3682	0.0696	11.2276
35	7.686	0.1301	111.435	0.0090	14.4983	0.0690	11.4319
40	10.286	0.0972	154.762	0.0065	15.0463	0.0665	12.3590
45	13.765	0.0727	212.744	0.0047	15.4558	0.0647	13.1413
50	18.420	0.0543	290.336	0.0035	15.7619	0.0635	13.7964
55	24.650	0.0406	394.172	0.0025	15.9906	0.0625	14.3411
60	32.988	0.0303	533.128	0.0019	16.1614	0.0619	14.7910
65	44.145	0.0227	719.083	0.0014	16.2891	0.0614	15.1601
70	59.076	0.0169	967.932	0.0010	16.3846	0.0610	15.4614
75	79.057	0.0127	1300.949	0.0008	16.4559	0.0608	15.7058
80	105.796	0.0095	1746.600	0.0006	16.5091	0.0606	15.9033
85	141.579	0.0071	2342.982	0.0004	16.5490	0.0604	16.0620
90	189.465	0.0053	3141.075	0.0003	16.5787	0.0603	16.1891
95	253.546	0.0040	4209.104	0.0002	16.6009	0.0602	16.2905
100	339.302	0.0030	5638.368	0.0002	16.6176	0.0602	16.3711

7% Interest Factors

n	F/P	P/F	F/A	A/F	P/A	A/P	A/G
1	1.070	0.9346	1.000	1.0000	0.9346	1.0700	0.0000
2	1.145	0.8734	2.070	0.4831	1.8080	0.5531	0.4831
3	1.225	0.8163	3.215	0.3111	2.6243	0.3811	0.9549
4	1.311	0.7629	4.440	0.2252	3.3872	0.2952	1.4155
5	1.403	0.7130	5.751	0.1739	4.1002	0.2439	1.8650
6	1.501	0.6664	7.153	0.1398	4.7665	0.2098	2.3032
7	1.606	0.6228	8.654	0.1156	5.3893	0.1856	2.7304
8	1.718	0.5820	10.260	0.0975	5.9713	0.1675	3.1466
9	1.838	0.5439	11.978	0.0835	6.5152	0.1535	3.5517
10	1.967	0.5084	13.816	0.0724	7.0236	0.1424	3.9461
11	2.105	0.4751	15.784	0.0634	7.4987	0.1334	4.3296
12	2.252	0.4440	17.888	0.0559	7.9427	0.1259	4.7025
13	2.410	0.4150	20.141	0.0497	8.3577	0.1197	5.0649
14	2.579	0.3878	22.550	0.0444	8.7455	0.1144	5.4167
15	2.759	0.3625	25.129	0.0398	9.1079	0.1098	5.7583
16	2.952	0.3387	27.888	0.0359	9.4467	0.1059	6.0897
17	3.159	0.3166	30.840	0.0324	9.7632	0.1024	6.4110
18	3.380	0.2959	33.999	0.0294	10.0591	0.0994	6.7225
19	3.617	0.2765	37.379	0.0268	10.3356	0.0968	7.0242
20	3.870	0.2584	40.996	0.0244	10.5940	0.0944	7.3163
21	4.141	0.2415	44.865	0.0223	10.8355	0.0923	7.5990
22	4.430	0.2257	49.006	0.0204	11.0613	0.0904	7.8725
23	4.741	0.2110	53.436	0.0187	11.2722	0.0887	8.1369
24	5.072	0.1972	58.177	0.0172	11.4693	0.0872	8.3923
25	5.427	0.1843	63.249	0.0158	11.6536	0.0858	8.6391
26	5.807	0.1722	68.676	0.0146	11.8258	0.0846	8.8773
27	6.214	0.1609	74.484	0.0134	11.9867	0.0834	9.1072
28	6.649	0.1504	80.698	0.0124	12.1371	0.0824	9.3290
29	7.114	0.1406	87.347	0.0115	12.2777	0.0815	9.5427
30	7.612	0.1314	94.461	0.0106	12.4091	0.0806	9.7487
31	8.145	0.1228	102.073	0.0098	12.5318	0.0798	9.9471
32	8.715	0.1148	110.218	0.0091	12.6466	0.0791	10.1381
33	9.325	0.1072	118.933	0.0084	12.7538	0.0784	10.3219
34	9.978	0.1002	128.259	0.0078	12.8540	0.0778	10.4987
35	10.677	0.0937	138.237	0.0072	12.9477	0.0772	10.6687
40	14.974	0.0668	199.635	0.0050	13.3317	0.0750	11.4234
45	21.002	0.0476	285.749	0.0035	13.6055	0.0735	12.0360
50	29.457	0.0340	406.529	0.0025	13.8008	0.0725	12.5287
55	41.315	0.0242	575.929	0.0017	13.9399	0.0717	12.9215
60	57.946	0.0173	813.520	0.0012	14.0392	0.0712	13.2321
65	81.273	0.0123	1146.755	0.0009	14.1099	0.0709	13.4760
70	113.989	0.0088	1614.134	0.0006	14.1604	0.0706	13.6662
75	159.876	0.0063	2269.657	0.0005	14.1964	0.0705	13.8137
80	224.234	0.0045	3189.063	0.0003	14.2220	0.0703	13.9274
85	314.500	0.0032	4478.576	0.0002	14.2403	0.0702	14.0146
90	441.103	0.0023	6287.185	0.0002	14.2533	0.0702	14.0812
95	618.670	0.0016	8823.854	0.0001	14.2626	0.0701	14.1319
100	867.716	0.0012	12381.662	0.0001	14.2693	0.0701	14.1703

8% Interest Factors

n	F/P	P/F	F/A	A/F	P/A	A/P	A/G
1	1.080	0.9259	1.000	1.0000	0.9259	1.0800	0.0000
2	1.166	0.8573	2.080	0.4808	1.7833	0.5608	0.4808
3	1.260	0.7938	3.246	0.3080	2.5771	0.3880	0.9488
4	1.360	0.7350	4.506	0.2219	3.3121	0.3019	1.4040
5	1.469	0.6806	5.867	0.1705	3.9927	0.2505	1.8465
6	1.587	0.6302	7.336	0.1363	4.6229	0.2163	2.2764
7	1.714	0.5835	8.923	0.1121	5.2064	0.1921	2.6937
8	1.851	0.5403	10.637	0.0940	5.7466	0.1740	3.0985
9	1.999	0.5003	12.488	0.0801	6.2469	0.1601	3.4910
10	2.159	0.4632	14.487	0.0690	6.7101	0.1490	3.8713
11	2.332	0.4289	16.645	0.0601	7.1390	0.1401	4.2395
12	2.518	0.3971	18.977	0.0527	7.5361	0.1327	4.5958
13	2.720	0.3677	21.495	0.0465	7.9038	0.1265	4.9402
14	2.937	0.3405	24.215	0.0413	8.2442	0.1213	5.2731
15	3.172	0.3153	27.152	0.0368	8.5595	0.1168	5.5945
16	3.426	0.2919	30.324	0.0330	8.8514	0.1130	5.9046
17	3.700	0.2703	33.750	0.0296	9.1216	0.1096	6.2038
18	3.996	0.2503	37.450	0.0267	9.3719	0.1067	6.4920
19	4.316	0.2317	41.446	0.0241	9.6036	0.1041	6.7697
20	4.661	0.2146	45.762	0.0219	9.8182	0.1019	7.0370
21	5.034	0.1987	50.432	0.0198	10.0168	0.0998	7.2940
22	5.437	0.1840	55.457	0.0180	10.2008	0.0980	7.5412
23	5.871	0.1703	60.893	0.0164	10.3711	0.0964	7.7786
24	6.341	0.1577	66.765	0.0150	10.5288	0.0950	8.0066
25	6.848	0.1460	73.106	0.0137	10.6748	0.0937	8.2254
26	7.396	0.1352	79.954	0.0125	10.8100	0.0925	8.4352
27	7.988	0.1252	87.351	0.0115	10.9352	0.0915	8.6363
28	8.627	0.1159	95.339	0.0105	11.0511	0.0905	8.8289
29	9.317	0.1073	103.966	0.0096	11.1584	0.0896	9.0133
30	10.063	0.0994	113.283	0.0088	11.2578	0.0888	9.1897
31	10.868	0.0920	123.346	0.0081	11.3498	0.0881	9.3584
32	11.737	0.0852	134.214	0.0075	11.4350	0.0875	9.5197
33	12.676	0.0789	145.951	0.0069	11.5139	0.0869	9.6737
34	13.690	0.0731	158.627	0.0063	11.5869	0.0863	9.8208
35	14.785	0.0676	172.317	0.0058	11.6546	0.0858	9.9611
40	21.725	0.0460	259.057	0.0039	11.9246	0.0839	10.5699
45	31.920	0.0313	386.506	0.0026	12.1084	0.0826	11.0447
50	46.902	0.0213	573.770	0.0018	12.2335	0.0818	11.4107
55	68.914	0.0145	848.923	0.0012	12.3186	0.0812	11.6902
60	101.257	0.0099	1253.213	0.0008	12.3766	0.0808	11.9015
65	148.780	0.0067	1847.248	0.0006	12.4160	0.0806	12.0602
70	218.606	0.0046	2720.080	0.0004	12.4428	0.0804	12.1783
75	321.205	0.0031	4002.557	0.0003	12.4611	0.0803	12.2658
80	471.955	0.0021	5886.935	0.0002	12.4735	0.0802	12.3301
85	693.456	0.0015	8655.706	0.0001	12.4820	0.0801	12.3773
90	1018.915	0.0010	12723.939	0.0001	12.4877	0.0801	12.4116
95	1497.121	0.0007	18701.507	0.0001	12.4917	0.0801	12.4365
100	2199.761	0.0005	27484.516	0.0001	12.4943	0.0800	12.4545

9% Interest Factors

n	F/P	P/F	F/A	A/F	P/A	A/P	A/G
1	1.090	0.9174	1.000	1.0000	0.9174	1.0900	0.0000
2	1.188	0.8417	2.090	0.4785	1.7591	0.5685	0.4785
3	1.295	0.7722	3.278	0.3051	2.5313	0.3951	0.9426
4	1.412	0.7084	4.573	0.2187	3.2397	0.3087	1.3925
5	1.539	0.6499	5.985	0.1671	3.8897	0.2571	1.8282
6	1.677	0.5963	7.523	0.1329	4.4859	0.2229	2.2498
7	1.828	0.5470	9.200	0.1087	5.0330	0.1987	2.6574
8	1.993	0.5019	11.028	0.0907	5.5348	0.1807	3.0512
9	2.172	0.4604	13.021	0.0768	5.9953	0.1668	3.4312
10	2.367	0.4224	15.193	0.0658	6.4177	0.1558	3.7978
11	2.580	0.3875	17.560	0.0570	6.8052	0.1470	4.1510
12	2.813	0.3555	20.141	0.0497	7.1607	0.1397	4.4910
13	3.066	0.3262	22.953	0.0436	7.4869	0.1336	4.8182
14	3.342	0.2993	26.019	0.0384	7.7862	0.1284	5.1326
15	3.642	0.2745	29.361	0.0341	8.0607	0.1241	5.4346
16	3.970	0.2519	33.003	0.0303	8.3126	0.1203	5.7245
17	4.328	0.2311	36.974	0.0271	8.5436	0.1171	6.0024
18	4.717	0.2120	41.301	0.0242	8.7556	0.1142	6.2687
19	5.142	0.1945	46.018	0.0217	8.9501	0.1117	6.5236
20	5.604	0.1784	51.160	0.0196	9.1286	0.1096	6.7675
21	6.109	0.1637	56.765	0.0176	9.2923	0.1076	7.0006
22	6.659	0.1502	62.873	0.0159	9.4424	0.1059	7.2232
23	7.258	0.1378	69.532	0.0144	9.5802	0.1044	7.4358
24	7.911	0.1264	76.790	0.0130	9.7066	0.1030	7.6384
25	8.623	0.1160	84.701	0.0118	9.8226	0.1018	7.8316
26	9.399	0.1064	93.324	0.0107	9.9290	0.1007	8.0156
27	10.245	0.0976	102.723	0.0097	10.0266	0.0997	8.1906
28	11.167	0.0896	112.968	0.0089	10.1161	0.0989	8.3572
29	12.172	0.0822	124.135	0.0081	10.1983	0.0981	8.5154
30	13.268	0.0754	136.308	0.0073	10.2737	0.0973	8.6657
31	14.462	0.0692	149.575	0.0067	10.3428	0.0967	8.8083
32	15.763	0.0634	164.037	0.0061	10.4063	0.0961	8.9436
33	17.182	0.0582	179.800	0.0056	10.4645	0.0956	9.0718
34	18.728	0.0534	196.982	0.0051	10.5178	0.0951	9.1933
35	20.414	0.0490	215.711	0.0046	10.5668	0.0946	9.3083
40	31.409	0.0318	337.882	0.0030	10.7574	0.0930	9.7957
45	48.327	0.0207	525.859	0.0019	10.8812	0.0919	10.1603
50	74.358	0.0135	815.084	0.0012	10.9617	0.0912	10.4295
55	114.408	0.0088	1260.092	0.0008	11.0140	0.0908	10.6261
60	176.031	0.0057	1944.792	0.0005	11.0480	0.0905	10.7683
65	270.846	0.0037	2998.288	0.0003	11.0701	0.0903	10.8702
70	416.730	0.0024	4619.223	0.0002	11.0845	0.0902	10.9427
75	641.191	0.0016	7113.232	0.0002	11.0938	0.0902	10.9940
80	986.552	0.0010	10950.574	0.0001	11.0999	0.0901	11.0299
85	1517.932	0.0007	16854.800	0.0001	11.1038	0.0901	11.0551
90	2335.527	0.0004	25939.184	0.0001	11.1064	0.0900	11.0726
95	3593.497	0.0003	39916.635	0.0000	11.1080	0.0900	11.0847
100	5529.041	0.0002	61422.675	0.0000	11.1091	0.0900	11.0930

10% Interest Factors

n	F/P	P/F	F/A	A/F	P/A	A/P	A/G
1	1.100	0.9091	1.000	1.0000	0.9091	1.1000	0.0000
2	1.210	0.8265	2.100	0.4762	1.7355	0.5762	0.4762
3	1.331	0.7513	3.310	0.3021	2.4869	0.4021	0.9366
4	1.464	0.6830	4.641	0.2155	3.1699	0.3155	1.3812
5	1.611	0.6209	6.105	0.1638	3.7908	0.2638	1.8101
6	1.772	0.5645	7.716	0.1296	4.3553	0.2296	2.2236
7	1.949	0.5132	9.487	0.1054	4.8684	0.2054	2.6216
8	2.144	0.4665	11.436	0.0875	5.3349	0.1875	3.0045
9	2.358	0.4241	13.579	0.0737	5.7950	0.1737	3.3724
10	2.594	0.3856	15.937	0.0628	6.1446	0.1628	3.7255
11	2.853	0.3505	18.531	0.0540	6.4951	0.1540	4.0641
12	3.138	0.3186	21.384	0.0468	6.8137	0.1468	4.3884
13	3.452	0.2897	24.523	0.0408	7.1034	0.1408	4.6988
14	3.798	0.2633	27.975	0.0358	7.3667	0.1358	4.9955
15	4.177	0.2394	31.772	0.0315	7.6061	0.1315	5.2789
16	4.595	0.2176	35.950	0.0278	7.8237	0.1278	5.5493
17	5.054	0.1979	40.545	0.0247	8.0216	0.1247	5.8071
18	5.560	0.1799	45.599	0.0219	8.2014	0.1219	6.0526
19	6.116	0.1635	51.159	0.0196	8.3649	0.1196	6.2861
20	6.728	0.1487	57.275	0.0175	8.5136	0.1175	6.5081
21	7.400	0.1351	64.003	0.0156	8.6487	0.1156	6.7189
22	8.140	0.1229	71.403	0.0140	8.7716	0.1140	6.9189
23	8.953	0.1117	79.543	0.0126	8.8832	0.1126	7.1085
24	9.850	0.1015	88.497	0.0113	8.9848	0.1113	7.2881
25	10.835	0.0923	98.347	0.0102	9.0771	0.1102	7.4580
26	11.918	0.0839	109.182	0.0092	9.1610	0.1092	7.6187
27	13.110	0.0763	121.100	0.0083	9.2372	0.1083	7.7704
28	14.421	0.0694	134.210	0.0075	9.3066	0.1075	7.9137
29	15.863	0.0630	148.631	0.0067	9.3696	0.1067	8.0489
30	17.449	0.0573	164.494	0.0061	9.4269	0.1061	8.1762
31	19.194	0.0521	181.943	0.0055	9.4790	0.1055	8.2962
32	21.114	0.0474	201.138	0.0050	9.5264	0.1050	8.4091
33	23.225	0.0431	222.252	0.0045	9.5694	0.1045	8.5152
34	25.548	0.0392	245.477	0.0041	9.6086	0.1041	8.6149
35	28.102	0.0356	271.024	0.0037	9.6442	0.1037	8.7086
40	45.259	0.0221	442.593	0.0023	9.7791	0.1023	9.0962
45	72.890	0.0137	718.905	0.0014	9.8628	0.1014	9.3741
50	117.391	0.0085	1163.909	0.0009	9.9148	0.1009	9.5704
55	189.059	0.0053	1880.591	0.0005	9.9471	0.1005	9.7075
60	304.482	0.0033	3034.816	0.0003	9.9672	0.1003	9.8023
65	490.371	0.0020	4893.707	0.0002	9.9796	0.1002	9.8672
70	789.747	0.0013	7887.470	0.0001	9.9873	0.1001	9.9113
75	1271.895	0.0008	12708.954	0.0001	9.9921	0.1001	9.9410
80	2048.400	0.0005	20474.002	0.0001	9.9951	0.1001	9.9609
85	3298.969	0.0003	32979.690	0.0000	9.9970	0.1000	9.9742
90	5313.023	0.0002	53120.226	0.0000	9.9981	0.1000	9.9831
95	8556.676	0.0001	85556.760	0.0000	9.9988	0.1000	9.9889
100	13780.612	0.0001	137796.123	0.0000	9.9993	0.1000	9.9928

11% Interest Factors

n	F/P	P/F	F/A	A/F	P/A	A/P	A/G
1	1.110	0.9009	1.000	1.0000	0.9009	1.1100	0.0000
2	1.232	0.8116	2.110	0.4739	1.7125	0.5839	0.4740
3	1.368	0.7312	3.342	0.2992	2.4437	0.4092	0.9306
4	1.518	0.6587	4.710	0.2123	3.1024	0.3223	1.3698
5	1.685	0.5935	6.228	0.1606	3.6959	0.2706	1.7923
6	1.870	0.5346	7.913	0.1264	4.2305	0.2364	2.1975
7	2.076	0.4817	9.783	0.1022	4.7121	0.2122	2.5860
8	2.305	0.4339	11.859	0.0843	5.1462	0.1943	2.9585
9	2.558	0.3909	14.164	0.0706	5.5371	0.1806	3.3145
10	2.839	0.3522	16.722	0.0598	5.8893	0.1698	3.6545
11	3.152	0.3173	19.561	0.0511	6.2066	0.1611	3.9789
12	3.498	0.2858	22.713	0.0440	6.4922	0.1540	4.2876
13	3.883	0.2575	26.212	0.0382	6.7499	0.1482	4.5823
14	4.310	0.2320	30.095	0.0332	6.9818	0.1432	4.8616
15	4.785	0.2090	34.405	0.2091	7.1906	0.1391	5.1268
16	5.311	0.1883	39.190	0.0255	7.3790	0.1355	5.3789
17	5.895	0.1696	44.501	0.0225	7.5489	0.1325	5.6183
18	6.544	0.1528	50.396	0.0198	7.7018	0.1298	5.8444
19	7.263	0.1377	56.939	0.0176	7.8394	0.1276	6.0578
20	8.062	0.1240	64.203	0.0156	7.9631	0.1256	6.2582
21	8.949	0.1117	72.265	0.0138	8.0749	0.1238	6.4487
22	9.934	0.1007	81.214	0.0123	8.1759	0.1223	6.6289
23	11.026	0.0907	91.148	0.0110	8.2665	0.1210	6.7972
24	12.239	0.0817	102.174	0.0098	8.3479	0.1198	6.9549
25	13.586	0.0736	114.413	0.0087	8.4218	0.1187	7.1045
26	15.080	0.0663	127.999	0.0078	8.4882	0.1178	7.2449
27	16.739	0.0597	143.079	0.0070	8.5477	0.1170	7.3752
28	18.580	0.0538	159.817	0.0063	8.6014	0.1163	7.4975
29	20.624	0.0485	178.397	0.0056	8.6498	0.1156	7.6119
30	22.892	0.0437	199.021	0.0050	8.6941	0.1150	7.7218
31	25.410	0.0394	221.913	0.0045	8.7329	0.1145	7.8199
32	28.206	0.0355	247.324	0.0040	8.7689	0.1140	7.9156
33	31.308	0.0319	275.529	0.0036	8.8005	0.1136	8.0019
34	34.752	0.0288	306.837	0.0033	8.8292	0.1133	8.0833
35	38.575	0.0259	341.590	0.0029	8.8550	0.1129	8.1586
40	65.001	0.0154	581.826	0.0017	8.9509	0.1117	8.4655
45	109.530	0.0091	986.639	0.0010	9.0082	0.1110	8.6777
50	184.565	0.0054	1688.771	0.0006	9.0416	0.1106	8.8182

12% Interest Factors

n	F/P	P/F	F/A	A/F	P/A	A/P	A/G
1	1.120	0.8929	1.000	1.0000	0.8929	1.1200	0.0000
2	1.254	0.7972	2.120	0.4717	1.6901	0.5917	0.4717
3	1.405	0.7118	3.374	0.2964	2.4018	0.4164	0.9246
4	1.574	0.6355	4.779	0.2092	3.0374	0.3292	1.3589
5	1.762	0.5674	6.353	0.1574	3.6048	0.2774	1.7746
6	1.974	0.5066	8.115	0.1232	4.1114	0.2432	2.1721
7	2.211	0.4524	10.089	0.0991	4.5638	0.2191	2.5515
8	2.476	0.4039	12.300	0.0813	4.9676	0.2013	2.9132
9	2.773	0.3606	14.776	0.0677	5.3283	0.1877	3.2574
10	3.106	0.3220	17.549	0.0570	5.6502	0.1770	3.5847
11	3.479	0.2875	20.655	0.0484	5.9377	0.1684	3.8953
12	3.896	0.2567	24.133	0.0414	6.1944	0.1614	4.1897
13	4.364	0.2292	28.029	0.0357	6.4236	0.1557	4.4683
14	4.887	0.2046	32.393	0.0309	6.6282	0.1509	4.7317
15	5.474	0.1827	37.280	0.0268	6.8109	0.1468	4.9803
16	6.130	0.1631	42.753	0.0234	6.9740	0.1434	5.2147
17	6.866	0.1457	48.884	0.0205	7.1196	0.1405	5.4353
18	7.690	0.1300	55.750	0.0179	7.2497	0.1379	5.6427
19	8.613	0.1161	63.440	0.0158	7.3658	0.1358	5.8375
20	9.646	0.1037	72.052	0.0139	7.4695	0.1339	6.0202
21	10.804	0.0926	81.699	0.0123	7.5620	0.1323	6.1913
22	12.100	0.0827	92.503	0.0108	7.6447	0.1308	6.3514
23	13.552	0.0738	104.603	0.0096	7.7184	0.1296	6.5010
24	15.179	0.0659	118.155	0.0085	7.7843	0.1285	6.6407
25	17.000	0.0588	133.334	0.0075	7.8431	0.1275	6.7708
26	19.040	0.0525	150.334	0.0067	7.8957	0.1267	6.8921
27	21.325	0.0469	169.374	0.0059	7.9426	0.1259	7.0049
28	23.884	0.0419	190.699	0.0053	7.9844	0.1253	7.1098
29	26.750	0.0374	214.583	0.0047	8.0218	0.1247	7.2071
30	29.960	0.0334	241.333	0.0042	8.0552	0.1242	7.2974
31	33.555	0.0298	271.293	0.0037	8.0850	0.1237	7.3811
32	37.582	0.0266	304.848	0.0033	8.1116	0.1233	7.4586
33	42.092	0.0238	342.429	0.0029	8.1354	0.1229	7.5303
34	47.143	0.0212	384.521	0.0026	8.1566	0.1226	7.5965
35	52.800	0.0189	431.664	0.0023	8.1755	0.1223	7.6577
40	93.051	0.0108	767.091	0.0013	8.2438	0.1213	7.8988
45	163.988	0.0061	1358.230	0.0007	8.2825	0.1207	8.0572
50	289.002	0.0035	2400.018	0.0004	8.3045	0.1204	8.1597

13% Interest Factors

n	F/P	P/F	F/A	A/F	P/A	A/P	A/G
1	1.130	0.8850	1.000	1.0000	0.8850	1.1300	0.0000
2	1.277	0.7831	2.130	0.4695	1.6681	0.5995	0.4695
3	1.443	0.6931	3.407	0.2935	2.3612	0.4235	0.9188
4	1.631	0.6133	4.850	0.2062	2.9745	0.3362	1.3480
5	1.842	0.5428	6.480	0.1543	3.5173	0.2843	1.7573
6	2.082	0.4803	8.323	0.1202	3.9976	0.2502	2.1469
7	2.353	0.4251	10.405	0.0961	4.4226	0.2261	2.5172
8	2.658	0.3762	12.757	0.0784	4.7987	0.2084	2.8683
9	3.004	0.3329	15.416	0.0649	5.1316	0.1949	3.2013
10	3.395	0.2946	18.420	0.0543	5.4262	0.1843	3.5162
11	3.836	0.2607	21.814	0.0458	5.6870	0.1758	3.8135
12	4.335	0.2307	25.650	0.0390	5.9175	0.1690	4.0932
13	4.898	0.2042	29.985	0.0334	6.1218	0.1634	4.3573
14	5.535	0.1807	34.883	0.0287	6.3024	0.1587	4.6048
15	6.254	0.1599	40.417	0.0247	6.4625	0.1547	4.8377
16	7.067	0.1415	46.672	0.0214	6.6037	0.1514	5.0548
17	7.986	0.1252	53.739	0.0186	6.7290	0.1486	5.2587
18	9.024	0.1108	61.725	0.0162	6.8399	0.1462	5.4492
19	10.197	0.0981	70.749	0.0141	6.9382	0.1441	5.6272
20	11.523	0.0868	80.947	0.0124	7.0249	0.1424	5.7923
21	13.021	0.0768	92.470	0.0108	7.1018	0.1408	5.9461
22	14.714	0.0680	105.491	0.0095	7.1695	0.1395	6.0880
23	16.627	0.0601	120.205	0.0083	7.2296	0.1383	6.2203
24	18.788	0.0532	136.831	0.0073	7.2828	0.1373	6.3428
25	21.231	0.0471	155.620	0.0064	7.3298	0.1364	6.4558
26	23.991	0.0417	176.850	0.0057	7.3719	0.1357	6.5623
27	27.109	0.0369	200.841	0.0050	7.4085	0.1350	6.6580
28	30.634	0.0326	227.950	0.0044	7.4410	0.1344	6.7468
29	34.616	0.0289	258.583	0.0039	7.4699	0.1339	6.8290
30	39.116	0.0256	293.199	0.0034	7.4957	0.1334	6.9054
31	44.201	0.0226	332.315	0.0030	7.5182	0.1330	6.9745
32	49.947	0.0200	376.516	0.0027	7.5381	0.1327	7.0375
33	56.440	0.0177	426.463	0.0023	7.5563	0.1323	7.0983
34	63.777	0.0157	482.903	0.0021	7.5717	0.1321	7.1509
35	72.069	0.0139	546.681	0.0018	7.5855	0.1318	7.1996
40	132.782	0.0075	1013.704	0.0010	7.6342	0.1310	7.3877
45	244.641	0.0041	1874.165	0.0005	7.6611	0.1305	7.5088
50	450.736	0.0022	3459.507	0.0003	7.6752	0.1303	7.5808

14% Interest Factors

n	F/P	P/F	F/A	A/F	P/A	A/P	A/G
1	1.140	0.8772	1.000	1.0000	0.8772	1.1400	0.0000
2	1.300	0.7695	2.140	0.4673	1.6467	0.6073	0.4673
3	1.482	0.6750	3.440	0.2907	2.3216	0.4307	0.9129
4	1.689	0.5921	4.921	0.2032	2.9138	0.3432	1.3371
5	1.925	0.5194	6.610	0.1513	3.4331	0.2913	1.7400
6	2.195	0.4556	8.536	0.1172	3.8886	0.2572	2.1217
7	2.502	0.3996	10.730	0.0932	4.2883	0.2332	2.4834
8	2.853	0.3506	13.233	0.0756	4.6389	0.2156	2.8246
9	3.252	0.3075	16.085	0.0622	4.9463	0.2022	3.1462
10	3.707	0.2697	19.337	0.0517	5.2162	0.1917	3.4493
11	4.226	0.2366	23.045	0.0434	5.4529	0.1834	3.7336
12	4.818	0.2076	27.271	0.0367	5.6603	0.1767	3.9997
13	5.492	0.1821	32.089	0.0312	5.8425	0.1712	4.2494
14	6.261	0.1597	37.581	0.0266	6.0020	0.1666	4.4819
15	7.138	0.1401	43.842	0.0228	6.1421	0.1628	4.6989
16	8.137	0.1229	50.980	0.0196	6.2649	0.1596	4.9006
17	9.277	0.1078	59.118	0.0169	6.3727	0.1569	5.0883
18	10.575	0.0946	68.394	0.0146	6.4675	0.1546	5.2631
19	12.056	0.0829	78.969	0.0127	6.5505	0.1527	5.4247
20	13.744	0.0728	91.025	0.0110	6.6230	01.510	5.5729
21	15.668	0.0638	104.768	0.0095	6.6872	0.1495	5.7119
22	17.861	0.0560	120.436	0.0083	6.7431	0.1483	5.8386
23	20.362	0.0491	138.297	0.0072	6.7921	0.1472	5.9551
24	23.212	0.0431	158.659	0.0063	6.8353	0.1463	6.0629
25	26.462	0.0378	181.871	0.0055	6.8729	0.1455	6.1607
26	30.167	0.0331	208.333	0.0048	6.9061	0.1448	6.2514
27	34.390	0.0291	238.499	0.0042	6.9353	0.1442	6.3348
28	39.205	0.0255	272.889	0.0037	6.9609	0.1437	6.4109
29	44.693	0.0224	312.094	0.0032	6.9832	0.1432	6.4800
30	50.950	0.0196	356.787	0.0028	7.0028	0.1428	6.5429
31	58.083	0.0172	407.737	0.0025	7.0200	0.1425	6.6004
32	66.215	0.0151	465.820	0.0022	7.0348	0.1422	6.6514
33	75.485	0.0132	532.035	0.0019	7.0482	0.1419	6.6997
34	86.053	0.0116	607.520	0.0017	7.0597	0.1417	6.7421
35	98.100	0.0102	693.573	0.0014	7.0701	0.1414	6.7829
40	188.884	0.0053	1342.025	0.0008	7.1048	0.1408	6.9286
45	363.679	0.0027	2590.565	0.0004	7.1230	0.1404	7.0175
50	700.233	0.0014	4994.521	0.0002	7.1327	0.1402	7.0714

15% Interest Factors

n	F/P	P/F	F/A	A/F	P/A	A/P	A/G
1	1.150	0.8696	1.000	1.0000	0.8696	1.1500	0.0000
2	1.323	0.7562	2.150	0.4651	1.6257	0.6151	0.4651
3	1.521	0.6575	3.473	0.2880	2.2832	0.4380	0.9071
4	1.749	0.5718	4.993	0.2003	2.8550	0.3503	1.3263
5	2.011	0.4972	6.742	0.1483	3.3522	0.2983	1.7228
6	2.313	0.4323	8.754	0.1142	3.7845	0.2642	2.0972
7	2.660	0.3759	11.067	0.0904	4.1604	0.2404	2.4499
8	3.059	0.3269	13.727	0.0729	4.4873	0.2229	2.7813
9	3.518	0.2843	16.786	0.0596	4.7716	0.2096	3.0922
10	4.046	0.2472	20.304	0.0493	5.0188	0.1993	3.3832
11	4.652	0.2150	24.349	0.0411	5.2337	0.1911	3.6550
12	5.350	0.1869	29.002	0.0345	5.4206	0.1845	3.9082
13	6.153	0.1625	34.352	0.0291	5.5832	0.1791	4.1438
14	7.076	0.1413	40.505	0.0247	5.7245	0.1747	4.3624
15	8.137	0.1229	47.580	0.0210	5.8474	0.1710	4.5650
16	9.358	0.1069	55.717	0.0180	5.9542	0.1680	4.7523
17	10.761	0.0929	65.075	0.0154	6.0472	0.1654	4.9251
18	12.375	0.0808	75.836	0.0132	6.1280	0.1632	5.0843
19	14.232	0.0703	88.212	0.0113	6.1982	0.1613	5.2307
20	16.367	0.0611	102.444	0.0098	6.2593	0.1598	5.3651
21	18.822	0.0531	118.810	0.0084	6.3125	0.1584	5.4883
22	21.645	0.0462	137.632	0.0073	6.3587	0.1573	5.6010
23	24.891	0.0402	159.276	0.0063	6.3988	0.1563	5.7040
24	28.625	0.0349	184.168	0.0054	6.4338	0.1554	5.7979
25	32.919	0.0304	212.793	0.0047	6.4642	0.1547	5.8834
26	37.857	0.0264	245.712	0.0041	6.4906	0.1541	5.9612
27	43.535	0.0230	283.569	0.0035	6.5135	0.1535	6.0319
28	50.066	0.0200	327.104	0.0031	6.5335	0.1531	6.0960
29	57.575	0.0174	377.170	0.0027	6.5509	0.1527	6.1541
30	66.212	0.0151	434.745	0.0023	6.5660	0.1523	6.2066
31	76.144	0.0131	500.957	0.0020	6.5791	0.1520	6.2541
32	87.565	0.0114	577.100	0.0017	6.5905	0.1517	6.2970
33	100.700	0.0099	664.666	0.0015	6.6005	0.1515	6.3357
34	115.805	0.0086	765.365	0.0013	6.6091	0.1513	6.3705
35	133.176	0.0075	881.170	0.0011	6.6166	0.1511	6.4019
40	267.864	0.0037	1779.090	0.0006	6.6418	0.1506	6.5168
45	538.769	0.0019	3585.128	0.0003	6.6543	0.1503	6.5830
50	1083.657	0.0009	7217.716	0.0002	6.6605	0.1501	6.6205

20% Interest Factors

n	F/P	P/F	F/A	A/F	P/A	A/P	A/G
1	1.200	0.8333	1.000	1.0000	0.8333	1.2000	0.0000
2	1.440	0.6945	2.200	0.4546	1.5278	0.6546	0.4546
3	1.728	0.5787	3.640	0.2747	2.1065	0.4747	0.8791
4	2.074	0.4823	5.368	0.1863	2.5887	0.3863	1.2742
5	2.488	0.4019	7.442	0.1344	2.9906	0.3344	1.6405
6	2.986	0.3349	9.930	0.1007	3.3255	0.3007	1.9788
7	3.583	0.2791	12.916	0.0774	3.6046	0.2774	2.2902
8	4.300	0.2326	16.499	0.0606	3.8372	0.2606	2.5756
9	5.160	0.1938	20.799	0.0481	4.0310	0.2481	2.8364
10	6.192	0.1615	25.959	0.0385	4.1925	0.2385	3.0739
11	7.430	0.1346	32.150	0.0311	4.3271	0.2311	3.2893
12	8.916	0.1122	39.581	0.0253	4.4392	0.2253	3.4841
13	10.699	0.0935	48.497	0.0206	4.5327	0.2206	3.6597
14	12.839	0.0779	59.196	0.0169	4.6106	0.2169	3.8175
15	15.407	0.0649	72.035	0.0139	4.6755	0.2139	3.9589
16	18.488	0.0541	87.442	0.0114	4.7296	0.2114	4.0851
17	22.186	0.0451	105.931	0.0095	4.7746	0.2095	4.1976
18	26.623	0.0376	128.117	0.0078	4.8122	0.2078	4.2975
19	31.948	0.0313	154.740	0.0065	4.8435	0.2065	4.3861
20	38.338	0.0261	186.688	0.0054	4.8696	0.2054	4.4644
21	46.005	0.0217	225.026	0.0045	4.8913	0.2045	4.5334
22	55.206	0.0181	271.031	0.0037	4.9094	0.2037	4.5942
23	66.247	0.0151	326.237	0.0031	4.9245	0.2031	4.6475
24	79.497	0.0126	392.484	0.0026	4.9371	0.2026	4.6943
25	95.396	0.0105	471.981	0.0021	4.9476	0.2021	4.7352
26	114.475	0.0087	567.377	0.0018	4.9563	0.2018	4.7709
27	137.371	0.0073	681.853	0.0015	4.9636	0.2015	4.8020
28	164.845	0.0061	819.223	0.0012	4.9697	0.2012	4.8291
29	197.814	0.0051	984.068	0.0010	4.9747	0.2010	4.8527
30	237.376	0.0042	1181.882	0.0009	4.9789	0.2009	4.8731
31	284.852	0.0035	1419.258	0.0007	4.9825	0.2007	4.8908
32	341.822	0.0029	1704.109	0.0006	4.9854	0.2006	4.9061
33	410.186	0.0024	2045.931	0.0005	4.9878	0.2005	4.9194
34	492.224	0.0020	2456.118	0.0004	4.9899	0.2004	4.9308
35	590.668	0.0017	2948.341	0.0003	4.9915	0.2003	4.9407
40	1469.772	0.0007	7343.858	0.0002	4.9966	0.2001	4.9728
45	3657.262	0.0003	18281.310	0.0001	4.9986	0.2001	4.9877
50	9100.438	0.0001	45497.191	0.0000	4.9995	0.2000	4.9945

25% Interest Factors

n	F/P	P/F	F/A	A/F	P/A	A/P	A/G
1	1.250	0.8000	1.000	1.0000	0.8000	1.2500	0.0000
2	1.563	0.6400	2.250	0.4445	1.4400	0.6945	0.4445
3	1.953	0.5120	3.813	0.2623	1.9520	0.5123	0.8525
4	2.441	0.4096	5.766	0.1735	2.3616	0.4235	1.2249
5	3.052	0.3277	8.207	0.1219	2.6893	0.3719	1.5631
6	3.815	0.2622	11.259	0.0888	2.9514	0.3388	1.8683
7	4.768	0.2097	15.073	0.0664	3.1611	0.3164	2.1424
8	5.960	0.1678	19.842	0.0504	3.3289	0.3004	2.3873
9	7.451	0.1342	25.802	0.0388	3.4631	0.2888	2.6048
10	9.313	0.1074	33.253	0.0301	3.5705	0.2801	2.7971
11	11.642	0.0859	42.566	0.0235	3.6564	0.2735	2.9663
12	14.552	0.0687	54.208	0.0185	3.7251	0.2685	3.1145
13	18.190	0.0550	68.760	0.0146	3.7801	0.2646	3.2438
14	22.737	0.0440	86.949	0.0115	3.8241	0.2615	3.3560
15	28.422	0.0352	109.687	0.0091	3.8593	0.2591	3.4530
16	35.527	0.0282	138.109	0.0073	3.8874	0.2573	3.5366
17	44.409	0.0225	173.636	0.0058	3.9099	0.2558	3.6084
18	55.511	0.0180	218.045	0.0046	3.9280	0.2546	3.6698
19	69.389	0.0144	273.556	0.0037	3.9424	0.2537	3.7222
20	86.736	0.0115	342.945	0.0029	3.9539	0.2529	3.7667
21	108.420	0.0092	429.681	0.0023	3.9631	0.2523	3.8045
22	135.525	0.0074	538.101	0.0019	3.9705	0.2519	3.8365
23	169.407	0.0059	673.626	0.0015	3.9764	0.2515	3.8634
24	211.758	0.0047	843.033	0.0012	3.9811	0.2512	3.8861
25	264.698	0.0038	1054.791	0.0010	3.9849	0.2510	3.9052
26	330.872	0.0030	1319.489	0.0008	3.9879	0.2508	3.9212
27	413.590	0.0024	1650.361	0.0006	3.9903	0.2506	3.9346
28	516.988	0.0019	2063.952	0.0005	3.9923	0.2505	3.9457
29	646.235	0.0016	2580.939	0.0004	3.9938	0.2504	3.9551
30	807.794	0.0012	3227.174	0.0003	3.9951	0.2503	3.9628
31	1009.742	0.0010	4034.968	0.0003	3.9960	0.2503	3.9693
32	1262.177	0.0008	5044.710	0.0002	3.9968	0.2502	3.9746
33	1577.722	0.0006	6306.887	0.0002	3.9975	0.2502	3.9791
34	1972.152	0.0005	7884.609	0.0001	3.9980	0.2501	3.9828
35	2465.190	0.0004	9856.761	0.0001	3.9984	0.2501	3.9858

30% Interest Factors

n	F/P	P/F	F/A	A/F	P/A	A/P	A/G
1	1.300	0.7692	1.000	1.0000	0.7692	1.3000	0.0000
2	1.690	0.5917	2.300	0.4348	1.3610	0.7348	0.4348
3	2.197	0.4552	3.990	0.2506	1.8161	0.5506	0.8271
4	2.856	0.3501	6.187	0.1616	2.1663	0.4616	1.1783
5	3.713	0.2693	9.043	0.1106	2.4356	0.4106	1.4903
6	4.827	0.2072	12.756	0.0784	2.6428	0.3784	1.7655
7	6.275	0.1594	17.583	0.0569	2.8021	0.3569	2.0063
8	8.157	0.1226	23.858	0.0419	2.9247	0.3419	2.2156
9	10.605	0.0943	32.015	0.0312	3.0190	0.3312	2.3963
10	13.786	0.0725	42.620	0.0235	3.0915	0.3235	2.5512
11	17.922	0.0558	56.405	0.0177	3.1473	0.3177	2.6833
12	23.298	0.0429	74.327	0.0135	3.1903	0.3135	2.7952
13	30.288	0.0330	97.625	0.0103	3.2233	0.3103	2.8895
14	39.374	0.0254	127.913	0.0078	3.2487	0.3078	2.9685
15	51.186	0.0195	167.286	0.0060	3.2682	0.3060	3.0345
16	66.542	0.0150	218.472	0.0046	3.2832	0.3046	3.0892
17	86.504	0.0116	285.014	0.0035	3.2948	0.3035	3.1345
18	112.455	0.0089	371.518	0.0027	3.3037	0.3027	3.1718
19	146.192	0.0069	483.973	0.0021	3.3105	0.3021	3.2025
20	190.050	0.0053	630.165	0.0016	3.3158	0.3016	3.2276
21	247.065	0.0041	820.215	0.0012	3.3199	0.3012	3.2480
22	321.184	0.0031	1067.280	0.0009	3.3230	0.3009	3.2646
23	417.539	0.0024	1388.464	0.0007	3.3254	0.3007	3.2781
24	542.801	0.0019	1806.003	0.0006	3.3272	0.3006	3.2890
25	705.641	0.0014	2348.803	0.0004	3.3286	0.3004	3.2979
26	917.333	0.0011	3054.444	0.0003	3.3297	0.3003	3.3050
27	1192.533	0.0008	3971.778	0.0003	3.3305	0.3003	3.3107
28	1550.293	0.0007	5164.311	0.0002	3.3312	0.3002	3.3153
29	2015.381	0.0005	6714.604	0.0002	3.3317	0.3002	3.3189
30	2619.996	0.0004	8729.985	0.0001	3.3321	0.3001	3.3219
31	3405.994	0.0003	11349.981	0.0001	3.3324	0.3001	3.3242
32	4427.793	0.0002	14755.975	0.0001	3.3326	0.3001	3.3261
33	5756.130	0.0002	19183.768	0.0001	3.3328	0.3001	3.3276
34	7482.970	0.0001	24939.899	0.0001	3.3329	0.3001	3.3288
35	9727.860	0.0001	32422.868	0.0000	3.3330	0.3000	3.3297

Appendix C   Effective Interest Rates

r	Compounding frequency					
	Semi-annually $\left(1+\dfrac{r}{2}\right)^2 -1$	Quarterly $\left(1+\dfrac{r}{4}\right)^4 -1$	Monthly $\left(1+\dfrac{r}{12}\right)^{12} -1$	Weekly $\left(1+\dfrac{r}{52}\right)^{52} -1$	Daily $\left(1+\dfrac{r}{365}\right)^{365} -1$	Continu-ously $\left(1+\dfrac{r}{\infty}\right)^{\infty} -1$
1	1.0025	1.0038	1.0046	1.0049	1.0050	1.0050
2	2.0100	2.0151	2.0184	2.0197	2.0200	2.0201
3	3.0225	3.0339	3.0416	3.0444	3.0451	3.0455
4	4.0400	4.0604	4.0741	4.0793	4.0805	4.0811
5	5.0625	5.0945	5.1161	5.1244	5.1261	5.1271
6	6.0900	6.1364	6.1678	6.1797	6.1799	6.1837
7	7.1225	7.1859	7.2290	7.2455	7.2469	7.2508
8	8.1600	8.2432	8.2999	8.3217	8.3246	8.3287
9	9.2025	9.3083	9.3807	9.4085	9.4132	9.4174
10	10.2500	10.3813	10.4713	10.5060	10.5126	10.5171
11	11.3025	11.4621	11.5718	11.6144	11.6231	11.6278
12	12.3600	12.5509	12.6825	12.7336	12.7447	12.7497
13	13.4225	13.6476	13.8032	13.8644	13.8775	13.8828
14	14.4900	14.7523	14.9341	15.0057	15.0217	15.0274
15	15.5625	15.8650	16.0755	16.1582	16.1773	16.1834
16	16.6400	16.9859	17.2270	17.3221	17.3446	17.3511
17	17.7225	18.1148	18.3891	18.4974	18.5235	18.5305
18	18.8100	19.2517	19.5618	19.6843	19.7142	19.7217
19	19.9025	20.3971	20.7451	20.8828	20.9169	20.9250
20	21.0000	21.5506	21.9390	22.0931	22.1316	22.1403
21	22.1025	22.7124	23.1439	23.3153	23.3584	23.3678
22	23.2100	23.8825	24.3596	24.5494	24.5976	24.6077
23	24.3225	25.0609	25.5863	25.7957	25.8492	25.8600
24	25.4400	26.2477	26.8242	27.0542	27.1133	27.1249
25	26.5625	27.4429	28.0731	28.3250	28.3901	28.4025
26	27.6900	28.6466	29.3333	29.6090	29.6796	29.6930
27	28.8225	29.8588	30.6050	30.9049	30.9821	30.9964
28	29.9600	31.0796	31.8880	32.2135	32.2976	32.3130
29	31.1025	32.3089	33.1826	33.5350	33.6264	33.6428
30	32.2500	33.5469	34.4889	34.8693	34.9684	34.9859
31	33.4025	34.7936	35.8068	36.2168	36.3238	36.3425
32	34.5600	36.0489	37.1366	37.5775	37.6928	37.7128
33	35.7225	37.3130	38.4784	38.9515	39.0756	39.0968
34	36.8900	38.5859	39.8321	40.3389	40.4722	40.4948
35	38.0625	39.8676	41.1979	41.7399	41.8827	41.9068

Index